# 4th International Conference on Advanced Materials and Chemical Engineering (AMCE 2024)



Journal of Physics: Conference Series Volume 2961

Guangzhou, China
25-27 October 2024

ISBN: 979-8-3313-1488-0
ISSN: 1742-6588

**Printed from e-media with permission by:**

Curran Associates, Inc.
57 Morehouse Lane
Red Hook, NY 12571

proceedings.com

**Some format issues inherent in the e-media version may also appear in this print version.**



Institute of Physics
Dirac House, Temple Back
Bristol BS1 6BE UK

Phone: 44 1 17 929 7481
Fax: 44 1 17 920 0979

techtracking@iop.org

**Additional copies of this publication are available from:**

Curran Associates, Inc.
57 Morehouse Lane
Red Hook, NY 12571 USA
Phone: 845-758-0400
Fax: 845-758-2633
Email: curran@proceedings.com
Web: www.proceedings.com

# TABLE OF CONTENTS

# Preface

The 2024 4th International Conference on Advanced Materials and Chemical Engineering (AMCE 2024) successfully convened from October 25 to 27, 2024, in Guangzhou, China. This annual conference continues to serve as a pivotal platform for researchers, scientists, and industry professionals to exchange innovative ideas, discuss recent trends, and foster collaborations in the fields of advanced materials and chemical engineering.

During the conference, esteemed professors were invited as keynote speakers to address speeches, including Professor Xianyou Wang from Xiangtan University, China and Professor Jingang Yu from Central South University, China. Their presentations provided invaluable insights into the latest advancements and challenges in advanced materials and chemical engineering, setting the tone for the rest of the conference. Moreover, the knowledge shared by these experts surely enlightened attendees and catalyzed further research and exploration.

The conference provided a comprehensive exploration of current developments in advanced materials and chemical engineering. It featured a series of keynote speeches that offered deep dives into pioneering research and emerging technologies. In addition to these talks, a wide range of oral presentations provided attendees with insights into breakthrough research in the current field. Complementing these were the poster displays, which offered a setting for presenters to showcase their work and engage in one-on-one discussions with interested peers.

AMCE 2024 has fostered extensive discussions and exchanges, with the collaborations established among attendees poised to lead to impactful research endeavors. Looking ahead, we hope the insights and knowledge shared at this conference will continue to inspire innovation in the fields of advanced materials and chemical engineering.

This conference proceedings represents the hard work of the authors and is also a recognition of their achievements. The papers accepted were gone through rigorous review process by 2-3 experts in related fields. The topics researched cover but not limited to nano materials and technology, smart materials, artificial band gap materials, all-oxide heterostructures and devices, high-performance computing in material design, data physical chemistry, electrochemistry and biosensors, etc.

Gratitude is extended to the anonymous referees who assessed the suitability of the articles for publication in the Journal of Physics: Conference Series. The invaluable feedback and constructive criticism provided by these referees ensured the high quality and scientific rigor of the work presented in this conference proceeding. A special thanks also goes to all the authors for their time and genuine efforts in preparing their contributions. Their dedication is a testament to the collaborative spirit that is essential to advancing knowledge in our field. Sincere appreciation is also extended to Guangdong University of Petrochemical Technology for their support, which played a pivotal role in the successful organization of this conference.

The Committee of AMCE 2024

Journal of Physics: Conference Series **2961** (2025) 011001
doi:10.1088/1742-6596/2961/1/011001

# Committee Member

**Conference General Chair**
Prof. Xianyou Wang, Xiangtan University, China
Prof. Saidur Rahman, Sunway University, Malaysia

**Technical Program Committee Chair**
Prof. Jingang Yu, Central South University, China
Prof. Guanghui Yue, Xiamen University, China
Assoc. Prof. CHING YERN CHEE, University of Malaya, Malaysia

**Publication Chair**
Prof. Jinbao Zhang, Xiamen University, China

**Organizing Committee**
Prof. Chunxiang Cui, Hebei University of Technology, China
Prof. Wenbo Liu, Sichuan University, China
Prof. Dongfang Yang, Guizhou Minzu University, China
Prof. Xuegong Yu, Zhejiang University, China
Prof. Sheng-Rui Jian, I-Shou University, China
Prof. Ir. Dr. Ts. Abdul Talib bin Din, Universiti Teknikal Malaysia Melaka, Malaysia
Prof. Chongwen Zhou, University of Science and Technology of China, China
Prof. Liang Li, Anhui University, China
Prof. Tao Yu, Tongji University, China
Dr. Hao Peng, Yangtze Normal University, China

**Technical Program Committee**
Prof. Dr. Yasuto Hijikata, Saitama University, Japan
Prof. Eckart H. Meiburg, University of California, Santa Barbara
Prof. DAVID BASSIR, French University of Technology, France
Prof. Volkan Kahya, Karadeniz Technical University, Turkey
Prof. David E. Esezobor, University of Lagos, Nigeria
A. Prof. Nour F. Attia, National Institute of Standards (NIS), Egypt
Prof. Jinyu Yang, Huizhou University, China
Prof. Sheng-Rui Jian, I-Shou University, China
Prof. Yuwu Zhong, Institute of Chemistry, Chinese Academy of Sciences, China
Prof. Xiaotao Hao, Shandong University, China
Prof. Guangyu Ren, Wuhan University, China
Prof. Qiang Zhao, Nanjing University of Posts and Telecommunications, China
Prof. Jie Xu, Wuhan Textile University, China
Prof. Guangzhi Hu, Yunnan University, China
Prof. Hongen Wang, Yunnan Normal University, China
Assoc. Prof. Xiaomin Kang, University of South China, China
Assoc. Prof. Chao Chen, Taiyuan University of Technology, China
Researcher Ning Nie, Sinochem Holdings, China
Researcher Yang Zhao, Southeast University, China
Senior Engineer Bingjie Zhou, Tianjin University, China
Dr. Hao Peng, Yangtze Normal University, China

Journal of Physics: Conference Series **2961** (2025) 011002 doi:10.1088/1742-6596/2961/1/011002

# Peer Review Statement

All papers published in this volume have been reviewed through processes administered by the Editors. Reviews were conducted by expert referees to the professional and scientific standards expected of a proceedings journal published by IOP Publishing.

- **Type of peer review:** Single Anonymous
- **Conference submission management system:** Morressier
- **Number of submissions received:** 98
- **Number of submissions sent for review:** 80
- **Number of submissions accepted:** 53
- **Acceptance Rate (Submissions Accepted / Submissions Received × 100):** 54.1
- **Average number of reviews per paper:** 3
- **Total number of reviewers involved:** 10
- **Contact person for queries:**
  **Name:** Xuexia
  **Email:** xx.ye@keoaeic.org
  **Affiliation:** AEIC Academic Exchange Information Centre

Journal of Physics: Conference Series **2961** (2025) 012001 doi:10.1088/1742-6596/2961/1/012001

# Research on the coating parameters and properties of $Cr_3C_2$-NiCr coating sprayed by HVOF

**Caiyan Liu[1], Zhengping Mao[2] and Xiaojie Xu[1*]**

[1]School of Artificial Intelligence, Shanghai Normal University Tianhua College, Shanghai 201815, China

[2]Oerlikon Metco Surface Technology (Shanghai) Co. Ltd, Shanghai, 201814, China

[*]Corresponding author's e-mail: hitment@163.com

**Abstract:** The influence of certain coating parameters of HVOF (High-Velocity Oxy-Fuel) spraying on the microstructure and properties of $Cr_3C_2$-NiCr coating and the mechanism of microstructure formation of the coating is investigated. The results indicate that the characteristics of the high-temperature and high-velocity flame jet enable the acquisition of coatings with high hardness (>910 $HV_{0.3}$), low porosity (<0.8%), high density (>99.2%) and uniform microstructure distribution. The neutral salt spray test demonstrates that the coating has good corrosion resistance properties.

## 1. Introduction

Surface modification has always been a research hotspot in the field of material preparation and performance development. It not only enables the reuse of substrates but also improves the surface properties of materials while responding to the national development concept of "green economy and circular development" [1-2]. Common surface treatment methods include nitriding, electroplating, thermal spraying, laser cladding, welding, and surface quenching. Among these, thermal spraying technology stands out due to its diversity of methods, the versatility of coating preparation, and the cost-effectiveness of its applications. It can deposit various characteristic ceramic coatings and their composite coatings on metal substrates [3-4].

Chrome carbide cermet coatings are widely used in different industries such as machinery, aerospace, automotive, hydropower, and printing [5-6] for improving the wear resistance of component surfaces and repairing the working surfaces of friction pairs due to their high hardness, high wear resistance, and good toughness. The preparation of $Cr_3C_2$-NiCr cermet coatings using thermal spraying technology is an important field of scientific research. HVOF (High-Velocity Oxy-Fuel) technology is considered the best and most widely applied technology for preparing $Cr_3C_2$-NiCr cermet coatings because such coatings exhibit excellent bond strength and wear resistance properties [7-8].

In summary, $Cr_3C_2$-NiCr coatings play a significant role in enhancing the performance and reliability of industrial equipment due to their unique physical and chemical properties, particularly in applications where wear resistance, high-temperature resistance, and corrosion resistance are particularly required.

## 2. Experimental process

### *2.1. Preparation of thermal spraying powder*

The powder used in the experiment was chromium carbide/nickel-chromium cermet, with the composition of $Cr_3C_2$-25 ($Ni_{20}Cr$) and the trade name of Woka 7202, provided by Oerlikon Metco (Shanghai) Co., Ltd. The powder was prepared using agglomeration and sintering methods with good sphericity and flowability. The particle size distribution was between 15-45 μm and the apparent density was in the range of 2.4-3.0 g/cm$^3$.

### *2.2. Sample and coating preparation*

The substrate material was a stainless steel sheet with dimensions of 70 mm in length and 20 mm in width, and a thickness of 4 mm. The coating material was Woka 7202 powder and the coating was sprayed by WokaStar 610 gun with Oerlikon Metco's Multicoat system. Before the powder was added to the feeder, it was baked in an oven at approximately 100°C for 1 hour to remove any moisture vapor that might have accumulated on the surface of the powder in the atmosphere, thereby improving the flowability of the powder. The stainless steel substrate was blasted with 60 mesh white $Al_2O_3$ grit to achieve a rough surface and to increase the bond strength of the coating. The temperature of the substrate was controlled to below 100°C with a compressed air cooling process during coating deposition in order to get high-quality coating and avoid oxidation of substrate material. The relatively high surface speed of 50 m/min was used to control the application rate of the $Cr_3C_2$-NiCr coating and acceptable residual stress in the coating in order to obtain good bond strength of the coating. The detailed coating parameters for spraying Woka 7202 are shown in Table 1.

**Table 1.** Coating parameters of woka 7202.

| Barrel Length (mm) | Oxygen Flow (L/min) | Kerosene Flow (L/h) | Carrier Gas Flow (L/min) | Powder Feed Rate (g/min) | Surface Speed (m/min) | Rotation Speed (r/min) | Step Size (mm) | Gun Speed (mm/s) | Spray Distance (mm) |
|---|---|---|---|---|---|---|---|---|---|
| 152 | 880 | 25 | 5 | 75 | 50 | 106 | 5 | 8.83 | 330 |

### *2.3. Microstructure and coating performance testing*

The morphology of the powder and the cross-sectional microstructure of the coating were evaluated using a metallographic microscope. The microhardness of the coating was measured using a Vickers hardness tester with a 0.3 kg load for a dwell time of 15 seconds. Ten values were tested for microhardness, and the average value was taken as the final hardness value. The macro hardness of the coating was measured using a Rockwell hardness tester with a 15 kgf load for a dwell time of 5 seconds. Five values were tested for macro hardness, and the average value was taken as the final hardness value.

### *2.4. Salt spray test of the coating*

According to ISO 9227 standard, a neutral salt spray test using a 5% sodium chloride solution was conducted to evaluate the corrosion resistance of the $Cr_3C_2$-NiCr coating. The areas of the coating samples were encapsulated with resin only leaving the coating surface exposed. The samples were then placed in a salt spray chamber. At regular intervals, the samples were removed, and cleaned with deionized water, and the corrosion condition of the coating was observed.

## 3. Results and discussion

### *3.1 Microstructure and performance of the coating*

Figure 1 shows the cross-sectional metallographic photograph of the $Cr_3C_2$-25 ($Ni_{20}Cr$) powder. From the photograph, it can be observed that the distribution of the hard phase $Cr_3C_2$ and the metallic phase NiCr within the powder and the particle size distribution is relatively uniform. Each agglomerated powder particle contains both metallic and ceramic phases. The sphericity is good, which ensures flowability during the powder-feeding process. Since the powder was prepared using an agglomeration and sintering method, some pores were inevitably present in the particles.

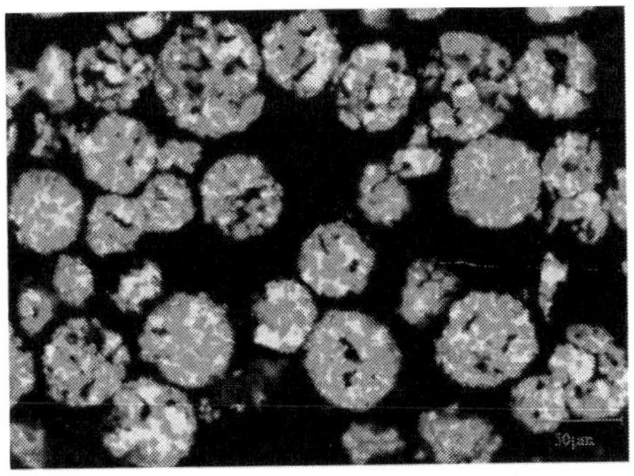

**Figure 1.** The morphology of $Cr_3C_2$-25 ($Ni_{20}Cr$) powder.

Figure 2 shows the metallographic structure of the coating prepared using the coating parameters in Table 1. From the microstructure of the coating, it can be seen that the coating is very dense with low oxides, which is mainly due to the high-speed particles forming a dense microstructure upon impact with the substrate. The metal binder NiCr and the hard phase $Cr_3C_2$ were distributed uniformly, exhibiting the typical lamellar structure of thermal sprayed coatings.

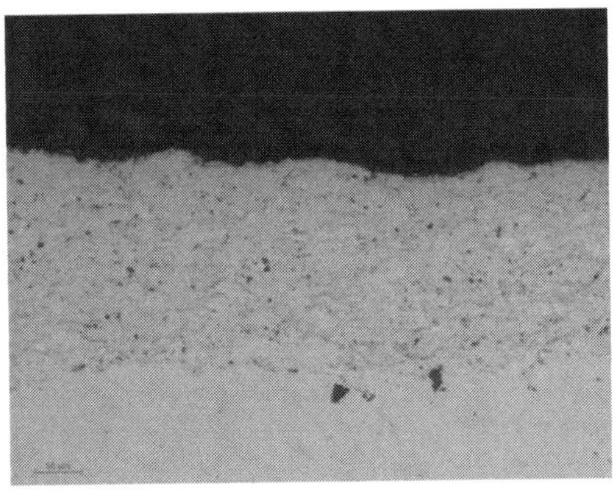

**Figure 2.** The cross-sectional morphology of $Cr_3C_2$-NiCr coating.

It is known that the particle velocity in the HVOF flame is very high (normally 600-1200 m/s) due to high gas flow and the special structure of the HVOF gun with a laval nozzle. This high-speed impact can cause particles to experience tremendous pressure at the moment of impact, forcing the particles to pack tightly together and reducing the formation of pores. In addition, the particles are heated to a molten or partially molten state during the flight of particles in the flame. When these molten or semi-molten particles strike the substrate, they are able to better fill the gaps between the particles, thereby forming a dense coating. The flattening of particles allows for better spreading and overlapping, which reduces the porosity within the coating. Due to the high flight speed of the particles during spraying and the short contact time with the air, the oxide content in the coating is relatively low, which can better maintain the original properties of the material. The dense coating helps to improve the hardness and wear resistance of the coating. The microstructure of HVOF sprayed coatings typically contains fine grains and fewer defects, a structure that is beneficial for increasing the hardness of the coating. At the same time, the retained $Cr_3C_2$ hard phase in the raw material has also greatly enhanced the hardness of the coating.

Journal of Physics: Conference Series **2961** (2025) 012001 doi:10.1088/1742-6596/2961/1/012001

As indicated in Table 2, the porosity of the coating is below 1%, specifically 0.6%. The microhardness is 910 HV0.3, and the macro hardness is 87.5 HR15N, which meets the general industry requirements for the density and performance of $Cr_3C_2$-NiCr coatings. Since the particle size distribution of Woka 7202 powder is between 15-45 μm, the as-sprayed coating surface is relatively rough, with a roughness of about Ra 6 μm. Depending on the specific coating specifications, it may be necessary to perform grinding and polishing if needed.

**Table 2.** The coating properties of $Cr_3C_2$-NiCr coating.

| Coating Thickness (μm) | Porosity (%) | Macro Hardness (HR15N) | Microhardness ($HV_{0.3}$) | Surface Roughness Ra (μm) |
|---|---|---|---|---|
| 210 | 0.6 | 87.5 | 910 | 6.04 |

*3.2. Salt spray test results of the coating*

Figure 3 shows the metallographic photos of the $Cr_3C_2$-NiCr coating after 48 hours and 96 hours of neutral salt spray testing in a 5% sodium chloride solution. From the surface condition of the coating, it can be observed that after 48 hours of salt spray corrosion, the coating still maintained its original morphology, demonstrating excellent corrosion resistance. When the salt spray test reached 96 hours, a few rust spots and some pores appeared on the surface. This is mainly due to the prolonged corrosion, which led to a certain chemical reaction between the iron elements in the metal matrix and the corrosive solution. The proportion of the corroded area was small, still reflecting the good corrosion resistance of the coating.

In the coating, chromium (Cr) is a key component that provides passivation. It can form a dense layer of chromium oxide ($Cr_2O_3$) in an oxidizing environment. This oxide layer has good stability and effectively prevents corrosive media such as oxygen and salt water from further penetrating into the interior of the coating, thereby protecting the underlying material from corrosion. Chromium carbide is not easily reactive with corrosive media and the very dense coating structure reduces the possibility of corrosion medium penetration, thus maintaining the integrity of the coating. In summary, the corrosion resistance of the Cr3C2-NiCr coating is the result of the combined effects of its chemical composition, microstructure, and coating preparation process.

**Figure 3.** The surface conditions of $Cr_3C_2$-NiCr coating after salt spray test with 48 hours (left) and 96 hours (right).

## 4. Conclusion

This paper mainly investigates the influence of certain coating parameters of HVOF spraying on the performance of $Cr_3C_2$-NiCr coatings and discusses the mechanism of how coating parameters affect the microstructure, porosity, microhardness, and distribution. The $Cr_3C_2$-NiCr coating prepared with these coating parameters exhibits good performance, meeting the needs of general industrial production for the wear resistance of component surfaces. The coating demonstrates good corrosion resistance after 48 hours and 96 hours of neutral salt spray testing.

**References**

[1] Medvedovski E. Advanced ceramics and coatings for erosion related applications in mineral and oil & gas production: A technical review[J]. International Journal of Applied Ceramic Technology, 2023,20(2):612-659.

[2] Singh S, Osi A, Joshi V V, et al. Development of fine WC-NiCr powder coatings by optimizing HVOF spray parameters[J]. International Journal of Refractory Metals and Hard Materials, 2024, 121:106667.

[3] Testa V, Morelli S, Bolelli G, et al. Corrosion and wear performances of alternative TiC-based thermal spray coatings[J]. Surface & Coatings Technology, 2022,438:128400.

[4] Qadir D, Sharif R, Nasir R, et al. A review on coatings through thermal spraying[J].Chemical papers, 2024, 78(1):71-91.

[5] Zhang D, Peng Z , Liu Z ,et al. Tribological characteristics of WC-$Cr_3C_2$-Ni cermet coatings under different wear parameters[J]. Journal of the Australian Ceramic Society, 2024, 60(3):871-884.

[6] Shi C, Liu S, Irfan, et al. Deposition mechanisms and characteristics of nano-modified multimodal $Cr_3C_2$-NiCr coatings sprayed by HVOF[J]. Reviews on Advanced Materials Science, 2022, 61(1):526-538.

[7] Murthy J K N, Venkataraman B. Abrasive wear behavior of WC-CoCr and $Cr_3C_2$-20(NiCr) deposited by HVOF and detonation spray processes[J]. Surface & Coatings Technology, 2006, 200(8):2642-2652.

[8] Encalada A I, Alidokht S A, Sharifi N, et al. Wear behavior of HVOF sprayed cobalt-based composite coatings reinforced with $Cr_3C_2$[J]. Wear, 2024, 546:205310.

Journal of Physics: Conference Series **2961** (2025) 012002 doi:10.1088/1742-6596/2961/1/012002

# Automatic recognition for cracks in aging rubber materials

**Yunhui Xu [1], Shifeng Wang [2], Hui Tu [1,*], Zhenghua Xin [3] and Haifeng Xu[3]**

[1] School of Materials Engineering, Xuzhou College of Industrial Technology, Xuzhou, Jiangsu, 221140, China

[2] School of Chemistry and Chemical Engineering, Shanghai Jiao Tong University, Shanghai, 200240, China

[3] School of Information Engineering, Suzhou University, Suzhou, Anhui, 234099, China

*Corresponding author's e-mail: LB21170010@cumt.edu.cn

**Abstract:** Rubber materials are extensively used in the industrial sector, where aging is a critical factor affecting the performance of various rubber materials in industrial applications. After aging, rubber surfaces exhibit crazing, a type of crack that currently relies on subjective visual observation. This method suffers from issues such as subjective judgment bias, difficulty in dynamic tracking, and quantification challenges. This study proposes an objective and effective method for identifying and analyzing the degree of erosion of rubber surfaces due to aging. The results of this research provide an automated recognition scheme for a deeper understanding of the rubber material aging process, making it possible to record the dynamic process of crack propagation and establish a relationship between morphology and performance. Furthermore, it provides process data for developing strategies to improve the aging resistance of rubber materials.

## 1. Introduction

Rubber materials are widely used in various industries due to their unique properties, such as elasticity, durability, and resistance to wear. Since they are often exposed to multiple environmental factors, this can lead to the surface cracking of rubber products and a significant reduction in their service life [1], accelerating their aging process. Aging is a common issue in the rubber industry [2, 3], with aging factors including ozone aging [4, 5], thermal aging [6], and ultraviolet aging [7], which can trigger chain reactions within the rubber matrix [8], leading to the formation of cracks and the loss of mechanical properties [9]. Accurate identification and analysis of various types of rubber aging are crucial for quality control, life prediction, maintenance planning, and the development of more durable rubber materials [10]. Although standards such as "GB/T 7762-2003 static tensile test for ozone resistance of vulcanized rubber or thermoplastic rubber" require the assessment of crack severity in the recording of experimental results, it is not easy to record the number of cracks per unit area and the average length of the ten largest cracks. Traditional methods for assessing rubber aging and crazing often rely on manual inspection or qualitative assessment, which can lead to inconsistencies in subjective judgment. Moreover, early-stage fine cracks cannot be effectively observed and recorded in a timely manner. Detecting early degradation is crucial for taking preventive measures. With the advancement of digital imaging and image processing technologies, this paper has developed an automated method for identifying rubber material aging cracks. The proposed method can automatically recognize and analyze ozone aging, thermal aging, and ultraviolet aging in rubber materials, objectively quantifying the extent of damage caused by aging to rubber products through the use of high-resolution imaging and advanced image processing algorithms. The introduction of this technology can change the way rubber aging is assessed, providing a more accurate, efficient, objective, and reliable evaluation method.

## 2. The proposed method

The traditional judgment of rubber cracks mainly relies on manual observation, such as visually inspecting for crazing after a specified period of aging or strain exposure and measuring the time at which crazing appears under any prescribed strain conditions. The former method is not easy to obtain ideal numerical results when faced with numerous and irregular cracks, while the latter requires constant observation to record the time at which cracks appear. Computer image processing has a wide range of applications in improving the quality of rubber crack images [11-13], including image enhancement techniques, feature extraction techniques, crack annotation, etc., which can achieve automated recognition and measurement, significantly saving labor and reducing subjective errors.

Firstly, the original image is subjected to image enhancement, which is done to make the image clearer and enhance the visibility of the feature areas while suppressing non-interesting features [14]. Secondly, brightness correction is performed. Since the photography of the rubber surface may include certain background areas, it is essential to eliminate the influence of the background. The rubber surface is black and generally darker than the background color. Therefore, this paper selects the bottom-hat operation for brightness correction. The bottom-hat operation, also known as the black-hat operation, is a type of closing operation in image morphology that can highlight the darker areas of an image [15]. Then, a 6×6 Gaussian low-pass filter and a replicate boundary padding filter are used for filtering treatment, with the aim of reducing image noise and highlighting the edges of the image. Subsequently, the image is binarized through an appropriate threshold to display the crack areas. Finally, the cracks are annotated. This step can calculate the number and length of cracks based on the distance and pixel count. The above processing steps are displayed in Figure 1.

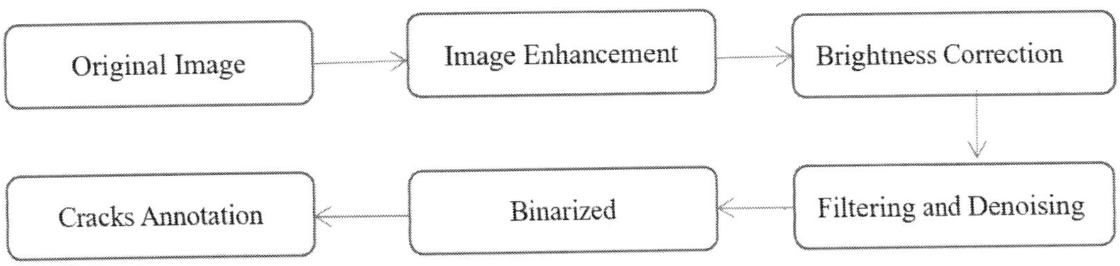


**Figure 1.** Process of rubber crack identification.

## 3. Results of the identification

The experiment selected rubber strips that had been aged in an ozone environment, including images with no cracks, a few cracks, and a large number of cracks. These images all have blurry backgrounds and uneven lighting conditions. The experimental results are shown in Figures 2 to 6. Additionally, the experimental diagrams from [16] and [17] were also recognized. The results indicate that the method presented in this paper can effectively identify cracks in various types of rubber.

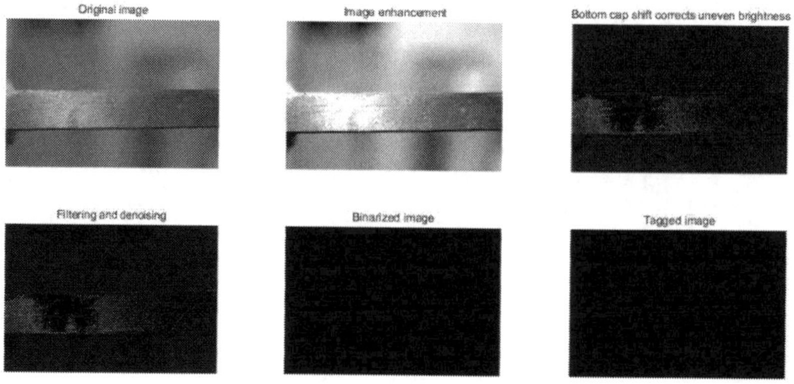


**Figure 2.** Crack-free specimen image recognition.

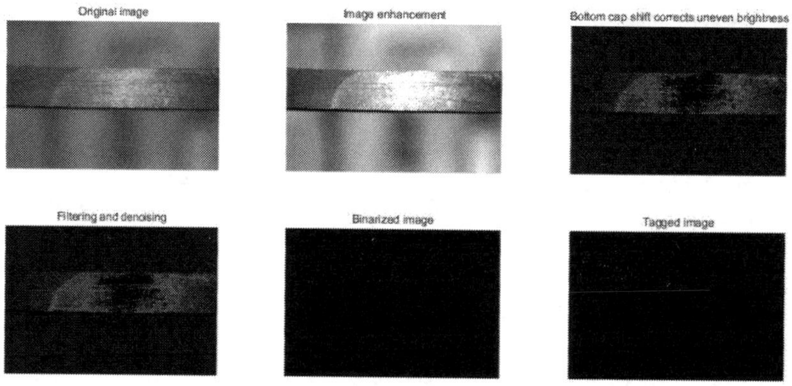


**Figure 3.** Crack-free specimen image recognition.

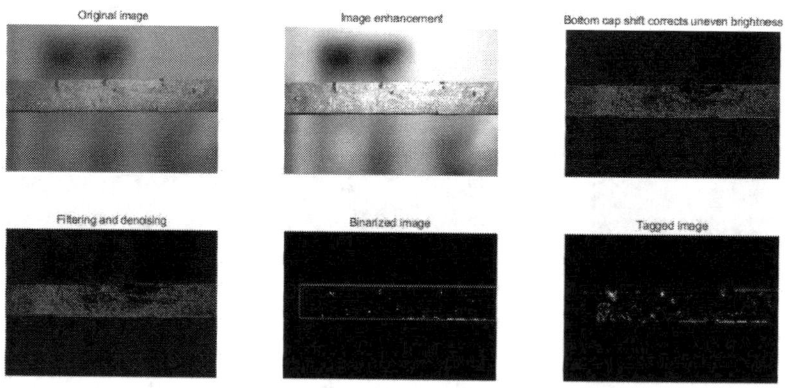


**Figure 4.** Recognition of specimen images with a few cracks.

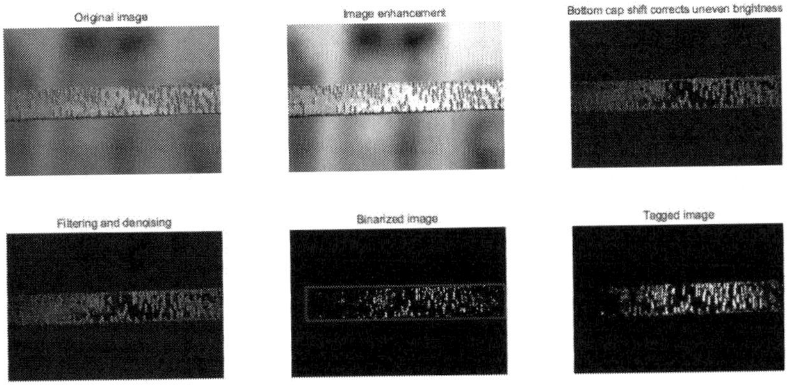


**Figure 5.** Recognition of specimen images with numerous cracks.

The rubber strips subjected to ozone aging follow the corresponding experimental standards. The method presented in this paper can effectively identify the number of cracks and the proportion of the overall rubber surface they occupy. It can also mark the specific locations of the oxidation cracks. The length and maximum width of the cracks can be calculated.

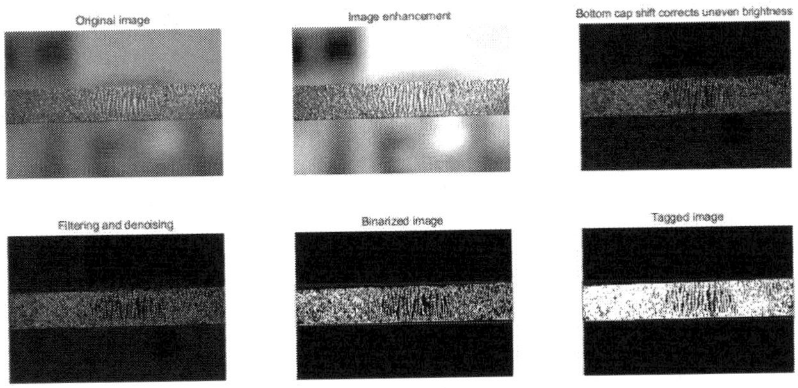


**Figure 6.** Recognition of specimen images with numerous cracks.

**Figure 7.** Original non-aged rubber image and recognition results.

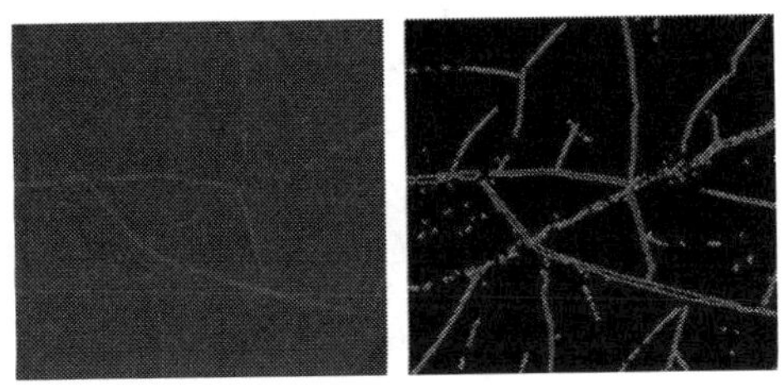

**Figure 8.** Irradiation with 1.35 $W{\cdot}m^{-2}$ intensity ultraviolet light at 40°C Celsius.

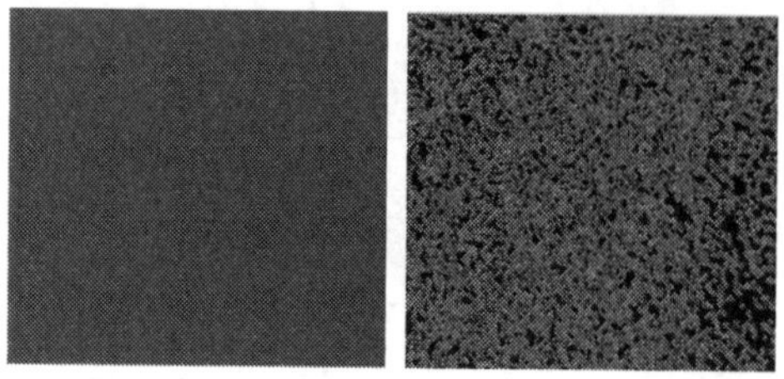

**Figure 9.** Thermal oxidative aging and recognition results in an 80°C environment.

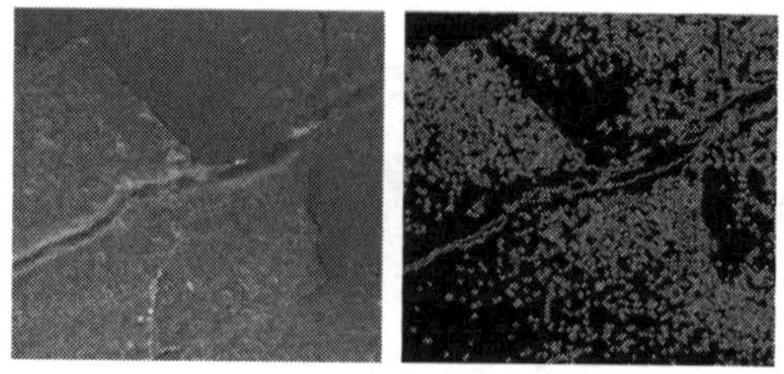

**Figure 10.** Thermal oxidative aging in an 80°C environment and irradiation aging under 1.35 W•$m^{-2}$ intensity ultraviolet light.

Under ultraviolet irradiation, the aging cracks in rubber products appear as irregular strips, while under thermal-oxidative aging, they appear as small flaky patches. When both factors act together, the cracks exhibit both strip-like and flaky characteristics. Figures 7 to 10 can identify a large number of crack conditions.

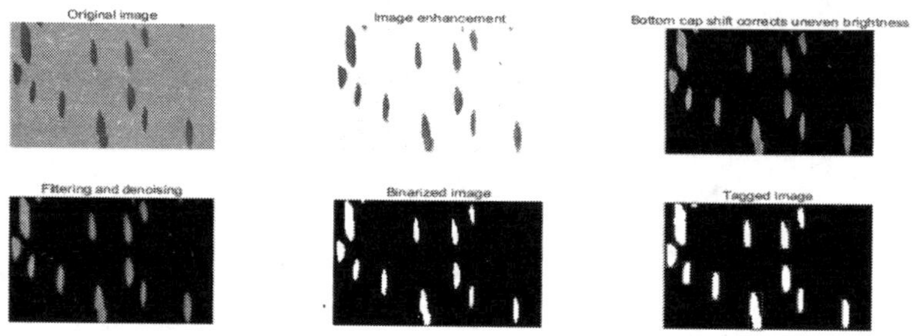


**Figure 11.** Rubber aging images and recognition results at a strain of 6%.

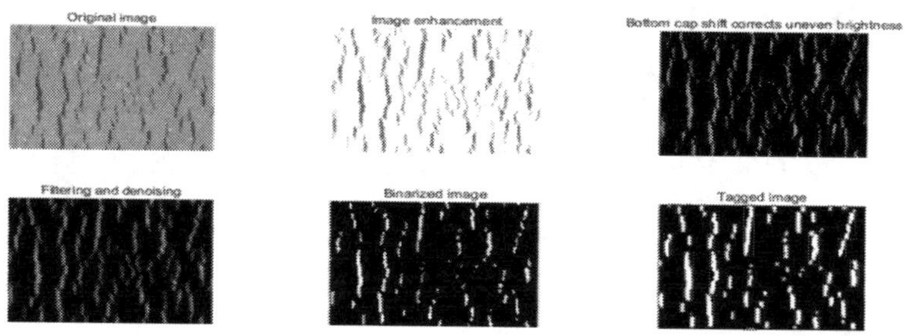


**Figure 12.** Rubber aging images and recognition results at a strain of 10%.

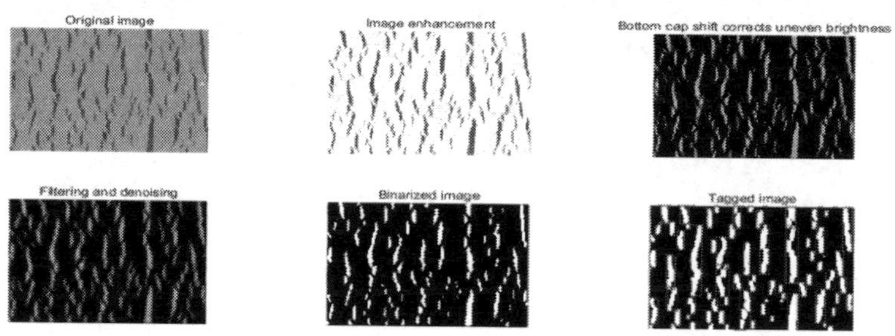


**Figure 13.** Rubber aging images and recognition results at a strain of 20%.

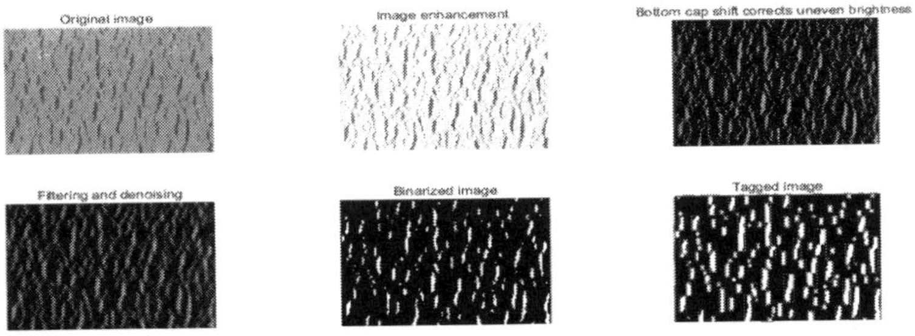


**Figure 14.** Rubber aging images and recognition results at a strain of 40%.

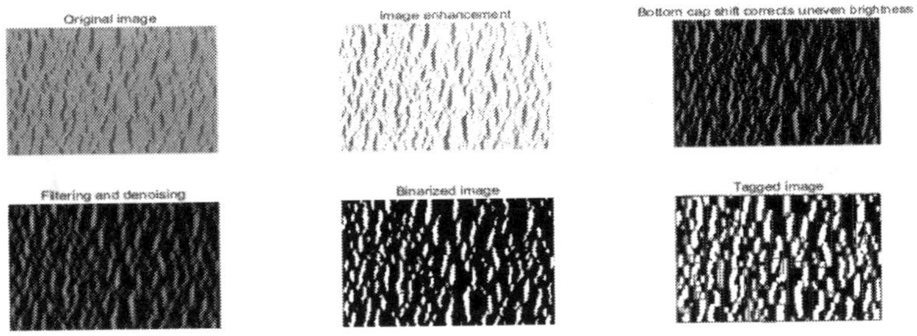


**Figure 15.** Rubber aging images and recognition results at a strain of 80%.

Figures 11 to 15 display the rubber crack conditions under different strain ratios. The method presented in this paper can effectively identify and mark the locations of the cracks. Combined with the magnification of the microscope and the size of the image, it can assist in calculating the number of cracks and their morphological characteristics.

## 4. Conclusion

In some assessment experiments of rubber performance, it is necessary to record the moment when visible crazing appears or to count the number of cracks that occur. This poses a challenge for manual vigilance in recording or manual counting. This paper proposes an automatic recognition method for the aging crazing of rubber products based on image processing, which can effectively identify cracks with various complex backgrounds, uneven lighting, and different types of aging. At the same time, the depth of the cracks does not increase linearly with the length of the cracks. There is still relatively little research on the relationship between crack depth and rubber performance. Therefore, in the future, we will continue to delve into the relationship between the depth of rubber product cracks and rubber performance, as well as the automated recognition method for rubber crazing based on video data, to achieve a fully automatic and comprehensive dynamic expansion data recording method for rubber cracks.

**Acknowledgments**

The work is supported by the National Natural Science Foundation of China, Mechanism and Application of Oxygen-Supply-Regulated Low-Temperature Efficient Mild Cracking of Rubber Powder (52173103), Jiangsu Province Higher Education Institutions Fundamental Research Program of Natural Science, and Micro-Oxygen Micro-Nano Recycling of Waste Rubber and Its High-Value Application in Tires (23KJA430015).

## References

[1] Gehling, Tobias, Jacopo Schieppati, Winoj Balasooriya, Roman Christopher Kerschbaumer, and Gerald Pinter. "Fatigue Behavior of Elastomeric Components: A Review of the Key Factors of Rubber Formulation, Manufacturing, and Service Conditions." Polymer Reviews 63, no. 3 (2023): 763-804. [CrossRef]

[2] Kunheng, Li, Chen Zhiyong, and Shi Wenku. "Rubber Aging Life Prediction Based on Interpolation and Improved Time-Temperature Superposition Principle." Materials Research Express 9, no. 1 (2022): 015301. [CrossRef]

[3] Treib, Caroline, Klara Loos, Michael Johlitz, and Alexander Lion. "Ozone Ageing: Experimental Methods on Pristine and Protected Natural Rubber." Continuum Mechanics and Thermodynamics 34, no. 6 (2022): 1563-77. [CrossRef]

[4] Liang, Meichen, Zhen Fu, Meng Guo, Xiuli Du, and Yiqiu Tan. "Effect of Ozone Aging on the Rheological and Chemical Properties of Crumb Rubber-Modified Bitumen." Road Materials and Pavement Design 25, no. 3 (2024): 529-46. [CrossRef]

[5] Tu, Y., W. Wang, M. Li, and L. Ding. "A Study on the Influence of Ozone on Aging of Htv Silicone Rubber." Paper presented at the 2009 IEEE Conference on Electrical Insulation and Dielectric Phenomena, 18-21 Oct. 2009. [CrossRef]

[6] Li, Zhou-Xian, Yi-Ran Kong, Xiang-Fei Chen, Ya-Jiang Huang, Ya-Dong Lv, and Guang-Xian Li. "High-Temperature Thermo-Oxidative Aging of Vulcanized Natural Rubber Nanocomposites: Evolution of Microstructure and Mechanical Properties." Chinese Journal of Polymer Science 41, no. 8 (2023): 1287-97. [CrossRef]

[7] Lin, Y., K. Chen, C. Heng, L. Dong, Z. Liu, and Y. Liu. "Gradient Degradation Behaviors of Silicone Rubber under Uv-B Radiation." Paper presented at the 2024 IEEE 7th International Electrical and Energy Conference (CIEEC), 10-12 May 2024. [CrossRef]

[8] Zhao, Wufan, Jing He, Peng Yu, Xueliang Jiang, and Liqun Zhang. "Recent Progress in the Rubber Antioxidants: A Review." Polymer Degradation and Stability 207 (2023): 110223. [CrossRef]

[9] Wang, Xiaolei, Hongwei Pan, Kun Yang, and Ping Zhang. "Cracking, Structural, and Mechanical Property Changes of Sibr and Related Elastomers During the Ozone Aging Process." Polymer Degradation and Stability 195 (2022): 109774. [CrossRef]

[10] Tayefi, Masoud, Mostafa Eesaee, Meysam Hassanipour, Said Elkoun, Eric David, and Phuong Nguyen-Tri. "Recent Progress in the Accelerated Aging and Lifetime Prediction of Elastomers: A Review." Polymer Degradation and Stability 214 (2023): 110379. [CrossRef]

[11] Hong, Shuxian, Chuan Kuang, Jianchao Zhang, Dongshuai Hou, Jinrui Zhang, Laibao Liu, and Biqin Dong. "Visual Analysis for Microscopic Cracking Propagation of Rubberized Concrete." Construction and Building Materials 265 (2020): 120599. [CrossRef]

[12] Mohan, Arun, and Sumathi Poobal. "Crack Detection Using Image Processing: A Critical Review and Analysis." Alexandria Engineering Journal 57, no. 2 (2018): 787-98. [CrossRef]

[13] Wang, JingWen, Chen Li, Xu Zhang, and YuJie Jiang. "Surface Crack Detection of Rubber Insulator Based on Machine Vision." Paper presented at the Intelligent Robotics and Applications, Cham, 2021//2021. [CrossRef]

[14] Panetta, K. A., E. J. Wharton, and S. S. Agaian. "Human Visual System-Based Image Enhancement and Logarithmic Contrast Measure." IEEE Transactions on Systems, Man, and Cybernetics, Part B (Cybernetics) 38, no. 1 (2008): 174-88. [CrossRef]

[15] Zhou, Hu, Dengdeng Shu, Chongjun Wu, Qingxia Wang, and Qibing Wang. "Image Illumination Adaptive Correction Algorithm Based on a Combined Model of Bottom-Hat and Improved Gamma Transformation." Arabian Journal for Science and Engineering 48, no. 3 (2023): 3947-60. [CrossRef]

[16] Zhou, Meng-Zhen, Hao-Ran Wang, Xing Guo, Yan-Chan Wei, and Shuangquan Liao. "Synergistic Effect of Thermal Oxygen and Uv Aging on Natural Rubber." 23, no. 1 (2023). [CrossRef]

[17] Treib, Caroline, Michael Johlitz, and Alexander Lion. "Crack Evaluation of ozone-Aged Elastomers." In Lectures Notes on Advanced Structured Materials 2, edited by Holm Altenbach, Leonhard Hitzler, Michael Johlitz, Markus Merkel and Andreas Öchsner, 103-21. Cham: Springer Nature Switzerland, 2024. [CrossRef]

# Preparation and microstructure of $BaTiO_3$ thin films by RF magnetron sputtering

**Zhengjie Zhang[1], Ruoxi Zhang[1], Feng Zhang[1,*], Xinguo Ma[1,*], Zhengwang Cheng[1] and Han Wang[1]**

[1] School of Chip Industry, Hubei University of Technology, Wuhan 430068, China

* Corresponding author's e-mail:fengz8186@hbut.edu.cn; maxg@hbut.edu.cn

**Abstract:** The preparation of high-quality films is a key challenge in building high-performance electro-optic modulators and achieving ultra-high-speed optical communication. In this research, we used magnetron sputtering to precisely prepare $BaTiO_3$ thin films on $SrTiO_3$ substrates, and systematically studied the effects of process parameters on their crystallinity and microstructure. Under optimized process conditions—a sputtering power of 60 W, chamber pressure of 1.0 Pa, argon-oxygen ratio of 6:1, substrate temperature of 350 °C, and a sputtering duration of 60 minutes—the thin film exhibited a <200> preferred orientation, a grain size of 38.7 nm, X-ray diffraction <200> half-peak full width is 0.288°, and demonstrated high crystalline quality and microstructural order. This research provides scientific evidence for the preparation of high-performance electro-optic modulator materials and has important academic and practical value for the development of Si-based photonic integration technology.

## 1. Introduction

The electro-optic modulators (EOMs) serve as essential devices for facilitating electro-optic signal conversion in optical communication systems and have garnered considerable attention in the field of optical communications in recent years[1]. The performance of EOMs is largely dependent on the electro-optic material employed. Among the numerous candidate materials, $BaTiO_3$ (BTO) stands out as one of the ideal materials for developing high-performance EOMs due to its exceptional photorefractive properties, high dielectric constant, and adjustable refractive index values[2]. The crystallinity of BTO thin films, as a key factor determining their electro-optic performance, yield, and production cost, has received considerable attention[3,4]. Specifically, the refractive index of BTO thin films is influenced by both material purity and crystal structure, resulting in a range of 1.6 to 2.4 within the visible light wavelength spectrum[5], offering substantial potential for applications in the field of electro-optic modulation.

The preparation process of BTO thin films significantly influences its growth orientation, surface roughness, grain size, thickness uniformity, defect concentration, and, ultimately, its electro-optic coefficient. Various methods have been employed for the synthesis of BTO thin films, including chemical vapor deposition (CVD)[6], pulsed laser deposition (PLD)[7], radio frequency magnetron sputtering (RF-MS)[8], sol-gel processing[9], and molecular beam epitaxy (MBE)[10]. While MBE and PLD exhibit advantages in producing high-quality, controlled epitaxial BTO thin films, their limitations in terms of slow deposition rates and small fabrication areas restrict their application primarily to laboratory research. In contrast, RF-MS stands out as the preferred industrial method for

Journal of Physics: Conference Series **2961** (2025) 012003 doi:10.1088/1742-6596/2961/1/012003

BTO thin film preparation due to its efficiency and capability for large-area deposition.

In this research, BTO films were prepared on $SrTiO_3$ (STO) substrate by RF-MS method, and the influences of sputtering pressure, argon-to-oxygen ratio, substrate temperature, and sputtering time on the crystallinity and microstructure of the films were investigated. Through systematic optimization of these process parameters, BTO thin films with exceptional crystallinity have been successfully obtained. This research not only provides a valuable reference for the preparation of high crystallinity perovskite films, but also establishes a solid foundation for the practical application of BTO films in EOMs and other optoelectronic devices.

## 2. Materials and methods

### *2.1. Sample preparation*

In this study, BTO thin films were prepared on an STO substrate with <100> crystallographic orientation using the RF-MS method. High-purity BTO target (diameter 2 inches, purity 99.99%) provided by Zhongnuo Xincai (Beijing) Technology Co., Ltd. Were employed. To guarantee substrate cleanliness, single-crystal STO substrates were first subjected to ultrasonic cleaning steps involving deionized water, acetone, and anhydrous ethanol. The cleaned STO substrates were dried under a nitrogen stream and then placed on the sample holder of a multi-target magnetron sputtering system (PD-500, Wuhan PDVACUUM Technology Co., Ltd.). The holder featured a heat radiation heating function to maintain the substrate temperature below 400°C during deposition and ensure uniform film growth through rotation.

In the chamber preparation phase, we first ensure that the vacuum is reduced to below 7.5 × 10-4 Pa to provide an almost impurity-free environment. Then, the STO substrate is degassed and cleaned with argon and oxygen plasma to further purify the surface. Finally, under a series of carefully designed process parameters, including argon-oxygen flow ratio (v), chamber pressure (P), sputtering power (W), substrate temperature (T), sputtering time (T) (see Table 1 for specific parameters), BTO films with different conditions were successfully deposited by RF magnetron sputtering.

**Table 1.** Specific process parameters for the preparation of BTO films using the RF-MS method.

| | v(sccm) | P(Pa) | W(W) | T(°C) | T(min) |
|---|---|---|---|---|---|
| | 3: 1 | | | | |
| Control group 1 | 6: 1 | 1.0 | 60 | 350 | 60 |
| | 9: 1 | | | | |
| | | 0.7 | | | |
| Control group 2 | 6: 1 | 1.0 | 60 | 350 | 60 |
| | | 1.3 | | | |
| | | | 40 | | |
| Control group 3 | 6: 1 | 1.0 | 50 | 350 | 60 |
| | | | 60 | | |
| | | | | 150 | |
| Control group 4 | 6: 1 | 1.0 | 60 | 250 | 60 |
| | | | | 350 | |
| | | | | | 45 |
| Control group 5 | 6: 1 | 1.0 | 60 | 350 | 60 |
| | | | | | 75 |

### *2.2. Testing and Characterization*

In this research, the ULtimaIV XRD system produced by the Japan RIKEN company was used to conduct X-ray diffraction (XRD) analysis on the BTO thin film to precisely elucidate its phase structure and crystal phase composition, while the intensity information of the diffraction peaks was used to evaluate its crystallinity. The microstructures, surface morphologies, and particle sizes of the BTO thin film were characterized using the Thermo Scientific Helios 5 CX SEM produced by the US FEI company. The test results provide a solid experimental basis for a deeper understanding of the crystallographic characteristics and microstructures of the BTO thin film and their impact on performance. They also provide guidance for subsequent material modification, device design, and performance optimization.

## 3. Results and discussion

### *3.1. Impact of sputtering power.*

Figure 1(a) shows the XRD patterns of BTO films prepared with different sputtering powers of 40 W, 50 W and 60 W. It is evident that, exclusive of the diffraction peaks attributed to the STO substrate, the remaining peaks correspond well with the standard diffraction peaks of BTO. This confirmation underscores that the fabricated BTO films consist of polycrystalline materials with a pristine perovskite structure and exhibit a distinct preferential orientation along the (200) direction.

Further analysis shows that the half-peak width (FWHM) of the BTO characteristic peak (200) decreases with the increase of sputtering power. Specifically, when the sputtering power is increased to 60W, the half-peak width of the characteristic peak reaches a minimum value of 0.318°, as shown in Figure 1 (b). This trend indicates that the increase in sputtering power is helpful in improving the crystallization quality of BTO films.

To quantitatively evaluate the impact of sputtering power on the grain size of BTO films, Equation (1) is given by:

$$D = \frac{0.9\lambda}{\beta\cos\theta} \tag{1}$$

In Equation (1), $\lambda$ is the X-ray wavelength (0.15406 nm), and $\beta$ is the half-peak width of the X-ray diffraction peak at $2\theta$ (FWHM).

The results indicate that the average grain size of BTO films prepared at different sputtering powers ranges from 15.6 nm to 29.0 nm, as illustrated in Figure 1(c). This study confirms the superior crystallinity of BTO films prepared at a sputtering power of 60W. The plausible mechanism behind the influence of sputtering power on the crystallization of BTO films may involve the acquisition of higher energy by sputtered particles as sputtering power increases, enabling easier migration and rearrangement on the substrate, thereby facilitating the crystallization growth of the films and the augmentation of grain size. However, the specific mechanisms underlying these effects necessitate further research and validation.

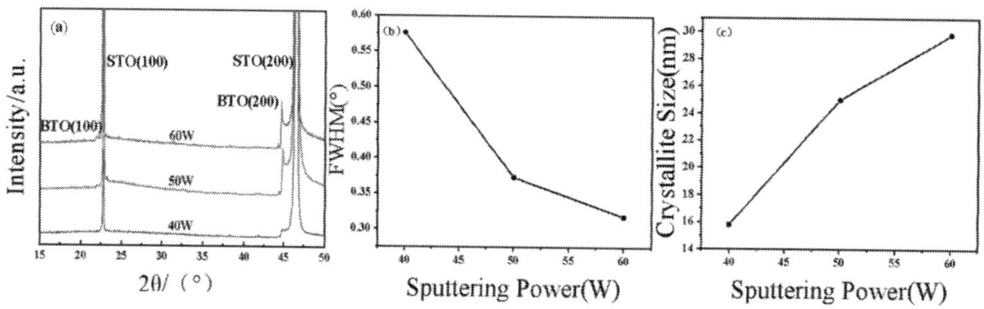


**Figure 1.** XRD pattern (a), bandwidth (b) and grain size (c) of BaTiO3 films deposited under controlled sputtering power.

### *3.2. Effect of sputtering pressure.*

Figure 2 (a) shows the XRD patterns of BTO films with different sputtering pressures (0.7 Pa, 1.0 Pa and 1.3 Pa). The XRD pattern analysis shows that the prepared BTO films all have good crystallinity, and when the sputtering pressure is 1.0 Pa, the prepared BTO films show a minimum FWHM of 0.36°

(as shown in Figure 2 (b)), and the corresponding films' average grain size reaches a maximum of 27.6 nm (as shown in figure 2 (c)). Sputtering pressure is one of the key factors affecting the microstructure of BTO films. The sputtering pressure is controlled at 1.0 Pa, and the crystal quality and grain growth of BTO films reached the optimal state. The internal mechanism may be that the sputtering particles have a more suitable collision frequency and energy distribution during flight, which promotes the crystal growth and grain size increase of the films.

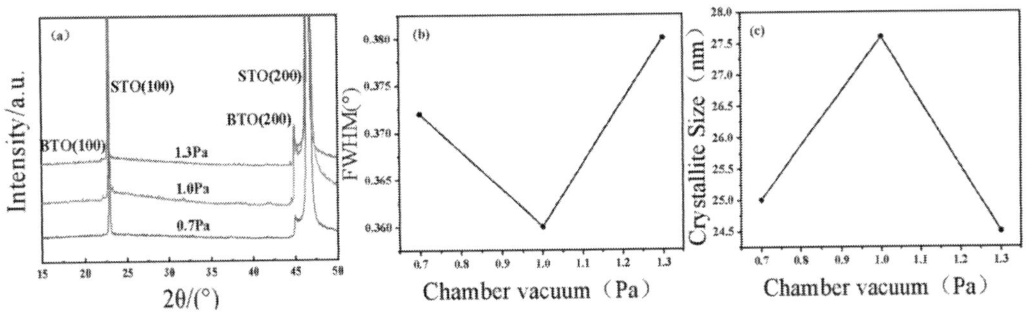


**Figure 2.** XRD pattern (a), XRD peak bandwidth (b) and grain size (c) of barometric BaTiO3 film under controlled sputtering.

### *3.3. Influence of argon-to-oxygen ratio.*

Figure 3(a) displays the XRD patterns of BTO films with different sputtered $Ar/O_2$ ratios of 3:1, 6:1 and 9:1. Analysis of the XRD reveals that when the $Ar/O_2$ ratio is 6:1, the prepared BTO film exhibits a minimum FWHM of 0.358° (as shown in Figure 3(b)), corresponding to a maximum average grain size of 25.8 nm (as illustrated in Figure 3(c)).

The experimental results indicate that the sputtered $Ar/O_2$ ratio is one of the key factors regulating the growth characteristics of BTO films. Under conditions with an $Ar/O_2$ ratio of approximately 6:1, the crystal quality and grain growth of BTO films reach the optimal state. We speculate that the underlying mechanism may be that the effective mixing of oxygen and argon at this ratio provides a suitable oxidation atmosphere and sputtering conditions for film growth, thus promoting the crystal growth and grain size increase of thin films.

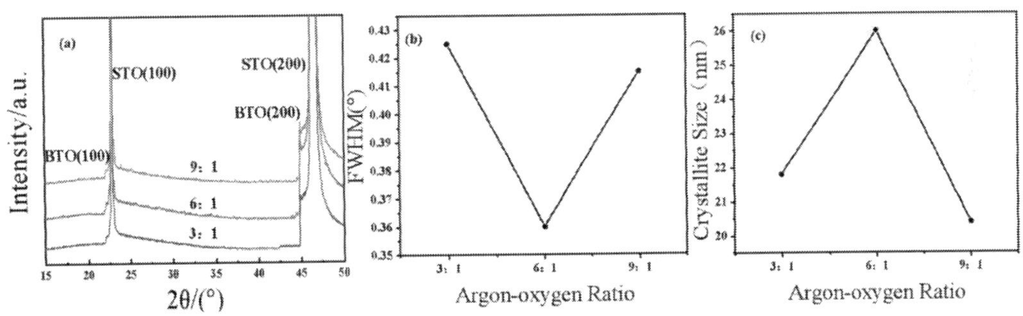


**Figure 3.** XRD pattern (a), bandwidth (b) and grain size (c) of BaTiO3 film deposited under a controlled argon-oxygen flow rate.

### *3.4. Influence of base temperature.*

The substrate temperature is a key factor in regulating the nucleation and growth of thin films, which affects the microstructure of BTO films. Figure 4(a) shows XRD patterns of films deposited at different substrate temperatures with barium titanate. It was found that as the substrate temperature rises, the FWHM of BTO film at (200) preferred orientation showed a decreasing trend (Figure 4(b)). When the substrate temperature reached 350 °C, FWHM reached the minimum value of 0.373°, and the corresponding average grain size of BTO film reached the maximum value of 25 nm (Figure 4(c)).

The results show that a higher substrate temperature is conducive to improving the crystal quality of the film and reducing the lattice defects. The mechanism is that when the substrate temperature is low, it is difficult for surface particles to obtain enough kinetic energy from the lattice vibration of the substrate material for effective migration, resulting in premature condensation of particles and insufficient nucleation and growth. In contrast, a suitable substrate temperature can ensure that the particles get enough energy to migrate and rearrange, thereby optimizing the crystalline structure of the film, increasing the grain size, and improving the overall performance of the film.

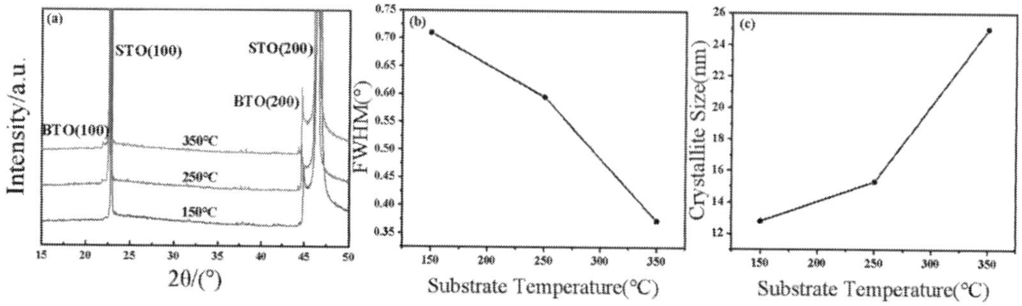


**Figure 4.** XRD pattern (a), bandwidth (b) and grain size (c) of BaTiO3 film deposited under controlled substrate temperature.

*3.5. Impact of sputtering time.*

Figure 5(a) shows the XRD pattern of BTO films deposited at different sputtering times (45 min, 60 min and 75 min). The XRD pattern analysis shows that when the sputtering time is 60 min, the prepared BTO film presents a minimum of 0.288° at the FWHM of (200) preferred orientation (as shown in Figure 5(b)). The average grain crystal size of the corresponding film reaches a maximum of 38.7 nm (as shown in Figure 5(c)).

The experimental results show that sputtering time has a significant effect on the crystallinity and microstructure of BTO films. The internal mechanism is that a sputtering time that is too short may lead to insufficient film thickness and insufficient deposition and crystallization of particles. If the sputtering time is too long, it may cause excessive particle bombardment, resulting in surface damage and a decrease in crystal quality. Therefore, there is an optimal sputtering time, so that the film can achieve the best crystal quality and grain growth while ensuring sufficient thickness. In this study, the sputtering time of 60 min was proved to be the optimal choice for the preparation of high-performance BTO films.

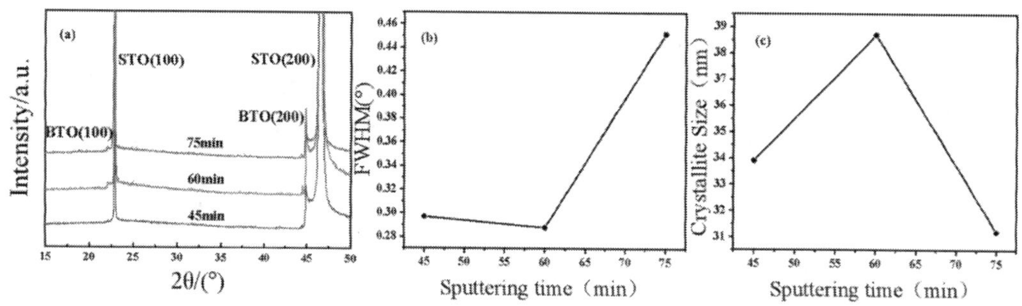


**Figure 5.** XRD pattern (a), bandwidth (b) and grain size (c) of BaTiO3 films deposited under controlled sputtering time.

*3.6. Structure and morphological characteristics of optimized BaTiO3 <200> films.*

Based on the research results from 3.1 to 3.5, the optimal process parameters were determined, including sputtering power of 60 W, sputtering pressure of 1.0 Pa, Ar/O2 ratio of 6:1, sputtering

Journal of Physics: Conference Series **2961** (2025) 012003 doi:10.1088/1742-6596/2961/1/012003

temperature of 350 °C, 60 minutes sputtering time to prepare BTO films. The XRD pattern is shown in Figure 6 (a). The FWHM at (200) preferred orientation is 0.288°, and the corresponding average grain size is 38.7 nm, indicating that the film has high crystallinity and good orientation. SEM was further used to characterize the surface morphology and cross-section of the film. Figure 6 (b) shows the surface morphology of the prepared BTO film. It was found that the surface of the film was smooth and flat, with good crystallization, uniform grain size and compact arrangement, showing excellent flatness. The profile of Figure 6 (c) clearly indicates that there is an apparent interface between the STO substrate and BTO film, the interface of the BTO film is flat, and the thickness is about 276 nm, which further verifies the stability and controllability of the deposition process.

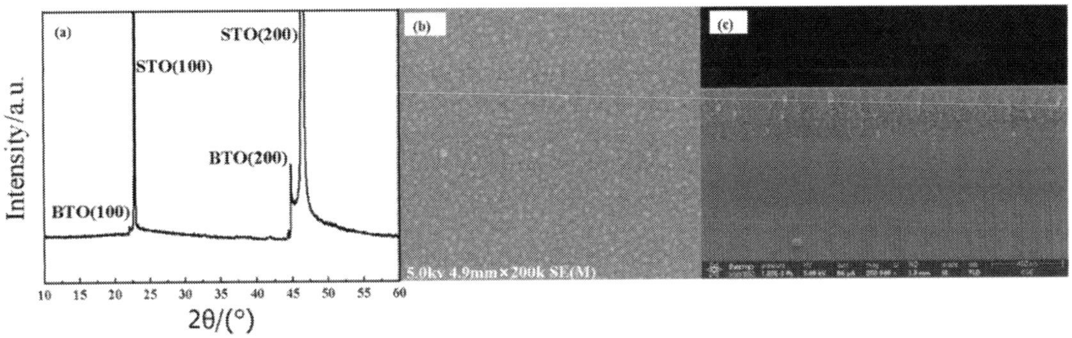


**Figure 6.** XRD pattern of BTO thin film prepared with optimized process parameters (a) SEM (b) surface and SEM (c) cross-section.

## 4. Conclusion

By systematically exploring and optimizing the RF-MS deposition parameters of BTO films, this research aims to reveal the impact of these key parameters on the crystallinity, grain size, and microstructure of BTO films. The results indicate that when the sputtering power is 60 W, the chamber vacuum is 1.0 Pa, the flow rate ratio of $Ar/O_2$ is 6:1 sccm, the substrate temperature is 350°C, and the sputtering time is 60 minutes. The BTO films exhibit high crystallinity, large grain size, and excellent surface smoothness. This research finding is of great significance for the application of BTO films in the field of EOMs, providing a reliable experimental and theoretical basis for subsequent research and applications.

**Acknowledgments**

This work has been supported by various funding and logistics. The Key Laboratory of Inorganic Functional Materials and Devices, Chinese Academy of Sciences, is an open project (approval number: "Preparation of ultra-high electro-optical coefficient $BaTiO_3$ single crystal film and electro-optical modulation characteristics at temperature field" (KLIFMD202302); research on the structure and mechanism of organic ferroelectric thin film integrated flexible sensor for touch/fingerprint recognition supported by the Doctoral Research Start-up Fund project (XJ2021008501).

**References**

[1] Palodiya V, Raghuwanshi S K (2016) Performance Study of optical Modulator based on electro-optic effect. C. Journal of Physics: Conference Series. IOP Publishing. 735(1).

[2]Jiang, Beibei, et al (2019) $BaTiO_3$ at the nanoscale: controlled synthesis and dielectric and ferroelectric properties. *Chemical Society Reviews*. 48.4, 1194-1228.

[3]Tewatia K, Sharma A, Sharma M, et al (2021) Factors affecting morphological and electrical properties of $BaTiO_3$: A brief review. J. Materials Today: Proceedings. 44, 4548-4556.

[4]Cheng Z, Wang A, Bo H, et al (2023) Preparation and structural investigation of ultra-uniform Mo films on a $Si/SiO_2$ wafer by the direct-current magnetron sputtering method. J. Crystal Growth & Design. 23(2), 1014-1022.

Journal of Physics: Conference Series **2961** (2025) 012003 doi:10.1088/1742-6596/2961/1/012003

[5]Cao Y, Liang H, Lin H L, et al (2023) Engineering refractive index contrast in thin film $BaTiO_3$-on-insulator. J. Nano Letters. 23(16), 7267-7272.
[6]Reinke, Michael, et al (2017) Low temperature epitaxial $BaTiO_3$ thin film growth in high vacuum CVD. *Advanced Materials Interfaces* 4.18. 1700, 116.
[7]Saroukhani, Zahra, et al (2015) Effect of working pressure and annealing temperature on microstructure and surface chemical composition of barium strontium titanate films grown by pulsed laser deposition. *Bulletin of Materials Science.* 38, 1645-1650.
[8]Li J, Zhang C C, Wang Y L, et al (2017) Radio frequency (RF) magnetron sputtered barium strontium titanate (BST) thin film.J. Key Engineering Materials. 727, 942-946.
[9]Sharma, Savita, et al (2015) Ultraviolet radiation detection by $BaTiO_3$ thin films grown by sol–gel hydrothermal method. *Sensors and Actuators A. Physical* 230, 175-181.
[10]Lin, Edward L., et al (2019) Atomic layer deposition of epitaxial ferroelectric $BaTiO_3$ on Si (001) for electronic and photonic applications. *Journal of Applied Physics.* 126,6.

# Study on the properties of silicone-modified basalt fiber composites under damp heat aging

**Wenhao Lu[1], Xiaoxing Wei[1], Xiang Xiao[1], Jinyin Zhang[1], Zhixiong Huang[2], Peng Wu[3]* and Hecheng Liu[3]**

[1]Electric Power Research Institute of Uhv Transmission Company of China Southern Power Grid Co., Ltd. Guangzhou, China
[2]China Southern Power Grid Company Limited, Guangzhou, China
[3]Hebei Key Laboratory of Green and Efficient New Electrical Materials and Equipment North China Electric Power University, Baoding, China

* Corresponding author: peng_k60@163.com

**Abstract:** The humid and hot environment is a key factor that threatens the reliability of basalt fiber-reinforced polymer (BFRP). In this work, a sizing agent with silicone-modified epoxy resin emulsion was used to study the capability of silicone-modified basalt fiber composites under damp heat aging. The test results show that the modified composite BFRP-SER has good resistance to damp heat aging, and its flexural strength and breakdown field strength decrease by 30.1% and 35.2%, respectively. This is because chemical bonds such as Si-O bonds and C-O bonds have high bond energy, and more energy is required to break these chemical bonds at high temperatures, resulting in a lower degree of hydrolysis of BFRP-SER and hindering the penetration of the medium, thus maintaining performance for a longer period of time.

## 1. Introduction

Basalt fiber (BF) is a green glass fiber made by melting pure natural basalt at high temperatures. Its production process does not require additives, and it does not cause environmental pollution or produce carcinogens, hence being called "the green strategic resource of the 21st century" by the scientific community. Basalt fiber has excellent resistance to high temperatures, corrosion, high tensile strength, and stiffness, so it is widely used in the aerospace, national defense equipment, and construction engineering fields. Additionally, basalt fiber also possesses good insulation and dielectric properties, making it promising for applications in the field of electrical insulation, such as making insulator core rods, insulating pull rods, etc[1-2]. Figure 1 shows the composite mandrel in production.

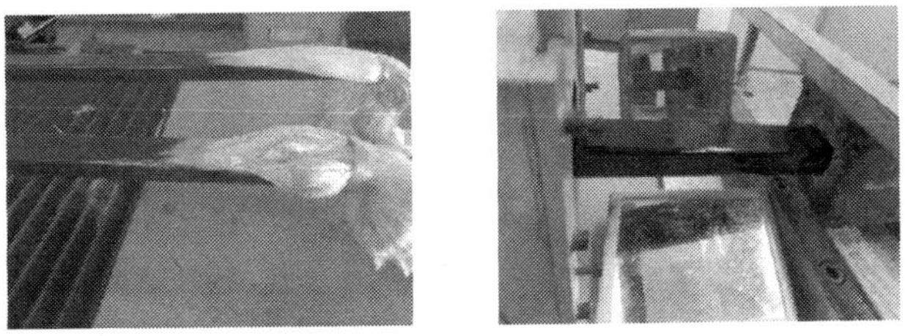

**Figure 1.** Composite mandrels.

IOP Publishing
doi:10.1088/1742-6596/2961/1/012004

However, basalt fiber has a smooth surface, strong chemical inertness, and poor wettability, which leads to a relatively high porosity in basalt fiber-reinforced composites and low adhesion strength in the BFRP. Therefore, it is common to modify the basalt fiber to change its surface roughness and chemical reactivity, thereby enhancing the adhesion strength in the BFRP[3-4].

Basalt fiber sizing agents primarily consist of many types of agents. Among these, the film-forming agent accounts for the largest proportion. Therefore, the selection of the agent and the design of sizing formulations are crucial for enhancing the functionality of BFRP, and as a result, scholars have conducted extensive research in this area. Wei et al. used an epoxy resin/$SiO_2$ hybrid sizing agent for the modification of basalt fiber (BF), increasing the tensile strength of BF bundles and interlaminar shear strength of BFRP[5]. Xu et al. prepared basalt fiber reinforced cement-based (BFRC) composites using three BF. The results showed that the sizing agents greatly influence the properties of the BFRC, with the sizing agent containing KH-560 showing the best improvement effect on the BFRC interface[6].

This article uses weather-resistant silicone-modified epoxy resin emulsion as the film-forming agent component, which is formulated into a wetting agent slurry to coat and modify basalt fibers. The basalt fiber reinforced composite BFRP-SER and the commercial control BFRP-CK are prepared, and the performance of the silicone-modified basalt fiber composites under wet heat aging is studied.

## 2. Experiment

### *2.1. Materials*

The materials utilized in this study include: basalt fiber plain fabric, 300g $m^{-2}$, with a single filament diameter of 22μm (provided by Sichuan Aerospace Tuoxin Basalt Industrial Co., Ltd, China); bisphenol A epoxy resin (DGEBA) with an epoxy value of 0.51mol/100g (Zhejiang Qiyi Electrical Technology Co., Ltd, China); the curing agent is methylhexahydrophthalic anhydride (MHHPA), and the epoxy curing accelerator is DMP-30 (Zhejiang Qiyi Electrical Technology Co., Ltd, China); silicone-modified epoxy resin (SER) emulsion, with an epoxy value of 0.06-0.16mol/100g (Shenzhen Jipeng Silicon Fluorine Materials Co., Ltd, China); hydrogenated bisphenol A epoxy resin (HBPA) with an epoxy value of 0.42-0.45mol/100g (Shanghai Yueyi Chemical Co., Ltd, China); the silane coupling agent is KH-560 (Dongwan Kangjin New Material Technology Co., Ltd, China); lubricant NBR-1090, antistatic agent NBR-1270 (Nanjing Fiberglass Research Institute, China).

### *2.2. Sample preparation*

*2.2.1 Preparation of sizing agent.* We measure a certain amount of deionized water in a beaker and slowly add the coupling agent while stirring the solution until the coupling agent is completely hydrolyzed, with no apparent oily substances present. Then, it is transferred into a beaker, and we add the lubricant, antistatic agent, and fully hydrolyzed coupling agent, mix the mixture thoroughly, and stir in a magnetic stirrer for about one hour to obtain the required sizing agent. The component ratios of the sizing agents are shown in Table 1.

**Table 1.** Component ratios of the sizing agent.

| Film-forming agent | Coupling agent | Lubricant | Antistatic agent |
|---|---|---|---|
| SER | KH-560 | NBR-1090 | NBR-1270 |
| 5 wt% | 1 wt% | 0.4 wt% | 0.1 wt% |

*2.2.2 Surface modification of basalt fiber.* Basalt fiber plain weave was positioned in a Soxhlet extractor filled with acetone and treated for 12 hours to remove the commercial wetting agent from the fiber surface. It was then cleaned five times using deionized water and put into an oven to dry for 10 hours. The acetone-treated fiber cloth is designated as BF-CK. The dried fiber plain was immersed in the wetting agent for 1 hour and then dried at 80°C for 5 hours. The modified fiber plain is designated as BF-SER.

*2.2.3 Preparation of BFRP.* The bisphenol A epoxy resin, methylhexahydrophthalic anhydride, and DMP-30 are mixed in a mass ratio of 100:80:0.5 to prepare a resin mixture, which is then mixed in a planetary agitator for 2 minutes and defoamed. The commercial fiber fabric BF-CK and the modified fiber fabric BF-SER are placed in a metal mold, and the prepared resin mixture is added until the fabric is completely impregnated. The metal mold is then placed in a vulcanizer and hot pressed at 140°C and 10MPa for 40 minutes. The resulting BFRP is post-curing at 140°C for 10 hours. The prepared BFRP samples are respectively named BFRP-CK and BFRP-SER.

### *2.3. Test Method*

*2.3.1 Damp heat aging test design.* Artificial accelerated aging is carried out using a constant temperature water bath method, with a 0.1% NaCl solution at 100°C as the aging medium to accelerate the degradation rate of the composite specimens. The aging duration is set at 30, 60, 90, and 120 hours, and the surface of the sample is wiped after sampling.

The test samples are cut for subsequent testing. Since the properties of BFRP are significantly affected by the arrangement of fibers, in this study, the BFRP is sliced along the orientation of the parallel fiber bundles to reduce the influence of fiber arrangement on the test results.

*2.3.2 Performance testing.* Fourier transform infrared spectroscopy (FT-IR): Infrared spectroscopy analysis of the basalt fiber, both pre- and post-modification, was performed using the Nicolet Summit FTIR spectrometer. The scanning range was set from 400 to 4000 $cm^{-1}$, with a resolution of 4 $cm^{-1}$, and each scan was conducted 32 times.

Bending test: Referring to the standard GB/T 1449-2005, the three-point bending method was used to conduct at least five bending tests on 40mm × 15mm × 1mm (length × width × height) composites. The average of the final results was taken, and the testing speed was adjusted to 10 mm/min.

Breakdown strength: The AC breakdown voltage test is conducted according to IEC60243-1 with a sample thickness of 1±0.1mm and a test frequency of 50Hz. The breakdown data are statistically analyzed using the Weibull distribution model, with a minimum of 15 valid samples.

Leakage current: Sample composites of dimensions 80mm × 1mm × 30mm (length × width × height) using a digital leakage current measurement device DM-3068. The test voltage is 10kV, with a voltage ramp rate of 0.5kV $s^{-1}$.

Dielectric loss factor: According to the GB/T 1409-2006 standard, we employ a fully automatic precision dielectric loss tester (YG9100, Shanghai Yanggao Electric Co., Ltd., China) to measure the tangent of dielectric loss for a 1.5mm thick sample at power frequency. The testing conditions include a temperature of 25°C, a voltage frequency of 50Hz, and a test voltage of 2kV.

## 3. Result and discussion

### *3.1. FT-IR*

Figure 2 displays the FT-IR spectra of basalt fibers before and after modification. The results indicate a substantial increase in the number of absorption peaks following treatment with the wetting agent. Both fiber types show a prominent absorption peak near 3440 $cm^{-1}$, associated with the stretching vibration of hydroxyl groups (-OH) on the basalt fiber surface. After modification with the wetting agent, this peak is significantly intensified, suggesting an increase in hydrogen bond formation either between the wetting agent and the fiber surface or among the molecules of the wetting agent.

Additionally, the treatment with the sizing agent results in the emergence of new functional groups on the fiber surface. The absorption peaks observed at 3100-3000 $cm^{-1}$ and 700 $cm^{-1}$ in sample BF-SER correspond to the C-H bond vibrations of the benzene ring present in the SER sizing agent. Peaks at 2932 $cm^{-1}$ and 2871 $cm^{-1}$ are due to the stretching vibrations of $CH_2$ and $CH_3$ groups, respectively. The peak at 1729 $cm^{-1}$ is associated with the carboxyl group in the C=O double bond, while the peak at 1260 $cm^{-1}$ indicates the symmetric stretching vibration of epoxy groups from the epoxy film-forming agent. Finally, the absorption peak at 1100 $cm^{-1}$ is characteristic of the Si-O bond found in both the coupling agent KH-560 and the SER sizing agent. This analysis confirms that both sizing agents have effectively bonded to the fiber surface.

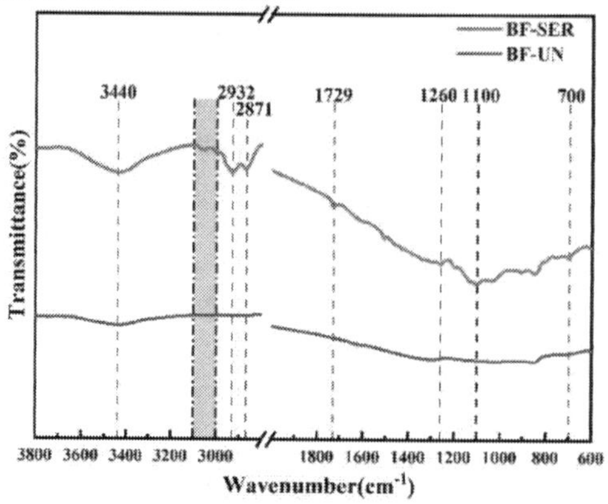


**Figure 2.** FT-IR spectra of basalt fiber samples.

*3.2. Bending strength*

The bond strength at the interface between the fiber and the matrix plays a crucial role in the flexural strength of the composite. When subjected to load, the difference in stiffness between the fiber and the matrix can lead to significant local stress and strain concentrations in the matrix, resulting in microcracks. If the interface bonding between fibers and matrix is poor, the efficiency of load transfer to the fibers will decrease, leading to increased component deformation and more microcracks, accelerating the deterioration process of the composites[7].

Figure 3 presents the bending strengths of the two composites. Before aging, BFRP-CK exhibits an average bending strength of 612 MPa, while BFRP-SER shows an average of 674 MPa. After 120 hours of aging, the average bending strengths of both materials decrease to 333 MPa and 471 MPa, with reductions of 45.6% and 30.1%, respectively. It is noteworthy that during the aging period of 90 to 120 hours, the bending strength drops significantly, decreasing by 15.2% and 10.2% of the initial strength. This is because, in the early stages of aging, the composites have fewer pores and damages, and the degree of hydrolysis is also low, resulting in a less noticeable decline in bending strength. In the later stages of aging, local interfacial failure can hinder the effective transfer of loads from the matrix to the fibers and significantly reduce the flexural strength of the composite.

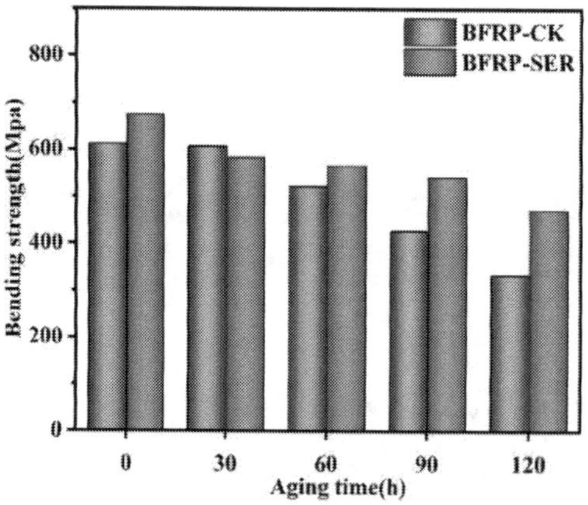


**Figure 3.** Bending strength of composites.

*3.3. Breakdown field strength*

Defects or voids in composites can raise the local electric field strength, influencing the breakdown field strength. Thus, the insulation performance of the samples is linked to the number of defects present in the sizing agent coating. In addition, during the process of wet heat aging, due to the diffusion migration and capillary action of the medium, invading media such as $Na^+$, $Cl^-$, $H_2O$, etc., form conductive channels under the electric field, causing significant damage to the insulation performance[8].

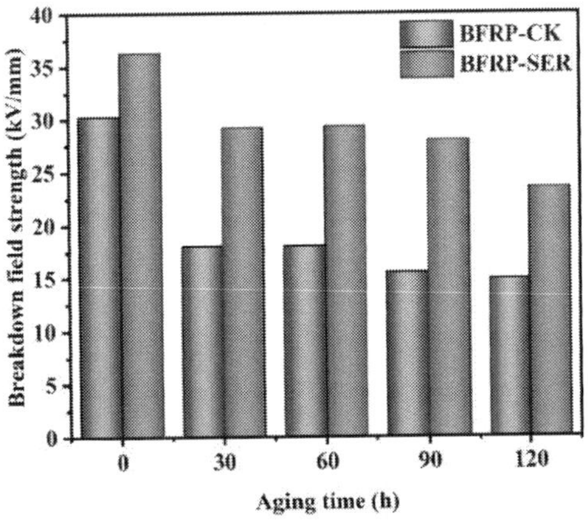


**Figure 4.** Breakdown field strength of composites.

The breakdown field strength test results of the composites are shown in Figure 4. Before aging, both materials have high insulation strength, with breakdown field strengths of BFRP-CK and BFRP-SER being 30.29 kV/mm and 36.31 kV/mm, respectively. However, as wet-heat aging progresses, the breakdown field strength of BFRP-SER remains relatively high (above 28.01 kV/mm) for a longer duration, while that of BFRP-CK rapidly decreases to around 15 kV/mm. Analyzing this in conjunction with the bending strength after aging, the BFRP-SER samples experience relatively low bending strength loss during the aging period of 0 to 90 hours. This indicates that the interface defects caused by wet-heat aging are also relatively few in BFRP-SER, allowing it to better resist medium intrusion and maintain a higher insulation strength over a longer period.

### *3.4. Leakage current*

The leakage current in composites is closely tied to the quality of interface bonding. Poor bonding can create pathways for charge transport, leading to higher leakage currents. Conversely, strong interface bonding inhibits the formation of these pathways, reducing leakage current. Additionally, moisture within the composites can significantly alter the internal electric field, further increasing leakage current.

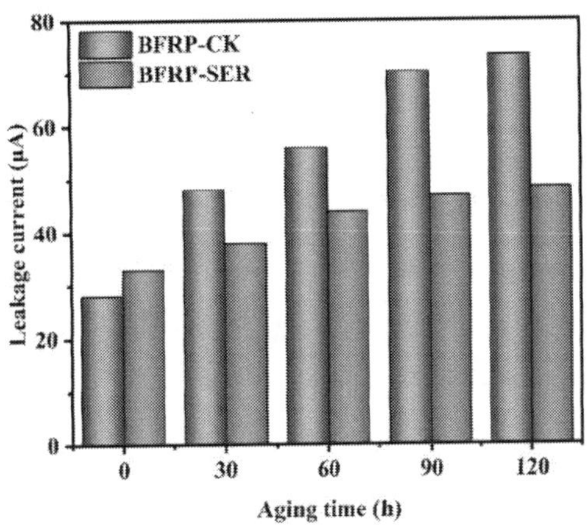


**Figure 5.** Leakage current of Composites.

The leakage current test results of the composites are shown in Figure 5. Before aging, the leakage currents of BFRP-CK and BFRP-SER are 28 μA and 33 μA, respectively. This is because, prior to aging, the interface bonding of both BFRP-CK and BFRP-SER is relatively good, which suppresses the formation of charge transport pathways, resulting in relatively low leakage currents. As aging progresses, mediums such as $Na^+$, $Cl^-$, and $H_2O$ gradually invade the coating of the wetting agent, leading to an increase in leakage current.

During the aging process, the leakage current of BFRP-SER increases at a slower rate, reaching 48 μA after 120 hours of aging, which is an increase of 31.3%. In contrast, the leakage current of BFRP-CK rises rapidly during aging, ultimately reaching 73 μA, with an increase rate of 162.2%. The results indicate that saline corrosion significantly raises the leakage current of the insulation materials. Thanks to the high bond energy Si-O bonds in SER, BFRP-SER can maintain a relatively intact interface structure during wet-heat aging, which hinders moisture penetration and suppresses the increase in leakage current.

*3.5. Dielectric loss*

Excessive dielectric loss can cause internal insulation heating of composites in an electric field, which can increase the temperature of the insulator and accelerate its damage. Therefore, it is very important to measure the dielectric loss tangent of insulating materials. In addition, due to the significantly higher dielectric constant of water molecules compared to the reinforcing materials and matrix, the increase in the dielectric constant of composites during damp heat aging is closely related to the absorption of water molecules.

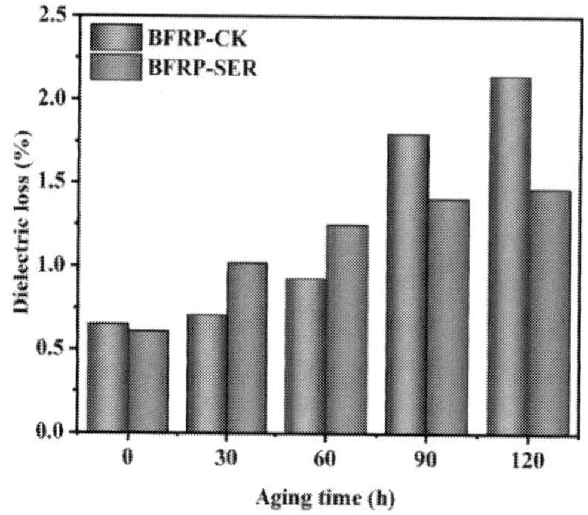


**Figure 6.** Dielectric loss of composites.

Figure 6 shows the change in the dielectric loss of two materials during wet-heat aging. Before aging, the dielectric of BFRP-CK and BFRP-SER is 0.65% and 0.61%, respectively. After 120 hours of aging, the dielectric of both materials increases to 2.15% and 1.47%, representing increases of 230.8% and 141.0%, respectively. It showed that the dielectric of both composite materials has significantly increased. This is due to the fact that after the onset of wet-heat aging, mediums such as $Na^+$, $Cl^-$, and $H_2O$ begin to infiltrate the BFRP, which can cause interfacial polarization phenomena.

## 4. Conclusion

This paper presents the preparation of silicon-modified basalt fiber composites and explores their changes in mechanical and electrical properties under wet-heat aging conditions. The conclusions of this study contribute to the promotion of BFRP in the power system. The conclusions are as follows:

1) In the initial state, the SER wetting agent coating allows for good bonding between the fiber layer and the resin layer. The bending strength and breakdown field strength of BFRP-SER are higher

by 10.1% and 19.9% compared to BFRP-CK, while the leakage current and dielectric loss are 85.1% and 93.8% of those of BFRP-CK, respectively.

2) During wet-heat aging, the wetting agent coating undergoes varying degrees of hydrolysis, causing damage to the interfacial bonding of the composite material. Additionally, the infiltration of mediums such as $Na^+$, $Cl^-$, and $H_2O$ can also affect the performance of the BFRP. After aging, the bending strength and breakdown field strength of the SER-emulsion-treated samples decreased by 30.1% and 35.2%, respectively, while the leakage current and dielectric loss increased by 31.3% and 141.0%, all of which are better than those of the commercially available BFRP-CK.

3) The superior moisture-heat resistance of BFRP-SER is attributed to the Si-O bonds in SER, which are similar to inorganic silicate structures. These chemical bonds, such as Si-O and C-O bonds, possess high bond energies, requiring more energy to break them at high temperatures. This leads to a lower degree of hydrolysis in BFRP-SER and hinders the penetration of mediums, allowing it to maintain performance over a longer period.

**References**

[1] LIU Yunpeng, ZHANG Mingjia, YU Hong, MA Long, FU Xiaotao, LIU Hechen, Aging Characteristics and Mechanism of Continuous Basalt Fibers Used in Composite Insulation Materials[J]. High Voltage Engineering:1-9.

[2] Torres J P, Hoto R, Andrés J, García-Manrique J A. Manufacture of Green-Composite Sandwich Structures with Basalt Fiber and Bioepoxy Resin[J]. Advances in Materials Science and Engineering, 2013, 2013: e214506.

[3] Khandelwal S, Rhee K Y. Recent advances in basalt-fiber-reinforced composites: Tailoring the fiber-matrix interface[J]. Composites Part B: Engineering, 2020, 192: 108011.

[4] Kim S-H, Lee J-H, Kim J-W, Lee S-Y, Park S-J. Interfacial Behaviors of Basalt Fiber-Reinforced Polymeric Composites: A Short Review[J]. Advanced Fiber Materials, 2022, 4(6): 1414–1433.

[5] Wei B, Cao H, Song S. Surface modification and characterization of basalt fibers with hybrid sizings[J]. Composites Part A: Applied Science and Manufacturing, 2011, 42(1): 22–29.

[6] Xu X. Effect of different fiber sizing on basalt fiber-reinforced cement-based materials at low temperature: From macro mechanical properties to microscopic mechanism[J]. Construction and Building Materials, 2023.

[7] Sreekala M S, George J, Kumaran M G, Thomas S. The mechanical performance of hybrid phenol-formaldehyde-based compo-sites reinforced with glass and oil palm fibres[J]. Composites Science and Technology, 2002, 62(3): 339–353.

[8] Liu Y, Wang W, Liu H, Zhang M, Liu J, Qi J. Blending Modification of Alicyclic Resin and Bisphenol A Epoxy Resin to Enhance Salt Aging Resistance for Composite Core Rods[J]. Polymers, 2022, 14(12): 2394–2407.

# Studies on properties of coral aggregate strengthened by granulated blast furnace slag (GBFS) and concrete prepared by strengthened coral aggregates

**Yijie Huang*** **and Hongggang Liu**

Shandong University of Science and Technology, Qingdao, Shandong, 266590, China

*Corresponding author's e-mail:302huangyijie@163.com

**Abstract:** To enhance the mechanical, durability and deformation properties of coral coarse aggregate concrete (CAC), alkali-activated granulated blast furnace slag (GBFS) paste (AGP) coated CCA is used to strengthen coral coarse aggregate (CCA). The strengthened coral aggregate concrete (SCAC) was prepared by strengthened CCA (SCCA). The influences of the strengthening parameters (molarity of sodium hydroxide (Mol) and modulus of sodium silicate (Ms) of AGP) on the microstructures and properties of SCCA and SCAC were studied by axial compression tests and scanning electron microscopy. The results showed that microstructures and properties of SCAC were greatly improved after using SCCA. Compared to CAC, the cubic compressive strength ($f_{cu}$) of SCAC increased by 23.74% on average. With the increase of Ms, the $f_{cu}$ of SCAC first increased and then decreased, while the influence of Mol was negligible. The deformation properties of SCAC also varied with the change of strengthening parameters. With the increase of Ms of AGP, the peak strain ($\varepsilon_0$) of SCAC increased, and the maximum increase in $\varepsilon_0$ was 28.32% in comparison to CAC. The decline of the stress-strain curve of SCAC became gentle, and its curve curvature was enhanced, which caused the energy dissipation capacity and ductility of SCAC to be 122.76% and 15.28% greater than those of CAC, respectively.

## 1.Introduction

Adopting coral aggregate concrete (CAC) in marine engineering reduces construction costs and time and utilizes local marine resources. Thus, the use of CAC in marine engineering is promising. However, the properties of coral coarse aggregate (CCA), such as its lightweight and porous structure, can cause great reductions in mechanical properties, deformability, and durability of CAC, which limits the development of the properties in CAC [1]. Therefore, enhancing the properties and microstructures of CCA is critical for the sustainable development of CAC.

Physical and chemical methods are generally used to improve CCA's characteristics (high porosity, low strength, brittleness, and poor durability). However, the improvement of CCA through physical methods is negligible. Therefore, chemical methods are always adopted. Wang et al. [2] modified CCA with acetic acid and found that the compressive strength ($f_{cu}$) of specimens prepared by strengthened CCA increased by 60MPa. Yao et al. [3] obtained C45 strength grade CAC by preparing with strengthened CCA, which was modified by hydrochloric acid solution and sodium silicate. The chemical methods mentioned above can enhance the strength of CAC, while its effects on ductility and energy dissipation of specimens are still negligible due to the properties of CCA.

In order to enhance the microstructure of CCA, some studies have coated aggregates with cementitious material. Yodsudjai et al.[4] found that the mechanical and deformation properties of CAC greatly increased by coating CCA with cement mortar. Wang et al. [5] modified CCA by geopolymer paste after using acid, causing the $f_{cu}$ of SCAC to increase by 34.5%. Chu et al. [6] adopted the magnesium sulfate cement paste to strengthen CCA, which changed the CCA microstructure and enhanced the specimen's strength (31.8%).

According to the above analyses, coated aggregate with cementitious material is an available method to enhance the microstructure and properties of CCA and CAC. Nevertheless, little research in this area has been conducted, and the deformability and energy dissipation capacity of SCAC prepared by strengthened CCA (SCCA) need to be further investigated. This study adopted the cementitious material (i.e., alkali-activated granulated blast furnace slag (GBFS)) coated aggregates method because of the low carbon footprint, recycling, and perfect erosion resistance. The microstructure and the properties of SCCA and SCAC were studied through compression tests and scanning electron microscopy (SEM), which provided technical support for its application in marine engineering.

## 2.Research program

*2.1.Ral materials*

In this study, SCAC and CAC were prepared. The raw materials (Figure 1) used in the test included Ordinary Portland cement (PO 42.5), CCA, sea sand (SS), fresh water (FW), and seawater (SW). The aggregate properties were evaluated per the standard codes [7] [8] (Table 1). The alkali-activator was prepared by mixing the sodium silicate and sodium hydroxide [9].

(a)

(b)

**Figure 1.** Coarse aggregates and fine aggregate.

**Table 1.** Properties of aggregates.

| Material | Size (mm) | $\omega_a$ [a] (%) | $\rho_b$ [b] (kg/m$^3$) | $\rho_a$ [c] (kg/m$^3$) | $S_t$ [d] (MPa) | Fineness modulus (%) | Clay content (%) | Shell content (%) |
|---|---|---|---|---|---|---|---|---|
| CCA | 5.0-31.5 | 16.40 | 890 | 1845 | 1.51 | - | 12.1 | - |
| Sea sand | 0.15-4.75 | - | 3190 | 2657 | - | 3.2 | 3.01 | 6.38 |

[a] $\omega_a$ is the water absorption of CCA.
[b] $\rho_b$ is the bulk density.
[c] $\rho_a$ is the apparent density.
[d] $S_t$ is the cylinder compression strength.

*2.2.Strengthened CCA*

*2.2.1 Preparation of strengthened CCA.* The alkali-activated GBFS paste (AGP) coated aggregates method was adopted in this test. Six groups of SCCA were prepared to investigate the influences of the strengthening parameters (molarity of sodium hydroxide (Mol) and modulus of sodium silicate

(Ms) of AGP) on the microstructures and the properties of SCCA and SCAC. The preparation of the SCCA was as follows (Figure 2):

First, the alkali-activator was prepared by mixing the sodium silicate and sodium hydroxide solution, which was sealed and stored for 24 hours. Second, AGP was prepared by mixing the alkali activator and GBFS. Third, CCA was mixed with AGP for approximately 3 min to ensure that AGP covered the CCA adequately. Next, CCA was vibrated for 15s to eliminate the excess paste of CCA. Finally, SCCA were separated and cured under standard conditions (temperature: 20 ± 2 ◦C) for 28d.

The SCCA were named by the following guidelines "C-Mol-Ms". For example, C-4-0.4 stands for CCA strengthened by AGP with Mol of 3mol and Ms of 0.4.

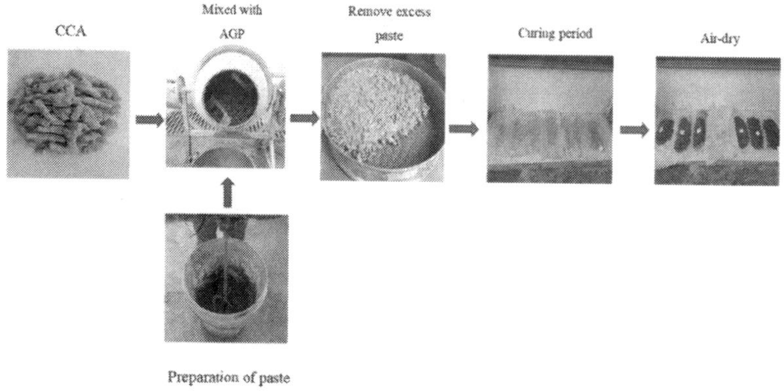


**Figure 2.** The fabrication procedures of the SCCA.

*2.2.2 Microstructures and properties of the SCCA.* The properties of CCA were remarkably enhanced after strengthening (Table 2). The water absorption and porosity ($\omega_a$ and $p$) of the SCCA decreased by 41.72 % and 17.49 % on average, compared to CCA, respectively, and the density and tube compressive strength ($\rho_a$ and $S_t$) of SCCA increased by 4.96 % and 229.97 % on an average, respectively. This is because the pores of the aggregates were filled by AGP (Figure 3) and formed a coating covering the surface of the aggregate (Figure 4). AGP with possessing adequate strength combined with the CCA formed a new strong particle. Additionally, the properties of SCCA were modified with changes in strengthening parameters (Table 2).

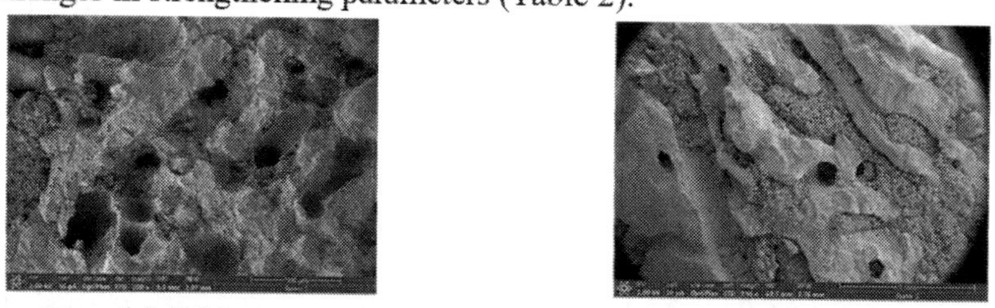

(a) CCA (b) C-3-0.4

**Figure 3.** SEM of CAA (a) and C-3-0.4 (b).

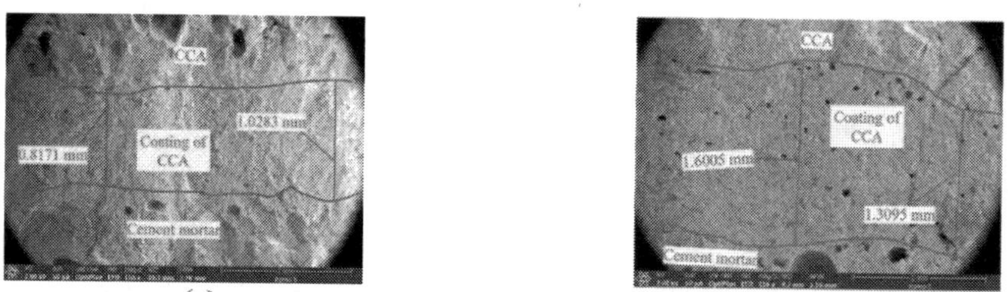


(a) (b)

**Figure 4.** The coating thickness of the C-3-0.4 (a) and C-4-0.4 (b).

**Table 2.** Physical properties of the SCCA.

| Aggregates | Mol | Ms | $\omega_a$ (%) | $p$ (%) | $\rho_a$ (kg/m$^{-3}$) | $S_t$ (MPa) | $\delta$ (mm) |
|---|---|---|---|---|---|---|---|
| CCA | - | - | 16.40 | 35.47 | 1845 | 1.51 | - |
| C-3-0.4 | 3 | 0.4 | 10.81 | 31.11 | 1942 | 3.43 | 0.92 |
| C-3-0.6 | 3 | 0.6 | 9.60 | 29.32 | 1955 | 3.66 | 1.08 |
| C-3-0.8 | 3 | 0.8 | 10.32 | 30.16 | 1950 | 3.52 | 1.26 |
| C-4-0.25 | 4 | 0.25 | 11.14 | 31.44 | 1923 | 3.25 | 1.35 |
| C-4-0.4 | 4 | 0.4 | 9.81 | 30.99 | 1928 | 3.48 | 1.46 |
| C-4-0.6 | 4 | 0.6 | 12.58 | 32.67 | 1919 | 3.06 | 1.62 |

*2.3. Specimen design and testing*

*2.3.1 Specimen design and preparation.* Six types of SCAC along with a group of CAC were prepared to systematically investigate the effect of strengthening parameters (i.e., Mol and Ms of AGP) on the properties of SCAC. The concrete mixing ratio was kept constant (Table 3). The detailed preparation of SCAC was as follows. Firstly, sea sand, cement, and SCCA were mixed for 3 minutes. After that, the mixing solution was added and mixed for 2 minutes. Then the mixture was poured into molds and vibrated for 1 minute. Finally, specimens were cured under standard conditions. The specimens were named based on "concrete type - aggregate type" (Table 3). For example, SCAC-C-3-0.4 denotes SCAC prepared with C-4-0.4 SCCA.

**Table 3.** Mixing ratio of concrete.

| Specimen | SW (kg/m$^3$) | SS (kg/m$^3$) | CCA (kg/m$^3$) | SCCA (kg/m$^3$) | Cement (kg/m$^3$) | Additive (kg/m$^3$) | Slump (mm) |
|---|---|---|---|---|---|---|---|
| CAC | 306.861 | 992 | 806 | - | 517 | 1.293 | 93 |
| SCAC-C-3-0.4 | 262.129 | 992 | - | 806 | 517 | 4.653 | 94 |
| SCAC-C-3-0.6 | 252.376 | 992 | - | 806 | 517 | 4.912 | 93 |
| SCAC-C-3-0.8 | 258.179 | 992 | - | 806 | 517 | 5.170 | 97 |
| SCAC-C-4-0.25 | 264.788 | 992 | - | 806 | 517 | 5.429 | 97 |
| SCAC-C-4-0.4 | 254.069 | 992 | - | 806 | 517 | 5.687 | 95 |
| SCAC-C-4-0.6 | 276.395 | 992 | - | 806 | 517 | 5.946 | 99 |

*2.3.2 Test system.* The macro properties (strength, stress-strain curve and ductility, etc.) and the microstructure of SCAC were tested by a TAW-2000 electro-hydraulic servo testing machine (Figure 5(a)) and Apreo S HiVac scanning electron microscope (SEM) system (Figure 5(b)) respectively based on a standard code [10].

(a)

(b)

**Figure 5.** Test system (TAW-2000 and Apreo S HiVac SEM system).

## 3.Test results and analysis

### *3.1.Failure process and pattern*

The failure process of CAC and SCAC was similar, but the differences were not negligible. The development of cracking in CAC was rapid, and the failure of CAC was brittle when the axial stress ($\sigma$) reached the peak stress ($\sigma_{max}$). The cracks of SCAC developed slowly, and $\sigma$ gradually decreased with increasing deformation compared to CAC. Furthermore, the angle ($\alpha$) between the loading direction and the main crack of concrete was changed after using SCCA [11]. The average α of CAC and SCAC was 39° and 32°, respectively. The types of aggregates had an influence on the failure pattern. The failure surface of SCAC was rough, and the SCCA were seldom destroyed (Figure 6) compared to CAC:

(a) (b) (c) (d)

**Figure 6.** Typical failure surface of concrete (CAC, SCAC-C-3-0.6, SCAC-C-3-0.8, and SCAC-C-4-0.6).

### *3.2.Cubic compressive strength*

Compared to CAC, the cubic compressive strength ($f_{cu}$) of SCAC was enhanced (Table 4). In general, $f_{cu}$ of SCAC increased by 10% on average compared with CAC, owing to the AGP with possessing adequate strength combined with CCA forming a new strong particle, which resulted in the improvement in $f_{cu}$ of SCAC. Furthermore, $f_{cu}$ of SCAC varied with changes in strengthening parameters of aggregates (i.e., Mol and Ms of AGP).

With increasing Ms of APG, the $f_{cu}$ of SCAC first increased and then decreased. Compared to CAC, $f_{cu}$ of SCAC-C-3-0.4, SCAC-C-3-0.6, and SCAC-C-3-0.8 were 20.90%, 35.43%, and 27.82% higher, respectively. This is because $p$ and $\omega_a$ of SCCA decreased, $S_t$ and coating thickness of aggregates ($\delta$) increased with increasing Ms of APG, and the concrete microstructure was improved (Figure 7), causing a higher $f_{cu}$ of SCAC. Furthermore, the influence of Mol on $f_{cu}$ of SCAC was negligible (Figure 8). The test results indicated that the difference in $f_{cu}$ of SCAC-C-3-0.4 and that of SCAC-C-4-0.4 were negligible.

It was also observed that the $f_{cu}$ of SCAC and CAC developed fast (Table 4). The $f_{cu}$ of SCAC and CAC on days 7 and 14 were 87.80% and 94.67% of the 28-day $f_{cu}$ on average, respectively. The reason is that cement hydration was promoted because of SW and SS's high $Cl^-$ content.

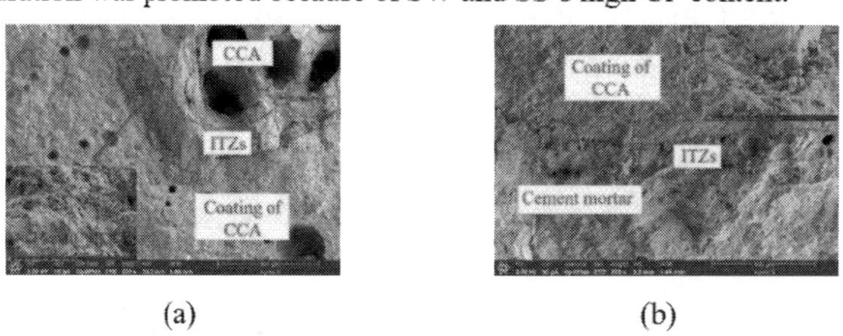


(a) (b)

**Figure 7.** Interfacial transition zones of SCAC-C-3-0.6.

**Table 4.** The properties of SCAC.

| Specimen | $f_{cu}$ (MPa) | | | $f_c$ (MPa) | | | $E_c$ | $\mu$ | $\eta$ | $\varepsilon_0$ |
|---|---|---|---|---|---|---|---|---|---|---|
| | 7d | 14d | 28d | 7d | 14d | 28d | ($10^4$MPa) | | | |
| CAC | 33.07 | 35.86 | 37.03 | 26.27 | 28.49 | 31.40 | 2.45 | 1.08 | 0.0487 | 0.00194 |
| SCAC-C-3-0.4 | 39.53 | 42.87 | 44.77 | 31.20 | 34.40 | 37.33 | 2.56 | 1.26 | 0.1076 | 0.00234 |
| SCAC-C-3-0.6 | 43.83 | 47.55 | 50.15 | 37.52 | 40.41 | 42.85 | 2.78 | 1.35 | 0.1509 | 0.00245 |
| SCAC-C-3-0.8 | 41.12 | 44.41 | 47.33 | 36.00 | 38.06 | 40.45 | 2.71 | 1.32 | 0.1299 | 0.00244 |
| SCAC-C-4-0.25 | 40.44 | 41.73 | 43.46 | 31.41 | 34.09 | 35.41 | 2.52 | 1.18 | 0.0682 | 0.00213 |
| SCAC-C-4-0.4 | 42.19 | 45.30 | 48.72 | 35.57 | 38.17 | 41.34 | 2.63 | 1.30 | 0.1436 | 0.00249 |
| SCAC-C-4-0.6 | 33.65 | 37.44 | 40.49 | 27.74 | 30.52 | 33.36 | 2.47 | 1.06 | 0.0507 | 0.00206 |

*3.3.Prismatic compressive strength*
The changes in prismatic compressive strength ($f_c$) and $f_{cu}$ in SCAC were similar. With increasing Ms of APG, the $f_c$ of SCAC first increased and then decreased. Compared to CAC, $f_c$ of SCAC-C-3-0.4, SCAC-C-3-0.6, and SCAC-C-3-0.8 were 20.90%, 35.43%, and 27.82% higher, respectively. The $f_c$ of SCAC-C-3 was 9.55% higher than that of SCAC-C-4. This is attributed to the fluidity of AGP with sodium hydroxide molarity of 3mol, which is higher, resulting in the lower porosity and higher strength of aggregates. It was also observed that the ratio of $f_c$ to $f_{cu}$ ($f_c$ / $f_{cu}$) changed after using SCCA. In general, the $f_c$ / $f_{cu}$ of SCAC was approximately 0.84. Due to the constraint between the concrete ends and the loading plate of the device being reduced by SCCA, the $f_c$ / $f_{cu}$ improved.

*3.4.Stress-strain curve*
The stress-strain curves ($\sigma$-$\varepsilon$ curves) of CAC and SCAC are shown in Figure 9, which consists of initial ascending, quasi-linear plastic, and descending stages. Nonetheless, the differences were negligible, and the feature of $\sigma$-$\varepsilon$ curves changed with a variation in strengthening parameters of aggregates.

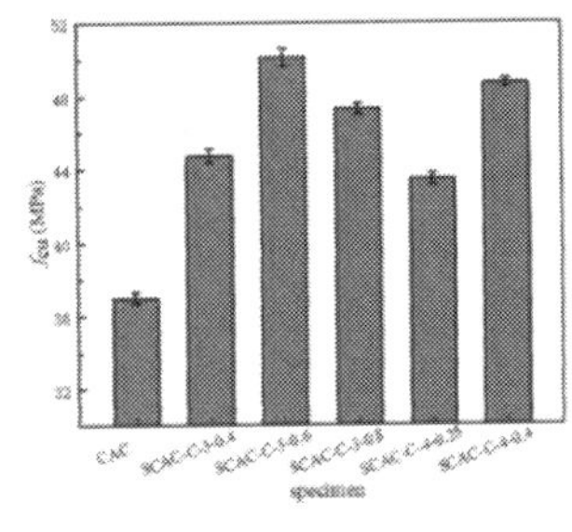

**Figure 8.** $f_{cu}$ of SCAC.

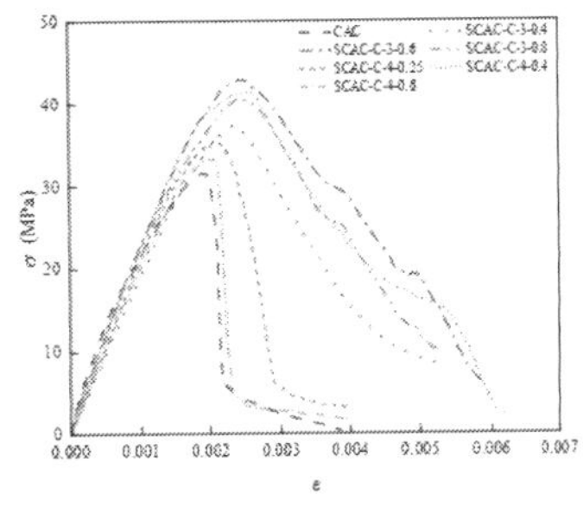


**Figure 9.** σ-ε curves.

*3.4.1 Initial ascending branch.* During the initial ascending branch (before 0.4 peak stress ($\sigma_{max}$)), $\sigma$ approximately linearly increased with increasing $\varepsilon$. The slope of the SCAC curve (elastic modulus ($E_c$)) in this stage was 6.60% higher on average in comparison with CAC (Table 4), and it increased with the increase of Mol of AGP. The reason is the effects of the coarse aggregates [12]. With the increase of Mol of AGP, the properties of SCCA were improved (lower porosity, higher coating thickness, and higher strength compared to CCA), increasing $E_c$.

*3.4.2 Quasi-linear plastic stage.* With increasing $\varepsilon$, the $\sigma$ nonlinearly increased during the quasi-linear plastic stage. The curve curvature of SCAC was slightly improved, causing an increase in the peak strain ($\varepsilon_o$). In general, $\varepsilon_o$ of SCAC was 19.50% higher than that of CAC (Table 4), owing to the improvement in the properties of SCCA, and the ITZ of SCAC caused the plastic deformation properties of SCAC to be greater. Furthermore, the strengthening parameters (i.e., Mol and Ms of AGP) could change $\varepsilon_o$, and it first increased and then decreased with increasing Ms, while the influence of Mol was negligible. The maximum enhancements in $\varepsilon_o$ of SCAC was 28.35% (SCAC-C-4-0.4).

*3.4.3 Descending stage.* After $\sigma_{max}$, the curve of CAC dropped suddenly. However, the curve of SCAC dropped slowly due to the adoption of SCCA, indicating that the deformation properties and energy dissipation of SCAC improved. The ductility ($\mu$) and energy dissipation capacity ($\eta$) were adopted to reflect these improvements (Equation (1) and Equation (2)) [13].

$$\mu = \frac{\varepsilon_{0.8}}{\varepsilon_0} \quad (1)$$

$$\eta = \int_0^{\varepsilon_{0.8}} \sigma \, d\varepsilon \quad (2)$$

where $\varepsilon_{0.8}$ denotes the strain corresponding to σ dropped to 80 % $\sigma_{max}$.

Compared with CAC, $\mu$ and $\eta$ of SCAC were 15.28% and 122.76% greater on average, respectively, owing to the enhanced ITZs and strong particle structure effectually restricting the initiation and improvement of cracks and enhancing the deformation properties of concrete. The $\mu$ and $\eta$ of SCAC changed with the variations in strengthening parameters of aggregates. In general, the $\mu$ and $\eta$ of SCAC first increased and then decreased with increasing Ms, while the influence of Mol was negligible. Compared with CAC, $\mu$ of SCAC-C-3-0.4, SCAC-C-3-0.6, and SCAC-C-3-0.8 were 16.67%, 25.00%, and 22.22% higher, respectively, and the $\eta$ of SCAC-C-3-0.4, SCAC-C-3-0.6 and SCAC-C-3-0.8 were 120.94%, 209.86%, and 166.74% higher, respectively. This can be attributed to the increasing Ms of APG, the properties of SCCA are improved, and ITZs become dense, causing the increase in the $\mu$ and $\eta$.

## 4.Conclusion

(1) AGP-coated CCA efficiently improved the properties of aggregates. Compared to the CCA, the porosity of SCCA decreased by 17.49 %, while its strength increased by 229.97 % on average, respectively. This caused a relatively ductile failure of SCAC.

(2) The adoption of SCCA resulted in enhancements in the strength and deformation properties of SCAC. Compared with CAC, the $f_{cu}$, $\mu$ and $\eta$ of SCAC increased by 23.74%, 15.28% and 122.76% on average, respectively. Furthermore, the properties of SCCA also changed with strengthening parameters.

(3) The $\sigma$-$\varepsilon$ curves of CAC and SCAC were similar, while the slope of the SCAC curve increased by 6.60%, and εo of SCAC increased by 19.50%.

## Acknowledgements

Throughout the writing of this dissertation, I have received a great deal of support and assistance. I would particularly like to acknowledge my team members for their wonderful collaboration and patient support. I would also like to thank my tutor for their valuable guidance throughout my studies, who provided me with the tools I needed to choose the right direction and successfully complete my dissertation. In addition, I would like to thank my parents for their wise counsel and sympathetic ear. You are always there for me.

## References

[1] Yang Z Y, Zhan X W, Zhu H, Zhang B, Lu F, and Dong Z Q 2024 Mesoscopic investigation of the matrix pores and ITZ effects on the mechanical properties of seawater sea-sand coral aggregate concrete. *J. Journal of Building Engineering* 90:109375. https://doi.org/10.1016/j.jobe.2024.109375

[2] Wang A G, Lyu B C, Zhu Y C, Liu K W, Guo L P, and Sun D S 2020 A gentle acid-wash and pre-coating treatment of coral aggregate to manufacture high-strength geopolymer concrete. *J. Construction and Building Materials* 121780. https://doi.org/J.CONBUILDMAT.2020.121780

[3] Yao Y, Wang Z D, Wang L, et al. The Invention Relates to a Method for Preparing Coarse Aggregate from Coral Reef and its Concrete. CN105731897A[P]. 2016.07.06. (in Chinese)

[4] W. Yodsudjai, N. Otsuki, T. Nishida, and R. Onitsuka 2003 Study on strength and durability of concrete using low quality coarse aggregate from circum-pacific region. *J. Doboku Gakkai Ronbunshu* 746. https://doi. org/10.2208/jscej.2003.746_103.

[5] Wang A G, Wang X Y, and Sun D S 2021 Preparation method of chemically etched coral aggregate and geopolymer-based ultra-high strength coral concrete, CN113200697. 2021-08-03

[6] Chu Y J, Wang A G, Zhu Y C, Wang H, Liu K W, Ma R, GUO L P, and Sun D S 2021 Enhancing the performance of basic magnesium sulfate cement-based coral aggregate concrete through gradient composite design technology. *J. Composites Part B 227 109382.* https://doi.org/10.1016/j.compositesb.2021.109382

[7] Standardization Administration of China. Specification of Pebble and Crushedstone for Building: GB/T 14685-2011[S]. Beijing, China, 2011.

[8] Standardization Administration of China. Lightweight aggregates and its test methods – Part 2: Test methods for lightweight aggregates: GB/T 17431.2-2010. Beijing, China, 2010.

[9] Wang C R 2019 Experimental study on mechanical properties and durability of alkali-activated slag recycled concrete[D]. Harbin: Harbin Institute of Technology

[10] Ministry of Housing and Urban-Rural Development of the People's Republic of China. Standard for test methods of concrete physical and mechanical properties: GB/T 50081-2019. Beijing, China, 2019.

[11] Guo Z H 2013 Principle of reinforced concrete [M]. Beijing: Peking University Press.

[12] YUAN Y F 2015 Mix Design and Property of Coral Aggregate Concrete[D]. Nanjing: Nanjing University of Aeronautics and Astronautics.

[13] Chen Z P and Liang Y H 2020 Experimental study of axial compression performance of seawater mixed coral sand aggregate concrete. *J. Journal of Guangxi University ( Natural Science Edvition)* 45(05): 967-977. https://doi.org/10.13624/j.cnki.issn.1001-7445.2020.0967

Journal of Physics: Conference Series **2961** (2025) 012006 doi:10.1088/1742-6596/2961/1/012006

# Easily prepared and high-performance AgNPs as SERS substrates for trace thiram detection

**Yongquan Wei, Mimi Qin*, Yansheng Liu, Xiaobo Jia, Jin Zhou and Guofu Wang**

Guangxi University of Science and Technology, No.2, Wenchang Road, Liuzhou City, 545000, Guangxi, China.

*Corresponding author's e-mail: qinmi-328.cool@163.com

**Abstract:** Surface-enhanced Raman scattering (SERS) technology, known for its ultra-high sensitivity and molecular specificity, is widely used for material detection. In this study, a high-performance SERS substrate was successfully fabricated by depositing a 10 nm silver film on a silicon wafer under low-vacuum conditions, systematically optimizing the annealing temperature. At the optimal annealing temperature, the silver film underwent particle formation, creating a high-density "hotspot" distribution, which significantly enhanced the SERS sensitivity. With rhodamine 6G (R6G) as a beta-molecule, the SERS performance of the nanostructure was evaluated, with the optimized substrate demonstrating the detection LOD achieved $10^{-8}$ M. A linear relationship was established between the intensity of the Characteristic Raman peak of thiram at 1384 $cm^{-1}$ and the logarithm of its concentration, achieving a Correlation value is 0.986 and the LOD achieved $10^{-9}$ M.

## 1. Introduction

Today, the variety of pesticides has become increasingly diverse, with thiram being widely used as a fungicide. However, thiram residues tend to persist after application, posing significant risks to human health and raising substantial safety concerns [1]. Prolonged exposure to low concentrations of thiram can result in liver and neurological damage, as well as immune suppression. Therefore, the development of highly efficient and sensitive detection techniques is of paramount importance [2].

Surface-enhanced Raman scattering (SERS) technology has demonstrated tremendous potential in substance analysis and environmental monitoring due to its exceptional sensitivity and molecular specificity [3-5]. SERS has emerged as a powerful analytical technique that is applied in fields such as chemistry and physics. SERS's unique capabilities include single-molecule detection, distinguishing different substances based on their characteristic vibrational fingerprint spectra, and providing spatially resolved information at the nanoscale. These features make SERS an effective tool for tracking environmental pollutants, biomedical diagnostics, and monitoring chemical reactions. The non-contact and non-destructive detection mode of SERS ensures the integrity of the analyte during testing, allowing for clearer identification of its composition and origin. Compared to conventional Raman spectroscopy, SERS amplifies Raman signals significantly through the LSPR effect of noble metal nanostructures, enabling the detection of substances at extremely low concentrations. However, achieving large-scale application of SERS remains challenging, as SERS substrates have high signal intensity and reproducibility, which is crucial for pesticide detection [6].

In this study, we analyzed the structural evolution of silver films under different annealing temperatures and designed a cost-effective, easily fabricated spherical silver nanoparticle (AgNPs) substrate. This AgNPs substrate exhibits a higher density of "hotspots," leading to stronger SERS enhancement. Using R6G as a biotechnology probe, we validated the SERS substrate's sensitivity, uniformity, and reproducibility. Furthermore, the optimized substrate was employed for gradient detection of thiram, achieving a detection limit of $1\times10^{-9}$ M. The results indicate that the AgNPs substrate holds significant potential for biosensors and environmental monitoring applications.

## 2. Materials and methods

### *2.1. Materials*
First, R6G, 95% purity, was purchased from Aladdin Biochemical Technology Co., Ltd, thiram 97%, purchased from Beijing Inokai Co., and the Silver target, 99.99%, was purchased from the Fuzhou Infixon Optoelectronics Technology Co., Ltd.

### *2.2. Preparation of AgNPs*
The blank Si/SiO2 substrates are first ultrasonically cleaned multiple times in ethanol and then air-dried. Using thermal deposition under a vacuum pressure of $5\times10^{-4}$, a 10 nm layer of silver is uniformly deposited onto the Si/SiO2 surface at a rate of 0.1 Å/s. The samples are then placed on a heating stage at preset temperatures and annealed at various temperatures under atmospheric pressure.

### *2.3. Apparatus and characterization*
Characterization is performed using a scanning electron microscope (Zeiss Sigma 300). The R6G is collected by the Raman spectrometer (RST2-301-SMS) excited by the 532 nm laser installations, with the acquisition settings of the grating at 1800, the integration time of 3 s, and the laser power maintained at 1 mW. A 785 nm laser source is used for thiram testing with an integration time of 5 s and the power maintained at 10 mW.

## 3. Results & discussion

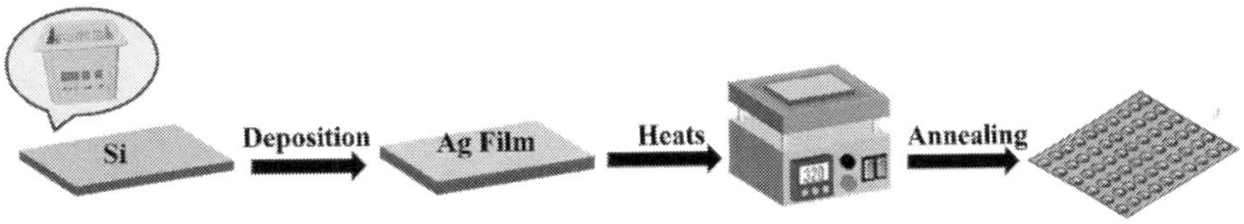


**Figure 1.** Preparation of AgNPs as substrates for SERS detection.

Figure 1. illustrates the preparation method of the AgNPs substrate, with the detailed procedure described in the experimental section. The AgNPs are formed by thermally treating a 10 nm thick silver film. By varying the annealing temperature while keeping the annealing time constant, the density and average diameter of the AgNPs can be controlled. Characterization through SEM is presented in Figures 2(a) - 2(e), which show a series of large-area images with temperatures ranging from 290°C to 350°C. Figures 2(f) - 2(j) correspond to the detailed small-area images. Many densely packed AgNPs can be found in these detailed images, with the particle shapes appearing generally uniform. Initially, the particle diameters are relatively small, and aggregation is minimal. As the temperature increases, the diffusion rate of silver atoms accelerates, causing the particles to fuse into larger nanoparticles with clearer boundaries between them and increased surface roughness. Some larger, irregular particle structures also emerge. As shown in Figures 2(k) - 2(o), the diameter of the AgNPs gradually increases. At 320°C, the particle uniformity reaches its peak, and the formation of strong SERS "hotspot" regions between the particles is more likely, significantly enhancing the Raman intensity of the SERS spectroscopy. This will facilitate the detection of other target substances using this SERS substrate.

Journal of Physics: Conference Series **2961** (2025) 012006 doi:10.1088/1742-6596/2961/1/012006

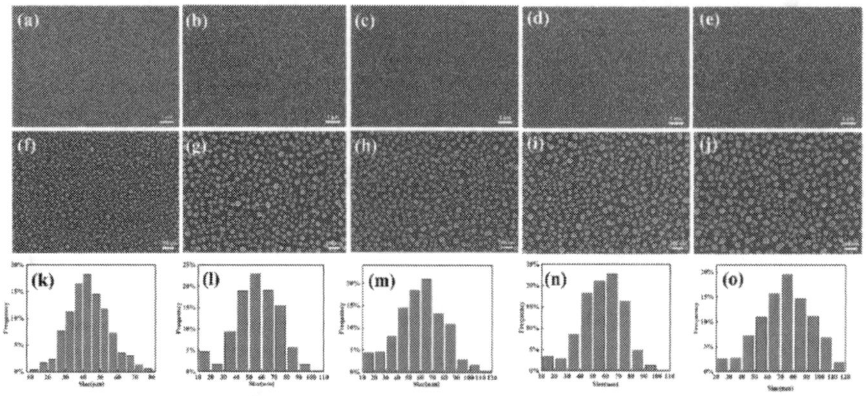


**Figure 2.** (a) - (e) show a wide range of SEM images at annealing temperatures of 290°C, 305°C, 320°C, 335°C, and 350°C at 45 minutes; (f) - (j) show the corresponding small-range SEM images; and (k) - (o) show the particle size distribution data for the corresponding small-range SEM characterization.

To evaluate the SERS activity of AgNPs at different annealing temperatures, R6G ($10^{-5}$ M) is used as a probe molecule to analyze the SERS performance of the substrates. In Figure 3(a), the SERS signal of AgNPs substrates prepared at various annealing temperatures is obtained, with the Raman characteristic peaks of R6G observed at 612, 774, 1123, 1178, 1307, 1359, 1504, 1574, and 1648 $cm^{-1}$. By the spectral data, the Raman characteristic signal at 612 $cm^{-1}$ corresponds to C-C-C in-plane vibrations, while the peaks at 774, 1123, and 1178 $cm^{-1}$ originate from vibrations of the C-H bonds. The remaining characteristic peaks originate from the C-C stretching vibrations of the aromatic ring. Figure 3(b) indicates that the SERS performance is strongest at 320°C for 45 minutes, suggesting this is the optimal annealing condition. Subsequently, we conduct the limit of detection (LOD) analysis for the optimized substrate using R6G, achieving a detection LOD of $10^{-8}$ M, as shown in Figure 3(c). Figure 3(d) illustrates the linear relationship following the serial dilution of R6G, with a calculated $R^2$ of 0.961, indicating a strong linear correlation between the SERS intensity and the concentration's logarithm.

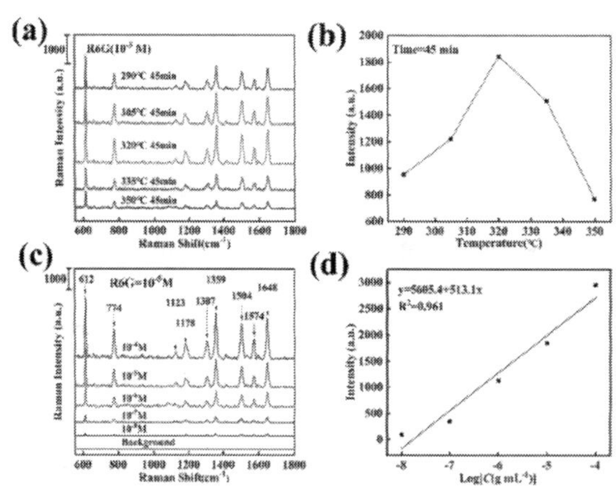


**Figure 3.** (a) SERS spectra under different annealing temperatures, (b) Intensity variation curve at 612 $cm^{-1}$, (c) Optimized limit spectral data of R6G concentration gradient on AgNPs substrate, and (d) Correlation image composed of the logarithm of intensity and concentration of R6G at 612 $cm^{-1}$.

To demonstrate the high performance of the prepared substrate, a $1\times10^{-5}$ M R6G solution is drop-cast onto the optimized AgNPs substrate, and 20 points are selected for analysis. In Figure 4(a), the RSD of the SERS signal at the 612 $cm^{-1}$ characteristic peak is calculated to be 4.9%. To further verify the effectiveness of the substrate, it is stored under ambient conditions and tested every two days. In Figure 4(b), the intensity fluctuations are minimal, indicating that the SERS substrate exhibits good stability.

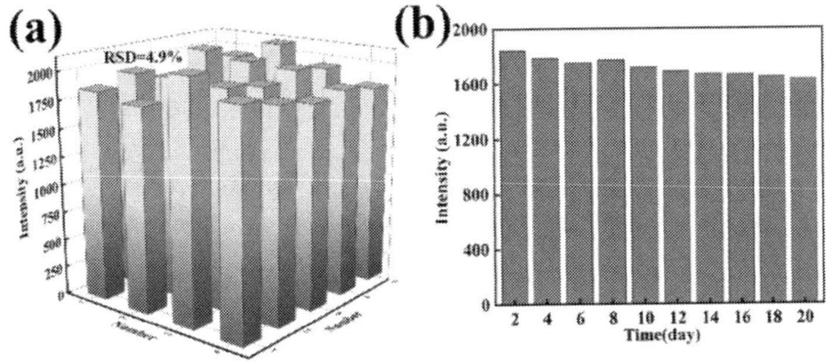


**Figure 4.** (a) shows the intensity variation map of 20 randomly selected points on the optimized AgNPs substrate at the 612 $cm^{-1}$ peak, while (b) presents the intensity changes of the AgNPs substrate measured at different time intervals.

The spectra of the thiram sample are subjected to gradient dilution to demonstrate the detection capability of the prepared biosensor for quantifying thiram samples. In Figure 5(a), the detection limit for trace amounts of thiram reaches $1\times10^{-9}$ M. A comparison with the literature is conducted to analyze the characteristic peaks, where 561 $cm^{-1}$ corresponds to the stretching of S-S bonds, 943 $cm^{-1}$ is attributed to the vibration of C-H bonds, 1149 $cm^{-1}$ corresponds to the stretching of C-N bonds and the wagging vibration of $CH_3$, and 1382 $cm^{-1}$ corresponds to both C-N bond stretching and $CH_3$ symmetric deformation. Additionally, 1511 $cm^{-1}$ is related to the stretching of C-N bonds and $CH_3$ deformation. To further analyze the relationship between spectral intensity and thiram concentration, the Raman peak intensity at 1382 $cm^{-1}$ is plotted against the logarithm of the concentration, and a linear correlation is calculated. As shown in Figure 5(b), $R^2$ is 0.986, showing a strong linear relationship between the SERS intensity and the logarithm of the concentration. Such results indicate that the detection of thiram using this substrate is effective.

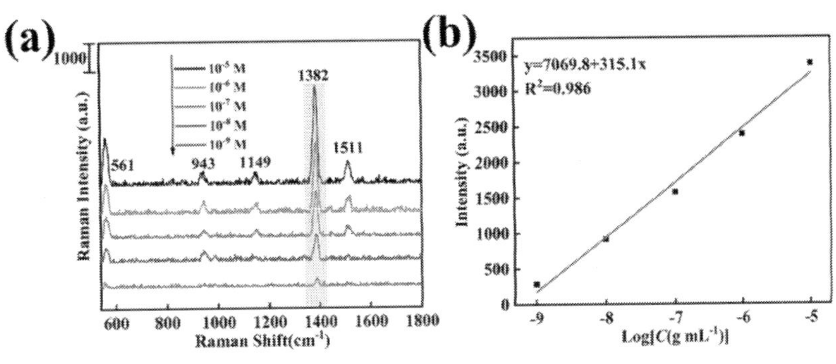


**Figure 5.** (a) presents the SERS signal of thiram at concentrations ranging from $10^{-5}$ to $10^{-9}$ M, while (b) shows the correlation between the intensity of the 1382 $cm^{-1}$ and the logarithm of the concentration.

## 4. Conclusions

In this study, we fabricate AgNPs nanostructures with enhanced SERS performance using a high-temperature annealing process. The SERS performance of AgNPs at different annealing temperatures is investigated to determine the optimal conditions for preparing SERS substrates. Based on the SERS spectra, we explore the detection limit of the optimized AgNPs substrate and demonstrate the uniformity and stability of the SERS signal. Using the optimized AgNPs substrate, we detect thiram under varying concentration conditions, achieving a favorable detection limit. This indicates that the SERS substrate holds great potential for pesticide detection, providing a valuable reference for developing new technologies in environmental monitoring and food safety.

## References

[1] T. Wu, S. Yang, W. Feng, K. Li, Y. Li, L. Wang, et al., Recyclable Fe3O4@SiO2@PEI-DTC@Au-Ag nanocomposites as a SERS platform for sensitive detection of thiram on fruit surfaces, Microchemical Journal, 198(2024).

[2] S. Chakraborty, V. Awasthi, R. Goel, S. K. Dubey, Label-free detection of thiram in trace quantity in fruits and vegetables using chemically made SERS-active substrate, Optical Materials, 154(2024).

[3] C. Zong, M. Xu, L. J. Xu, T. Wei, X. Ma, X. S. Zheng, et al., Surface-Enhanced Raman Spectroscopy for Bioanalysis: Reliability and Challenges, Chem Rev, 118(2018) 4946-80.

[4] A. X. Wang, X. Kong, Review of Recent Progress of Plasmonic Materials and Nano-Structures for Surface-Enhanced Raman Scattering, Materials (Basel), 8(2015) 3024-52.

[5] K. K. Strelau, A. Brinker, C. Schnee, K. Weber, R. Möller, J. Popp, Detection of PCR products amplified from DNA of epizootic pathogens using magnetic nanoparticles and SERS, Journal of Raman Spectroscopy, 42(2011) 243-50.

[6] P. K. Sharma, J. S. Kumar, V. V. Singh, U. Biswas, S. S. Sarkar, S. I. Alam, et al., Surface plasmon resonance sensing of Ebola virus: a biological threat, Anal Bioanal Chem, 412(2020) 4101-12.

# Study on crack extension behavior of epoxy resin composites under typhoon conditions

**Ruodong Huang [1,*], Fusheng Zhou [1], Chao Gao [1] and Hao Yang [2]**

[1] National Engineering Research Center of UHV Technology and Novel Electrical Equipment Basis, Kunming, 651705, China
[2] School of Electronics and Information, Xi'an Polytechnic University, Xi'an, 710048, China

*Corresponding author's e-mail: Huangruodong999@126.com

**Abstract:** For the problem of accelerated expansion of epoxy resin cracks in tub insulators under typhoon conditions, this paper establishes an accelerated evolution model of epoxy resin crack defects based on the maximum circumferential stress criterion, simulates the crack expansion process of epoxy resin materials under different typhoon levels, grasps the characteristic parameters of the crack expansion behaviors under the influence of the three major typhoon levels, and achieves the prediction of the crack expansion path and fatigue life of the epoxy resin materials under extreme environments. The results show that the type I stress intensity factor at the crack tip of the epoxy resin material increases linearly with the crack extension length, the type II stress intensity factor reaches the peak at the early stage of crack extension and then decreases to 0, the type III stress intensity factor tends to float up and down to 0, and the fatigue life decreases linearly with the increase in typhoon level; the results of the paper have an important theoretical application value for the prediction of crack extension of tub insulators under the action of typhoon loads. The results of this paper have important theoretical application value for carrying out crack prediction of basin insulators under typhoon loading.

## 1. Introduction

Epoxy-based insulating materials are widely used in gas-insulated metal-enclosed switchgear (GIS) and gas-insulated transmission pipelines (GIL) due to their excellent electrical and mechanical properties [1-2]. However, due to many factors such as production process, transportation and installation, and daily operation and maintenance, a variety of defects of epoxy-based insulators inside GIL/GIS have been observed [3-4], especially in extreme typhoon climatic conditions, which can lead to the damage and failure of basin insulators, and ultimately induce insulating flashover and breakdown accidents [5]. Therefore, it is of great significance to realize the simulation of the development of crack defects in epoxy resin composites under extreme climates to ensure the safe and stable operation of the equipment.

At present, the problems in the scalability of cracks in metallic materials have attracted extensive attention and research from scholars in related fields at home and abroad. Taking DMWJ metal welded joints as the research object, Yuan et al. selected four center crack-containing specimens with different constraints, and numerically simulated different vertical seismic loads by applying five seismic intensities and three vertical seismic loads under actual earthquakes through the method of mass acceleration application [6]. Song established a finite element model of the steel structure, obtained the dangerous parts of the platform through the static analysis under extreme storm conditions, and calculated the number of load cycles of single cracks and multi-cracks under extreme storm conditions by using the Paris and NASGRO crack expansion rate model, respectively, and obtained the expansion life of single cracks and multi-cracks through the conversion of the ambient load cycle [7]. Wang et al. used a stochastic crack extension model to analyze the wind vibration fatigue problem of towering structures and also considered the probability distribution of mean wind speed and the influence of time correlation on the crack extension, simulated the relevant time course samples of the mean wind speed by the ZMNL method, and derived the probability distribution of the crack length according to the Monte-Carlo method [8]. Li et al. carried out the establishment of a steel material model according to the structural characteristics of a typical steam pipe section of a certain type of ship and analyzed the stress response and displacement response of the steel material under the impact of waves at 45°, 90° and 180° heading angles, determined the weak link of the ship's steam pipe section structure under the impact of the waves as the stress concentration area, and finally calculated the change of the crack size of the steam pipe section during the operation period based on the Paris formula [9].

In summary, scholars have studied the effects of extreme conditions such as seismic loads, wind loads, and wave loads on the performance of metallic materials. However, there are relatively few literatures that consider the crack extension of epoxy resin composites by extreme pulsating wind loads. Therefore, the crack extension model of the GIS basin insulator is now analyzed to predict the stress intensity factor at the crack tip and the fatigue life of crack extension of epoxy resin composites under extreme pulsating wind conditions using the joint simulation software of ABAQUS and FRANC3D.

## 2. Model building

### *2.1. Modeling of epoxy crack defects*

In this paper, FRANC3D, a crack analysis software, and ABAQUS, a finite element analysis software, are selected to extend the analysis of epoxy resin specimens containing cracks [10-11]. The epoxy resin specimens without defects are constructed in ABAQUS with length, width, and height of 0.2 mm, 0.4 mm, and 0.05 mm, respectively. The material parameters are set up in ABAQUS with a density of $1.2\times10^3$ kg/m$^3$, Young's modulus of 3340 MPa, and a Poisson's ratio of 0.31 [12]. The crack extension parameters are set up in FRANC3D with C of $3.35 \times 10^{-8}$, and m being 5.54 [13]. The maximum circumferential stress criterion and the Paris formula [14] are used to solve for life. The direction of crack expansion and step of expansion are shown in Figure. 1.

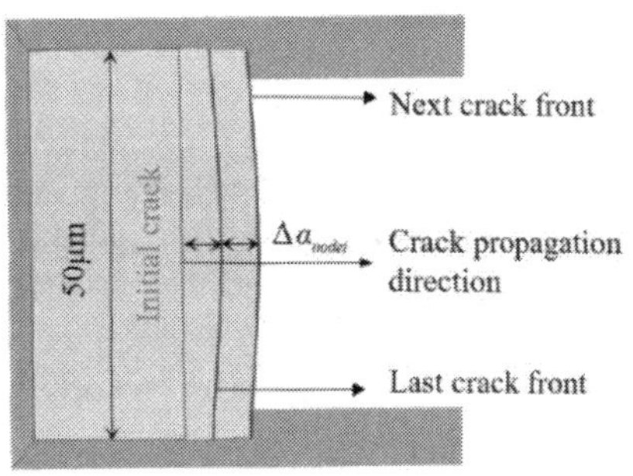


**Figure 1.** Schematic diagram of extension step.

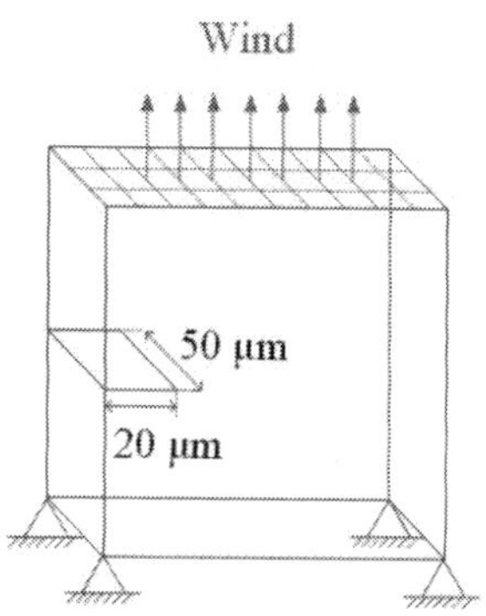


**Figure 2.** 3D view of crack tip loading.

Due to the different typhoon classes, including typhoons, strong typhoons, and super typhoons, the basin insulators are subjected to different typhoon loads [15]. In this paper, the trends of the stress intensity factor at different locations of the crack tip of the tub insulator subjected to different typhoon levels and wind directions, as well as the variation curves of the crack extension life under this typhoon condition are investigated respectively. The crack is located at the left middle edge of the specimen, with a size of 0.05 mm*0.02 mm. The lower surface of the epoxy resin specimen is fixed, and three different levels of typhoon loads are applied on top of the epoxy resin specimen, which is subjected to three different stresses on the epoxy resin specimen as shown in Figure 2.

### *2.2. Calculation of wind load*

According to a large number of measured data [16], the average wind and pulsating wind are superimposed on each other to form a natural wind, the average wind is generally used for the calculation of static analysis of the structure, and the pulsating wind is used for the calculation of dynamic analysis of the structure.

1) Average wind

According to the American Petroleum Institute (API) standard [17], the wind load acting on the platform can be described as:

$$F_W = K_g K \alpha A u^2 \tag{1}$$

In the formula: $F_W$ is the wind force, N; $K_g$ is the height coefficient of wind pressure; $K$ is the shape coefficient of the structure; $\alpha$ is the wind pressure coefficient; $A$ is the area of the structure subjected to wind pressure, m$^2$; $u$ is the wind speed, m/s.

Due to the installation location of GIS basin insulators, according to API quasi [13], it can be seen that the platform wind pressure height coefficient is 1.37; and due to the basin insulators being similar to the structure of the round basin, the platform structural shape coefficient is 0.5, and the wind pressure coefficient is 0.613; according to the different wind direction, the platform to withstand the wind pressure of the area is different.

2) Pulsating wind

The Davenport wind speed spectrum coincides with the pulsating wind speed spectrum in the South China Sea, so the Davenport wind speed spectrum is chosen to simulate the pulsating wind spectrum [18], which is given by:

$$S_\upsilon(n) = \overset{-2}{\upsilon_{10}} \frac{4kx^2}{n(1+x^2)^{4/3}} \tag{2}$$

According to the geographic location of the platform, the average wind speed under extreme storm conditions is 64 m/s, without considering the effect of vertical pulsating wind, the simulation space is 60 sample points, the time step is 0.1 s, the frequency step is 0.01 Hz, the ground roughness coefficient is 0.00129, and the simulation time is one wave period, i.e., 12 s [19]. Matlab programming software and harmonic synthesis method are used to obtain the pulsating wind speed time scale, as shown in Figure 3.

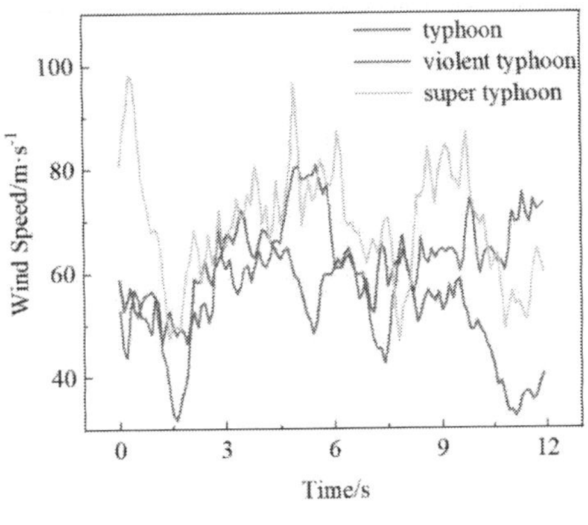


**Figure 3.** Pulsating wind time course curves for different classes.

According to GB50009-2012 Code for Structural Loads of Buildings [20], the instantaneous wind load is calculated as follows:

$$F_s = \frac{1}{2}\rho u_s u_z A \upsilon^2 \tag{3}$$

In Formula (3): $F_s$ is the instantaneous wind load, N; $\rho$ is the air density, 1.235 kg/m$^3$; $v$ is the instantaneous wind speed, m/s; $u_s = 1.1$ is the body shape coefficient; $u_z = 1.25$ is the coefficient of variation of the wind pressure height.

## 3. Calculation of characteristic parameters of crack extension behavior

The type I stress intensity factor is simulated under typhoon, violent typhoon, and super typhoon conditions, and it can be seen from the trend in Figure 4 that the overall trend of the type I stress intensity factor at the crack tip of the epoxy resin under three types of typhoon conditions is relatively the same, which is growing linearly with the length of the crack extension, and the growth rate of the type I stress intensity factor is roughly the same. When the stress intensity factor at the crack tip reaches the set fracture toughness of 3 KPa·m$^{1/2}$, the fatigue extension stage ends. Under the same crack extension length, the greater the typhoon intensity is, the greater the type I stress intensity factor at the crack tip is, and the smaller the extension length of the fatigue extension stage is. The value of the type I stress intensity factor at the crack tip is typhoon<violent typhoon<super typhoon. From Figure 5, the simulation results of the type II stress intensity factor show that the type II stress intensity factor under different typhoon conditions reaches the peak value at the initial stage of crack extension (0 μm), and the magnitude decreases with typhoon, violent typhoon, and super typhoon in that order. With the increase in the crack extension length, the type II stress intensity factor under the three typhoon conditions is in a small range with a small fluctuation trend and reaches a minimum value of about 0 MPa·m$^{1/2}$ at the extension of 5 μm. From the simulation results of the type III stress intensity factor in Figure 6, it can be seen that the type III stress intensity factor under different typhoon levels is near 0 MPa·m$^{1/2}$ at the initial stage of crack extension, and then fluctuates around ± 1.5 KPa·m$^{1/2}$. Under strong typhoon conditions, the crack tip-out type III stress intensity factor shows a large fluctuation when the crack extends to 20 μm. When the crack is about to end the fatigue extension stage, the type III stress intensity factor is relatively stable in a smaller floating region and reaches the minimum value under different accelerations at the end of the fatigue crack extension.

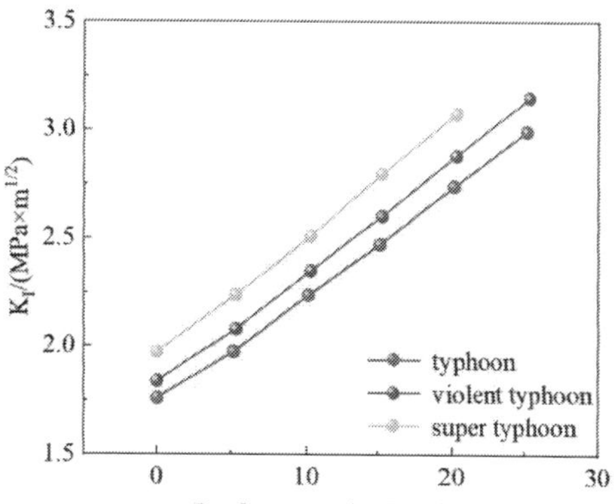


**Figure 4.** Comparison curves of type I stress intensity factor under different typhoon levels.

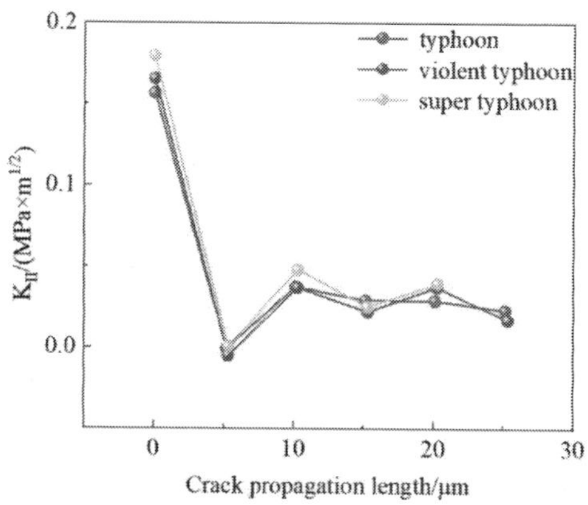


**Figure 5.** Comparison curves of type II stress intensity factor under different typhoon levels.

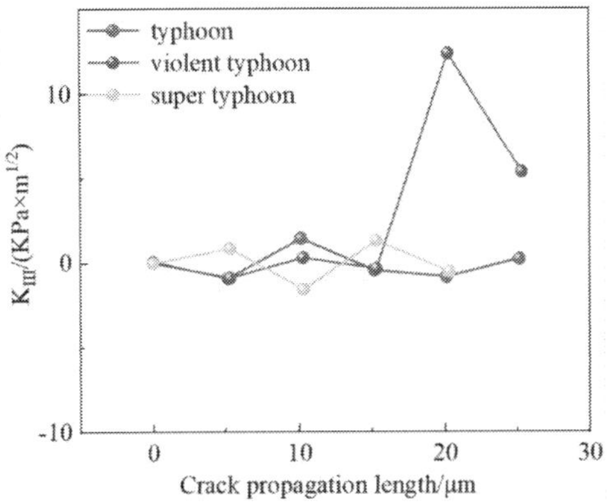


Figure 6. Comparison curves of type III stress intensity factor under different typhoon levels.

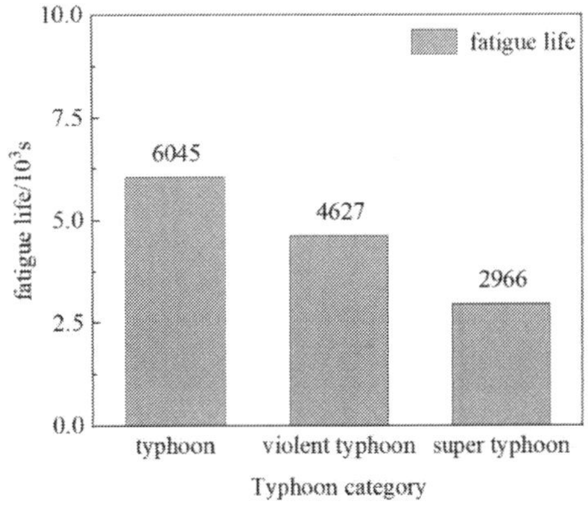


Figure 7. Fatigue life of epoxy resin under different typhoon levels.

The prediction of fatigue crack extension life of epoxy composites containing defects is performed under three levels of typhoon, violent typhoon, and super typhoon. It can be observed that the life of tub insulators containing crack defects decreases in order under the extreme environment of typhoons, violent typhoons, and super typhoons. According to the simulation results in Figure 7, the fatigue life value of the basin insulator affected by a super typhoon is 2966 s, which is 56% less than the fatigue life value of the basin insulator affected by a violent typhoon (4627 s), and 104% less than the fatigue life value of the basin insulator affected by the typhoon (6045 s).

## 4. Conclusion

To investigate the effect of typhoon level (typhoon, violent typhoon, super typhoon) on the extended behavior of basin insulators with crack defects, it can be seen that the crack tip of the epoxy resin material, type I stress intensity factor, increases linearly with the length of crack extension, type II stress intensity factor reaches a peak in the early stage of crack extension and then decreases to 0, and the type III stress intensity factor tends to fluctuate up and down to 0. The extended life of the tub insulator with crack defects is reduced with the increase in the crack tip in different grades of pulsating wind time course. Under different levels of pulsating wind duration, the extended life of basin insulators with crack defects tends to decrease with the increase in typhoon intensity. Under typhoon loading, its life is 6045 s; under violent typhoon conditions, its life decreases to 4627 s; and under super typhoon, its life is 2966 s.

**Funding statement**

This work was supported by the National Engineering Research Center of UHV Technology and Novel Electrical Equipment Basis (No. NERCUTNEEB-2022-KF-07).

## References

[1] DU B X, YAO H, LIANG H C, et al. Electric field relaxation of basin spacer under variable temperature gradient in DC-GIL/GIS [J]. Transactions of China Electrotechnical Society, 2024, 39(9): 2851-2859.

[2] SHA W Y, LUO Y, MA Penghuan. Study on surface discharge and electric field characteristics of GIS basin insulator with different metal particle defects [J]. Insulation Materials, 2023, 56(9): 89-93.

[3] SUN H B, ZHOU Z Y, LÜ Y T, et al. Research on defect detection method of basin insulator based on low-frequency dual probe ultrasonic [J]. Insulation Materials, 2023, 56(1): 110-114.

[4] HAN S, GAO F, LIAO S Z, et al. Review on surface defects and diagnosis methods of basin-type insulator in GIS [J]. Insulation Materials, 2022, 55(2): 12-22.
[5] ZHANG G B, ZHAO H Y, YANG W, et al. Joint diagnosis and application of GIS defects based on partial discharge and x-ray imaging detection [J]. High Voltage Apparatus, 2022, 58(9): 197-202, 220.
[6] YUAN H W, YANG J, CUI W, et al. Numerical simulation of fracture behavior of 52M nickel-based alloy DMWJ under different vertical seismic loads [J].Mechanical Strength, 2022, 44(06): 1469-1475.
[7] SONG L Z. Study on crack extension of pile leg of conduit rack platform under extreme storm condition [D]. Northeast Petroleum University, 2021.
[8] WANG S C, YANG X Q. Random crack extension analysis of wind vibration fatigue in towering structures [J]. Steel Structure, 2007, (07): 43-47.
[9] LI C S, LIU F J, WU Y. Study on dynamic response and crack extension of ship steam pipeline under wave impact [J]. Thermal Power Engineering, 2024, 39(06): 191-198
[10] ZHAO H, LI Y X, WANG Y Q, et al. Study on crack extension of crane truss at fatigue hazardous location based on ABAQUS-FRANC3D joint simulation [J/OL]. Industrial Building, 1-15[2024-08-05].
[11] XING B D, GAO D, HU Z H, et al. Three-dimensional crack extension analysis of typical monolithic structures based on Franc3D [J]. Computer Aided Engineering, 2023, 32(03): 15-19+25.
[12] LI Z, ZHONG L M, WANG W X, et al. Study on in-service crack initiation and failure fracture mechanism of tub insulators based on Al_2O_3/epoxy composites [J]. Insulation Materials, 1-8[2024-08-05].
[13] Duan Q, Song J, Menouillard T, et al. Element-local level set method for three-dimensional dynamic crack growth [J]. International Journal for Numerical Methods in Engineering, 2009, 80(12): 1520-1543.
[14] LI Y, FENG Y Z, DING J W, et al.Parameterization of initial cracks in Franc3D crack extension calculation [J]. Science and Technology Innovation, 2020, (18): 33-34.
[15] ZHU B R, CHEN G M, LIN H, et al. Numerical simulation of the whole process of conduit rack platform collapse under strong typhoon [J]. Ocean Engineering, 2015, 33(02): 97-104.
[16] Cao Y H, Xiang H F, Zhou Y. Simulation of stochastic wind velocity field on long-span bridges [J]. Journal of Engineering Mechanics, 2000, 126(1): 1-6.
[17] American Petroleum Institute. Planning, designing and constructing fixed offshore platforms-working stress design: API RP 2AWSD 22nd Edition [S]. Washington, DC: API, 2014.
[18] Liao Y D, Wang L, Zhang P C, et al. Dynamic response analysis of offshore wind turbine structures under combined environmental loads [J]. Water Resources and Hydropower Engineering, 2021, 52(3): 206-216.
[19] Zhang P, Cao J F. Safety evaluation of jacket offshore platform in service under strong typhoon [J]. Low-Temperature Architecture Technology, 2020, 42(12): 60-64.
[20] Ministry of Housing and Urban-Rural Development of the People's Republic of China. Load code for the design of building structures: GB 50009—2012[ S]. Beijing: China Architecture & Building Press, 2012.

doi:10.1088/1742-6596/2961/1/012008

# Oxygen-vacancy-abundant amorphous layer and Dual-doping strategy enabling superior electrochemical performance in Li-rich materials

**Deyuan Liu** [1,*]

[1] School of Materials and Energy, University of Electronic Science and Technology of China, West High-Tech Zone, Chengdu 611731, China

[*]Corresponding author's e-mail: dyliu.uestcsme@foxmail.com

**Abstract**: To meet the rapid development demands of high-energy Li-ion batteries, this paper presents a strategy for a Co- and Ni-free single-crystal Li-rich cathode via a high-speed temperature rise sintering method. By constructing an oxygen-vacancy-abundant amorphous layer on the surface and inducing dual-ion doping in the bulk phase, this research aims to optimize the irreversible oxygen loss, voltage decay, and poor cycling stability of single-crystal Li-rich Mn-based materials. Testing results indicate that the prepared material exhibits an initial discharge capacity of 342.1 mAh $g^{-1}$ and capacity retention of 76.2% after 800 cycles, and the gas generation and voltage decay are also significantly reduced.

## 1. Introduction

The rapid development of electric vehicles has diversified the demand for advanced cathode materials. Achieving high specific energy, low cost, and environment-friendly cathodes has gradually become the mainstream. In recent years, polycrystalline Li-rich Mn-based (PC-LRM, $Li_{1.2}Mn_{0.54}Ni_{0.13}Co_{0.13}O_2$) materials with high theoretical specific capacity (≥250 mAh $g^{-1}$) due to anionic redox reactions have garnered attention [1, 2]. However, the extensively studied PC-LRM cathode has gradually failed to meet commercial requirements due to its low compacted density and the addition of toxic Co [3, 4]. Co-free PC-LRM materials exhibit poor battery stability due to transition metal (TM) dissolution [5, 6]. Therefore, developing single-crystal Co- and Ni-free LRM (SCNF-LRM) materials is imperative. The process flow of single-crystal cathode is complicated, and particle size control is the key point. Under Co- and Ni-free conditions, irreversible anionic oxygen redox reactions lead to low initial Coulombic efficiency (ICE) and cycling performance [7, 8]. The irreversible loss of surface oxygen and voltage decay also continually hinder the commercial application of SCNF-LRM materials.

Based on this, this paper proposes a synthesis strategy for SCNF-LRM materials combining coating and doping. Coating an amorphous oxygen-vacancy-abundant Li-Nb-P oxide layer inhibits the irreversible loss of surface elements. Doping $Nb^{5+}$ and $P^{5+}$ in the bulk phase reduces the $Li^{+}$ migration barrier and stabilizes the material's crystal structure. The synthesized SCNF-LRM material exhibits excellent rate and cycling performance, with low voltage decay at a rate of 5 C (1 C = 250 mA $g^{-1}$). In situ, differential electrochemical mass spectrometry (in situ DEMS) indicates that the material has extremely low gas generation and possesses good commercial application prospects.

## 2. Experiment

### *2.1. Material synthesis*

The precursor was synthesized by the co-precipitation method and sintered to prepare single-crystal Li-rich Mn-based materials (denoted as LRM) [3]. No Ni or Co is added during the SCNF-LRM precursor preparation. In the sintering process, LRMNP precursor, $Li_2CO_3$, and $NbP_2O_7$ were mixed in a molar ratio of 1.97:1.3:0.03 and sintered to obtain the modified sample (denoted as LRMNP). The sintering process will be detailed later.

### *2.2. Sample characterization*

Morphology, mapping, and particle size analysis of LRNMP were characterized by using SEM (Hitachi Regulus8100). The focused ion beam transmission electron microscope (FIB-TEM, Helios G4 PFIB HXe) was used for sectioning, and SAED and EELS were performed for characterization. The crystal structure was analyzed using an X-ray powder diffractometer (Xrd, Ultima IV), and elemental analysis was conducted via X-ray photoelectron spectroscopy (XPS, Thermo ESCALAB 250XI). Elemental loss analysis after cycling was performed by using an inductively coupled plasma optical emission spectrometer (ICP-OES, PerkinElmer ICP 2100). Gas production was tested by using in situ DEMS (QAS 100 Li Plus). First-principles (DFT) calculations were performed by CASTEP software, with elemental vacancy formation energy based on the plane-wave pseudopotential method, and the $Li^+$ migration barrier was calculated by the PBE function.

### *2.3. Electrochemical measurements*

The cathode slurry (~15.0 mg $cm^{-2}$) comprised active material (95%), conductive carbon black (3%), and polyvinylidene fluoride (2%). The anode slurry (~9.0 mg $cm^{-2}$) consisted of graphite (80%), conductive carbon black (15%), and polyacrylic acid (5%). The pouch cells were assembled in an Ar-filled glove box using a Celgard 2500 separator. The electrolyte (1 M $LiPF_6$) was a mixture of ethylene carbonate (EC) and dimethyl carbonate (DMC) in a 3:7 volume ratio. Rate and cycle tests were performed on pouch cells using the Land battery test system (CT3004A).

## 3. Results and discussion

Figure 1 (a-c) shows the SEM images of LRMNP. The morphology of LRMNP appears as single-crystal particles, showing no significant difference from conventional single-crystal cathode, indicating that the modification does not disrupt the original morphology. Figure 1 (d) reveals the impact of rapid temperature rise sintering on the particle size distribution. Traditional single-crystal cathode material sintering adopts a pre-sintering process, typically heating at a rate of 10 °C $min^{-1}$ to 450°C for 10 hours, followed by heating to 950°C at the same rate and sintering for 25 hours to obtain a material grain size distribution ranging from 1-5 μm in descending order. The rapid temperature rise sintering method used in the modification eliminates the pre-sintering, heating to 950°C at a rate of 50 °C $min^{-1}$, sintering for 5 hours, cooling to room temperature (RT), and then sintering at 950°C for 20 hours at the same heating rate, and LRMNP single-crystal particles with a ~3 μm particle distribution can be obtained. Figure 1 (e) is the FIB-TEM image of LRMNP, revealing that LRMNP shows a core-shell structure. The interior of the material exhibits a layered structure, while the exterior is covered by a ~2 nm thick shell. The SAED image shows a pattern combining ring- and spot-like features, further indicating that the shell is amorphous. To explore the element distribution of the core-shell structure, EELS testing was conducted. Figure 1 (f) indicates that the amorphous shell is rich in Li-Nb-P oxides with oxygen vacancies, and the distribution of oxygen vacancies gradually decreases from the surface to the interior of the bulk phase, which is beneficial for inhibiting irreversible oxygen loss in the material. The distribution of Nb and P within the material shows elemental doping within the bulk phase. The Nb-O bond is more stable than the Mn-O bond, which is conducive to suppressing element loss and structural collapse in cycling. Figure 1 (g) is the mapping diagram of LRMNP, indicating that LRMNP has a uniform element distribution.

**Figure 1.** (a-c) SEM, (d) Particle size distribution, (e) FIB-TEM, (f) EELS, and (g) EDS mapping images of samples.

To investigate the impact of modification on the crystal structure of the material, XRD testing was conducted and refined, as shown in Figure 2 (a, b). The XRD patterns of the samples all exhibit a layered R-3m structure, and there are $Li_2MnO_3$ superlattice diffraction peaks between 20° and 25°. The high peak intensities of (003) and (104) indicate the successful synthesis of the two-phase solid solution material. The XRD pattern of LRMNP does not show significant changes compared to LRM, suggesting that the modification does not significantly disrupt the crystal structure of LRM. LRMNP has a higher $I_{(003)}/I_{(104)}$ peak intensity ratio than LRM, indicating that LRMNP has a more complete layered structure. The higher c/a value of LRNMP compared to LRM suggests that Nb doping expands the interplanar spacing of the material, enabling faster $Li^+$ diffusion. The modified LRMNP material has lower Li/Ni mixing, which will achieve more reversible $Li^+$ intercalation and good cycle stability. The modified LRMNP has a higher proportion of the C2/m phase, which provides a basis for more anionic redox reactions in the material. Figure 2 (c-e) shows the XPS test results of Nb and Mn in LRM and LRMNP, indicating that $Nb^{5+}$ is successfully doped into LRMNP. The Mn $2p_{1/2}$ and Mn $2p_{3/2}$ peaks of LRMNP shift to higher binding energies compared to LRM, indicating that doping increases the valence state of Mn. The proportion of $Mn^{4+}$ in LRMNP is 73%, while it is only 57% in

LRM. The lower proportion of $Mn^{3+}$ will bring a lower Jahn-Teller effect to LRMNP, and reduce the lattice.

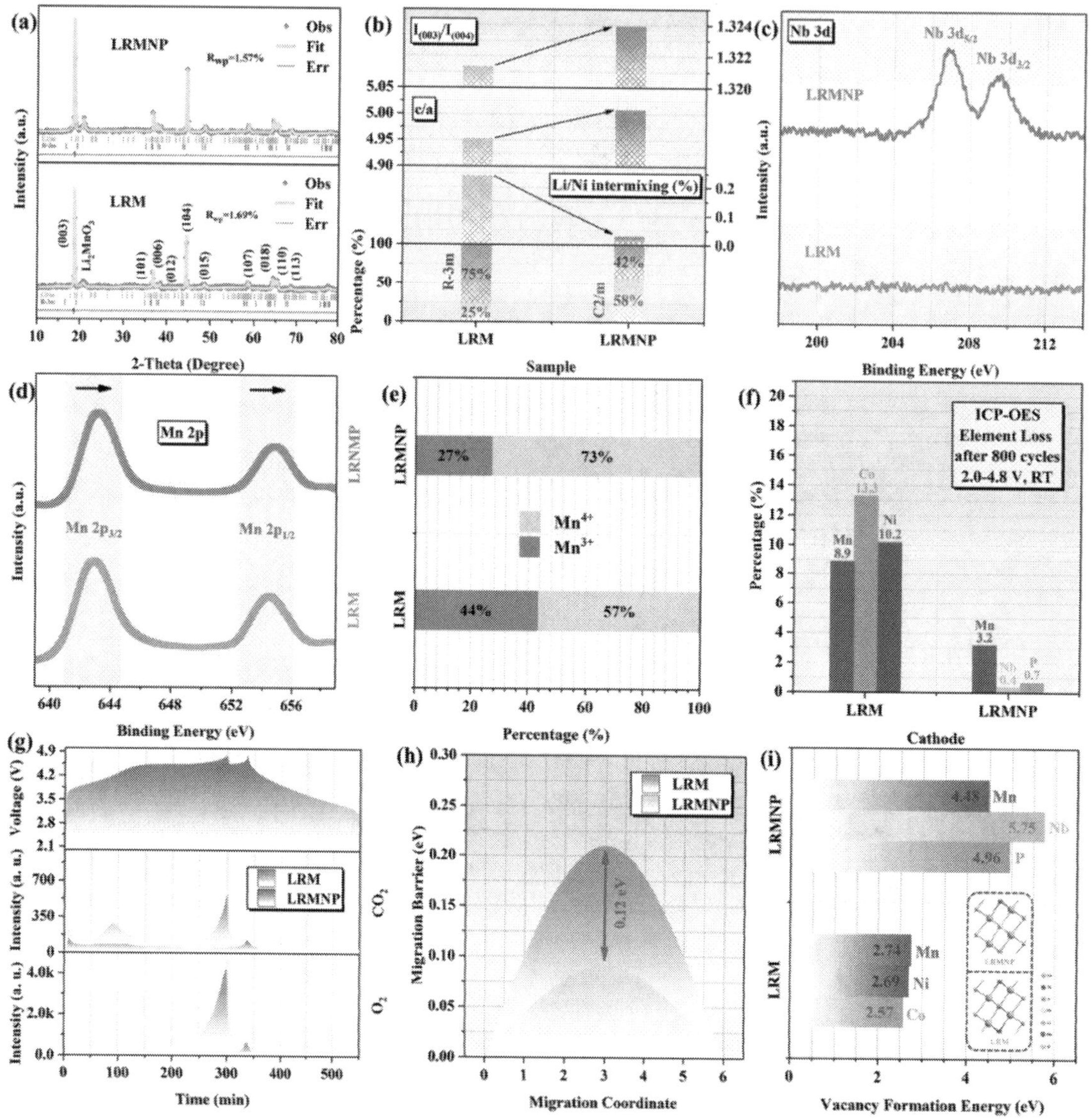

**Figure 2.** (a, b) XRD Rietveld refinement, (c-e) XPS, (f) ICP-OES, (g) In-situ DEMS, and (h, i) DFT images of samples.

Distortion of the material during charging and discharging brings more stable cycling and lower voltage decay. Figure 2 (f) uses ICP-OES to analyze the element loss in LRM and LRMNP in pouch cells before and after long cycles. Benefiting from the more stable surface and bulk structures brought by the coating and doping, the Mn element loss in LRMNP is only 3.2%, and Nb and P are stable in the material (the loss ratios are both below 1%). The unstable structure of LRM results in higher metal element loss, with Mn, Co, and Ni losses of 8.9%, 13.3%, and 10.2%, respectively. Figure 2 (g) reveals the gas generation during charging and discharging. LRMNP has lower gas production compared to LRM, indicating that P doping and the amorphous coating with oxygen vacancy on the surface assist LRMNP in achieving more stable anionic redox reactions, which also lays the foundation for the further commercial application of the cathode. Figure 2 (h) calculates the $Li^+$

migration barriers of LRM and LRMNP. Benefiting from the larger interplanar spacing brought by doping, the $Li^+$ migration barrier of LRNMP is significantly lower than that of LRM. Figure 2 (i) calculates the vacancy formation energies of elements in LRM and LRMNP. Mn, Nb, and P in LRMNP are higher.

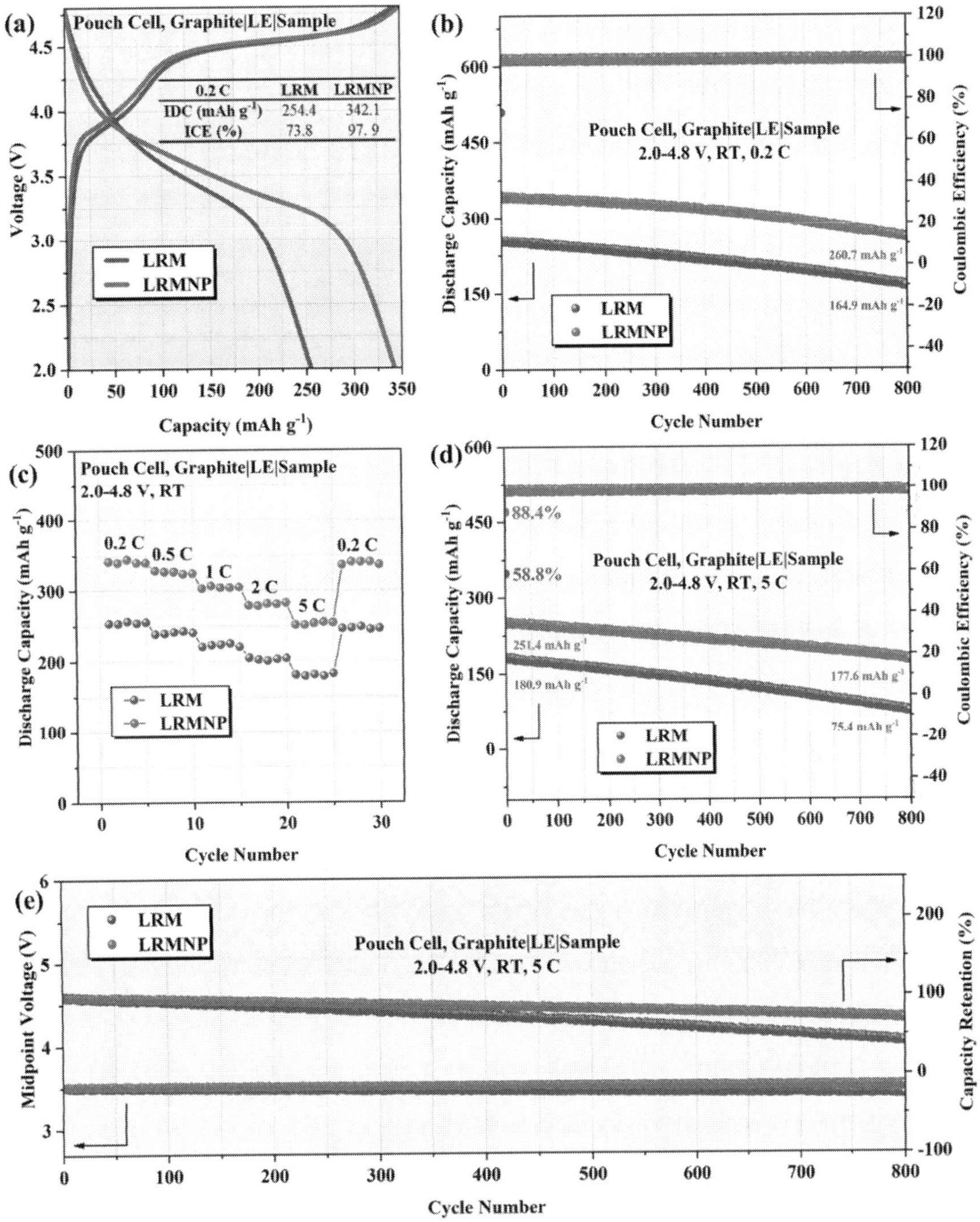


**Figure 3**. (a) Initial charge and discharge, (b, d) Cycling performance, (c) Rate capability, and (e) Voltage decay curves of samples.

Vacancy formation energies other than Mn, Co, and Ni in LRM indicate that LRMNP has a stable crystal structure. Figure 3 (a) shows the first charge-discharge curve of the material at a current density of 0.2C. The ICE of LRMNP is 97.9%, and the initial discharge capacity (IDC) is 342.1 mAh $g^{-1}$, while LRM is only 73.8% and 254.4 mAh $g^{-1}$. Figure 3 (b-d) reveals that LRMNP has good cycle performance at both high and low rates. LRMNP still retains a specific discharge capacity of 177.6 mAh $g^{-1}$ after 800 cycles at a current density of 5C. Figure 3 (e) summarizes the midpoint voltage of LRMNP and LRM at a high rate, with LRMNP showing lower voltage decay. The excellent electrochemical performance of LRMNP consists well with the characterization results of its more stable crystal structure.

## 4. Conclusion

To meet the performance, cost, and environmental requirements of high-energy-cathode, this paper presents a synthesis strategy for Co- and Ni-free single-crystal Li-rich material based on rapid temperature-rise sintering. Characterization results from FIB-TEM, EELS, XRD, XPS, in situ DEMS, and DFT calculations indicate that the synthesized single-crystal cathode has a core-shell structure, with an amorphous shell rich in oxygen vacancies on the surface and a co-doped bulk phase. The amorphous shell inhibits the irreversible loss of oxygen from the material surface, while element doping stabilizes the crystal structure of the material. LRMNP exhibits an IDC of 251.4 mAh $g^{-1}$ and an ICE of 88.4% at a current density of 5C, with extremely low voltage decay.

**Acknowledgments**

This work was financially supported by the Department of Science and Technology of Sichuan Province (2020ZHCG0035).

## References

[1] Brucker A., Duran A., Sullivan N. P., and Mistry A. (2024). Electrifying off-road vehicles: is 1, 00 0 [Wh/kg] enough? ACS Energy Lett., 9: 4, 053-8. https://doi.org/10.1021/acsenergylett.4c0 1276.

[2] Shen C., Hu L., Duan Q., Liu X., Huang S., Jiang Y., Li W., Zhao B., Sun X., and Zhang J. (2023). Understanding lattice oxygen redox behavior in lithium-rich manganese-based layered oxides for lithium-ion and lithium-metal batteries from reaction mechanisms to regulation strategies. Adv. Energy Mater., 13: 2, 302, 957. https://doi.org/10.1002/aenm.202302957.

[3] Zhang Y., Yin C., Qiu B., Chen G., Shang Y., and Liu Z. (2022). Revealing Li-ion diffusion kinetic limitations in micron-sized Li-rich layered oxides. Energy Stor. Mater., 53: 763-73. https://doi.org/10.1016/j.ensm.2022.10.008.

[4] Cui S. L., Gao M. Y., Li G. R., and Gao X. P. (2022). Insights into Li-rich Mn-based cathode materials with high capacity: from dimension to lattice to atom. Adv. Energy Mater., 12: 2, 003, 885. https://doi.org/10.1002/aenm.202003885.

[5] Park N. R., Zhang M., Han B., Li W., Qian K., Nguyen H., Kumakura S., and Meng Y. S. (2024). Understanding boron chemistry as the surface modification and electrolyte additive for Co-fr ee lithium-rich layered oxide. Adv. Energy Mater., 2, 401, 968: https://doi.org/10.1002/aenm. 202401968.

[6] Zhao H., Lam W. Y. A., Sheng L., Wang L., Bai P., Yang Y., Ren D., Xu H., and He X. (2022). Cobalt-free cathode materials: families and their prospects. Adv. Energy Mater., 12: 2, 103, 894. https://doi.org/10.1002/aenm.202103894.

[7] Pei Y., et al. (2022). Fluorinated rocksalt cathode with ultra-high active Li content for lithium-ion batteries. Angew. Chem. Int. Ed., 61: e202212471. https://doi.org/10.1002/anie.202212471.

[8] Huang Y., et al. (2024). Integrated rocksalt-polyanion cathodes with excess lithium and stabilized cycling. Nat. Energy, https://doi.org/10.1038/s41560-024-01615-6.

# Preparation of amino acid-based polyesteramides and study on poly-ε-caprolactone/amino acid-based polyesteramides fiber film

**Lu Bai[1], Mingguang Hu[1], Xiaojian Li[2], Nan Liu[2], Haijun Zhou[1], Fen Zhang[1], Yantao Li[1] and Xiaoqi Chen[1*]**

[1]Institute of Energy Resources, Hebei Academy of Sciences, Shijiazhuang, Hebei Province, 050000, China

[2]Hebei Huami New Material Technology Co., Ltd., Xingtai, Hebei Province, 055150, China

[*]Corresponding author's e-mail: cxq198508@163.com

**Abstract:** A series of PCL/AA-PEAs nanofibrous films with different amino acid-based polyesteramides contents were successfully synthesized by interfacial polymerization using L-phenylalanine, 1,3-propanediol, and sebacic dichloride as the main raw materials and complexed with poly-ε-caprolactone. A series of PCL/AA-PEAs nanofibrous films with different amino acid-based polyesteramide contents were prepared using electrostatic spinning technology. The effects of different ratios on the performance of the fibrous films were investigated. The results show that with the increase of the dosage of amino acid-based polyesteramide, the tensile strength of the PCL/polyesteramide fiber film firstly increases and then decreases, the glass transition temperature increases and the dosage is 70/30, the diameter of the PCL/polyesteramides fiber film is 309-684 nm, the tensile strength reaches a maximum of 5.2±0.9 MPa, and the $T_g$ is 31.5°C which is lower than the human body temperature 37°C.

## 1. Introduction

Aliphatic polyesters contain ester bonds in their molecular structure, which makes them widely used in tissue engineering materials with good biodegradability, excellent mechanical properties, and other advantages. Commonly used biodegradable polymer materials mainly include poly-ε-caprolactone (PCL), poly (lactic acid) (PLA), polyurethane (PU), etc. However, this type of material's degradation process produces acid, which harms the human body and hinders its wider application.

Polyesteramides (PEAs) can achieve fine control of the molecular structure of PEAs by selecting different monomers and reaction conditions, thereby giving the material different physical and chemical properties. Due to the presence of ester bonds and amide bonds in its repeating structural units, it has adjustable amphiphilicity, excellent biodegradability, and biocompatibility, as well as good mechanical strength and thermal properties, and shows great application potential in the fields of tissue engineering, drug controlled release, biosensors, etc. [1-4].

The introduction of amino acid structures into the main chain of PEAs can make their degradation products nontoxic, thus reducing the inflammatory reactions caused by implanted materials.[5-6] In this study, L-phenylalanine, 1,3-propanediol, and sebacic dichloride were used as the main raw materials to synthesize amino acid-based polyesteramides (AA-PEAs) material, which was compounded with PCL to form an electrostatic fibrous film by using electrostatic spinning technology. The effect of the polymer's molecular structure on the performance was also investigated, as well as the effect of AA-PEAs on the performance of electrostatic fiber films.

## 2. Experiment

### *2.1. Materials*

L-phenylalanine was provided by Shanghai Aladdin Biochemical Technology Co.. 1.3-Propanediol and Decanedioyl chloride were from Shanghai McLean Biochemical Technology Co.. p-Toluenesulfonic acid monohydrate was provided by Alfa Aesar (China) Chemical Co.. Dichloromethane was provided by Sigma Aldrich (Shanghai) Trading Co.. Sodium carbonate anhydrous, ethyl acetate, etc., were provided by Tianjin Yongda Chemical Reagent Co..

### *2.2. Instruments*

Vacuum drying oven (Shanghai Senxin Experimental Instrument). Electrostatic Spinning Instrument (E05, Foshan Lite-On Precision Measurement & Control Technology Co.). Microcomputer control electronic universal testing machine (Shenzhen Wanji Testing Equipment Co.). Scanning Electron Microscope(Inspect-S50, FEI, USA). Infrared Spectrum Analyzer (Spectrum 100 (FT-IR) PerkinElmer Co.). Circulating water vacuum pump (SHZ-95B Gongyi Yuhua Instrument Co.).

### *2.3. Preparation of phenylalanine malonyl ester*

L-phenylalanine, p-toluenesulfonic acid monohydrate, and 1,3-propanediol were added to a four-necked flask with a water separator in the ratio of 2:2:1 in substance, heating was started by adding aqueous solvent, and the temperature was raised until the solvent refluxed until no water was produced. The mixture was then cooled to room temperature and filtered, purified by passing through a large amount of water and dried in a vacuum drying oven at room temperature to finally obtain a solid white product.

### *2.4. Synthesis of AA-PEAs*

Phenylalanine propanediol and sodium carbonate were placed in pure water at a molar ratio of 1:2 until completely dissolved. Sebacyl chloride in an equal molar ratio to phenylalanine propanediol was dissolved in anhydrous dichloromethane and then slowly added dropwise to the above aqueous solution, stirred at room temperature for 20 h, and vacuum filtered for 2 h to remove dichloromethane to obtain crude AA-PEAs, which were purified using a tetrahydrofuran/ethyl acetate solvent system and dried in a vacuum drying oven at room temperature for 24 h to obtain AA-PEAs.

### *2.5. Preparation of poly-ε-caprolactone/amino acid-based polyesteramides (PCL/AA-PEAs) electrostatic fiber films*

PCL/AA-PEAs electrostatic fiber films were prepared using an electrostatic spinning technique using PCL (Mn=80,000 Da) and compounding different masses of AA-PEAs. The different ratios of PCL/AA-PEAs were dissolved in trichloromethane/dimethyl sulfoxide ($CHCl_3$/DMSO) mixed solvent and stirred until completely dissolved, with a solvent concentration of 150 mg/mL, a voltage of 18-20 KV, a receiving distance of 18 cm, a 10 mL syringe, and a propulsion rate of 1.2 mL/h. The electrostatic spinning films were obtained by vacuum drying for more than 48 h at room temperature.

### *2.6. Characterization*

*2.6.1 Infrared spectroscopy (FI-IR).* The molecular structures of phenylalanine malonyl esters and AA-PEAs were tested using a Spectrum 100 infrared spectrometer.

*2.6.2 Microscopic morphology.* PCL/AA-PEAs electrostatic fiber film's microscopic morphology was observed using an Inspect-S50 scanning electron microscope from FEI, USA.

Journal of Physics: Conference Series **2961** (2025) 012009 doi:10.1088/1742-6596/2961/1/012009

*2.6.3 Thermal weight loss (TGA).* The thermal stability of AA-PEAs was determined using a TGA-Q50 thermogravimetric analyzer under the following conditions: nitrogen atmosphere, the heating rate at 20°C/min, and the test range between 30-700°C.

*2.6.4 Glass transition temperature ($T_g$).* The $T_g$ of phenylalanine malonate and AA-PEAs were determined using a DSC214 differential thermal scanning calorimeter under the following conditions: nitrogen atmosphere, heating rate at 20°C/min, and test range between 30-300°C.

## 3. Results and discussion

### *3.1. Infrared spectral analysis*

Firstly, phenylalanine and propylene glycol were synthesized to obtain phenylalanine malonyl ester by esterification reaction, and then amino acid-based polyesteramide was synthesized with sebacic dichloride by interfacial polycondensation reaction. The process flow is shown in Figure 1. The molecular structures of phenylalanine malonyl ester and amino acid-based polyesteramide were tested by using infrared spectroscopy, respectively, and the results are shown in Figure 2.

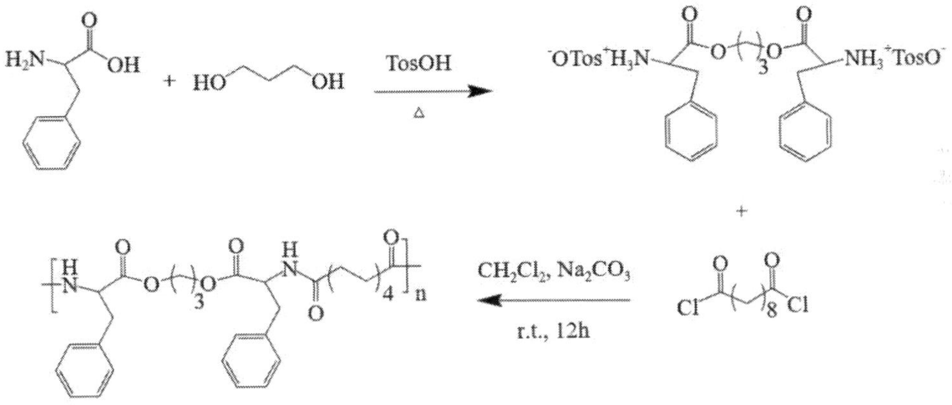


**Figure 1.** AA-PEAs synthesis process flow.

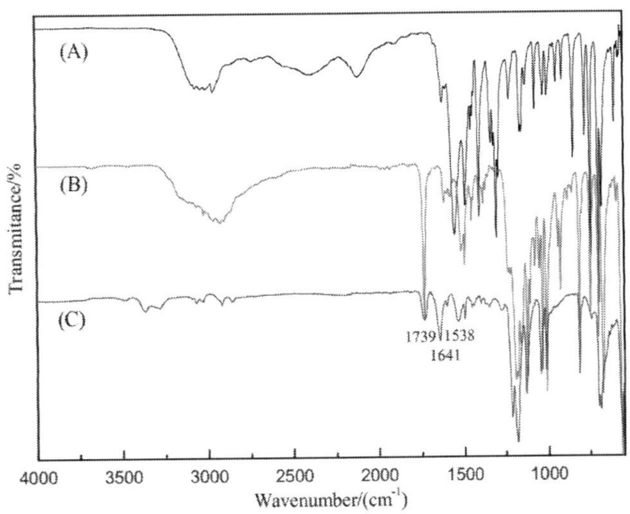


**Figure 2.** Infrared spectra.

As shown in Figure 2, spectra (A), (B), and (C) are the infrared spectra of phenylalanine, phenylalanine malonyl ester, and amino acid-based polyesteramide, respectively. A comparison of spectra (A) and (B) shows that the characteristic absorption peak of C=O at 1739 $cm^{-1}$ in spectra (B) indicates that phenylalanine malonyl ester was successfully synthesized by esterification reaction. In spectrum (C), it can be seen that the characteristic absorption peak of C=O appeared at 1739 $cm^{-1}$, and the characteristic absorption peak of the amide group appeared at both 1641 $cm^{-1}$ and 1538 $cm^{-1}$. In contrast, in spectrum (B), only the characteristic absorption peak of C=O of the ester bond appeared. There was no characteristic absorption peak of the amide, which indicated that phenylalanine-based polyesteramide was successfully synthesized.

### *3.2. Synthesis of phenylalanine malonyl ester*

Phenylalanine malonyl ester was synthesized by esterification reaction, which is an equilibrium reversible reaction. Selecting a suitable solvent with water is beneficial to improve the efficiency of the reaction. The effect of reaction time on the yield of phenylalanine malonyl ester was investigated in the reaction system of toluene and xylene under the other conditions, which remained unchanged, as shown in Figure 3.

As shown in Figure 3, no matter whether toluene or xylene was chosen as the aqueous solvent, with the increase of reaction time, the yield of phenylalanine malonyl ester gradually increased and then stabilized; among them, xylene as the aqueous solvent was more conducive to the reaction than toluene to improve the efficiency of the reaction and reduce the reaction time. In the xylene reaction system, when the reaction temperature 12 h, the yield reached 80.8%. Continuing to extend the reaction time, the yield gradually increased, but the growth rate was low, with 20 h almost reaching the maximum value of 85.2%. Therefore, the solvent with water was chosen as xylene, and the reaction time was 20 h.

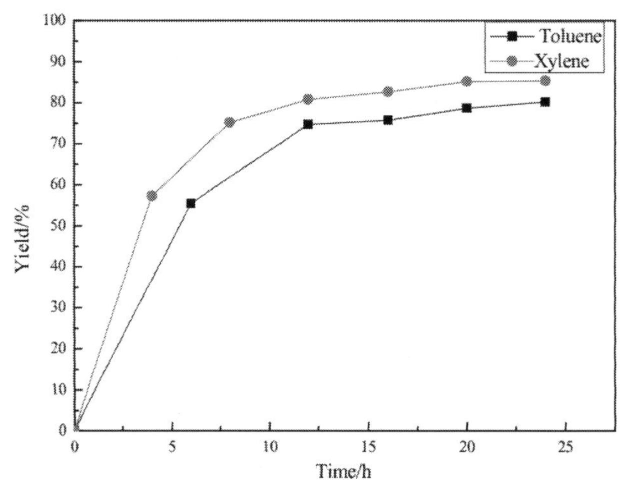


**Figure 3.** Effect of reaction time on the yield of phenylalanine malonate.

### *3.3. Synthesis of amino acid-based polyesteramides*

Phenylalanine malonyl ester and sebacic dichloride were subjected to the interfacial polymerization reaction to obtain AA-PEAs, but the products contained impurities such as unreacted monomers and a large number of salts, which affected the properties of the materials. The AA-PEAs were purified using a tetrahydrofuran/ethyl acetate solution system and vacuum-dried to obtain pure AA-PEAs. The thermal properties, residual monomer content, water content, and other properties of the polymers were also determined, and the results are shown in Table 1.

**Table 1.** Properties of amino acid-based polyesteramide

| Test | Performance |
|---|---|
| Appearance | White powder |
| $T_g$/°C | 37.7 |
| $Td_{5\%}$/°C | 303.1 |
| $Td_{10\%}$/°C | 319.3 |
| Residual monomer/% | 0.1 |
| Total solvent residue/% | 0.1 |
| Water content/% | 0.18 |

As shown in Table 1, the tetrahydrofuran/ethyl acetate solution system can effectively remove impurities and obtain high-purity AA-PEAs. The polymer's glass transition temperature is 37.7°C, and the temperature of 5% thermal weight loss is 303.1°C, showing good thermal stability.

*3.4. Characterization of PCL/polyesteramide fiber film*

As shown in Table 2, PCL/AA-PEAs with different mass ratios were dissolved in a solution consisting of $CHCl_3$/DMSO to prepare an electrospinning solution. A series of PCL/AA-PEAs electrostatic fiber films were prepared using an electrospinning apparatus, and their morphology and $T_g$ were measured. The results are shown in Figure 4 and Table 2.

As shown in Figure 4 and Table 2, the fibers of pure PCL and PCL/AA-PEAs electrostatic fiber films are randomly distributed without beads, and the intersections are entangled. The addition of AA-PEAs broadened the size distribution of PCL/AA-PEAs electrostatic fiber membrane fibril diameters, but they were all less than 900 nm. The tensile strength of the pure PCL fiber film was 2.5±0.6 MPa, and with the increase of the dosage of amino acid-based polyesteramide, the tensile strength of the PCL/AA-PEAs electrostatic fiber film firstly increases and then decreases, and the glass transition temperature then increases. When the dosage of PCL/polyesteramide was 70/30, the tensile strength of PCL/polyesteramide fiber film reached a maximum of 5.2±0.9 MPa, and the $T_g$ was 31.5°C, which is lower than the human body temperature of 37°C. PCL/AA-PEAs electrostatic fiber film has the flexibility and mechanical strength to be used as a postoperative barrier film.

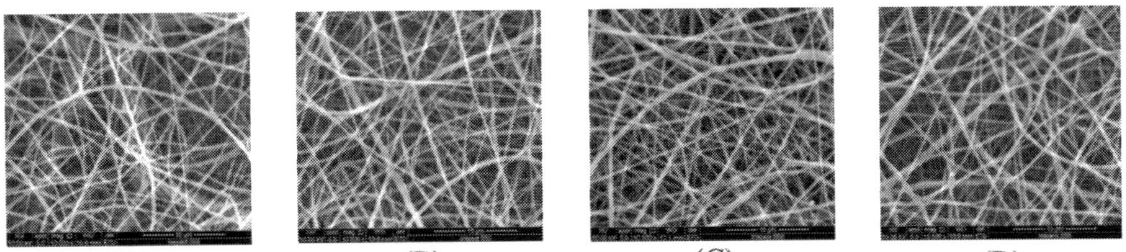

(A) (B) (C) (D)

**Figure 4.** SEM image of fiber film.

**Table 2.** Performance of PCL/AA-PEAs electrostatic fiber film.

| Serial number | PCL | PEA | Concentrations /(g/mL) | Fiber diameter /nm | Tensile strength /MPa | $T_g$/°C |
|---|---|---|---|---|---|---|
| A | 100 | 0 | 0.15 | 383~567 | 2.5±0.6 | -61.2 |
| B | 80 | 20 | 0.15 | 373~823 | 4.4±1.2 | 28.4 |
| C | 70 | 30 | 0.15 | 309~684 | 5.2±0.9 | 31.5 |
| D | 60 | 40 | 0.15 | 314~835 | 1.9±1.7 | 33.8 |

## 4. Conclusion

Phenylalanine malonyl monomer was successfully synthesized by using L-phenylalanine, p-toluenesulfonic acid monohydrate compound, and 1,3-propanediol as the main raw materials, and xylene as the aqueous solvent with a reaction time of 20 h.

By means of interfacial polycondensation, the reaction of phenylalanine malonyl monomer with sebacoyl chloride successfully yielded an amino acid-based polyesteramide, and the polymer showed good thermal stability with a glass transition temperature of 37.7°C and a heat loss of 5% of the weight of the polymer of 303.1°C. The polymers were also successfully applied in the reaction of the phenylalanine malonyl monomer with decanoyl chloride.

PCL/polyesteramide fiber film was prepared by compounding amino acid-based polyesteramide with PCL through electrostatic spinning technology. The electrospun fibers were randomly distributed without beading, there were entanglements in fiber crossings, and the diameters of the fibers were all less than 900 nm, reaching the nanoscale. With the increase in the amount of amino acid-based polyesteramide added, the tensile strength of PCL/polyesteramide fiber film first increased and then decreased, and the glass transition temperature increased accordingly. When the dosage of PCL/polyesteramide is 70/30, the diameter of PCL/polyesteramide fiber film is 309~684 nm, the tensile strength reaches a maximum of 5.2±0.9MPa, and $T_g$ is 31.5°C, which is lower than the human body temperature of 37°C.

## Acknowledgement

This work was financially supported by the Hebei Academy of Sciences Science and Technology Project (Project No.24706), Xingtai City key research and development plan (Project No.2023ZZ041), and Basic Research Expenses System Pilot Project of the Hebei Academy of Sciences (Project No.2024PF01).

## Reference

[1]Zhang H and Liu X. (2002) Synthesis of biodegradable polyester amides and their effects on drug release.synthetic chemistry, 10(5):3.
[2]He Z and Wu X. (2024) Synthetic methods of polyesteramide. Energy Chemical Industry, 45(2):14-20.
[3]Winnacker M and Rieger B. (2016) Biobased polyamides: recent advances in basic and applied research. Macromolecular Rapid Communications, 37(17):1391-1413.
[4]Haung C, Xu S, Zhao J, Jiang S, and Yang W. (2010) Synthesis and characterization of polyester amide from aliphatic diamide diol. Journal of polymers, (5):574-581.
[5]Cui X, Wang Y, Han L, Sun Q, and Song X. (2020) Preparation and properties of polylactic acid/polyesteramide composite nanofiber film[J]. Engineering Plastics Application, 48(3):14-21.
[6]Winnacker M and Rieger B. (2016) Poly(ester amide)s: Recent insights into synthesis, stability and biomedical applications. Polymer Chemistry, 46(7):7039-7046.

# Study of polyamine species modulating the properties of polyurea microencapsulated wall materials

**Yongjun He [1,2], Jingyun Feng [2,3], Xin Yi [2,3*], Furu Kang [2,3] and Jun Deng [2,3]**

[1] Intelligent Policing and National Security Risk Management Laboratory, Sichuan Police College, 186, Longtouguan, Rd., Luzhou, Sichuan, 646000, PR China
[2] College of Safety Science and Engineering, Xi'an University of Science and Technology, Xi'an, Shaanxi 710054, China
[3] Key Laboratory of Coal Fire Disaster Prevention in Shaanxi Province, Xi'an, Shaanxi 710054, China

*Corresponding author's e-mail: yixin05@163.com

**Abstract:** This study explored the impact of various polyamines on polyurea shell performance and identified optimal candidates for high-performance microcapsules. Isophorone diisocyanate (IPDI) was combined with six polyamines (EDA, DAB, HMD, PPD, DETA, and TETA) to synthesize crosslinked polyurea shells via solvent-assisted precipitation polymerization. The polyurea coatings were characterized using ultraviolet-visible spectrophotometry, SEM, FTIR, X-ray diffraction, and thermogravimetric analysis to assess polymerization, microstructure, composition, and crystallinity. Results showed that polyamine type significantly influenced polyurea properties. Diamine-based polyureas had higher initial thermal stability, while triamine (DETA) and tetraamine (TETA) coatings exhibited better thermal degradation resistance. Tri- and tetraamine-based coatings also had uniform spherical morphologies and lower crystallinity, with TETA-derived polyurea demonstrating excellent thermal stability and low crystallinity, making it ideal for high-performance microcapsules. These findings guide the selection of suitable polyurea coatings for microcapsule applications, advancing phase change material technology.

## 1. Introduction

Phase Change Materials (PCMs) are crucial in thermal energy management due to their high energy storage density and phase change reversibility [1-3]. However, they often face volume changes and leakage issues that limit their practical applications. To mitigate these problems, microencapsulation technology creates microencapsulated phase change materials (MPCMs) by enclosing PCMs in mechanically strong wall materials, enhancing stability and effectiveness in real-world applications [4-6].

The selection of wall material is of paramount importance for the performance of MPCMs, as different materials impart disparate properties. Polyurea is an optimal choice due to its favorable thermal stability, chemical resistance, and mechanical strength [7-8]. The morphology, structure, and cross-linking of polyurea walls exhibit significant impacts on MPCM performance, influencing thermal insulation, mechanical strength, and service life at elevated temperatures [9].

Research has concentrated on three areas: the impact of varying monomers and chain extenders on polyurea structure and microcapsule characteristics [10-11], the thermal stability of polyurea for high-temperature applications [12-13], and the interaction between polyurea walls and phase-change materials [14-15]. However, existing studies lack a comprehensive analysis of the thermal stability of polyurea itself, which is influenced by its molecular structure, cross-linking, and molecular weight.

To address this, isophorone diisocyanate (IPDI) was copolymerized with various polyamines (EDA, DAB, HMD, PPD, DETA, and TETA) in order to synthesize polyurea walls with different molecular structures. The thermal decomposition characteristics of the polyurea samples were analyzed using thermogravimetric analysis (TG-DTG), with particular attention paid to key temperature points and weight loss processes. Furthermore, the study investigated the relationship between polyurea's thermal stability and MPCM performance, providing theoretical guidance for the design of optimal polyurea microcapsules.

## 2. Preparation and characterization of polyamine polyurea wall materials

### *2.1. Polyamine polyurea wall material preparation*

In order to prepare the polyurea wall material in white powder form, the following experimental procedure should be followed: The experimental setup was designed to maintain a constant temperature of 30°C using a thermostat system comprising six 500 mL flasks placed in a water bath. Each flask was filled with 90 grams of acetonitrile, which served as the solvent. However, acetone or methyl n-butyl fluorophenyl ether could be used as alternatives, provided that the equivalent mass was maintained. A magnetic stirrer was set at 300 rpm for mixing, while 6.0 ml of IPDI was titrated at a rate of 20 ml/h, and six distinct polyamines were simultaneously added at a rate of 10 ml/h to maintain a 1:1 molar ratio of NCO to $NH_2$ groups. The polymerization reaction was conducted at 30°C for a period of 24 hours. Subsequently, the mixture was subjected to centrifugation at 10,000 rpm for the purpose of solid-liquid separation. The precipitate was then washed on five occasions and subsequently dried in an oven at 70°C for a period of 12 hours, resulting in the production of a white polyurea powder. The reaction equation between IPDI and polyamines is illustrated in Figure 1.

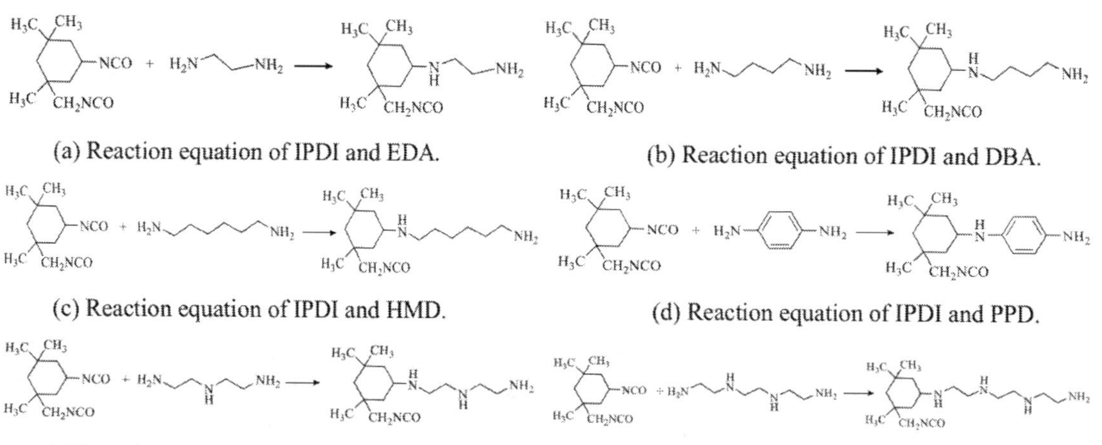

(a) Reaction equation of IPDI and EDA. (b) Reaction equation of IPDI and DBA.

(c) Reaction equation of IPDI and HMD. (d) Reaction equation of IPDI and PPD.

(e) Reaction equation of IPDI and DETA. (f) Reaction equation of IPDI and TETA.

**Figure1.** Equation for the formation of polyurea by the reaction of IPDI with polyamines.

### *2.2. Polyamine Polyurea Wall Characterization*

Morphological analysis: SEM was used to examine the surface morphology of polyamine polyurea walls. A gold coating was applied to the non-conductive samples, and SEM images were taken at various magnifications.

Infrared spectral analysis: FTIR was used to identify functional groups in the polyamine polyurea material. Samples were prepared by mixing with KBr (1:150 mass ratio) and analyzed using a Thermo Scientific Nicolet iS20 infrared spectrometer within the range of 400-4000 $cm^{-1}$.

Crystallinity analysis: XRD was performed using a Bruker D8 Advance diffractometer with CuKα radiation (λ = 0.154 nm) over a 2θ range of 5-50°, with crystal structure and crystallinity determined via Bragg's equation.

Thermal stability: Thermal stability was assessed using a Setaram Setline TGA within a temperature range of 30-800°C under an $N_2$ atmosphere (50 mL/min) at a heating rate of 10°C/min.

## 3. Results and discussion

### *3.1. Analysis of the polymerization degree of polyamine polyurea wall materials.*

*3.1.1 Determination of the optimal polymerization time.* This paper examines the changes in amine content in polyurea from IPDI and EDA polymerization, and the effect of solvents—acetonitrile, acetone, and methyl nonafluorobutyl ether. Figure 2 shows that as polymerization progresses, molecular weight increases while surface amine content decreases due to urea bond formation, with more amine groups becoming embedded in the polymer particles.

The polymerization process, as shown in Figure 2, is divided into three phases: pre-polymerization stage, intermediate stage, and late stage. In the pre-polymerization stage, amino acid content decreased rapidly, especially in acetonitrile. After 6 hours, the decline slowed, with a gentler decrease in acetone and methyl n-fluorobutyl ether systems. Around 24 hours, amino acid content stabilized at $5.80\times10^{-4}$, $4.79\times10^{-4}$, and $5.92\times10^{-4}$ mol/g in acetonitrile, acetone, and methyl n-fluorobutyl ether, respectively. The fastest polymerization occurred in acetonitrile, reflecting its higher polarity and solubility. Turbidization times further confirmed solvent effects, with times of 43, 35, and 31 seconds for acetonitrile, acetone, and methyl n-fluorobutyl ether. After 24 hours, the polymerization reaction was complete across all solvents.

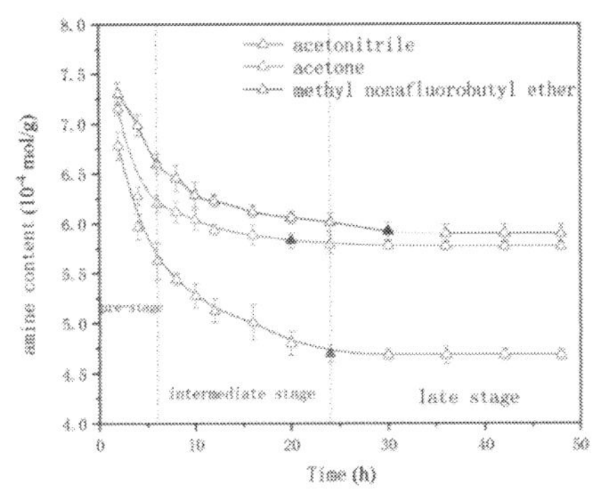


**Figure 2.** Variation of amino content in IPDI and EDA polyurea products with polymerization time.

*3.1.2 Polymerization degree of polyamine polyurea wall materials.* Acetonitrile was used as the solvent to polymerize IPDI with six polyamines (EDA, DAB, HMD, PPD, DETA, and TETA) for 24 hours, yielding various polyurea wall materials. As shown in Figure 3, the amine group contents varied significantly among the polyureas, with EDA-derived polyurea having the highest amine content ($4.69\times10^{-4}$ mol/g) and PPD-derived polyurea the lowest ($3.15\times10^{-4}$ mol/g). This establishes a hierarchy of effectiveness: EDA > DAB > DETA > TETA > HMD > PPD.

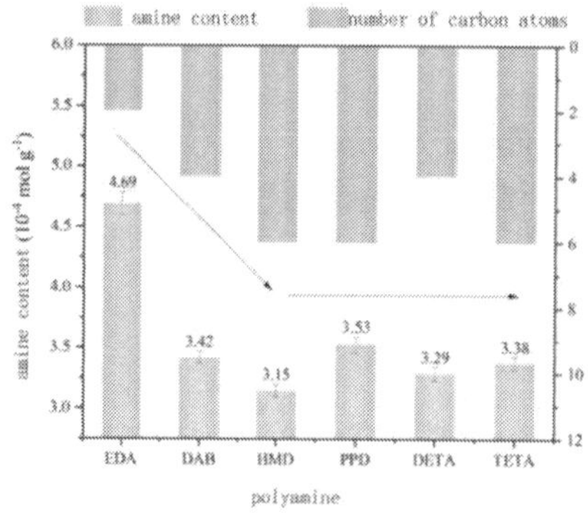

**Figure 3.** Surface amine content of six polyurea wall materials.

The high reactivity of EDA is attributed to its simple diamine structure, while PPD's benzene ring causes steric hindrance, reducing its reaction efficiency with IPDI. Although HMD has a higher amine content than PPD, its flexible chain structure limits its amino utilization compared to EDA. In the dibasic aliphatic amine series, the decrease in amine content with increasing carbon atoms is due to greater chain flexibility and increased spatial barriers, affecting chain-terminal amine reactivity and cross-linking density. Despite DETA and TETA having more amine groups, their similar amine content is likely due to comparable conversion rates and network structures.

*3.2. Morphological analysis of polyamine polyurea wall materials*

Microscopic observation of polyurea walls formed by reacting IPDI with six diamines in acetonitrile revealed varied morphologies, including flocculent, blocky, and flaky structures, influenced by the diamines' molecular characteristics (Fig. 4). EDA's smaller size leads to a flocculent structure, DAB's longer chain creates blocky morphology, and HMD, despite its length, also forms blocky structures. PPD's benzene ring promotes a flaky morphology, illustrating the impact of molecular structure on polymer morphology.

The polyurea walls formed by tertiary amine DETA and tetrameric amine TETA displayed uniform spherical structures. TETA-produced polyurea, with particle sizes around 100 nm, showed enhanced homogeneity due to increased cross-linking points and molecular flexibility from its longer aliphatic chains. This promotes the formation of spherical oligomers. Additionally, TETA's polyamino structure increases intermolecular hydrogen bonding and cross-linking density, resulting in stable, uniform particle sizes.

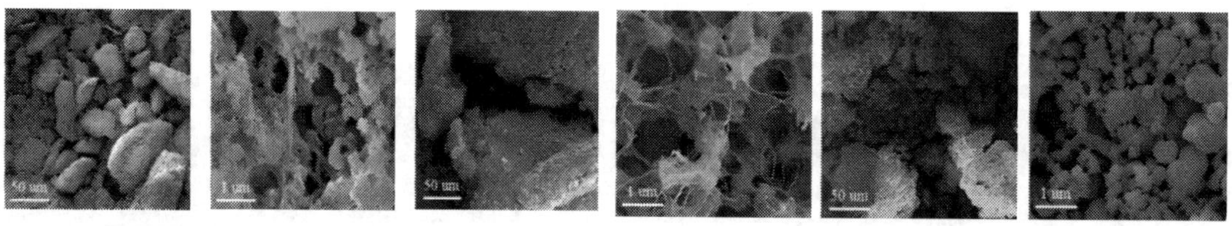

(a) EDA (b) DAB (c) HMD

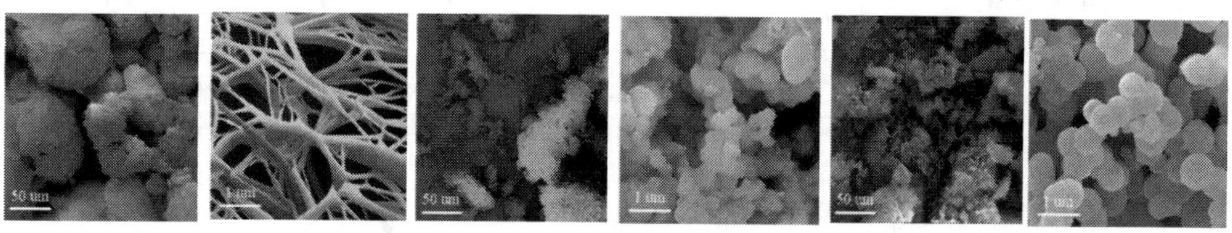

(d) PPD (e) DETA (f) TETA

**Figure 4.** Microscopic morphology of polyurea walls prepared from IPDI and polyamines.

Journal of Physics: Conference Series **2961** (2025) 012010 doi:10.1088/1742-6596/2961/1/012010

### *3.3. Infrared characterization of polyamine polyurea wall materials*

FTIR analysis of the polyurea walls synthesized from IPDI with different polyamines revealed key functional groups (Figure 5). A strong absorption peak at 3320 cm$^{-1}$ corresponds to NH stretching in the urea group, while the peak at 2930 cm$^{-1}$ is due to $CH_2$ stretching. Additional peaks were observed at 1670 cm$^{-1}$ for CO stretching, 1540 cm$^{-1}$ for NH bending, 1230 cm$^{-1}$ for CN stretching, and 670 cm$^{-1}$ for CH out-of-plane bending in the benzene ring.

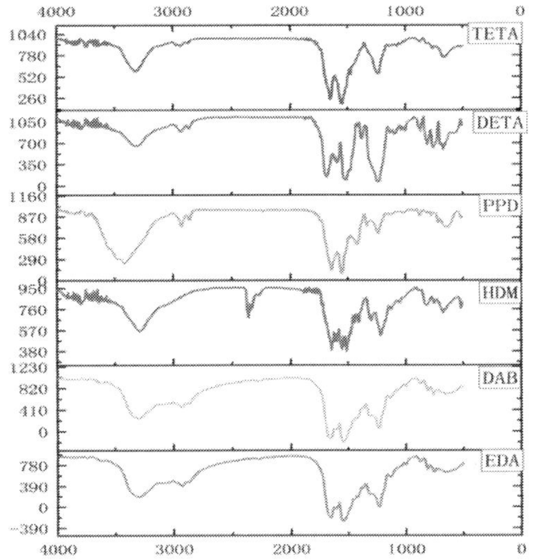


**Figure 5.** FTIR curves of polyurea walls prepared with IPDI and polyamines.

Hydrogen bonding between the carbonyl group and NH in polyurea is significant. Although theoretical predictions suggest multiple NH stretching peaks, a strong 3320 cm$^{-1}$ peak lacks distinct secondary peaks due to NH in both ordered and disordered hydrogen bonds and free NH stretching. Carbonyl absorption peaks shift to lower wavenumbers, with disordered bonds at 1669 cm$^{-1}$ and ordered bonds at 1649 cm$^{-1}$. FTIR analysis of polyureas from IPDI and various polyamines shows distinct NH stretching and carbonyl absorption. HMD and PPD-based polyureas exhibit lower polymerization, while TETA-based polyureas show stronger NH stretching and carbonyl peaks due to additional amino groups.

### *3.4. Crystallinity analysis of polyamine polyurea wall materials*

X-ray diffraction data revealed varying crystallinity in the six polyureas (Figure 6), with differences in peak location, number, and intensity. Polyureas synthesized from aliphatic diamines (EDA, DAB, HMD) showed diffuse scattering peaks near $2\theta = 20°$, with crystal plane spacing related to chain length. EDA had the highest crystallinity due to its short chain, while DAB and HMD showed decreasing crystallinity. Polyureas from aromatic diamine PPD exhibited multiple diffuse peaks, attributed to the unique crystalline structure induced by the benzene ring.

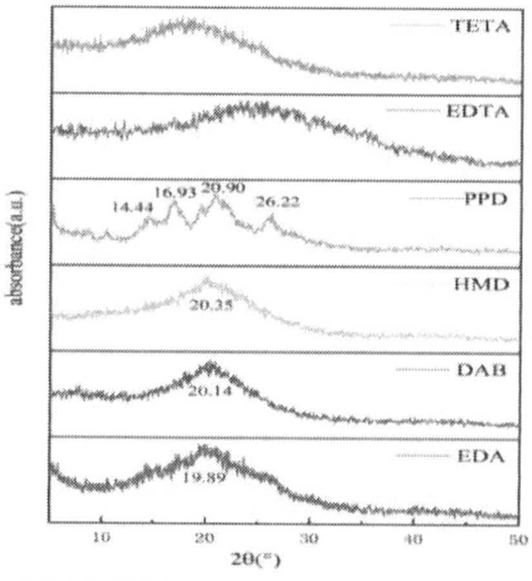


**Figure 6.** XRD diffractogram of polyurea wall prepared by IPDI and polyamines.

The polyureas prepared with DETA and TETA showed weak diffraction peaks, indicating low crystallinity. This is due to the long chains and branching structures, which increase chain entanglement and steric hindrance, preventing ordered crystal formation. The additional hydrogen bonding from amino and amide groups, along with flexible methylene chains, further reduces crystallization. Among them, TETA polyurea had the lowest crystallinity, making it ideal for microcapsules with uniform spherical structures and excellent encapsulation properties.

*3.5. Thermal stability analysis of polyamine polyurea wall materials*

*3.5.1 Analysis of characteristic temperature points.* The thermal stability of polyurea materials is influenced by several factors, including the presence of rigid groups, flexible groups, the number of cross-linking points, molecular weight and size, and the nature of intermolecular forces within the molecular structure. The use of different amine chain extenders leads to the formation of polyurea materials with different molecular structures and cross-linking densities, which in turn affect their thermal stability. To compare the thermal stability of polyurea wall materials prepared from IPDI and different polyamines, a heat loss analyzer was used to assess mass changes from 30 to 800°C, resulting in TG and DTG curves as shown in Figure 7. The TG/DTG curve for IPDI and EDA shows six characteristic temperature points in Figure 8: T1 (growth rate temperature), T005 (5% mass loss), T2 (tangent method temperature), Tmax (maximum mass loss), T3 (degradation slowdown), and T4 (end of degradation). Figure 9 illustrates that T1, T005, and T2 increase with carbon atoms in diamine-based polyureas, indicating improved initial thermal stability. In contrast, tertiary amine (DETA) and tetrameric amine (TETA) polyureas show lower temperatures, likely due to their loose aliphatic structures. DETA, DAB, and TETA exhibit the highest Tmax, indicating the fastest decomposition rate. Thermal stability decreases with rising temperature, and mass loss slows down to T4 due to the difficulty of decomposing the cross-linked structure and carbonization. For diamine-prepared polyureas, T3 and T4 show increased stability with more carbon atoms.

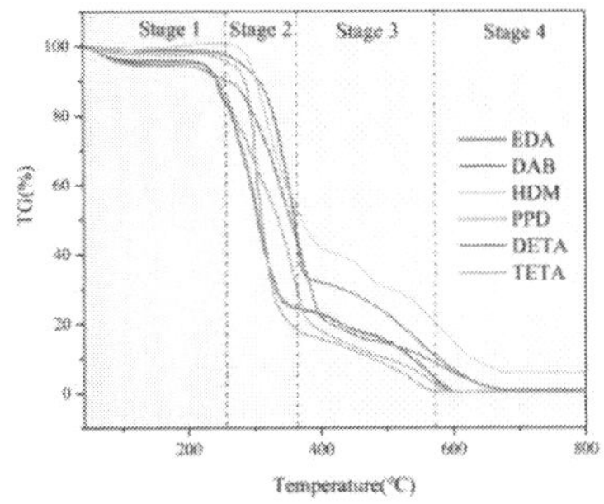


(a) TG

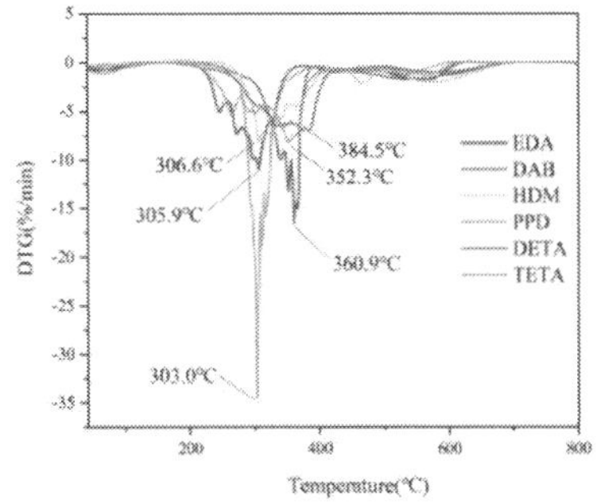


(b) DTG

**Figure 7.** TG/DTG curves of polyurea walls prepared with IPDI and polyamines.

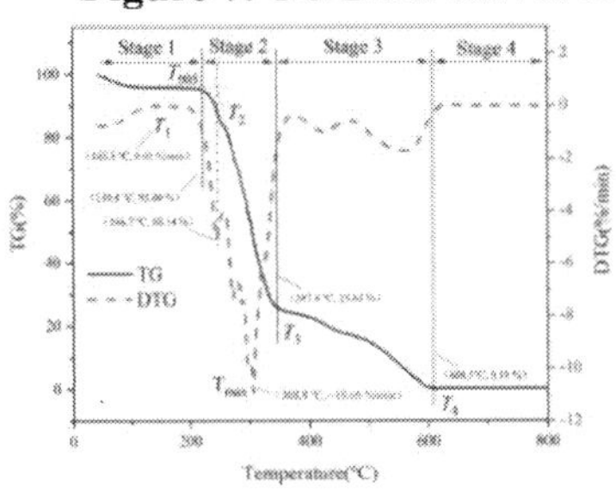


**Figure 8.** TG/DTG curves of polyurea walls prepared by IPDI and EDA.

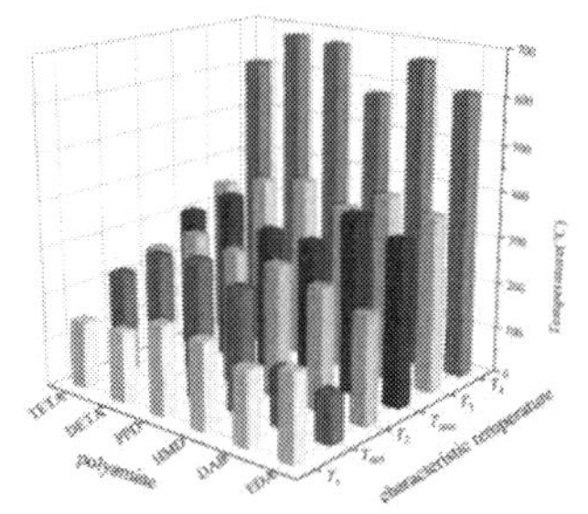


**Figure 9.** Variation of the characteristic temperature of polyurea walls prepared with IPDI and polyamines.

*3.5.2 Division of weight loss process stages.* Based on the TG/DTG curves and characteristic temperature points of six polyurea wall materials, their weight loss process is divided into four stages:

1) Initial Mass Loss Stage (Stage 1): A 5% mass loss occurs, mainly from the evaporation of water, low molecular weight volatiles, or residual solvents. The high specific surface area and porosity of polyurea lead to water adsorption, which evaporates upon heating.

2) Rapid Decomposition Stage (Stage 2): This stage features significant polymer backbone breakdown, free radical reactions, and cross-linking, resulting in major structural changes. Polyureas made with EDA, DAB, HMD, and PPD show high onset temperatures and good stability, while DETA and TETA exhibit lower onset temperatures, possibly due to less cross-linking or weaker bonds.

3) Slow Decomposition Stage (Stage 3): Characterized by slower degradation, mainly involving carbonization and thermal degradation. Most polymer backbones are degraded, leading to the formation of thermally stable aromatic structures that slow weight loss. DETA and TETA show narrower weight loss temperature intervals, indicating better thermal stability.

4) Final Decomposition Stage (Stage 4): All polyurea materials are completely degraded, resulting in inorganic and residual carbonaceous materials.

## 4. Conclusion

(1) A series of cross-linked polyurea wall materials were successfully prepared by solvent-assisted precipitation copolymerization using IPDI and various polyamine chain extenders (EDA, DAB, HMD, PPD, DETA, and TETA).

(2) The structure of polyamines significantly affects the microscopic morphology and crystallinity of polyurea. Diamine-based polyureas exhibit diverse morphologies (flocculent, flaky, and blocky), while tertiary/quaternary amine systems produce uniform spherical particles with reduced crystallinity due to more flexible chain segments.

(3) Diamine-prepared polyureas show higher initial thermal stability, with increasing $T_1$, $T_{005}$, and $T_2$ characteristic temperatures as carbon atom numbers rise. In contrast, tertiary/quaternary amine

polyureas exhibit higher thermal degradation stability and narrower weight loss temperature intervals, indicating their potential as microencapsulation materials.

(4) Future studies should focus on optimizing the performance of different polyurea microcapsule wall materials for specific applications, enhancing the practical use and development of high-performance microcapsules.

**Acknowledgment:**

This work was supported by National Natural Science Foundation of China projects (52074218, 52074215); Key international cooperation projects of Shaanxi Provincial Department of Science and Technology (No. 2023GHZD-31); General projects of Luzhou Science and Technology Bureau (No. 2021-SYF-42); The Intelligent Policing and National Security Risk Management Laboratory Autonomous Youth Project (No. ZHZZQN2301).

**References**

[1] WANG C J, WANG L Q, MA J, et al. Research progress of carbon matrix composite phase change materials[J]. Chemical Industry and Engineering Progress, 2023,42(12): 6383-6398.

[2] YANG H Y, WANG S Y, WANG X, et al. (2020). Wood-based composite phase change materials with self-cleaning superhydrophobic surface for thermal energy storage[J]. Applied Energy, 2020,261:114481.

[3] WU L M, LIU Q X, WANG XL, et al. Review on Phase Change Energy Storage Materials[J]. Materials reports, 2021,35(S1): 501-506.

[4] ZHANG J, CAI Y B, WEI Q F, et al. Preparation and characterization of n-octadecane@urea formaldehyde resin phase change energy storage microcapsule[J]. New Chemical Materials, 2017,45(12): 57-59.

[5] LIU W, LI Z M, LIU M Y, et al. Review of high-temperature phase change heat storage material preparation and applications[J]. Energy Storage Science and Technology, 2023,12(02): 398-430.

[6] PARK S K, CHO Y J, PARK S Y, et al. Polyurea Microcapsules with Different Phase Change Material for Thermochromic Smart Displays[J]. Chemistry Letters, 2019,48(11): 1343-1346,

[7] LI Z J, GONG X, WANG C Y, et al. Preparation of n-hexadecane-polyurea phase change microcapsule and study on its thermal performance[J]. Modern Chemical Industry, 2022,42(09): 170-174.

[8] YANG L K. Research progress in preparation methods of coal/PANI composites[J]. Journal of Xi'an University of Science and Technology, 2013,33(06): 715-718.

[9] CHANG Z J, WANG K, WU X H, et al. Review on the preparation and performance of paraffin-based phase change microcapsules for heat storage[J]. Journal of Energy Storage, 2022,46: 103840.

[10] WANG Q Y, LU S F, SHI W Z, et al. Preparation and properties of polyurea/polyurethane shell microencapsulated essence by interfacial polymerization[J]. Chemical Industry and Engineering Progress, 2022,41(08): 4432-4440.

[11] ZHANG Y, ZHAO D, JIAO X, et al. Fabrication and Characterization of Microcapsulated Dementholized Peppermint Oil with Polyurea Shell[J]. Acta Ploymerica Sinica, 2016(01): 46-52.

[12] Fortuniak W, Slomkowski S, Chojnowski J, et al. Synthesis of a paraffin phase change material microencapsulated in a siloxane polymer[J]. Colloid and Polymer Science, 2013,291: 725-733.

[13] YU S F, LUO W S. Preparation and Characterization of Microencapsulated Paraffin/Polyurea Phase Change Materials[J]. Journal of Materials Engineering, 2015,43(07): 100-104.

[14] MOGHADDAM M K, MORTAZAVI S M, KHAUAMIAN T. Preparation of calcium alginate microcapsules containing n-nonadecane by a melt coaxial electrospray method[J]. Journal of Electrostatics, 2015,73: 56-64.

[15] HE L N, WAN X, LIU Y K, et al. Polyurea Microcapsules Prepared by Interfacial Polymerization[J]. Plastics, 2013,42(05): 87-90+102.

Journal of Physics: Conference Series **2961** (2025) 012011 doi:10.1088/1742-6596/2961/1/012011

# Analysis of fire test methods and airworthiness requirements for materials in the cabin of civil aircraft

**Weixing Huang** [1,*], **Tonghui Yang** [2] **and Zhengye Lv** [3]

[1, 2, 3] Shanghai Aircraft Design and Research Institute, State Key Laboratory of Airliner Integration Technology and Flight Simulation, Shanghai 315211, China

[*] Corresponding author's e-mail: huangweixing1@comac.cc

**Abstract**: The safety of aircraft cabin materials is related to the safety of passengers and aircraft. With the development of science and technology, the types of materials are becoming more and more abundant, with frequent applications in the aircraft cabin, cargo compartment, and other locations. A large number of these materials improve the economic comfort of the aircraft. When a fire occurs, the performance of the materials involved can significantly influence the fire's progression, which in turn affects the safety of passengers and the aircraft. Additionally, it can impact emergency evacuation procedures. To enhance safety in aircraft research, development, and manufacturing, it is essential to understand the flame-retardant properties of materials used in aircraft. Scientific methods must be employed to verify the fireproofing capabilities of these materials, and data-driven evaluations should assess their flame-retardant properties. This is crucial to ensure the safety of passengers and the aircraft itself. In the event of a fire, effective materials can slow the spread of flames, allowing passengers to evacuate promptly and increasing their chances of survival. By analyzing the cabin material fire test verification method and airworthiness requirements, the aircraft cabin material fire requirements are outlined to provide guidelines for the commercial aircraft cabin material fire verification work, to guide the aircraft design process of cabin material selection, and to accelerate the pace of airworthiness certification.

## 1. Introduction

Aircraft cabin has many nonmetallic materials, decorations, and equipment, located in the passenger cabin, cargo hold, cockpit, and other locations, which are beautiful and easy to use but also greatly reduce the weight of the aircraft [1-2]. However, the extensive use of these materials also increases the risk of aircraft fire. When a fire occurs in the aircraft cabin, it will produce heat, smoke, etc. The space inside the aircraft is limited. The heat generated cannot dissipate. When concentrated, it accelerates the combustion of materials, significantly jeopardizing passenger safety.

The cabin safety research and technology group CRSTG found that in aircraft crashes, 40% of passengers survived, but many died in the ensuing fire. To ensure the safety of passengers and aircraft, aircraft cabin materials are required to meet fire safety requirements and to improve survivability in the event of a fire. All airworthiness agencies have stipulated requirements for the safety standards of aircraft cabin materials [3,4]. To ensure that the fire safety of aircraft meets the airworthiness requirements, the FAA stipulates the use of fire test methods to verify that the simulated cabin materials meet the specified performance standards when they are exposed to a heat source or fire environment. The test verifies and determines that the various materials meet the specified performance standards under the test conditions. This paper introduces the combustion test methods, test equipment, and standard method requirements for cabin materials in FAR25 Appendix FI to FVII [5-10].

## 2. Introduction to the test methods

### *2.1. Vertical bunsen burner test for cabin and cargo compartment materials*

There are two types of vertical combustion tests: 12 s and 60 s vertical combustion tests. The ignition time is 12 s and 60 s, respectively, to determine the flame retardancy of the material. The specimen is clamped and placed vertically into the test box according to the requirements. The door of the test box is closed and maintained throughout the test period. When the burner is pushed into the specified position of the test, the timing device will be synchronized to record the ignition time. The burner will be pushed out of the test position after the required ignition time is reached. During the test process, each specimen needs to be observed for test phenomena such as dripping, odor, smoke, etc. At the same time, after the ignition time is reached, if the specimen continues to burn, the timing device needs to be activated to record the flame burning time on the specimen until the flame goes out and stops the timing, i.e., the flame ignition time. If there are drops in the test process and continue to burn, you need to start the timer to record the burning time of the drops. We remove the test specimen with a steel ruler and other tools to measure the length of the burn.

### *2.2. Horizontal bunsen burner test for cabin, cargo compartment, and miscellaneous materials*

For the horizontal Bunsen burner test, we ignite the specimen for 15 seconds. The specimen must be clamped securely and placed horizontally into the test box. We close the door of the test box and maintain the entire test period. We push the burner into the test at the same time as the designated position to start the timing device. In order to achieve the required time that the ignition time meets the requirements, the burner is out of the test position. We observe and record data for each specimen during the test.

- (1) If the flame is self-extinguishing before the 38 mm marking line, the burning rate of the specimen is zero.
- (2) If the flame front passes through the 38mm marking line, we start timing. When the flame is extinguished, we stop timing, record the time, measure and record the combustion distance l1, and calculate the combustion rate according to Formula (1):

$$v = \frac{l_1 - 38}{t_1} \tag{1}$$

where:
v is the burning rate, mm/min.
$l_1$ is the burning distance, mm.
$t_1$ is the flame spread time, min.

### *2.3. 45-Degree bunsen burner test for cargo compartment liners and waste stowage compartment materials*

The ignition time of the 45° combustion test is 30 seconds to determine the flame-retardant property of the material. The specimen is clamped and placed into the test box at 45° according to the requirements. The door of the test box is closed and maintained throughout the test period. A timing device is activated to record the ignition time while the burner is pushed into the test position. The burner is pushed out of the test position after the required ignition time has been reached. The test phenomena are observed and recorded during the test of each specimen. Once the required ignition time is reached and the burner is removed from the specimen, it is crucial to monitor whether the specimen continues to burn. If it does, the duration of the combustion should be recorded until the flame is extinguished. This recorded time is known as the flame ignition time. If the specimen does not burn or is extinguished, it is necessary to observe and record the negative ignition time. We observe whether the flame penetration phenomenon occurs in each test specimen and make a record.

### *2.4. 60-degree Bunsen burner test for electric wire*

The ignition time of the 60°combustion test is 30 s. We put the specimen into the test box at 60° according to the requirement, closed the door of the test box, and kept it during the whole test. When the burner is pushed into the specified test position, the timing device will be synchronized to record the ignition time. After the required ignition time, the burner will be pushed out of the test position. The phenomena occurring during the test are observed and recorded for each specimen. When the required ignition time is reached and the burner is removed from the specimen, if the specimen continues to burn, it is necessary to record the time that the combustion continues to occur until the flame is extinguished, i.e., the flame ignition time. If a drop falls during the test and continues to burn, we start a timer to record the time the drop burns while it is falling. We remove the test specimen and measure the burned length with a tool such as a diamond size and record it.

### *2.5. Oil burner test for seat cushions*

This test method is used to evaluate the flame retardant and weight loss characteristics of aircraft seat cushions when exposed to high-intensity, open flame to demonstrate compliance with the requirements of FAR 25.853. An oil burner was used to generate a flame with a temperature of (1038±56)°C and a heat flow value of (11.9±0.6)W/cm$^2$ to ignite the specimen for 2 min. The burner was turned off and removed from the specimen after reaching the 2-minute time limit. The four burned lengths of each specimen were measured and recorded at 7 min. The test phenomena of each specimen were observed and recorded. The weight of each specimen was recorded and calculated in accordance with Formula (2) to calculate the weight loss percentage of each specimen.

$$P = \frac{m_1 - m_2}{m_1} \times 100\% \tag{2}$$

where:
P is the percentage of weight loss.
$m_1$ is the Pre-test weight, Kg.
$m_2$ is the after-test weight, Kg.

### *2.6. Oil burner test for cargo liners*

This test method utilizes a high-intensity, open flame to evaluate the ability of aircraft cargo hold liner materials to resist flame penetration to demonstrate compliance with the requirements of FAR 25.855. An oil burner was used to generate a flame with a temperature of (927±56) °C and a heat flow value of (9.1±0.6) W/cm$^2$ to ignite the specimen for 5 min. The burner was turned off and removed from the specimen after the ignition time was reached. During the test, we observe and record the test phenomenon of each specimen, whether and when it burns through, and the temperature at 102 mm above the horizontal specimen [11].

### *2.7. Heat release rate test for cabin materials*

This test method is used for the determination of the heat release rate of materials to demonstrate compliance with the requirements of FAR 25.853. The specimen is placed in a test chamber with constant airflow, exposed to a radiation value of (3.5±0.05) W/cm$^2$ and a length of 25 mm flame. During the test, the surface of the specimen subjected to radiation is in a vertical state. The test is conducted for 5 min. The data acquisition system collects the voltage of the thermoelectric stack and records it, which is used for calculating the rate of heat release at any moment in accordance with Formula (3).

$$R_t = K_h (V_t - V_0) \tag{3}$$

where:
$R_t$ is the heat release rate at time t, kW/m$^2$.
$V_t$ is the thermopile millivolt at time t, mV.
$V_0$ is the thermopile millivolt reading at the baseline, mV.

### *2.8. Smoke test for cabin materials*

This test method is used to determine the smoke production characteristics of aircraft cabin materials to show that the materials meet the requirements of FAR 25.853. The specimens are exposed to a radiant oven with a heat flow value of (2.5 ± 0.05) W/cm$^2$ for 4 min of test. The percentage of light transmission is collected and recorded throughout the test. The maximum specific optical density D of each specimen is calculated for the 4-min test time according to Formula (4).

$$^{4}D_{m} = \left(\frac{V}{LA}\right)\log_{10}\left(\frac{100}{^{4}T_{m}}\right) \tag{4}$$

where:

V is the chamber volume, which is 0.510 m$^3$.

L is the length of the light path, which is 0.914 m.

A is the exposed specimen area, which is 0.00424 m$^2$.

$^4$T is the minimum percent light transmission for 4 minutes.

log10(100/4Tm) is the maximum optical density for 4 minutes.

### *2.9. Test method to determine the flammability and flame propagation characteristics of thermal/acoustic insulation materials*

This test method is used to evaluate the flammability and flame spread characteristics of thermal/acoustic insulation materials exposed to radiant heat sources and flames. The heat flow value of the radiant plate is (1.7±5%) W/cm$^2$ thermal radiation, and the burner produces a flame of 127 mm long for the test ignition for 15 s. During the test, the test phenomenon and flame ignition time of each specimen are observed and recorded, and the charred length of each specimen is measured and recorded.

### *2.10. Test method to determine the burn-through resistance of thermal/acoustic insulation materials*

This test method is used to evaluate the burn-through resistance of thermal/acoustic insulation materials for aircraft when exposed to a high-intensity, open flame. An oil burner was used to generate a flame with a temperature of (1038±56)°C and a heat flow value of (18.2±0.9)W/cm$^2$ to ignite the test specimen for 4 min. Then, the burner was turned off, and the specimen was removed from the test specimen after the ignition time was reached. During the test, we observe and record the test phenomenon of each specimen, whether it is burned through the time of occurrence, and the heat flow value of the heat flow sensor at 102 mm on the back of the specimen.

### *2.11. Test method of noxious gas produced by burning cabin nonmetallic material installed in the fuselage for civil aircraft*

The testing method specified in this document refers to the method of determining the concentration of specific gases in the smoke components produced by the combustion of cabin materials in accordance with the steps specified in FAR 25 FV for 4 minutes under radiation heat and/or ignition source conditions [12].

## 3. Airworthiness requirements

### *3.1. Vertical bunsen burner test for cabin and cargo compartment materials airworthiness requirements*

12 s vertical bunsen burner test requires the specimen to self-extinguish. The average drip flame time is not greater than 5 s. The average burn length of the specimen is not greater than 203mm. The average flame time of the specimen out of the burner is not greater than 15 s.

60 s vertical bunsen burner test requires the specimen to self-extinguish. The average drip flame time is not more than 3 s. The average burn length of the specimen is not more than 152 mm. The average flame time of the specimen is not more than 15 s after pushing out the burner.

The vertical Bunsen burner test for cabin and cargo compartment materials' airworthiness requirements is shown in Table 1.

**Table 1.** Vertical bunsen burner test method requirements.

| Ignition time/s | Average flame time/s | Average burn length/mm | Average drip flame time/s |
|---|---|---|---|
| 60 | ≤15 | ≤152 | ≤3 |
| 12 | ≤15 | ≤203 | ≤5 |

*3.2. Horizontal bunsen burner test for cabin, cargo compartment, and miscellaneous materials airworthiness requirements*

Horizontal bunsen burner test requirements specimen average burn rate according to the type of material is not greater than 102 mm/min or 64mm/min, as shown in Table 2.

**Table 2.** Horizontal bunsen burner test method requirements.

| Ignition time/s | Average burn rate mm/min |
|---|---|
| 15 | ≤64 |
| 15 | ≤102 |

*3.3. 45-Degree bunsen burner test for cargo compartment liners and waste stowage compartment materials airworthiness requirements*

45-Degree bunsen burner test requirements do not allow any specimen flame penetration. The average glow time is not more than 10 s. The average flame time in the burner pushed away from the specimen is not more than 15 s, as shown in Table 3.

**Table 3.** 45-Degree bunsen burner test method requirements.

| Ignition time/s | Average flame time/s | Flame penetration of the specimen | Glow time/s |
|---|---|---|---|
| 30 | ≤15 | No | ≤10 |

*3.4. 60-degree Bunsen burner test for electric wire airworthiness requirements*

45-degree bunsen burner test requirements specimen is self-extinguishing. The average drip flame time is not more than 3 s. The average burn length of the charred specimen is not more than 76 mm. The average flame time is not more than 30 s after the launch of the burner, as shown in Table 4.

**Table 4.** 45-Degree bunsen burner test method requirements.

| Ignition time/s | Average flame time/s | Average burn length/mm | Average drip flame time/s |
|---|---|---|---|
| 30 | ≤30 | ≤76 | ≤3 |

*3.5. Oil burner test for seat cushions airworthiness requirements*

The average percentage of weight loss of the test specimen is not more than 10%, of which at least 2/3 of the specimen weight loss is not more than 10%. At least 2/3 of the specimen burn length shall not reach the edge of the seat cushion on the opposite side of the burner from one side of the burner. The burn length shall not be more than 432 mm, as shown in Table 5.

**Table 5.** Oil burner test for seat cushion requirements.

| Ignition time/min | Average percentage weight loss | Burn length of 2/3 of the total number of samples/mm |
|---|---|---|
| 2 | ≤10% | ≤432 |

*3.6. Oil burner test for cargo liners' airworthiness requirements*

Any specimen exposed to the flame for 5 min must not experience burn-through at a distance of 102 mm from the upper surface. This is applicable as long as the highest measured temperature does not exceed 203 °C, as indicated in Table 6.

**Table 6.** Oil burner test for cargo liner requirements.

| Ignition time/min | Burn through | The temperature at 102mm above the horizontal specimen/°C |
|---|---|---|
| 5 | No | ≤203 |

### *3.7. Heat release rate test for cabin materials airworthiness requirements*
The average value of the maximum heat release rate during the 5-minute test period shall not exceed 65 $kW/m^2$. The average value of the total heat release in the first 2 min shall not exceed 65 $kW \cdot min/m^2$, as shown in Table 7.

**Table 7.** Heat release rate test requirements.

| Ignition time/min | Maximum heat release rate during 5min test $kW/m^2$ | The average value of total heat release in the first 2min $kW \cdot min/m^2$ |
|---|---|---|
| 5 | ≤65 | ≤65 |

### *3.8. Smoke test for cabin materials airworthiness requirements*
The average $^4D_m$ during the 4-minute test will not exceed 200, as shown in Table 8.

**Table 8.** Smoke test requirements.

| Ignition time/min | $^4D_m$ during the 4-minute |
|---|---|
| 4 | ≤200 |

### *3.9. Test method to determine the flammability and flame propagation characteristics of thermal/acoustic insulation materials' airworthiness requirements*
For each specimen in the burner flame center line from the left, the flame spread distance is not greater than 51 mm, and the specimen after-flame time of each specimen cannot be greater than 3 s, as shown in Table 9.

**Table 9.** Test method to determine the flammability and flame propagation characteristics of thermal/acoustic insulation materials requirements.

| Ignition time/s | Flame propagation/mm | After-flame time/s |
|---|---|---|
| 15 | ≤51 | ≤3 |

### *3.10. Test method to determine the burn-through resistance of thermal/acoustic insulation materials' airworthiness requirements.*
Each of the three sets of insulation blanket test samples must not allow fire or flame penetration in less than 4 minutes. Each of the three sets of insulation blanket test samples must not allow more than 2.27 $W/cm^2$ on the cold side of the insulation samples at a point 305 mm from the face of the test rig, as shown in Table 10.

**Table 10.** Test method to determine the flammability and flame propagation characteristics of thermal/acoustic insulation materials requirements.

| Ignition time/min | Ignition or not | Whether to burn through | The cold side of the specimen from the surface of the test device at 305mm heat flow value $W/cm^2$ |
|---|---|---|---|
| 4 | No | No | ≤2.27 |

### *3.11. Test method of noxious gas produced by burning cabin nonmetallic material installed in the fuselage for civil aircraft*
According to HB7066, the toxic gases generated by the combustion of nonmetallic materials inside civil aircraft cabins mainly include CO, $SO_2$, $NO_x$, HF, HCN, and HCL, as shown in Table 11.

**Table 11.** Noxious gas requirements.

| Gas species | HF | HCL | HCN | $SO_2$ | CO | $NO_x$ |
|---|---|---|---|---|---|---|
| Requirement | $\leq 100\times 10^{-6}$ | $\leq 150\times 10^{-6}$ | $\leq 150\times 10^{-6}$ | $\leq 100\times 10^{-6}$ | $\leq 3500\times 10^{-6}$ | $\leq 100\times 10^{-6}$ |

## 4. Conclusion

This paper introduces fireproof verification test methods for cabin materials of commercial airplanes, and at the same time researches airworthiness requirements. The test verifies whether the materials meet the standard requirements, guides the design and selection of cabin materials, improves the safety of the airplane, and meets the airworthiness requirements.

## References

[1] Wu Yi, Chen Yinglong. Literature Review of Combustion Characteristics of Civil Aircraft Interior Typical Materials [J]. Journal of Civil Aviation Flight University of China. 2023. 07(4): 34.
[2] YANG Wuyi. Research on Design of Stowage Equipment and Airworthiness Method for Civil Aircraft [J]. Civil Aircraft Design & Research. 2018(2): 64-68.
[3] CCAR-25-R4 Airworthiness Standards: Transport Category Airplane [S]. Beijing: Civil Aviation Administration of China, 2011.
[4] FAR PART 25. AIRWORTHINESS STANDARDS: TRANSPORT CATEGORY AIRPLANES. 2003.
[5] William J. Hughes Technical Center. Aircraft Materials Fire Test Handbook: DOT/FAA/TC-17/55 [R]. US: Federal Aviation Administration, 2019.
[6] DU Yu, ZHANG Baoyan, WANG Tingting, Li Jia. Study on Properties of Flame Retardant Epoxy Resin and Composites for Civil Aircraft Interior Decoration [J]. Civil Aircraft Design & Research. 2019(4): 135.
[7] Li Pengyang. Discussion about Flammability Airworthiness Verification Method of Civil Aircraft Composite Material [J]. Civil Aircraft Design & Research. 2022(4): 147.
[8] Zhang Zhang, Zha Meng. Research on Application of Flame Retardant Composite in Plane Cabin [J]. NEW CHEMICAL MATERIALS. 2021, 49(11): 252-257.
[9] KUANG Lili. Airworthiness Requirements of Fire Prevention and Compliance Verification Test Method for Transport Airplane Interior Materials [J]. Civil Aircraft Design & Research. 2018, (1): 19-23.
[10] Federal Aviation Administration. Flammability Testing of Interior Materials: PS-ANM-25.853-01-R2 [S]. The US. Federal Aviation Administration. 2020.
[11] Lyu Zhengye, HUANG Weixing, HOU Zhenguo, DING Ku, LI Gen, WU Shuang, YONG Yao, HUANG Hongzhai, Differences in Buner Tests Between Seat Cushion and Seat Structure. Civil Aircraft Design & Research. 2020 (4): 139.
[12] Zhengye Lyu, Kun Ding, Songxing Han. Test Method of Toxicity Gas Produced by Burning Cabin Nonmetallic Materials Installed in the Fuselage for Civil Aircraft. HB 7066-2022, 2022-10-01.

# Self-assembled micelles prepared from amphiphilic Dextran-Jeffamine diblock copolymers

**Shuxin Wang[1, 2], Mengxiao Sun[1], Zhaoning Zeng[1], Dandan Wang[1], Xin Shen[1, 2, 4], Suming Li [3] and Feng Su [1, 2, 5]**

[1] College of Chemical Engineering, Qingdao University of Science and Technology, Qingdao, Shandong, 266042, China
[2] Institute of High Performance Polymers, Qingdao University of Science and Technology, Qingdao, Shandong, 266042, China
[3] Institut Européen des Membranes, IEM UMR 5635, Univ Montpellier, CNRS, ENSCM, Montpellier, France

Corresponding author's e-mail: [4] shenxin212@163.com; [5] sufengvip@126.com

**Abstract:** The creation of Dextran-Jeffamine (DEX-JEF) amphiphilic block polymers was achieved through the process of reductive amination. This synthesis involved reacting the semi-acetal terminal of the hydrophilic dextran (DEX) with the amine termini of Jeffamine (JEF), which could be in monoamine, diamine, or triamine forms. The result was a collection of block copolymers that included diblock, triblock, and triarm structures, each with distinct hydrophilic ratios and varying block lengths. These polymers were thoroughly analyzed by using nuclear magnetic resonance (NMR) spectroscopy and Fourier-transform infrared (FT-IR) spectroscopy to determine their structural characteristics. After the DEX-JEF polymer was self-assembled into micelles in a water medium, the micellar structure, dimensions, and zeta potential were analyzed by using Transmission Electron Microscopy (TEM) and Dynamic Light Scattering (DLS) techniques. Furthermore, the stability of the micelles was assessed by determining the Critical Micelle Concentration (CMC) through fluorescence spectroscopy measurements. Finally, the drug-carrying properties of the polymer micelles were also evaluated. The results showed that the micelle sizes ranged from 104.6-202.1 nm, CMC ranging from 0.04-0.28 mg/mL, and drug loading ranging from 4.71-7.16%. It is expected to be a potential drug carrier.

## 1. Introduction

Studies reveal that approximately 90% of novel chemical entities classified under the Biopharmaceutical Classification System's Classes II or IV exhibit low aqueous solubility [1]. These molecules, although potentially potent, did not enter the market due to their low bioavailability due to poor water solubility [2]. Therefore, searching for a drug carrier with good biocompatibility and improved drug bioavailability has received extensive attention.

Biocompatible polymers have been widely used in pharmacy as excipients in traditional drug formulations, where nanoparticles are expected to overcome the problems of unsatisfactory treatment and toxic side effects of traditional drugs [3]. Nanocarriers such as polymer micelles, polymer nanoparticles, liposomes, nanoemulsions, dendrimers, and polymer vesicles are emerging, and they have been proven to be intelligent carriers of nanomaterials drug delivery systems [4]. Nanoscale colloidal particles formed by self-assembly of amphiphilic polymers in water are polymer micelles. Above the critical micelle concentration (CMC), the hydrophobic part of the polymer chain gathers together to reduce the contact with water, thus forming a "core-shell" micelle structure [5]. The outer layer of the polymer micelle is composed of hydrophilic segments, whose function is to reduce the binding force between micelles. Hydrophobic segments can form a solid core of micelles in water, load drugs through chemical, physical, or electrostatic interactions, protect drugs from environmental effects during blood circulation, and control pharmacological activity through drug loading and release [6].

Polysaccharide is a kind of polymer material with excellent biocompatibility and without toxic side effects and biodegradation [7]. Polysaccharides-based nanocallers have been widely studied, including cellulose, dextran, chitosan, Pullulan, and so on. Dextran (DEX) is a polysaccharide of D-glucopyranose. In addition to its good water solubility, DEX has a variety of molecular weight products, which makes it easy to modify the chemical reactions.

The Jeffamine (JEF) series comprises a lineage of polyether amines that integrate poly (propylene oxide) (PPO), poly(ethylene oxide) (PEO), or a blend of PPO/PEO segments, each culminating with a primary amino group affixed to the terminal point of the polyether backbone. JEF exists in the form of monoamine, diamine, and triamine. In many cases in vivo, JEF is biocompatible [8]. Given its eco-friendly degradation properties and minimal toxicity to cells [9], JEF has found extensive application in various domains of biomedical research.

In this paper, a range of amphidetic block copolymers were produced through a process of reductive amination. The synthesis resulted in diblock, triblock, and triarm copolymers, each exhibiting varying degrees of hydrophilic-hydrophobic balance and distinct block lengths. These copolymers were thoroughly characterized by using NMR, FT-IR spectroscopy, and other techniques. Their self-assembling properties were then investigated through DLS, TEM, and CMC studies. The drug-loading properties of polymer micelles were also studied.

## 2. Materials and methods

### *2.1. Materials*

JEF M2005, JEF D2000, and JEF T5000 with molecular weights of 2000, 2000, and 5000 were produced by Huntsman Polyurethane Co., Ltd. (Shanghai, China). Sodium triacetoxyborohydride ($NaBH(OAc)_3$) and DEX with molecular weights of 5000 and 10000 were obtained from Shanghai Macklin Biochemical Co., Ltd. (Shanghai, China). Acquired from Sigma-Aldrich, the dimethyl sulfoxide underwent a drying process using calcium chloride ($CaCl_2$) and was subsequently redistilled before its utilization. L-929 mouse fibroblast cells were kindly supplied by Qilu Stem Cell Engineering Co., Ltd. (Shandong, China). It is pertinent to mention that all the chemical compounds utilized in this study were of analytical grade, sourced from Sigma-Aldrich.

### *2.2. Synthesis of DEX-JEF polymer*

DEX-JEF copolymer was synthesized by reductive amination. Specifically, a certain proportion of JEF M2005, JEF D2000, or JJEF T5000 and DEX (Mw=5000 or 10000), $NaBH(OAc)_3$ were put into a three-way flask, and 20 mL DMSO was added to dissolve. The process unfolded at ambient temperature, enduring 24 hours, followed by a 72-hour dialysis in uncontaminated water at the same temperature. Each six-hour interval marked water change, and subsequently, a 48-hour freeze-drying procedure led to the acquisition of a solid and snowy final compound. This course of events is depicted in Figure 1.

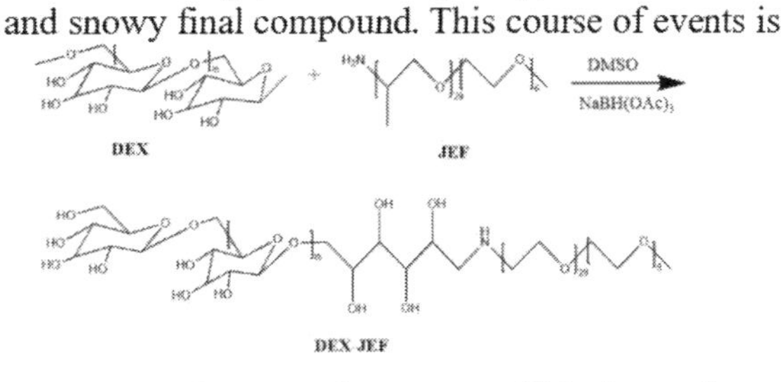


**Figure 1.** Synthesis route of DEX-JEF diblock copolymer.

### *2.3. Preparation of DEX-JEF polymer micelles*

The assembly of DEX-JEF polymeric micelles was accomplished through a straightforward method involving the dispersion of 10 mg of polymer in 10 mL of deionized water. This mixture was left to spontaneously form micelles via gentle agitation at ambient temperature for approximately 12 hours, culminating in the acquisition of a micellar dispersion. Subsequent purification was achieved by passing the solution through a 0.45 μm filtration unit, with the clarified micellar suspension being refrigerated at 4°C for storage.

For the construction of curcumin-encapsulated micelles, a multistep procedure was employed. Initially, a solution of empty micelles was formulated by dispersing 20 milligrams of a copolymer within 20 milliliters of purified water, serving as the foundational mixture. Thereafter, an aliquot of 50 microliters of acetone was utilized to solubilize 2 milligrams of curcumin, crafting a distinct curcumin-rich solution, which was carefully amalgamated into the micellar base under vigorous mixing conditions. The acetone was then permitted to naturally dissipate over an extended period, typically overnight, facilitating the entrapment of curcumin within the micelles. Any unbound curcumin was meticulously eliminated by filtration utilizing a 0.45 μm cellulose acetate membrane. The resultant micellar formulation underwent lyophilization and was kept chilled at 4°C for subsequent utilization.

### *2.4. Characterization*

A Bruker AVANCE 500 spectrometer with an operating frequency of 500 MHz was used for NMR analysis with tetramethylsilane as an internal reference and $CDCl_3$ DMSO or $D_2O$ as solvent. About 5 mg of the sample was added to the nuclear magnetic tube, and 600 μL solvent was added for nuclear magnetic analysis.

FT-IR analysis was carried out by utilizing a TENSOR 27 FT-IR spectrometer, a product of Bruker Optics from Germany. The procedure involved preparing the samples using the potassium bromide (KBr) disc technique. Each sample underwent 32 scans within the spectral range of 400 to 4000 $cm^{-1}$.

The measurement of the copolymer micelles' hydrodynamic dimension and zeta potential was carried out by utilizing a Nano-ZS90 nanosizer, a product of Malvern in Worcestershire, UK. The DLS analysis was performed at a consistent temperature of 25°C. In preparation for these assessments, the copolymer solutions were carefully filtered through a 0.45 μm cellulose acetate filter membrane.

Observation of micellar structure was carried out by using a JEM-1200EX TEM microscope operated at an acceleration voltage of 80 kV. The procedure involved dipping a copper grid into a 1.0 mg/mL micellar solution for 60 seconds and subsequently withdrawing it. A minute quantity of 2.0% phosphotungstic acid solution was applied to the copper grid for staining purposes. The sample was then left to air-dry at ambient temperature before the actual examination.

To ascertain the CMC of the synthesized copolymers, a method involving fluorescence spectroscopy with pyrene serving as the luminescent marker was employed [10]. A range of copolymer solution samples was concocted, spanning concentrations from $1.0\times10^{-3}$ to 5.0 mg/mL, while meticulously maintaining the pyrene concentration at a constant level of $2.0\times10^{-6}$ M. By utilizing a Perkin Elmer LS55 fluorescence spectrometer, the excitation spectra for these solutions were meticulously documented, with the excitation wavelength set precisely at 390 nm. The determination of the CMC was achieved by analyzing the point of intersection of the regression lines derived from plotting the ratio of intensities at wavelengths 339.2 and 336 nm against the natural logarithm of the copolymer concentration.

### *2.5. Drug-carrying properties of DEX-JEF polymer micelles*

A 2.0 mg/mL micellar formulation was prepared by reconstituting lyophilized micelles with a pH 7.4 solution. HPLC was employed to determine the drug incorporation level. The chromatographic setup entailed a mobile phase consisting of a 50/50 (v/v) blend of acetonitrile and water. A consistent thermal environment of 30°C was maintained for the column, and absorbance measurements took place at a wavelength of 425 nm. Every sample introduction amounted to a volume of 10 microliters, with the stream velocity calibrated to a pace of 1.0 milliliters per minute. A calibration curve, as illustrated in Figure 2, was derived from curcumin serial solutions varying from 0.001 to 0.08 mg/mL. We calculate the drug loading effect by using the following formulas.

$$DLC(\%)=\frac{\text{weight of loaded drug}}{\text{weight of polymeric micelles}}\times 100 \quad (1)$$

$$DLE(\%)=\frac{\text{weight of loaded drug}}{\text{therotical drug load}}\times 100 \quad (2)$$

## 3. Results and discussion

### *3.1. Synthesis of DEX-JEF copolymers*

The nuclear magnetic spectrum of JEF in $CDCl_3$ is shown in Figure 2A. The JEF monomer has signal peaks between 3.3-3.9 (g-k) and 1.11 (h). After the reduction amination reaction between JEF and DEX, the nuclear magnetism of the final polymer DEX-JEF is obtained, as shown in Figure 2B. Most of the characteristic peaks of DEX are concentrated in the range of 3.1-3.8, most of the characteristic peaks of JEF are concentrated in the range of 3.3-3.9, and there is a characteristic peak of methyl peak at 1.11. The nuclear magnetic results indicate the final synthesis of the polymer.

The chemical composition of DEX-PLA and its precursors was elucidated through FT-IR spectroscopy, as depicted in Figure 2C. JEF showed characteristic bands of C-H, C-O, C-N, and N-H at 2975, 1108, 1014, and 921 $cm^{-1}$, while DEX showed strong and wide bands of hydroxyl and carboxyl groups at 3450 $cm^{-1}$ and 1645 $cm^{-1}$. The DEX-JEF copolymer exhibits all the signature frequencies of its constituent elements, confirming the successful synthesis of the polymer.

### *3.2. DEX-JEF micelle formation*

Table 1 illustrates the outcomes of DLS analyses conducted to assess the micelle characteristics, specifically focusing on their size, size distribution, and Zeta potential. Regarding dimensions, the micelles exhibited an average diameter that varied between 104.6 nm and 202.1 nm, contingent upon the differing lengths of the DEX and JEF segments. The micelle size of the triarm copolymer is the smallest and that of the triblock copolymer is the largest. This can be attributed to differences in topology. On the other hand, the micellar size of $DEX_{10K}$-derived copolymers is larger than that of $DEX_{5K}$-derived copolymers, indicating that as the length of hydrophilic segments rises, the dimensions of the micelles expand. The polydispersity ranges from 0.12 to 0.28, indicating that the size distribution of micelles is narrow. The micelles' Zeta potential is virtually nil, echoing the neutral characteristics of the copolymer in question.

**Table 1.** DEX-JEF blank micelle and drug-loaded micelle properties.

| Sample | Blank micelles | | | | Drug-loading micelles | |
|---|---|---|---|---|---|---|
| | Size (nm) | PDI | Zeta (mV) | CMC (mg/mL) | DLC (%) | DLE (%) |
| $DEX_{5K}$-$JEF_{M2K}$ | 136.7±1.8 | 0.20 | -3.66±0.27 | 0.192 | 5.02±0.20 | 47.8±2.4 |
| $(DEX_{5K})_2$-$JEF_{D2K}$ | 179.5±2.2 | 0.12 | -3.06±0.21 | 0.245 | 5.72±0.16 | 54.1±1.7 |
| $(DEX_{5K})_3$-$JEF_{T5K}$ | 104.6±3.9 | 0.23 | -2.84±0.1 | 0.013 | 7.16±0.11 | 66.8±3.0 |
| $DEX_{10K}$-$JEF_{M2K}$ | 175.0±3.4 | 0.14 | -2.60±0.3 | 0.189 | 4.71±0.23 | 45.0±1.2 |
| $(DEX_{10K})_2$-$JEF_{D2K}$ | 202.1±2.5 | 0.18 | -4.33±0.45 | 0.237 | 5.31±0.19 | 50.4±3.2 |
| $(DEX_{10K})_3$-$JEF_{T5K}$ | 157.9±4.5 | 0.25 | -4.07±0.23 | 0.046 | 6.20±0.31 | 58.4±3.6 |

The micelle morphology observed is shown in Figure 2D. The shape of the six polymer micelles is spherical, and the size of the micelles is about 20-50 nm.

Figure 2E depicts the movement in the fluorescence excitation spectrum, which elucidates the micelle formation process. The CMC point was ascertained through the intersection of two linear regression lines drawn from the $I_{336.3}/I_{331.3}$ intensity ratio plotted against the polymer concentration graph. While the CMC for both the diblock and triblock copolymers hovered closely within the range of 0.189 to 0.245 mg/mL, showcasing negligible disparity, the triarm copolymers exhibited a markedly diminished CMC in comparison. The CMC of $(DEX_{5K})_3$-$JEF_{T5K}$ and $(DEX_{10K})_3$-$JEF_{T5K}$ were 0.013 and 0.046 mg/mL. To put it differently, the CMC of the triarm copolymer rises as the hydrophilic DEX segment's length increases. Lu et al. [11] prepared HPMC-JEF micelles with CMC between 0.03-0.24. Compared with DEX-JEF micelles, DEX-JEF micelles have lower CMC, that is, better stability.

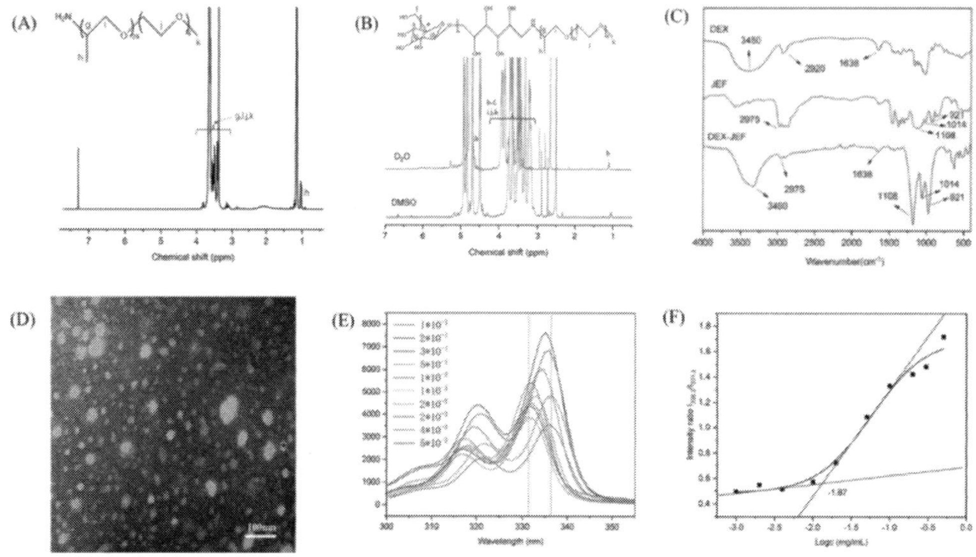


**Figure 2.** (A) NMR spectra of JEF in $CDCl_3$, (B) DEX-JEF in $D_2O$ and DMSO, (C) FT-IR spectra of DEX JEF and DEX-JEF, (D) TEM images of $DEX_{5K}$-$JEF_{M2K}$ micelles, (E) Fluorescence excitation spectra, and (F) $I_{336.3}/I_{331.3}$ vs logC plot of $(DEX_{5K})_3$-$JEF_{T5K}$ solutions.

### *3.3. Micellar drug-carrying properties*

Drug loading of DEX-JEF polymer micelles is shown in Table 1. The DLC values of DEX-JEF micelles with different molecular weights and structures ranged from 4.71 to 7.16%, and DLE values ranged from 45 to 66.8%, indicating that DEX-PLA micelles had a certain drug-carrying capacity. In the presence of the same hydrophilic group, the triblock copolymer has a higher drug loading capacity than the diblock copolymer, and the triarm copolymer has a higher drug loading capacity than the triblock copolymer.

## 4. Conclusion

In this paper, a biophilic amphiphilic block polymer was synthesized entirely through reductive amination, employing the hydrophilic DEX hemiacetal and the amino groups of monoamine, diamine, and triamine Jefamine as hydrophobic segments. The successful synthesis was confirmed via NMR and FT-IR spectroscopic analyses. Notably, all synthesized copolymers demonstrated the capacity to spontaneously form narrowly distributed spherical micelles. The micelle size of the triarm copolymer is the smallest and that of the triblock copolymer is the largest. The triarm copolymer in DEX-JEF micelles has lower CMC. In the presence of the same hydrophilic group, the drug encapsulation efficiency exhibits a higher value for the triblock copolymer in comparison to the diblock copolymer, and triarm copolymers are superior to triblock copolymers in drug loading capacity. DEX-JEF polymer micelle drug delivery system has a good effect on improving hydrophobic drug dissolution, prolonging drug efficacy, and enhancing bioavailability. They have important application value in the research of drug delivery systems.

**References**

[1]Wu C Y and Benet L Z (2010) Predicting Drug Disposition via Application of BCS: Transport/Absorption/Elimination Interplay and Development of a Biopharmaceutics Drug Disposition Classification System. Pharmaceutical Research, 106: 162–167.

[2]Fournier E, Dufresne M H, Smith D C, Ranger M and Leroux J C (2006) A Novel One-Step Drug Loading Procedure for Water-Soluble Amphiphilic Nanocarriers. Pharmaceutical Research, 23: 1948-1948.

[3]Zheng Y T, Oz Y, Gu Y M, Gu Y M, Ahamad N, Shariati K, Chevalier J, Kapur D and Annabi N (2024) Rational design of polymeric micelles for targeted therapeutic delivery. Nano Today, 55: 102147.

[4]Liu J Y, Li S Q, Wang J, Zhou J N and Chen H X (2023) Application of Nano Drug Delivery System (NDDS) in Cancer Therapy: A Perspective. Recent Patents on Anti-Cancer Drug Discovery, 18: 125-132.

[5]Udin S M, Maarof H, Rahman N A (2023) A Mini-Review on the Application of Computational Strategies in Multi-Armed Star-PEG Polymeric Micelles for Drug Delivery. Chemistryselect, 8: e202301022.

[6]Trimaille T and Verrier B (2023) Copolymer Micelles: A Focus on Recent Advances for Stimulus-Responsive Delivery of Proteins and Peptides. Pharmaceutics, 15: 2481.

[7]Gopinath V, Saravanan S, Al-Maleki A R, Ramesh M and Vadivelu J (2018) A review of natural polysaccharides for drug delivery applications: Special focus on cellulose, starch and glycogen. Biomedicine & Pharmacotherapy, 107: 96-108.

[8]Shachaf Y, Gonen-Wadmany M and Seliktar D (2010) The biocompatibility of Pluronic® F127 fibrinogen-based hydrogels. Biomaterials, 31: 2836-2847.

[9]Erdem A, Ngwabebhoh F A and Yildiz U (2016) Fabrication and characterization of soft macroporous Jeffamine cryogens as potential materials for tissue applications. RSC advances, 6: 111872-111881.

[10]Abbas R, Abolfazl K and Hossein B (2023) PCL-based hydrophobic chains grafted with two PEG-based hydrophilic branches: fluorescence and dynamic light scattering studies. Journal of Polymer Research, 30: 1-10.

[11]Lu A J, Petit E, Li S M, Wang Y D, Su F and Monge S (2019) Novel thermo-responsive micelles prepared from amphiphilic hydroxypropyl methyl cellulose-block-jeffamine copolymers. International Journal of Biological Macromolecules, 135: 38-45.

# Investigation of thermodynamic properties of sheet molding compound based on molecular dynamics simulations

**Wei Duan [1], Shaobin Li [2], Hangwei Cao [3], Xuefeng Zhao [1], Lu Pu [1], Haofei Sun [1], Xiongying Duan [3,*] and Zeli Ju [1]**

[1]Electric Power Research Institute of State Grid Shaanxi Electric Power Co., Ltd., Xi'an, Shaanxi, 710100, China
[2]State Grid Xi'an Power Supply Company, Xi'an, Shaanxi, 710199, China
[3]School of Electrical Engineering, Dalian University of Technology, Dalian, Liaoning, 116081, China

[*]Corresponding author's e-mail: xyduan@dlut.edu.cn

**Abstract:** Sheet Molding Compound (SMC) composed of unsaturated polyester resin (UPR) reinforced with sheets of glass fiber, has been widely used in fireproof support projects for tube corridors and tunnels. The crosslinking degree is a key factor affecting the internal structure of UPR. In this study, based on the molecular dynamic simulations, UPR models were constructed at varying degrees of cross-linking. Important intrinsic parameters, including the glass transition temperature and elastic modulus of the material, were obtained. It was found that both the thermodynamic and mechanical properties of the material improved with the increase of the crosslinking degree. Besides, a UPR/glass fiber interface model was developed at an 80% cross-linking degree to study the interfacial bonding at different temperatures. The microscopic compactness of the interface across various temperatures was further analyzed.

## 1. Introduction

Sheet molding compound (SMC), mainly composed of unsaturated polyester resin (UPR) [1] and glass fiber, offers advantages such as easy processing, low cost, and high production efficiency. It also provides electrical insulation, heat resistance, flame retardancy, and good mechanical properties, making it widely used in cable support structures within urban corridor tunnels and other fire-resistant materials in the electrical field. However, due to the aging of SMCs over long-term use, their mechanical and insulating properties have decreased substantially. While their thermodynamic properties can be measured from a macroscopic point of view through traditional test methods, they cannot fully explain the underlying mechanisms of these changes. Molecular dynamics simulations, on the other hand, allow for microscopic analysis of these materials and offer valuable insights into the aging mechanisms at the interface between UPR and glass fiber, which can provide knowledge support for the development of higher-performance SMCs.

In this study, molecular dynamics simulations were conducted using the Materials Studio (MS) software to construct a model of UPR with different crosslinking degrees. Important parameters including the glass transition temperature, shear modulus, and Young's modulus were calculated. Additionally, an interface model of the UPR and the glass fiber has been developed to calculate the binding energy and molecular diffusion coefficient at different temperatures. The binding energy at different temperatures for the interface was also calculated, to provide an in-depth understanding of the degradation mechanisms.

## 2. Modeling and validation of UPR

### *2.1. Crosslinking modeling*

In this study, UPR was selected for crosslinking with styrene. MS software was used for crosslinking modeling. The UPR model and styrene model were constructed according to the literature. As UPR is a homopolymer, its degree of polymerization will affect the performance of the UPR molecule. Therefore, the degree of polymerization should be considered when establishing the UPR molecule. According to the previous research, the homopolymer in the range of 10-15 can satisfy the calculation requirements. Thus, in the process of establishing the UPR molecule [2], the degree of polymerization was set to 10. Figure 1a) [3] shows the UPR molecular structure, Figure 1b) presents the UPR polymer chain, and Figure 2 depicts the styrene molecular structure.

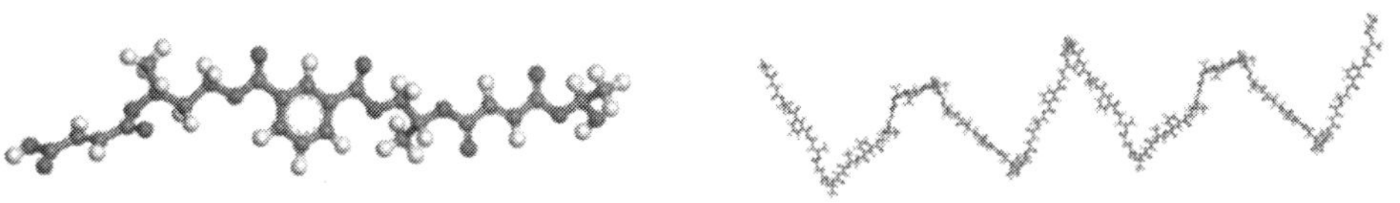

a) b)

**Figure 1.** Molecular model of a) UPR molecular structure and b) UPR polymer chain.

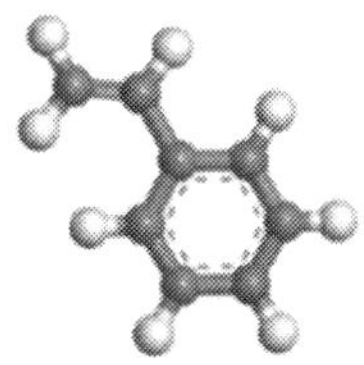

**Figure 2.** Molecular model of styrene.

According to the actual cross-linking process of the proportion of various substances, the MS Amorphous Cell module was used to place two UP polymer chains with a cross-linking degree of 10 and 40 styrene in a three-dimensional periodic box. The mass ratio of UP molecules to styrene molecules was 70:30. The density of the cross-linking system was set to 1.10 g/cm$^3$ according to the experimentally determined values. The initial amorphous model is shown in Figure 3a). Three chemical reactions are possible during the curing process: 1) styrene with polyester molecules; 2) styrene with styrene; and 3) polyester molecules with polyester molecules. Since the degree of polymerization has increased to 10 in this study, the main focus is on the reaction of styrene with polyester molecules.

An automated cross-linking script was developed using Perl language, enabling cross-linking within MS software. The cross-linking distance was set to 0.25-0.6 nm, with a step radius of 0.02 nm. The Forcefield was set to compass II, the charge method was set to group-based, and the temperature was set to 410K. Lower cross-linking UPR is unusable. Achieving higher levels is not possible. The crosslinking degree was set to 40%, 50%, 60%, 70%, and 80%. The 80% crosslinking degree model is shown in Figure 3b).

a) b)

**Figure 3.** Structure of a) initial amorphous structure of UP and b) UP with 80% crosslinking degree.

### *2.2. Molecular properties of UPR*

*2.2.1 Glass transition temperature.* The glass transition temperature ($T_g$) corresponds to the transition from the glassy state to the elastic state [4]. This inherent property of amorphous polymers represents the macroscopic embodiment of the transformation of the form of polymer movement [5], which directly affects the use of materials and process performance, making it a crucial parameter in polymer thermodynamics.

UPR with different degrees of crosslinking were annealed. The temperature-density curves during the annealing process were observed. These curves were fitted with the inflection points identified. Figure 4 shows the temperature-density diagram of the UPR under different crosslinking degrees. This phenomenon is analogous to that observed in epoxy resins, whereby a reduction in free volume percentage is accompanied by a decline in cohesive energy density as cross-linking increases.

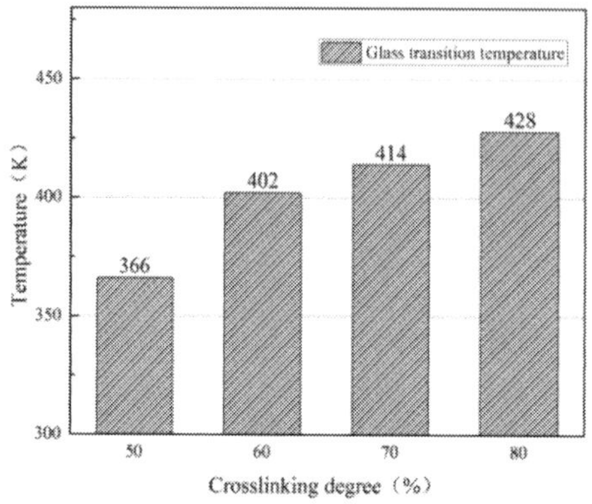


**Figure 4.** Glass transition temperatures of UPR with different crosslinking degrees.

*2.2.2 Modulus of elasticity.* The MS software permits the calculation of Young's modulus and shear modulus of a given material. Using this data, the stiffness matrix can be simplified as follows:

$$ij=\begin{bmatrix} \lambda+2\mu & \lambda & \lambda & 0 & 0 & 0 \\ \lambda & \lambda+2\mu & \lambda & 0 & 0 & 0 \\ \lambda & \lambda & \lambda+2\mu & 0 & 0 & 0 \\ 0 & 0 & 0 & \mu & 0 & 0 \\ 0 & 0 & 0 & 0 & \mu & 0 \\ 0 & 0 & 0 & 0 & 0 & \mu \end{bmatrix} \quad (1)$$

where $\lambda$ and $\mu$ are the elasticity constant, given as follows:

$$\lambda=\frac{1}{6}(C_{12}+C_{13}+C_{21}+C_{23}+C_{31}+C_{32}) \quad (2)$$

$$\mu=\frac{1}{3}(C_{44}+C_{55}+C_{66}) \quad (3)$$

The model of UPR under different crosslinking degrees was calculated at 300K. The results are shown in Table 1.

As shown in Table 1, compared to the model with a 40% crosslinking degree, the UPR with an 80% crosslinking degree shows an increase of 95.1% in Young's modulus and a 108.6% increase in the shear modulus. With the increase of the crosslinking degree, the three-dimensional mesh structure improved its mechanical properties. These improvements are mainly attributed to the increasing mesh structure, which enhances the overall rigidity and densification, thereby significantly improving the thermo-mechanical properties.

**Table 1.** A table with a modulus of elasticity of UPR at different crosslinking degrees.

| Modulus of elasticity \ Crosslinking degree | 40% | 50% | 60% | 70% | 80% |
|---|---|---|---|---|---|
| Young's modulus (GPa) | 1.237 | 1.215 | 1.306 | 1.835 | 2.413 |
| Shear modulus (GPa) | 0.465 | 0.502 | 0.494 | 0.768 | 0.970 |

## 3. Interface features

### *3.1. Interface modeling*

Since the filler material in SMC is usually glass fiber, amorphous silica was used to simulate the glass fiber surface in MS software. The $SiO_2$ cell was first imported into the MS software to supercell the crystal structure, and subsequent super-crystallization of the crystal structure; the model was then subjected to the Berendsen control system of pressure and temperature at a pressure of 1 atm. Firstly, the crystal is heated to obtain an amorphous cell structure, quenched to room temperature, and then slowly heated to fully melt to form a constant size [6]. It is annealed again to obtain amorphous silica with a density of 2.19g/cm$^3$, which is in accordance with the actual situation [7].

We intercept the $SiO_2$ surface and manually calculate the bond breaks for the stretched bonds within it to yield unsaturated silicon and oxygen atoms. Unsaturated atoms were grafted. The interface was modeled using the Build Layers feature. The resulting interface model is shown in Figure 5.

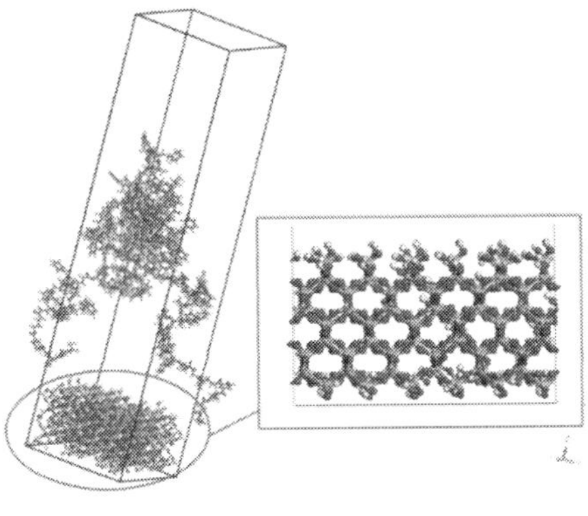

**Figure 5.** Interface model of UPR/glass fiber.

### *3.2. Interface binding energy*

In order to further investigate the microstructural changes of the model at different temperatures, molecular dynamics simulations were conducted using MS across a range of temperatures. The temperature range considered in this study was selected based on the real conditions: The lowest temperature in the coldest region of northeastern China can reach -45°C [8]. The maximum thermal decomposition temperature of unsaturated polyester resin is around 337°C [9], with an increment of 50k. Therefore, the temperature range was 228.15K to 628.15K. The calculation of the interface binding energy is shown in Figure 5. The formula for calculating the interfacial binding energy is given in Equation (4) [10].

$$E_{UPR\&glass} = (E_{UPR} + E_{glass}) - E_{UPR+glass} \tag{4}$$

where $E_{UPR\&glass}$ is the interfacial binding energy between UPR and amorphous silica, $E_{UPR}$ is the UPR energy, $E_{glass}$ is the glass fiber energy, and $E_{UPR+glass}$ is the total system energy.

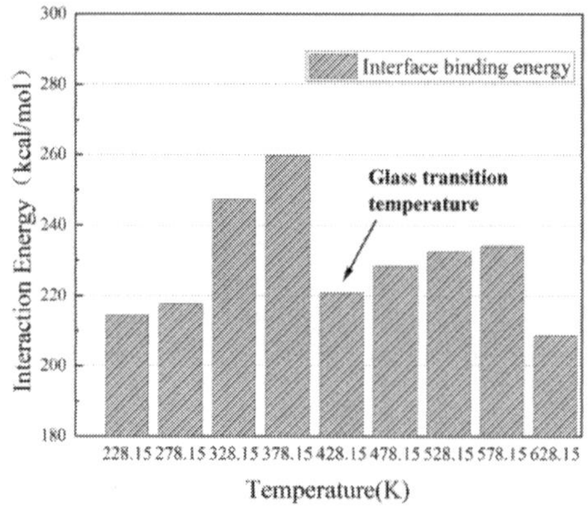


**Figure 6.** The interfacial binding energy of UPR/glass fiber at 80% crosslinking degrees at different temperatures.

Figure 6 shows the interfacial binding energy of UPR/glass fiber at 80% crosslinking at different temperatures. At the UPR/glass fiber interface without a coupling agent, the overall bonding is weak. With the increase in temperature, the interfacial bonding gradually strengthens, with the turning point observed at $T_g$. After $T_g$, the UPR transitions to the elastic state, which leads to a rapid decrease in interfacial bonding energy. This could be because in the glassy state, the molecular chains cannot be moved in a wide range, and the whole presents stability. However, in the high elastic state, the molecular chain segments can move and entangle with each other, increasing the degree of freedom. This movement fills the tiny cavities, reducing the molecular contact surface area, which will in turn weaken the interfacial bonding effect.

## 4. Conclusion

This paper presents a study of the cross-linking reaction of UPR, in which simulation models with varying degrees of cross-linking were constructed, and the glass transition temperature and elastic modulus were calculated. The interfacial binding energy was also calculated for a range of temperatures. It has been demonstrated that an elevated degree of crosslinking can facilitate enhanced performance of SMC, including heat resistance and favorable mechanical properties. In practical production, it is advised to augment the degree of crosslinking reaction. It was observed that elevated temperatures following the Tg resulted in a decline in interfacial bonding properties, leading to the formation of defects. Consequently, it is not recommended to place long-term SMCs in proximity to the Tg.

**Acknowledgments**

This work was supported by the State Grid Shaanxi Electric Power Co., Ltd. Technology Project "Quality Inspection Technology and Application of Fireproof Materials in Urban Tube Corridor Tunnel Project (5226KY24000K)".

**References**

[1] Boong G B, Goo S L, Jin H Y, et al. Preparation and Characteristic of Sheet Molding Compound using Unsaturated Polyester Resin with Low Profile Agent of Polystyrene [J]. Korean Chemical Engineering Research, 2012, 50(3): 588-593.

[2] Ceriani C, Fois E, Gamba A, et al. Dehydration dynamics of bikitaite: Part II. Ab initio molecular dynamics study [J]. American Mineralogist, 2004, 89(1): 102-109.

[3] Yang Gao. Study on Design, Performance and Dynamics of Unsaturated Polyester Resin Concrete [D]. Chang'an University, 2020. DOI: 10.26976/d.cnki.gchau.2020.000059.

[4] Du R, He L, Li P, et al. Polydopamine-Modified Al2O3/Polyurethane Composites with Largely Improved Thermal and Mechanical Properties [J]. Materials, 2020, 13(7): 1772-1772.

[5] Li P, Jia N, Pei X, et al. Effects of Temperature on Bending Properties of Three-Dimensional and Five-Directional Braided Composite [J]. Molecules, 2019, 24(21): 3977-3977.
[6] Jun X, Ke C, Meng Y, et al. Effect of temperature and water penetration on the interfacial bond between epoxy resin and glass fiber: A molecular dynamics study [J]. Journal of Molecular Liquids, 2022, 350
[7] Marko H B B B Š B K. Fibre density distribution in a layer of glass wool [J]. Glass Technology—European Journal of Glass Science and Technology Part A, 2015, 56(5): 145-152.
[8] Yumeng W, Jingyan M, Lijuan Z, et al. Spatial Distribution of, and Variations in, Cold Regions in China from 1961 to 2019 [J]. Sustainability, 2022, 14(1): 465-465.
[9] Saetiaw K, Atong D, Sricharoenchaikul V, et al. Thermal Decomposition and Kinetic Study on Different Types of Glass Fiber/Unsaturated Polyester Pipe Waste [J]. Materials Science Forum, 2010, 962(654-656): 2652-2655.
[10] S. R. Patel, S. W. Case. Durability of hygrothermally aged graphite/epoxy woven composite under combined hygrothermal conditions [J]. International Journal of Fatigue, 2002, 24(12): 1295-1301.

# Identification and regulation of catalytic active sites on material surface

**Shuqi Lu***

University of Sheffield, Sheffield, England, UK

* Corresponding author: shuqi5277@outlook.com

**Abstract:** In order to explore the identification and regulation of active sites on catalysts, various high-resolution characterization techniques combined with theoretical calculations were used to conduct an in-depth analysis of active sites on the surfaces of different types of catalytic materials. Through techniques such as X-ray photoelectron spectroscopy, scanning tunneling microscopy, and atomic force microscopy, the active sites of various catalysts were successfully identified, and their electronic and geometric structures were analyzed. The results indicate that the structural characteristics of active sites are closely related to catalytic performance, and precise regulation of these sites can significantly improve catalytic efficiency. In addition, the developed in-situ characterization technology provides new possibilities for real-time monitoring of catalytic processes, opening up new paths for designing and optimizing efficient catalysts.

## 1. Introduction

The active sites of a catalyst have a decisive impact on its catalytic performance, and precise identification and regulation are crucial to improving catalytic efficiency and selectivity. In recent years, with the rapid development of nanotechnology and surface science, researchers have been able to reveal the active sites and structural characteristics of catalyst surfaces at the atomic level. Especially with multi-scale characterization techniques, new perspectives and tools have been provided for catalyst design and optimization [1]. This study aims to systematically identify and regulate the catalytic active sites on the surface of materials through advanced characterization techniques combined with theoretical calculations and experimental methods.

### *1.1. Literature review*

In recent years, the recognition and regulation of active sites in catalysts have attracted widespread attention. Shao et al. [2] successfully identified active sites on nanodiamonds through various characterization techniques, providing an efficient catalytic system for the oxidation of organic pollutants. Wang et al. [3] used a combination of in-situ spectroscopy and theoretical calculations to reveal the influence of geometric active sites in crystalline cobalt oxide hydrides on oxygen evolution reactions. Ding et al. [4] determined the active sites for CO oxidation and water gas shift reactions on supported Pt catalysts through isotope labeling and kinetic analysis. Kumar et al. [5] reviewed the application of single-atom catalysts in electrocatalysis and proposed new strategies for activity descriptors and active site recognition. Wang et al. [6] conducted in-depth research on the influence of coordination environment on the active sites of heterogeneous catalysts, providing new ideas for rational design of high-performance catalysts. Yang et al. [7] successfully identified the active sites of oxygen reduction nanocatalysts using atomic scale characterization techniques, providing essential guidance for developing fuel cell electrode materials. These studies have laid the foundation for precisely identifying and regulating active sites. However, further exploration is needed to explore the structure-activity relationship between active sites and catalytic performance and how to synergistically enhance catalyst performance through multiple regulatory strategies.

## 2. Identification and regulation methods of multi-scale catalytic active sites

### *2.1. High-resolution surface characterization techniques*

This study employed various high-resolution surface characterization techniques to identify catalytic active sites. X-ray photoelectron spectroscopy (XPS) is used to analyze the valence states of surface elements with a resolution of up to 0.1 eV. Scanning tunneling microscope (STM) has achieved atomic level resolution (0.1 nm) for surface morphology imaging. Atomic force microscopy (AFM) is used to measure surface mechanical properties. X-ray absorption fine structure analysis (XAFS) analyzed local atomic structures, including near-edge structures (XANES) and extended fine structures (EXAFS). Electron paramagnetic resonance (EPR) is used to detect unpaired electrons. High-resolution transmission electron microscopy (HRTEM) achieved lattice imaging with a resolution of 0.1 nm. The comprehensive application of these technologies enables us to characterize active sites from multiple perspectives, such as electronic structure, geometric configuration, and chemical environment. For example, EXAFS analysis of Pt/CeO2 catalyst shows a Pt-O coordination number of 3.8±0.2 and a bond length of 2.01±0.01 Å, indicating that Pt mainly exists in the oxidized state. Table 1 summarizes the characteristics and applications of various characterization techniques.

**Table 1** High-Resolution surface characterization techniques and their applications.

| Technology | Resolving power | Main applications |
|---|---|---|
| XPS | 0.1 eV | Analysis of elemental valence states |
| STM | 0.1 nm | Atomic-level surface morphology |
| AFM | 0.1 nm | Surface mechanical properties |
| XAFS | 0.1 eV | Local atomic structure |
| EPR | $10^{-4}$ T | Unparalleled electronic detection |
| HRTEM | 0.1 nm | Lattice structure imaging |

### *2.2. Quantitative analysis methods for active sites*

A series of mathematical models and data processing methods have been developed to analyze active sites quantitatively. XPS peak fitting uses the Gaussian Lorentz mixture function to analyze the proportion of different chemical states quantitatively:

$$I(E) = I_0[\eta \cdot L(E) + (1-\eta) \cdot G(E)]$$

Among them, $\eta$ is a mixed parameter with a typical value of 0.3-0.7. STM image analysis utilizes two-dimensional fast Fourier transform (2D-FFT) and autocorrelation function to analyze surface atomic arrangement:

$$F(u,v) = \sum_{x=0}^{M-1} \sum_{y=0}^{N-1} f(x,y) \cdot e^{-j2\pi(\frac{ux}{M}+\frac{vy}{N})}$$

Among them, F (u, v) is the 2D-FFT result, and f (x, y) is the original STM image. We use the IFEFFIT software package for background subtraction, normalization, and Fourier transform of XAFS data and obtain coordination number (N) and bond length (R) information through EXAFS fitting:

$$x(k) = S_0^2 \cdot \frac{\Sigma(N \cdot F(k) \cdot e^{-2k^2\sigma^2} \cdot \sin(2kR + \delta(k)))}{kR^2}$$

Among them, x(k) is the EXAFS oscillation function, k is the photoelectron wave vector, and $S_0^2$ is the amplitude reduction factor. For example, for Pt/CeO2 catalyst, EXAFS fitting shows a Pt-O coordination number of 3.8±0.2 and a bond length of 2.01±0.01 Å, indicating that Pt mainly exists in the oxidized state. These quantitative analysis methods provide a reliable basis for a deeper understanding of the structure of active sites.

### *2.3. Active site regulation strategy*

Based on a deep understanding of the essence of active sites, the following multidimensional regulatory strategies are proposed: 1) Physical control: Nanoparticles of different sizes are synthesized using the solvothermal method. The particle size distribution can be precisely controlled by adjusting the precursor concentration, reaction temperature, and time. For example, Pt nanoparticles with a particle size of 3.5±0.2 nm can be obtained by reacting at 180°C for 4 hours. The morphology control adopts selective electrochemical etching, and applying a potential of 0.8 V in 0.1 M H2SO4 can preferentially expose the Pt {100} crystal plane. 2) Chemical control: Atomic layer deposition (ALD) technology is used to achieve atomic-level precision doping control, and a new gas-phase functionalization method is used to achieve precise control of surface properties while maintaining the integrity of the catalyst skeleton structure. If trimethylaluminum and water are used as precursors, 5 at% Al doping can be achieved by performing 50 ALD cycles at 250°C. 3) Composite regulation design: We design core-shell and heterostructure, such as using a layer-by-layer growth method to prepare Pt@Pd core-shell structure, controlling the Pd shell thickness to 2-3 atomic layers. Figure 1 shows the flowchart of multidimensional regulation strategies for active sites, covering a comprehensive range of regulation methods from atomic to nanoscale.

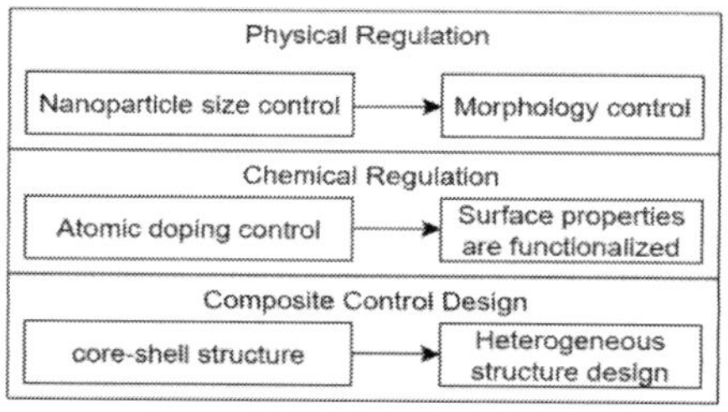


**Figure 1.** Flow chart of multidimensional regulation strategy for active sites.

### *2.4. In-situ/operating condition characterization techniques*

A series of in-situ characterization techniques have been developed to capture the dynamic changes of active sites during catalytic reactions: Environmental transmission electron microscopy (E-TEM) using microfluidic reaction cells achieves in-situ imaging with a resolution of 0.2 nm at 0.1 Pa pressure. In-situ XAS utilizes fast scanning XAFS technology, with a time resolution of up to 10 ms and an energy resolution of 0.1 eV. In-situ Raman spectroscopy combined with surface-enhanced Raman scattering (SERS) technology can achieve a detection limit of 10-15M. For example, in the CO oxidation reaction, we observed that the CO adsorption peak (2090 cm-1) on the surface of Pt nanoparticles gradually weakened with increasing temperature, while the intensity of the O-O vibration peak (877 cm-1) increased, revealing the reaction mechanism. The comprehensive application of these in-situ characterization techniques can establish a dynamic correlation between "structure electronic state reactivity", providing a powerful tool for a deeper understanding of catalytic reaction mechanisms.

## 3. Results and discussion

### *3.1. Identification and characterization results of active sites*

This study successfully identified the active sites of different types of catalysts through various advanced characterization techniques, as shown in Table 2.

**Table 2.** Active sites of different types of catalysts.

| Catalyst type | Characteristics of active sites | Main characterization methods | Catalytic performance indicators |
|---|---|---|---|
| Pt/C | Low coordination Pt atoms | EXAFS, STM | HER: $TOF = 0.5s^{-1}(\eta = 50mV)$ |
| TiO2 | Oxygen vacancy | EPR, XPS | Photocatalytic hydrogen production: $AQY = 15\%(\lambda = 365nm)$ |
| Cu/ZnO/Al 2O3 | Cu-Zn interface sites | In-situ XAFS, E-TEM | Methanol synthesis: $STY = 0.8g_{MeOH}/(g_{cat} \cdot h)$ |

Research has found a significant correlation between the number and properties of active sites and catalytic performance. For example, for Pt/C catalysts, EXAFS analysis revealed a linear correlation between the proportion of low coordination Pt atoms (coordination number<9) and the conversion frequency (TOF) of hydrogen evolution reaction (HER), as shown in Figure 2.

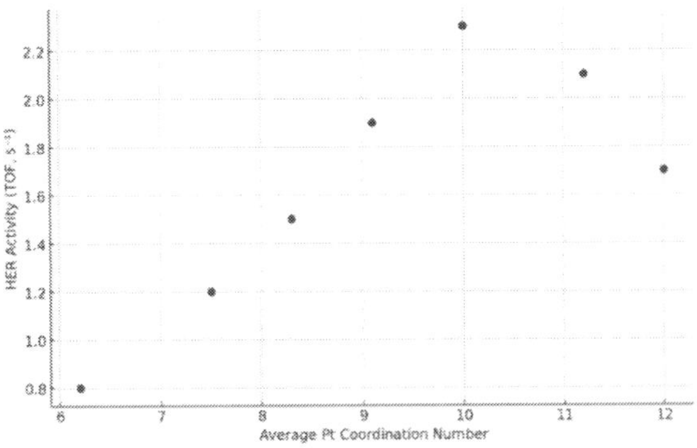


**Figure 2.** Relationship between average coordination number of Pt atoms and HER activity.

For $TiO_2$ photocatalysts, we quantitatively analyzed the oxygen vacancy concentration using EPR and XPS techniques. We found an exponential relationship between it and photocatalytic hydrogen production's apparent quantum yield (AQY). This discovery provides essential guidance for designing efficient photocatalytic materials.

*3.2. Analysis of the regulatory effect and mechanism of active sites*

The performance of various catalytic reactions has been successfully improved by precisely regulating the active sites. In terms of hydrogen production through water splitting, constructing a PtNi alloy shell reduced the HER overpotential from 78 mV to 28 mV (@ $10mA/cm^2$). In-situ XAS revealed that the unique electronic structure of the Pt Ni interface weakened the Pt-H bond and accelerated hydrogen desorption. In CO2 electroreduction, Fe doped N-graphene achieved 90% CO selectivity and 100-hour stability. DFT calculations indicate that the Fe-N4 site has the most moderate adsorption energy for COOH intermediates, effectively suppressing HER. In the partial oxidation reaction of methane, introducing CeO2 surface oxygen vacancies by plasma treatment increases the selectivity from 65% to 85%. In-situ EPR and XANES studies have found that an increase in the Ce3+/Ce4+ ratio (from 0.11 to 0.23) promotes the rapid migration of oxygen species. Figure 3 shows the effects of different control strategies on CO2 reduction performance. The composite control strategy (simultaneously controlling morphology and electronic structure) results in the most significant performance improvement, with a 2.8-fold increase in current density and a 15% increase in selectivity.

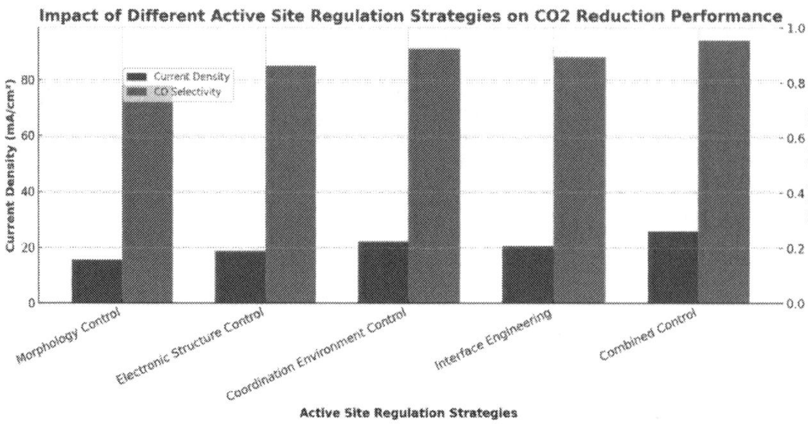


**Figure 3.** Effects of different active site regulation strategies on CO2 reduction performance.

### *3.3. Structure performance relationship and theoretical simulation*

A structure-activity relationship model is established based on experimental data and theoretical calculations to better understand the relationship between active site structure and catalytic performance. Taking CO2 electroreduction as an example, it was found that the activity of Fe-N-C catalyst is closely related to the Fe-N coordination structure. Using density functional theory (DFT) calculations, free energy diagrams of different Fe-N coordination models (FeN2, FeN3, FeN4) were constructed. It was found that the FeN4 configuration has the most moderate adsorption energy for the critical intermediate COOH (Δ G=-0.32 eV), which is beneficial for CO2 activation and suppresses HER side reactions. This theoretical prediction is consistent with the experimentally observed highest CO Faraday efficiency at the FeN4 site (91% @ -0.6 V vs. RHE). Furthermore, we establish a volcano curve model to describe catalytic activity:

$$\log(jCO) = a \cdot (\Delta G_{COOH} - \Delta G_0)^2 + b$$

Among them, jCO is the partial current density of CO, $\Delta G_{COOH}$ is the free energy of COOH adsorption, $\Delta G_0$ is the optimal adsorption energy, and a and b are the fitting parameters. This model can accurately predict the CO2 reduction activity of different M-N-C (M=Fe, Co, Ni) catalysts, as shown in Figure 4. The catalyst design method guided by this theory provides a new approach for developing high-performance CO2 reduction electrocatalysts.

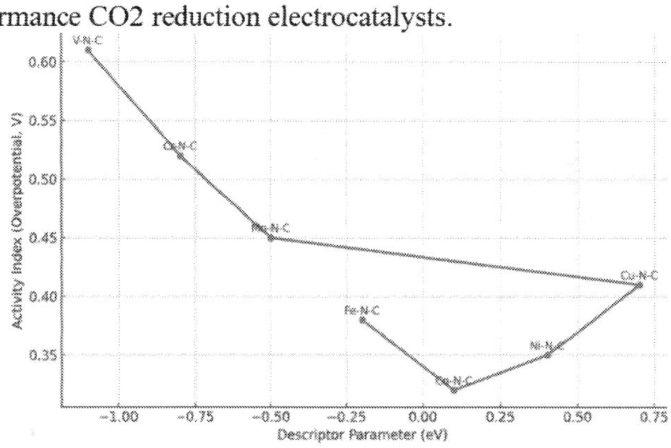


**Figure 4.** Volcanic curve of M-N-C catalyst parameterized model.

## 4. Conclusion

This study successfully revealed the active site characteristics and regulatory mechanisms of multiple types of catalysts through various high-precision characterization techniques, providing an in-depth scientific basis for understanding the relationship between catalytic activity and structure. Future research will focus on exploring the dynamic changes of active sites under different catalytic environments and their performance in actual catalytic reactions, especially in electrocatalysis and photocatalysis. By developing and optimizing in-situ characterization techniques, the catalytic process is expected to be monitored in real time, catalyst design can be further optimized, and efficient and sustainable catalytic reactions can be achieved. This has important guiding significance for the development of new high-performance catalytic materials.

**Reference**

[1] Shah S S A, Najam T, Bashir M S, et al. Identification of catalytic active sites for durable proton exchange membrane fuel cell: catalytic degradation and poisoning perspectives [J]. Small, 2022, 18(18): 2106279.

[2] Shao P, Tian J, Yang F, et al. Identification and regulation of active sites on nanodiamonds: establishing a highly efficient catalytic system for oxidation of organic contaminants [J]. Advanced functional materials, 2018, 28(13): 1705295.

[3] Wang S, Jiang Q, Ju S, et al. Identifying the geometric catalytic active sites of crystalline cobalt oxyhydroxides for oxygen evolution reaction [J]. Nature Communications, 2022, 13(1): 6650.

[4] Ding K, Gulec A, Johnson A M, et al. Identification of active sites in CO oxidation and water-gas shift over supported Pt catalysts [J]. science, 2015, 350(6257): 189-192.

[5] Kumar A, Vashistha V K, Das D K, et al. MNC-based single-atom catalysts for H2, O2 & CO2 electrocatalysis: activity descriptors, active sites identification, challenges and prospects [J]. Fuel, 2021, 304: 121420.

[6] Wang C, Wang Z, Mao S, et al. Coordination environment of active sites and their effect on catalytic performance of heterogeneous catalysts [J]. Chinese Journal of Catalysis, 2022, 43(4): 928-955.

[7] Yang Y, Zhou J, Zhao Z, et al. Atomic-scale identification of active sites of oxygen reduction nanocatalysts [J]. Nature Catalysis, 2024: 1-11.

# Preparation and performance study of modified GO/TPU nanocomposites

**Jiliang Xie* and Xia Sun**

Institute of Marine Research Development, Jiangsu Ocean University, Lianyungang 222005, Jiangsu, China

*Corresponding author's e-mail: jiliangxie@jou.edu.cn

**Abstract:** Thermoplastic polyurethane has a wide range of industrial applications, but its stiffness and strength limit its performance. Graphene oxide can be introduced as a nanofiller into thermoplastic polyurethane to enhance its properties, but it also has certain drawbacks. In order to improve the above situation, a modified graphene oxide/thermoplastic polyurethane nanocomposite material is proposed. In response to the poor dispersibility of traditional graphene, a study was conducted to modify it with isophorone diisocyanate. Based on this, the modified graphene oxide was introduced into thermoplastic polyurethane through a solution method to prepare the final composite material. The experimental results showed that the carbonyl hydrogen bonding content of the composite material reached 84.79%, and the degree of hydrogen bonding increased significantly; The tensile strength and fracture elongation of composite materials are 35.88 MPa and 569%, respectively. Compared with traditional materials, the tensile strength and fracture elongation have increased by 495% and 92.97%, respectively. Therefore, the modified graphene oxide proposed in the study can effectively improve the microstructure and macroscopic properties of TPU, which is of great significance for its performance enhancement and industrial applications.

## 1. Introduction

Thermoplastic Polyurethane (TPU) is a widely used industrial multi-block polymer that combines easy processing and flexibility[1]. TPU can customize its exclusive performance according to the diversified needs of modern technology and has wide applicability in the fields of coating, adhesive, fiber, foam, rubber, thermoplastic elastomer, and composite materials [2]. However, the traditional TPU has a lower proportion of hard segments, resulting in limited stiffness and strength, which limits its further development [3]. The existing methods usually add nanofillers to the TPU matrix to improve its performance [4]. Common nanofillers include clay, nano silica, cellulose nanocrystals, carbon nanotubes, graphene oxide (GO), reduced graphene oxide, etc. [5]. However, although these nanofillers perform well in enhancing polymer performance, they also have limitations such as poor dispersibility, insufficient compatibility with the matrix, complex preparation, and high cost, which to some extent also restrict the widespread promotion of TPU in large-scale industrial applications [6]. In view of this, in order to improve the performance of TPU, a modified GO material was proposed as the filler for TPU. To address the issue of poor dispersion of traditional GO, innovative research has been conducted using Isophorone Diisocyanate (IPDI) to modify GO; After obtaining IPDI modified GO dispersion (IPDI-GO), IPDI-GO was used as a filler to prepare TPU composite materials. Finally, an Isophorone Diisocyanate Graphene Oxide Thermoplastic Polyurethane (IPDI-GO-TPU) composite material was obtained. The research aims to prepare IPDI-GO-TPU and explore its performance, providing reference value for the application of TPU in sports equipment, protective equipment, electronic equipment casings, shoe materials, medical catheters, packaging materials, and other fields.

## 2. Preparation of modified GO/TPU nanocomposites

### *2.1. Preparation of modified GO based on isophorone diisocyanate*

Graphene is widely used in various fields due to its excellent properties. The hydroxyl, carboxyl, and epoxy groups on the surface of GO enable it to disperse well in water. However, GO tends to agglomerate in organic solvents, and ultrasound and mechanical stirring cannot solve the problem of poor dispersion [7]. Therefore, improving the dispersion of GO in polymers through chemical modification has become a mainstream approach. In view of this, the study first used IPDI to modify GO to improve its aggregation phenomenon in organic solvents. The experimental materials designed for GO modification include isophorone diisocyanate, analytical pure (Shanghai Aladdin Biochemical Technology Co., Ltd.); oxidized graphene slurry, industrial grade (Changzhou Sixth Element Material Technology Co., Ltd.); dibutyltin dilaurate (DBTDL), analytically pure (Shanghai Aladdin Biochemical Technology Co., Ltd.); N. N-Dimethylformamide (N, N-Dimethylformamide, N,N-DMF), analytically pure (China National Pharmaceutical Group Chemical Reagent Co., Ltd.). The experimental equipment involved in the preparation of modified GO mainly includes a collector-type constant temperature heating magnetic stirrer (DF-101S, Zhengzhou Ketai Experimental Equipment Co., Ltd.); vacuum drying oven, (DZF-6050, Qingdao Chuanghesheng Science and Education Instrument Equipment Co., Ltd.); desktop high-speed centrifuge, (TG16-WS, Hunan Xiangyi Laboratory Instrument Development Co., Ltd.); CNC ultrasonic cleaner, (KQ-800DE, Kunshan Ultrasonic Instrument Co., Ltd.).

The preparation process for IPDI-modified GO dispersion is as follows: First, the GO dispersion is poured into a three-necked flask. The three-necked flask needs to be equipped with a magnetic stirring device to ensure uniform mixing during the reaction process[8]. Next, 10 grams of IPDI and 0.02 grams of DBTDL are added to the three-necked flask, where DBTDL acts as a catalyst in this reaction. Then, the flask is placed on a constant temperature heating device and the temperature is adjusted to 80°C. At the same time, nitrogen is introduced to maintain an inert atmosphere and prevent side reactions caused by the presence of oxygen during the reaction process. Under a nitrogen atmosphere, continuous stirring is maintained through a magnetic stirring device to ensure that the reactants can come into full contact and mix evenly. To ensure that IPDI can fully react with oxygen-containing groups in GO and successfully modify GO, the entire reaction needs to be carried out continuously for 6 hours under these conditions. The mechanism of the reaction between IPDI and GO is shown in Figure 1.

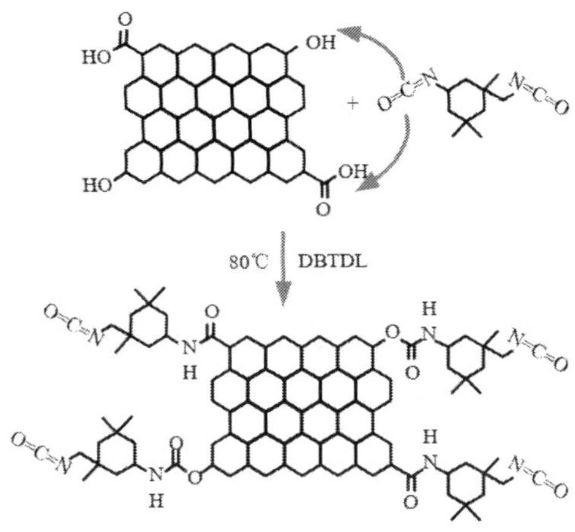


**Figure 1.** Reaction mechanism between IPDI and GO.

After the reaction is completed, heating and stirring are stopped, waiting for the mixture in the flask to cool to room temperature. Finally, the reaction system is centrifugally washed using N, N-DMF to remove unreacted IPDI and DBTDL. The specific operation is to pour the reacted mixture into a centrifuge, undergo high-speed centrifugation treatment, pour out the supernatant, and then add fresh anhydrous N, N-DMF solvent for washing. This process needs to be repeated at least three times to ensure that all impurities are completely removed and ultimately obtain pure IPDI-GO.

### *2.2. Preparation of TPU composite materials based on IPDI-GO*

TPU has the characteristics of easy processing and flexibility, and has adjustable physical properties due to the diversity of raw material selection, making it applicable in various fields [9-10]. TPU has excellent flexibility and elasticity, but its rigidity and strength are relatively low, which has significant limitations for its further application. The existing methods mainly improve the strength of TPU by adding nanofillers. However, the uneven dispersion and interface compatibility issues of nanofillers in the matrix still exist, leading to insignificant enhancement effects and potentially affecting other properties such as transparency and processability of the material [11-13]. Therefore, the study uses IPDI-GO as a nanofiller and prepares IPDI-GO-TPU through the solution method. IPDI-GO is first mixed with the soft segment, and the NCO on IPDI-GO can react with OH on the soft segment, making IPDI-GO uniformly dispersed in the soft segment. Afterward, hard segments are added for pre-polymerization reaction to obtain TPU nanocomposites with excellent mechanical properties. The hard segment type of TPU is 4,4 ` - diphenylmethane diisocyanate, the soft segment type is polytetrahydrofuran ether diol (molecular weight 1000), the chain extender is 1,4-butanediol, and the hard segment content is 35%. After determining the molar ratio of isocyanate groups to hydroxyl groups, the TPU equation can be calculated using Equation (1).

$$\begin{cases} Ch\% = \dfrac{mi+md}{mi+md+mg}\% \\ \mathrm{R} = \dfrac{n(NCO)}{n(OH)} = \dfrac{ni}{nd+ng} = \dfrac{mi/Mi}{md/Md+mg/Mg} \end{cases} \tag{1}$$

In Equation (1), $Ch\%$ represents the hard segment content; $mi$, $md$, and $mg$ respectively represent the mass of the hard segment, chain extender, and soft segment; $R$ represents the molar ratio of isocyanate groups to hydroxyl groups; $n(NCO)$ and $n(OH)$ represent the number of substances with isocyanate groups and hydroxyl groups; $ni$, $nd$, and $ng$ represent the number of substances in the hard segment, chain extender, and soft segment; $Mi$, $Md$, and $Mg$ represent the molar masses of the hard segment, chain extender, and soft segment, respectively [14-15]. The raw materials involved in the preparation of IPDI-GO-TPU include 4,4 ` - diphenylmethane diisocyanate (MDI), industrial grade (Wanhua Chemical Group Co., Ltd.); Polytetrahydrofuran ether diol (molecular weight 1000) (PTMG1000), industrial grade (BASF Corporation); 1,4-Butanediol (BDO), analytically pure (China National Pharmaceutical Group Chemical Reagent Co., Ltd.).

The synthesis step of TPU is to place PTMG1000 into a three-necked flask, vacuum, and remove water for 2 hours under conditions of 120°C and 300 r/min, then seal with nitrogen gas, heat and stir MDI at 50°C until clear, add dehydrated PTMG1000 and N, N-DMF, and stir at 80°C and 300 r/min. After repeating vacuum pumping and nitrogen injection three times, it is necessary to pre-polymerize in nitrogen for 2 hours. After pre-polymerization, vacuum defoaming is carried out, and BDO and DBTDL are added for chain expansion. After expanding the chain, the sample is poured into the mold and dried at 80°C for 48 hours to obtain a blank TPU.

The preparation of IPDI-GO-TPU first involves the reaction between IPDI-GO and PTMG1000, followed by pre-polymerization with MDI, and finally chain extension through BDO. The main steps

Journal of Physics: Conference Series 2961 (2025) 012015 doi:10.1088/1742-6596/2961/1/012015

are as follows: first, adding an appropriate amount of N, N-DMF to the IPDI-GO dispersion, and performing 2 hours of mechanical stirring and 1 hour of ultrasonic crushing to ensure uniform dispersion. Afterward, a quantitative amount of IPDI-GO dispersion is taken and placed in a vacuum oven for drying at 45°C. The solid mass after drying is weighed to determine the solid content of the modified GO dispersion, and then the amount of modified GO dispersion added to TPU is determined. The mass fraction of modified GO added in the study is 0.1wt%. Subsequently, the excess PTMG1000 is placed into a three-necked flask, and a mechanical stirring paddle is installed on a constant temperature magnetic stirrer. It is continuously vented using a vacuum pump to maintain a temperature of 120°C, a speed of 300 r/min, and dehydrated for 2 hours. An appropriate amount of IPDI-GO dispersion is added to the three-necked flask, PTMG1000 is added after removing moisture, and stirred with constant temperature magnetic stirring at 80°C for 2 hours to react with PTMG1000. During this process, nitrogen gas needs to be continuously introduced. After the reaction is completed, MDI is added to the heated magnetic stirrer at 50°C to make it clear and transparent. On this basis, the polymerized IPDI-GO is blended with PTMG1000, and the stirring blade speed is adjusted to 300 r/min and heated to 80°C. It is necessary to repeat three exhaust and nitrogen cycles, replace the air in the oven with nitrogen, and then perform 2 hours of pre-polymerization in a nitrogen atmosphere. On this basis, BDO and DBTDL are used as the main raw materials for chain extension, with BDO as the chain extender and DBTDL as the catalyst. Subsequently, the sample is injected into a polytetrafluoroethylene mold and dried at a constant temperature of 80°C for 48 hours. The solvent is removed and subjected to maturation treatment to obtain IPDI-GO-TPU composite material.

## 3. IPDI-GO-TPU material performance testing

The study uses Fourier transform infrared spectroscopy (FTIR), dynamic thermomechanical analysis (DMTA), and mechanical performance testing to investigate the dispersion of IPDI-GO in TPU and its impact on performance. FTIR uses the Brooke Vertex 70 spectrometer, DMTA uses the NETSCH GABO EPlexor 500N instrument, and mechanical properties use the Zwick Z005 tensile machine. Each test is conducted according to standard procedures to ensure the accuracy and reliability of the results.

Figure 2 shows the peaks of infrared carbonyl groups in TPU, GO-TPU, and IPDI-GO-TPU composites. Considering the complex hydrogen bonding mechanism of -NH in TPU, carbonyl groups are chosen as the research object. The degree of hydrogen bonding can be visually displayed by the ratio of the peak area of carbonyl hydrogen bonding to the sum of the peak areas of carbonyl-free peaks. Figures 2 (a) and 2 (b) indicate that due to the poor dispersion of GO in TPU and the breaking of the original hydrogen bonds by aggregation, the hydrogen bonding degree of GO-TPU composite material is lower than that of blank TPU. In Figure 2 (c), the hydrogen bonding degree of IPDI-GO-TPU composite is significantly improved compared to TPU and GO-TPU, indicating that IPDI-GO has good dispersion performance in TPU and will not damage the original hydrogen bonding structure. In addition, the surface of IPDI-GO can also form hydrogen bonds with carbonyl groups. In the composite system prepared using IPDI-GO as a template, the carbonyl hydrogen bond content reaches 84.79%. Therefore, after modifying TPU with diisocyanate, the degree of hydrogen bonding significantly increases, which can improve the performance of TPU composite materials.

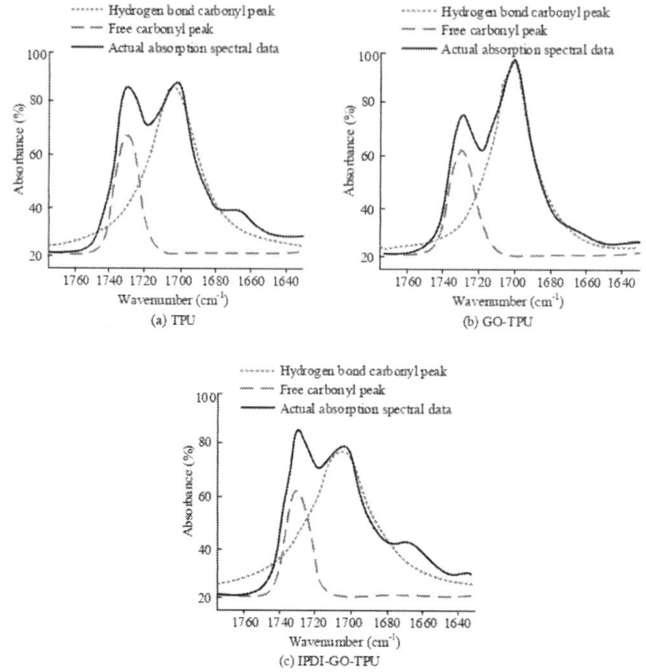


**Figure 2.** Peak separation of infrared carbonyl groups in TPU, GO-TPU, and IPDI-GO-TPU.

Figure 3 shows the DMTA curves of GO-TPU and IPDI-GO-TPU, Figure 3 (a) shows the storage modulus and Figure 3 (b) shows the loss factor. In Figure 3 (a), the IPDI-GO-TPU composite has a lower storage modulus than GO-TPU, indicating its lower elasticity. In Figure 3 (b), the IPDI-GO-TPU composite has a higher loss factor than the GO-TPU composite before the glass transition, while the loss factor of the IPDI-GO-TPU composite decreases after the glass transition. The main reason is that IPDI-GO first reacts with PTMG1000, consuming some hydroxyl groups, causing slight cross-linking of the composite material, thereby affecting its storage modulus and loss factor, and then pre-polymerizes with MDI, leading to the occurrence of this phenomenon. After the interaction between IPDI-GO and PTMG1000, some hydroxyl groups are consumed, causing a certain degree of cross-linking, which has a certain impact on the energy storage coefficient and loss coefficient of the material. In Figure 3 (b), the temperature corresponding to the peak loss coefficient is the glass transition temperature (Tg), and the Tg of IPDI-GO-TPU is lower than that of GO-TPU. The above results indicate that after adding IPDI-GO to TPU, although its energy storage coefficient decreases, its loss coefficient increases, and the glass phase transition temperature also changes. This is mainly due to mild cross-linking between IPDI-GO and PTMG1000. Therefore, IPDI-GO is a nanoscale filler that can effectively improve the microstructure and macroscopic properties of TPU.

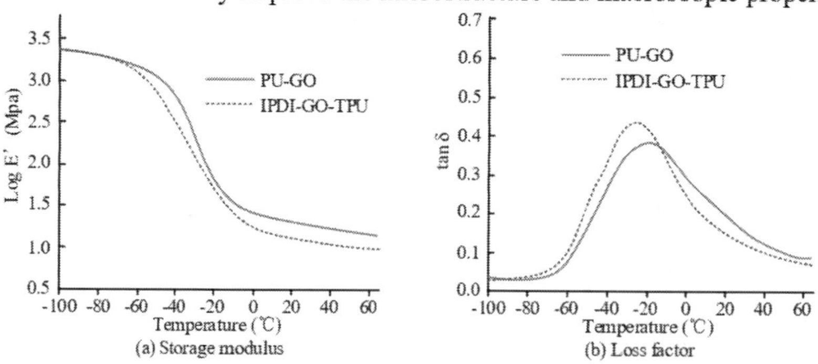


**Figure 3.** DMTA curves of GO-TPU and IPDI-GO-TPU.

Figure 4 shows the stress-strain relationship of TPU, GO-TPU, and IPDI-GO-TPU materials. As shown in the figure, the tensile strength and fracture elongation of GO-TPU are higher than those of the original TPU. In addition, during the stretching stage, GO-TPU exhibits a stress-yielding phenomenon. The reason is that GO has good mechanical properties, high specific surface area, and a two-dimensional layered structure, effectively enhancing compatibility with the matrix while preventing crack growth. In the figure, the mechanical properties of IPDI-GO-TPU are much better than GO-TPU, due to the chemical bonding between IPDI-GO and TPU, which improves the dispersion of the material in TPU and ensures the overall stability of the material. The tensile strength and elongation at break of IPDI-GO-TPU are 35.88 MPa and 569%, respectively. Compared with GO-TPU, IPDI-GO-TPU has a 495% and 92.97% increase in tensile strength and elongation at break, respectively. The above results indicate that IPDI-GO has a good dispersion effect in TPU, which significantly improves its mechanical properties. The aforementioned FTIR test results also indicate that IPDI-GO-TPU has a high hydrogen bonding strength, and the hydrogen bonding strength is positively correlated with its mechanical properties. Therefore, the determination of its mechanical properties is basically consistent with the FTIR results. Therefore, the introduction of IPDI-GO in TPU can significantly improve its microstructure, enhance its strength and toughness, and optimize its performance.

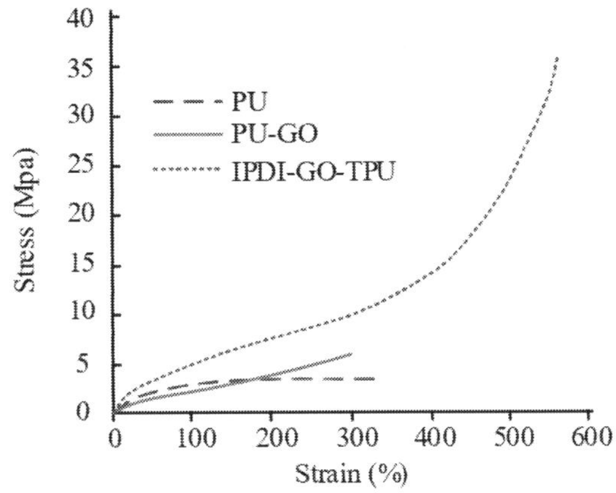


**Figure 4.** Stress-strain relationship of TPU, GO-TPU, and IPDI-GO-TPU.

## 4. Conclusion

In order to improve the performance of TPU, a modified GO, IPDI-GO, was prepared and introduced as a nanofiller into TPU. Finally, IPDI-GO-TPU composite material was synthesized. Related experiments were conducted to test the performance of IPDI-GO-TPU. The experimental results showed that the hydrogen bonding degree of the IPDI-GO-TPU composite was significantly improved compared to TPU and GO-TPU, with a carbonyl hydrogen bonding content of 84.79%. The IPDI-GO-TPU composite has a lower storage modulus and lower elasticity than GO-TPU; The IPDI-GO-TPU composite has a higher loss factor than the GO-TPU composite before glass transition, while the loss factor of the IPDI-GO-TPU composite decreases after glass transition. The tensile strength and elongation at break of IPDI-GO-TPU are 35.88 MPa and 569%, respectively. Compared with GO-TPU, IPDI-GO-TPU has a 495% and 92.97% increase in tensile strength and elongation at break, respectively. Therefore, introducing IPDI-GO into TPU can effectively enhance its microstructure, enhance its performance, and expand its industrial applications. The limitation of the study is that it failed to consider the impact of different filling amounts of IPDI-GO on TPU, which can be discussed in the future.

**Acknowledgement**
This work was supported by the Key Research and Development Program of Lianyungang (SF2334).

**Reference**

[1] Desai S M, Sonawane R Y, More A P. Thermoplastic polyurethane for three-dimensional printing applications: A review. Polymers for Advanced Technologies, 2023, 34(7): 2061-2082.
[2] Cui Z, Marcelle S, Zhao M, Wu J, Liu X, Si J, Wang Q. Thermoplastic polyurethane/titania/polydopamine (TPU/TiO 2/PDA) 3-D porous composite foam with outstanding oil/water separation performance and photocatalytic dye degradation. Advanced Composites and Hybrid Materials, 2022, 5(4): 2801-2816.
[3] Arifvianto B, Iman T N, Prayoga B T, Dharmastiti R, Salim U, Mahardika M, Suyitno. Tensile properties of the FFF-processed thermoplastic polyurethane (TPU) elastomer. The International Journal of Advanced Manufacturing Technology, 2021, 117(5): 1709-1719.
[4] Farzaneh A, Rostami A, Nazockdast H. Thermoplastic polyurethane/multiwalled carbon nanotubes nanocomposites: Effect of nanoparticle content, shear, and thermal processing. Polymer Composites, 2021, 42(9): 4804-4813.
[5] Liu C, Huang X, Wu Y Y, Deng X, Zheng Z, Xu Z, Hui D. Advance on the dispersion treatment of graphene oxide and the graphene oxide modified cement-based materials. Nanotechnology Reviews, 2021, 10(1): 34-49.
[6] Asghar F, Shakoor B, Fatima S, Munir S, Razzaq H, Naheed S, S. Butler I. Fabrication and prospective applications of graphene oxide-modified nanocomposites for wastewater remediation. RSC advances, 2022, 12(19): 11750-11768.
[7] Si J, Deng Y, Gong C, Cui Z, Wang Q. Preparation and characterization of deacetylated cellulose acetate/thermoplastic polyurethane nanofiber membranes modified with graphene oxide for methylene blue and Cr (VI) adsorption. Polymer Engineering & Science, 2023, 63(11): 3891-3905.
[8] Zhang J, Wang H, Sun W, Zhang Z, Li H, Zhang S, Sun J, Gu X. Surface modification on ammonium polyphosphate and its enhanced flame retardancy in thermoplastic polyurethane. Polymers for Advanced Technologies, 2021, 32(8): 2879-2886.
[9] Iqbal M S, Nazir M S, Ali Z, Iftikhar R, Hussain M, Imran S. Reduced graphene oxide coated poly-methyl methacrylate beads based thermoplastic polyurethane nanocomposites for gas sensing applications. Polymer-Plastics Technology and Materials, 2023, 62(6): 790-799.
[10] Wang J, Li J, Li H, Zhou H. Thermoplastic polyurethane (TPU) modifier to develop bimodal cell structure in polypropylene/TPU microcellular foam in presence of supercritical CO2. Journal of Vinyl and Additive Technology, 2021, 27(1): 127-136.
[11] Chen Y, Zhong W, Jia X, Hu D, Sun J, Peng Y, Yu J, Jiang X, Wang H, Zhao L. Microcellular Thermoplastic Polyurethane (TPU) with Multimodal Cell Structure Fabricated Based on Pressure Swing Strategy and Its Compressive Mechanical Properties. Industrial & Engineering Chemistry Research, 2024, 63(19): 8833-8845.
[12] Xu T, Shen W, Lin X, Xie Y, . Additively manufactured thermoplastic polyurethane (TPU) mold for concrete casting of complex structures. Rapid Prototyping Journal, 2022, 28(9): 1717-1730.
[13] Sabet M, Soleimani H, Mohammadian E, Hosseini S. The effect of graphene oxide on the mechanical, thermal characteristics and flame retardancy of polyurethane. Plastics, Rubber and Composites, 2021, 50(2): 61-70.
[14] Chen X, Zhang J, Chen X, Zhu Y, Liu X. Reduced graphene oxide-doped porous thermoplastic polyurethane sponges for highly efficient oil/water separation. ACS omega, 2023, 8(11): 10487-10492.
[15] Liu Q, Wang H, Li H, Sun J, Gu X, Zhang S. Constructing a novel synergistic flame retardant by hybridization of zeolitic imidazolate framework-67 and graphene oxide for thermoplastic polyurethane. Polymers for Advanced Technologies, 2022, 33(7): 2374-2385.

# Polyurethane porous elastomeric mixture composition design and aging properties

**Qizhuo Liu [1], Guilan Li [1], Jiaxin Wan [1], Wenchao Shi [3], Songtao Si [4] and Qi Wang [2]***

[1] Jiangxi Provincial Communications Investment Group Co., Ltd., Project Construction Management Company, Nanchang 330052, China
[2] Research Institute of Highway Ministry of Transport, Beijing 100088, China
[3] Chongqing Jiaotong University, Chongqing 400074, China
[4] Jiangsu Shuangning Engineering Consulting Co., Ltd., Nanjing 211200, China

* Corresponding author: wang.qi@rioh.cn

**Abstracts:** Polyurethane Porous Elastic Road Surface (PERS) has the characteristics of energy saving, emission reduction, carbon reduction, drainage, noise reduction, and de-icing. However, the aging properties of PERS have been rarely studied. The optimal gradation of the polyurethane mixtures has been designed in this paper. The concentration of polyurethane has been determined as 4.1%. Through scatter tests and segregation leakage tests, we analyzed the variation curves of the void ratio. Under aging conditions of thermo-oxidative, humid-heat, and without aging, through seepage test and wave tube test, the drainage function and noise performance have been studied in this paper. It has been found that under humid-thermal and thermo-oxidative aging procedures, the void ratio of mixtures increases by 1.2% and 1.4%, respectively. The sound absorption coefficient of mixtures was reduced by 7% and 10.9%, respectively. The water permeability coefficient of mixtures increased by 15.0% and 18.3%, respectively.

## 1. Introduction

Poroelastic road surface (PERS) has been known for its high strength, good drainage, and noise reduction [1~3]. The polyurethane mixtures have also been studied by some researchers. Liao et al. [4] proposed the rubber particle content significantly affected the mechanical properties of PERS mixtures, which determined the polyurethane content needed. And the high air void of PERS mixtures reduced water susceptibility noticeably. Lyu et al. [5] found that the optimal material composition of PERS is 5.5% polyurethane content, 10% rubber content, and 1.18 mm rubber particles, with a 60 mm thickness. Wang et al. [6] designed 20 PERS mixtures with different compositions and concluded the binder content and degree of compaction are the critical factors affecting the raveling resistance of PERS. Through finite element simulation and an indoor noise test, Wang [7] discovered that porous elastic polyurethane blends perform well in terms of noise reduction.

Based on the overview of the literature described above, it is worth noting that the composition design and the long-term aging performance of the polyurethane porous elastic mixtures have been rarely researched. Therefore, the grading for polyurethane mixtures based on the Marshall test proportion design method for asphalt mixtures has been designed in this paper. The optimum amount of polyurethane used in polyurethane porous elastic mixtures has been determined through the segregation leakage and scatter tests. Based on the Test Procedures for Asphalt and Asphalt Mixtures in Highway

Journal of Physics: Conference Series **2961** (2025) 012016 doi:10.1088/1742-6596/2961/1/012016

Engineering [8] and the experiments above, the effects of aging conditions of thermo-oxidative, and humid-heat aging of noise reduction performance and drainage function of PERS have also been studied in this paper.

## 2. Materials and testing

### *2.1. Materials*

The polyurethane was used as the adhesive of mixtures. The diameter of the mineral powder was less than 0.075mm, the diameter of fine aggregate ranged from 1.18mm to 0.15mm, and the diameter of coarse aggregate ranged from 16.0mm to 4.75mm. The aggregates of a diameter of 4.75 mm to 1.18 mm were replaced by the waste tire rubber granules with the same volume and particle size.

### *2.2. Experiments*

Referring to the range of PERS gradations identified in the Drainage Pavement Specification, a reasonable mineral gradation for PERS was determined by Marshall Stability and Target Void Ratio. To explore the optimal polyurethane content, the polyurethane contents were set to 3.0%, 3.5%, 4.0%, 4.5%, and 5.0%, respectively. The optimal content of polyurethane for PERS mixtures is obtained by deflection segregation leakage tests, scatter tests, and void ratio. The ratios of the polyurethane porous elastic compounds were established. Thus, the PERS mixtures were mixed at the 25°C temperature and then maintained in the 30°C temperature and 80% humidity curing environment for 48 hours. Then some of them experienced the hygrothermal treatment under 85°C temperature and an 80% humid-heat environment for 120 hours. Some of them experienced thermo-oxidative treatment under an 85°C temperature environment for 120 hours. Then, the standing wave tube tests and seepage tests were used to observe the variations in the water permeability and sound absorption coefficients of PERS mixtures after aging.

## 3. Design of material composition

### *3.1. Mineral gradation design*

According to the Drainage Pavement Specification, the gradation ranges of PERS are indicated in Table 1. The grading curve of stone material was determined first. Then, based on the volume replacement, the certain grading of stones was replaced with rubber particles. Within the top and lower limits gradation curves, three gradations of PERS were designed. Through the target void ratio and marshall stability, the gradation of PERS was ensured (Figure 1).

**Table 1.** Mineral composition gradation of PERS mixtures.

| Sieve size (mm) | 16.0 | 13.2 | 9.5 | 4.75 | 2.36 | 1.18 | 0.6 | 0.3 | 0.15 | 0.075 | Mineral powders |
|---|---|---|---|---|---|---|---|---|---|---|---|
| Gradation limit (%) | 100 | 100 | 71 | 30 | 20 | 17 | 14 | 12 | 9 | 5 | |
| Gradation median (%) | 100 | 95 | 55.5 | 20 | 14.5 | 12 | 10 | 8.5 | 6.5 | 4 | 3 |
| Grading Lower Limit (%) | 100 | 90 | 40 | 10 | 9 | 7 | 6 | 5 | 4 | 3 | |

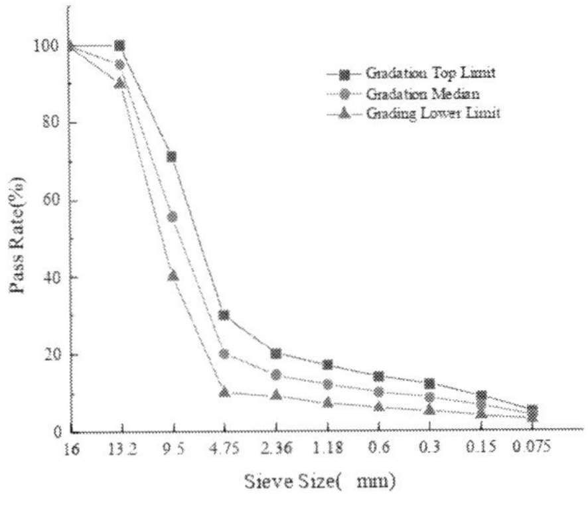


**Figure 1.** Grading curve of PERS.

### *3.2. Determination of the optimal polyurethane ratio*

The graphs from the analysis of segregation leakage tests and the polyurethane content are shown in Figure 2 and Figure 3.

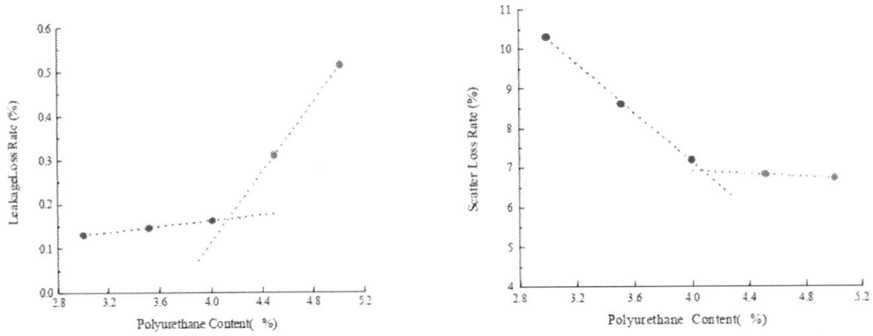


(a)Relationship between leakage loss rate and polyurethane content (b) Relationship between scatter loss rate and polyurethane content

**Figure 2.** Relationships among polyurethane content, the loss rate of scatter, and leakage loss rate.

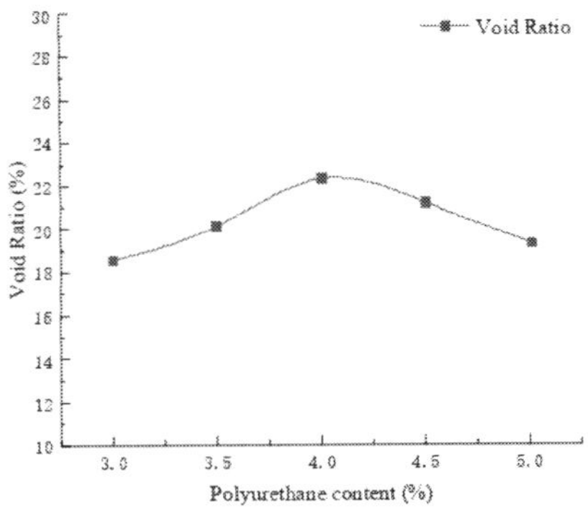


**Figure 3.** Relationship between the polyurethane content and the void ratio.

The optimum amount of polyurethane for PERS mixtures was determined by analyzing the inflection point of the leakage curve, the specified maximum value of leakage, the inflection point of the scatter loss rate curve, and the specified minimum rate of scatter loss. From Figure 2(a), the value of the intersection point of the leakage loss and polyurethane content curve was 4.08%. From Figure 2(b), the intersection value of the scatter loss rate and polyurethane content curve was 4.108%. It can be seen in Figure 3 that when the content of the polyurethane was close to 4.1%, the inflection point of the void ratio curve occurred. Generally, the optimum polyurethane content of PERS was 4.1%.

## 4. Aging properties

### *4.1. Noise reduction function*

The noise reduction function of PERS mixtures was studied after humid-heat aging and thermo-oxidative aging. The sound absorption coefficient at various frequencies (100Hz to 8000Hz) is shown in Figure 4.

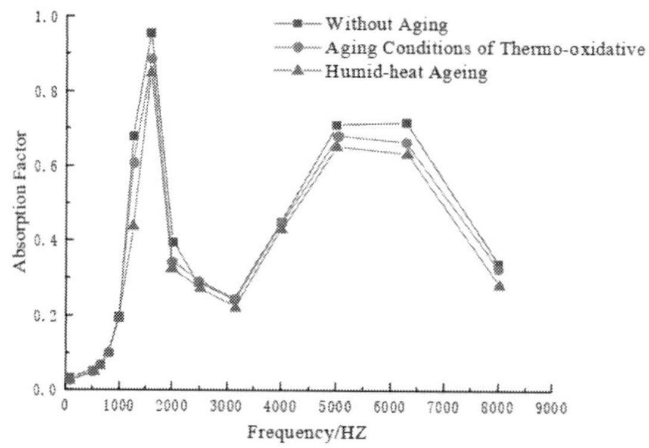


**Figure 4.** Variation of the sound absorption coefficient of PERS mixtures after aging.

It can be seen from Figure 4 that the acoustic absorption capabilities of PERS mixtures were degraded after aging. The peak sound absorption coefficients decreased by 7% and 10.9% respectively after thermo-oxidative aging and humid-heat aging. The high elastic impedance, large void ratio, and many connected voids of PERS mixtures result in the greater sound absorption property of PERS. The molecular structure of polyurethane was changed and rubber particles hardened after mixture aging, which resulted in the decreasing of elasticity and reduction of sound absorption property of PERS. Humid-heat aging has more effect on the acoustic absorption capabilities of PERS mixtures than thermo-oxidative aging.

### *4.2. Drainage function*

Table 2 shows that the void fraction of the PERS mixtures was increased after the thermo-oxidative aging and humid-heat aging. The void fraction of PERS mixtures increased by 1.2% and 1.4% after thermo-oxidative aging and humid-heat aging, respectively. The water permeability coefficient of aging mixtures increased with the void ratio increased. After thermo-oxidative aging, the water permeability coefficient of PERS mixtures increased from 5730 ml/min to 6592.6 ml/min. The water permeability coefficient of PERS mixtures increased by 18.3% after humid-heat aging. Due to the aging of the polyurethane and rubber particles in PERS mixtures, the volume contraction of rubber particles increases the void ratio and the seepage coefficient.

**Table 2.** Changes in void ratio and water permeability coefficient of PERS mixtures after aging.

| Type | Working condition | Void fraction (%) | Water permeability coefficient (ml/min) |
|---|---|---|---|
| PERS mixtures | Without aging | 21.2 | 5730 |
| | Thermo-oxidative aging | 22.4 | 6592.6 |
| | Humid-heat aging | 22.6 | 6776.3 |

## 5. Conclusion

The grading design method of PERS has been introduced in this paper. The optimal gradation of the PERS mixtures has been designed. Through seepage tests and wave tube tests, the drainage function and noise reduction performance have been researched. The main conclusions can be summarized as follows.

(1) Based on leakage tests and the scatter tests, the concentration of optimal polyurethane content was determined as 4.1%.

(2) The acoustic performance of PERS mixtures was reduced under the effect of aging. The peak acoustic absorption coefficient of PERS mixtures was 0.954. Under the effect of thermo-oxidative and humid-thermal aging, the peak acoustic absorption coefficient decreased by 7% and 10.9%, respectively.

(3) Compared with the PERS mixtures without aging, under aging conditions of humid-heat and thermo-oxidative, the void fraction of PERS mixtures increased by 1.2% and 1.4%, respectively. The water permeability coefficient under two aging conditions increased by 18.3% and 15.0%, respectively.

(4) In this study, drainage function and noise reduction performance of PERS under aging conditions of thermo-oxidative and humid-heat aging have been studied through the seepage test and wave tube tests. In the future, the mechanism of drainage function and noise reduction performance of PERS can be studied more comprehensively through the multi-field coupling analysis.

**Acknowledgments**

This research was supported by the Science and Technology Project of the Department of Transportation of Jiangxi Province (No. 2023C0003), the National Key R&D Program of China (No. 2021YFB2601200), and the Fundamental Research Funds for the Central Research Institute ((No. 2024-9020).

**References**

[1] Zhong K, Shao C, Sun M, Tian X (2024). Evaluation of the attenuation mechanism of the noise reduction performance of polyurethane porous elastic road surface [J]. Construction and Building Materials, 448:138247.

[2] Luo Y, Han S, Wu C, Zheng Y, Men C (2023). Laboratory evaluation on performance of Polyurethane Porous Elastic Mixture [J]. Construction and Building Materials, 397: 132399.

[3] Liao G, Fang X, Hu J, Wang H, Xu T, Chen J (2022). Improving the mechanical performance of poroelastic road surface with low polyurethane content through surface activation [J]. Construction and Building Materials, 323:126543.

[4] Liao G, Wang H, Xiong J, Chen J, Qi K (2020). Mechanical Properties of Poroelastic Road Surface with Different Material Compositions [J]. Journal of Materials in Civil Engineering, 32(9).

[5] Lyu L, Dong Y, Zhao D, Wen Y, Li R, Ren X, Pei J (2021). Mechanical and acoustic properties composition design and effects analysis of poroelastic road surface [J]. Journal of Materials in Civil Engineering, 33(10).

[6] Wang D, Schacht A, Leng Z, Leng C, Kollmann J, Oeser M (2017). Effects of material composition on mechanical and acoustic performance of poroelastic road surface (PERS) [J]. Construction and Building Materials, 135: 352-360.

[7] Wang C (2023). Research on the formation mechanism of vibration/pumping noise and evaluation of noise reduction function of porous elastic pavement [D]. Chongqing Jiaotong University.

[8] Ministry of Transport of the People's Republic of China (2000). Test specification of asphalt and asphalt mixture for highway engineering [M]. People's Traffic Press.

# Molecular cages for the selective detection of aromatic nitro compounds

**Bingyu Wang[1] and Fanzhi Yang[2*]**

[1]School of Chemistry and Chemical Engineering, Beijing Institute of Technology, Beijing, 100000, China

[2]Frontier Interdisciplinary Colleges, Beijing Institute of Technology, Beijing, 100000, China

Corresponding author's e-mail: 3120211262@bit.edu.cn

**Abstract**: With the widespread application of aromatic nitro compounds (ANCs) in multiple fields, such as pesticides, pharmaceuticals, high-energy explosives, and industrial dyes, the rapid detection of trace amounts is crucial for environmental safety and human health. In this article, a molecular cage CC7 was synthesized to detect 2,4-DNT based on the fluorescence quenching method, which exhibits high selectivity and sensitivity for 2,4-DNT with a detection limit of 0.53μM, providing a new method for the detection of energetic compounds.

## 1. Introduction

Aromatic nitro compounds (ANCs) are crucial for their detection in multiple fields due to their widespread applications in pesticides, pharmaceuticals, explosives, and dye manufacturing. The residues of ANCs in pesticides may affect human health through the food chain, and long-term exposure may lead to chronic poisoning and even cancer. Therefore, monitoring the residues of ANCs in food and agricultural products is crucial for food safety. In the pharmaceutical industry, ANCs, as intermediates in drug synthesis, are crucial for ensuring drug quality and efficacy. Rapid and sensitive detection methods help pharmaceutical companies control production processes and ensure consumer health. The application of ANCs in explosives poses a threat to public safety, and rapid detection is crucial for preventing terrorist activities and reducing casualties in emergency situations. In addition, ANCs in dye production may pollute water bodies and cause damage to the environment, so conducting ANCs testing on industrial emissions is crucial for environmental protection. In summary, rapid and sensitive detection of ANCs is of great significance for ensuring food safety, drug quality, public safety, and environmental protection, and is an effective means to improve quality of life and achieve environmental safety.

Currently, gas chromatography, Raman spectroscopy, ion mobility spectrometry, and fluorescence spectroscopy have been used to trace NACs explosives[1]. However, apart from fluorescence spectroscopy, other detection methods have some shortcomings, such as the inconvenience of carrying expensive instruments, many influencing factors, and low concentration of analytes, thus making fluorescence spectroscopy-based detection of explosives the preferred method.

In recent years, porous organic materials have been of interest due to their unique cavity structures, and they have been widely used in fields such as gas separation, energy storage, multiphase catalysis, drug delivery, and sensing[2]. Among them, molecular cages stand out due to their good solubility and ease of modification. Jaya Prakash's group[3] reported a fluorescent covalent organic cage (F-COC) molecule derived from triphenylamine, which has shown unique performance in various applications. They used the fluorescence quenching technology of F-COC to study the detection of nitric acid (PA). The observed fluorescence quenching is a static process, confirming that fluorescence quenching occurs in the ground state, showing high sensitivity and selectivity for PA. Bolse et al.[4] reported that a fluorescence quenching sensor was prepared using aerosol jet printing. The detection of nitrobenzene, 1,3-dinitrobenzene, and 2,4-DNT was achieved at the ppb level with a high classification rate using an array of six fluorescent spirochetes and homopolymers. Many materials have been applied in the detection of explosives, and porous molecular cages, as an emerging material, have very few applications in this field. Thus, it is necessary to develop a new, simple, fast, reliable, and highly sensitive polymer cage detection method.

## 2. Experiment

### *2.1. Synthesis and characterization of CC7*

According to the report by the Cooper group[5], the molecular cage CC7 was prepared as follows:

At room temperature, $CH_2Cl_2$ (100 mL) was slowly added to a 250 mL round-bottom flask containing tri(4-formylphenyl)amine (660 mg, 2 mmol) that had been thoroughly dried in an oven. After standing for 10 minutes, a solution of (1R,2R)-1,2-cyclohexanediamine (689 mg, 6 mmol) in $CH_2Cl_2$ (100 mL) was slowly added down the inner wall of the flask. The resulting mixture was left to stand at room temperature without stirring. After 72 hours, a clear orange-yellow solution with a large amount of yellow cubic crystals floating on top was observed. After filtration, the product was washed with DCM (3×5 mL) and air-dried to obtain pure CC7 product (741 mg) with a yield of 83%(The synthetic route is shown in Figure 1). And no further purification required CC7:$^1$H NMR (400 MHz, CDCl3) δ 8.36 (s, 12H), 8.24 (s, 12H), 7.54 (m, 48H), 7.05 (m, 48H), 3.55 (m, 24H), 1.93 – 1.72 (broad s, 72H), 1.50 (s, 24H). (Figure 2)

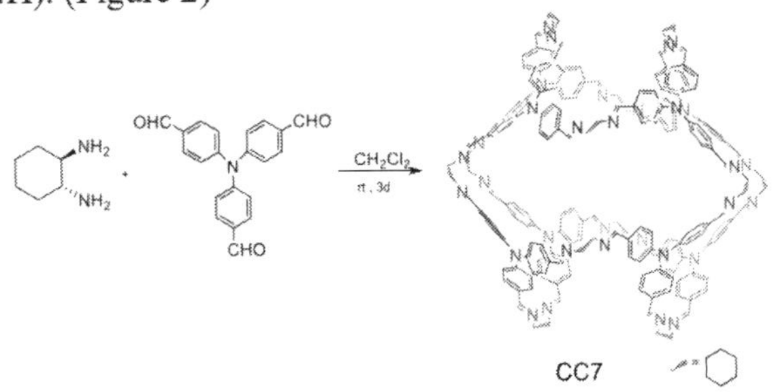


**Figure 1.** Synthetic route of CC7.

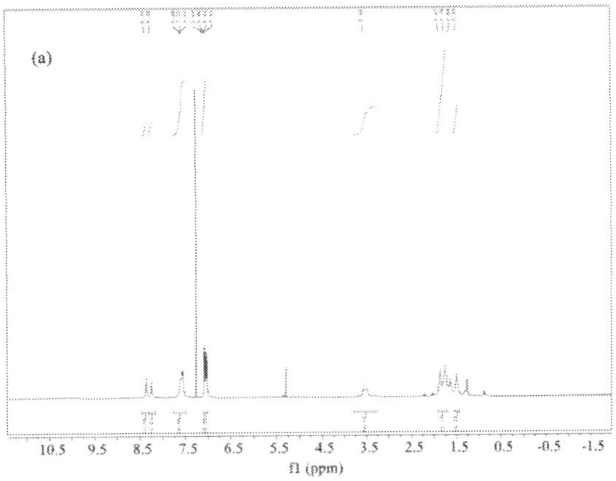


**Figure 2.** $^1$H NMR of CC7.

Journal of Physics: Conference Series **2961** (2025) 012017 doi:10.1088/1742-6596/2961/1/012017

In Figure 3(a), the peak at 1596 $cm^{-1}$ corresponds to the characteristic peak of the imine bond (-CH=N-), FT-IR: 2921.70, 2851.19, 1603.30, 1596.54, 1504.24, 1447.35, 1316.01, 1268.77, 1174.25, 1103.68, 1015.35,810.68, 749.54, 640.39, 612.42, 513.77, 412.41. The $^1$H NMR and FT-IR spectrum matches the literature reports, confirming the successful synthesis of the target product CC7.

As shown in the excitation spectrum of Figure 3(b), the maximum absorption value of CC7 is 360 nm, and the maximum emission wavelength is 420 nm.

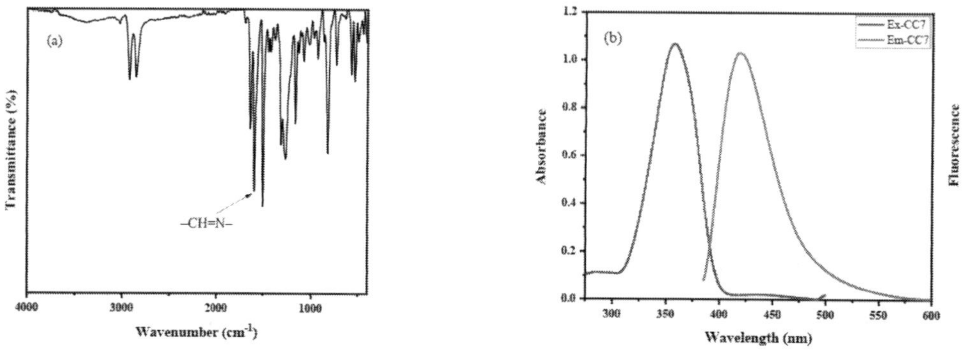


**Figure 3**(a). FT-IR of CC7,(b) the excitation spectrum of CC7.

*2.2. Fluorescence quenching experiment*

Fluorescence quenching titration is an experimental method used to study the interaction between fluorescent substances and quenchers. In this experiment, a certain amount of fluorescent material (in this case, a molecular cage) is first dissolved in DCM, and then the quencher (in this case,2,4-DNT) is gradually added. The extent of fluorescence quenching is analyzed by recording changes in fluorescence intensity.

The experimental steps are as follows:

1) 2ml of a $CH_2Cl_2$ solution of the molecular cage CC7 at a certain concentration was added into a 4ml cuvette.

2) It was titrated with a $CH_2Cl_2$ solution of 2,4-DNT at a concentration of 2.16 mM.

3) Under the same experimental conditions, we measured the fluorescence spectrum three times to ensure the stability of the obtained data. The excitation wavelength for the fluorescence spectrum was selected based on the absorbance values of the absorption spectrum, the measurement range of the fluorescence emission spectrum was set according to the excitation wavelength, and the slit width was set to 2.5 nm for both.

The data obtained from the experiment are calculated using the following equations:

Fluorescence quenching efficiency (η):

$$\eta = (I_0 - I)/I_0 \times 100\% \qquad \text{equation (1)}$$

where I0 represents the initial fluorescence intensity of the molecular cage, and I represents the fluorescence intensity after the addition of the quencher.

Quenching constant ($K_{SV}$):

$$I_0/I = 1 + K_{SV}[Q] \qquad \text{equation (2)}$$

where [Q] is the concentration of the quencher.

Detection limit (LOD):

$$LOD = 3\sigma/K_{SV} \qquad \text{equation (3)}$$

where σ represents the standard deviation, and KSV is the quenching constant.

## 3. Results and analysis

We performed fluorescence titration on cage CC7 using 2,4-DNT, which exhibited significant quenching effects. As shown in Figure 4(a), the fluorescence intensity decreased markedly with the increase of 2,4-DNT content in the cage solution. When the amount of added 2,4-DNT reached 30 equivalents, the quenching efficiency was 90%. Based on the fluorescence change data, a plot of the fluorescence intensity ratio ($I_0/I$) versus the concentration of 2,4-DNT was drawn (Figure 4b), which was linear. The exponential equation $I_0/I = 1 + K_{SV}[2,4\text{-DNT}]$ was applied for linear fitting over the range of [2,4-DNT] from 0 to 7μM ($R^2 = 0.9937$), and the quenching constant was calculated to be $K_{SV}[2,4\text{-DNT}] = 0.075\ M^{-1}$. The quenching constant indicates the strong quenching ability of 2,4-DNT on cage CC7, with a detection limit of 0.53μM.

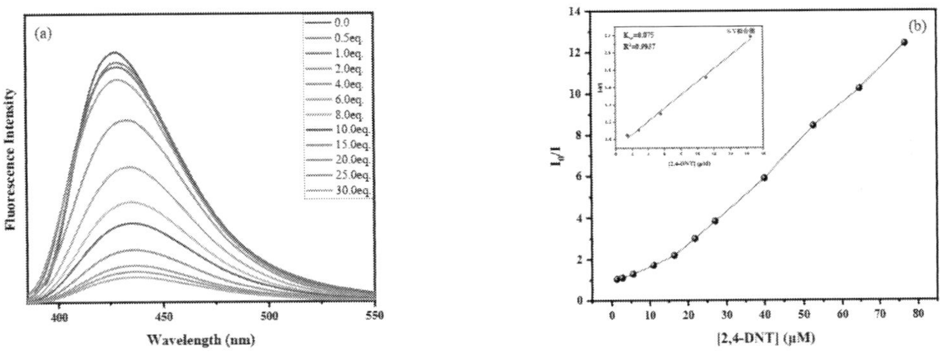


**Figure 4**(a). The change in fluorescence intensity of CC7 in DCM with incremental addition of 2,4-DNP. (b) Stern–Volmer plot $I_0/I$ versus [2,4-DNP] with CC7 in DCM, the inset shows linear fit in the range of 0-18μM.

Based on Figure 3(b), it can be seen that the quenching is static quenching. CC7 is a structure containing multiple π electrons. Generally, compared to the lowest unoccupied molecular orbital (LUMO) energy levels of aromatic nitro compounds, the conduction band orbital energy levels of polymers rich in π electrons are higher. When 2,4-DNT is added to a polymer-rich in π electrons, the excited electrons that have transitioned to the conduction band may transfer to the nearby LUMO orbital of the nitro compound due to its lower energy level, rather than returning to the ground state. Photoinduced electron transfer (PET) becomes a possible mechanism for quenching the fluorescence of the analyte.

## 4. Conclusion

We have developed a simple and rapid method for detecting 2,4-DNT. The molecular cage CC7 exhibits a quenching rate of 90% with 2,4-DNT, with a quenching constant $K_{SV}[2,4\text{-DNT}]$ of $0.075\ M^{-1}$ and a detection limit of 0.53 μM. This provides a new method and approach for the detection of explosives.

## References

[1] To, K. C., Ben-Jaber, S., & Parkin, I. P. (2020). Recent developments in the field of explosive trace detection. ACS nano, 14(9), 10804-10833.

[2] Song, J. H., & Kang, D. W. (2023). Hazardous nitroaromatic explosives detection by emerging porous solid sensors. Coordination Chemistry Reviews, 492, 215279.

[3] Gajula, R. K., Mohanty, S., Chakraborty, M., Sarkar, M., & Prakash, M. J. (2021) An imine linked fluorescent covalent organic cage: the sensing of chloroform vapour and metal ions, and the detection of nitroaromatics. *New Journal of Chemistry*, *45*(10), 4810-4822.

[4] Bolse, N., Eckstein, R., Habermehl, A., Hernandez-Sosa, G., Eschenbaum, C., & Lemmer, U. (2017). Reliability of aerosol jet printed fluorescence quenching sensor arrays for the identification and quantification of explosive vapors. ACS omega, 2(10), 6500-6505.
[5] Jelfs, K. E., Wu, X., Schmidtmann, M., Jones, J. T., Warren, J. E., Adams, D. J., & Cooper, A. I. (2011) Large self-assembled chiral organic cages: Synthesis, structure, and shape persistence. *Angewandte Chemie, 123*(45), 10841-10844.

Journal of Physics: Conference Series **2961** (2025) 012018 doi:10.1088/1742-6596/2961/1/012018

# Spin distribution and conductance variations in phenoxyl radical single-molecule junctions

**Fanxi Sun[1], Shengqing Jiang[1], Hanjun Zhang[1], Gaolu Zhu[1], Yangyang Pan[1], Yu Ji[1], Jiayu Li[1], Yonghao Zheng[1*] and Dongsheng Wang[1*]**

[1] School of Optoelectronic Science and Engineering, University of Electronic Science and Technology of China. Chengdu 610054, P.R. China

*Corresponding author's e-mail: zhengyonghao@uestc.edu.cn, and wangds@uestc.edu.cn

**Abstract:** This study explores the conductance characteristics and spin distributions in single-molecule junctions using phenoxyl radicals linked to bis(methylthio)triphenylene cores. Through a combination of UV-vis spectrophotometry, electron spin resonance (ESR) spectroscopy, and density functional theory (DFT) calculations, we examine how the structural modifications in phenoxyl radicals influence electron delocalization and molecular conductance. Initial tests confirmed the presence of radicals, and subsequent DFT analyses demonstrated significant shifts in electron spin densities affecting the conductive pathways. This study utilized scanning tunneling microscopy-break junction (STM-BJ) techniques to further elucidate the electronic properties of these radical-injected systems, highlighting the critical role of molecular architecture in defining conductive properties. Results indicate that the addition of phenyl rings leads to varying degrees of electron delocalization and corresponding changes in conductance, with a noted decrease in the HOMO-LUMO energy gap across different structures.

## 1. Introduction

Radicals, characterized by their unpaired electrons, are highly reactive due to their propensity for quickly gaining or losing electrons. These unpaired electrons are not anchored to specific atoms but are delocalized throughout the molecule, manifesting as a probability cloud [1]. This electron delocalization influences the radical's chemical behavior, leading to their involvement in reactions like oxidation [2], dimerization [3], and polymerization [4], significantly shortening their half-lives. Understanding this delocalization is critical as it impacts the radical's energy levels, spectral properties, and stability, factors that are fundamental in various molecular interactions and reaction mechanisms.

Phenoxyl radicals are a prime example of such reactive intermediates. They play significant roles in chemical synthesis [5], catalytic processes [6], and environmental chemistry [7], particularly in pollutant degradation [8]. The properties of phenoxyl radicals, affected by both solvent interactions and their molecular structure, provide valuable insight into the broader implications of radical behavior [9]. These radicals facilitate the exploration of electron transfer processes and stability, essential for advancements in chemical and material sciences.

Electron delocalization is also intimately related to spin density distribution [10]. In radical molecules, the presence of unpaired electrons leads to non-uniform spin densities, unlike non-radical molecules where paired electrons result in a uniform spin distribution and a net spin of zero [11]. The uniformity of spin density across a molecule often suggests greater stability, whereas concentrated densities indicate higher reactivity and lesser stability, due to the increased likelihood of chemical reactions [12]. These dynamics are typically analyzed using density functional theory (DFT), which provides a predictive framework for assessing molecular stability based on spin distributions [13].

Advancements in molecular electronics, where molecules serve as core components of electronic devices, further illustrate the importance of understanding these molecular behaviors [14]. Techniques like STM-BJ were used to study metal-molecule-metal junctions at the single-molecule level [15]. This methodology not only facilitates the detailed examination of molecular conductivity but also enables the controlled study of chemical reactions and kinetics at the molecular scale [16].

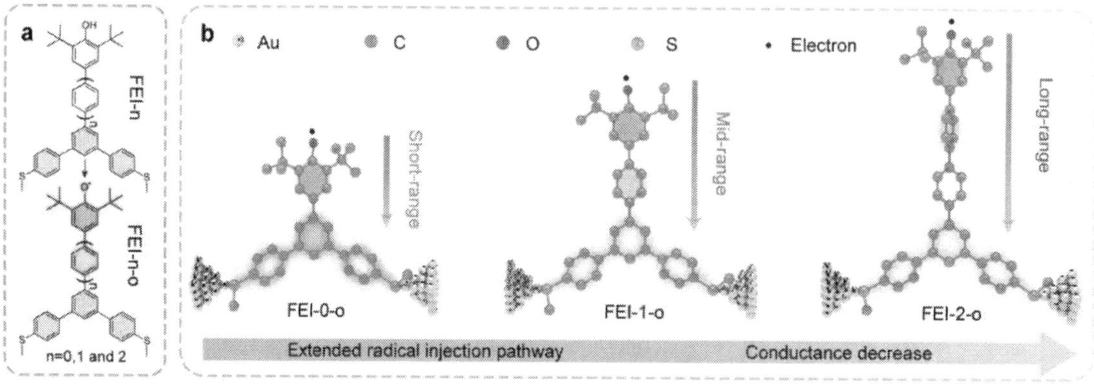


**Figure 1.** (a) Chemical structures of molecules before (FEI-n) and after (FEI-n-o) oxidation; (b) Schematic illustration of the molecular structure and conductance pathway variations for the bis(methylthio)triphenylene core-based phenoxyl radical injections at the single-molecule level using STM-BJ technique.

In our research, we specifically investigate a single-molecule conductive pathway using bis(methylthio)triphenylene cores, modified by varying the number of phenyl rings linking the core to phenoxyl radicals (Figure 1). This study utilizes UV-vis spectrophotometer and ESR spectroscopy to verify the existence of the radicals and employs DFT calculations to analyze their spin distributions. Through STM-BJ techniques, we explore how phenoxyl radical injection distances affect the electrical conductivity of the pathway, providing detailed insights into the influence of molecular architecture on electron transport properties, which is crucial for the development of molecular electronics.

## 2. Results and discussion

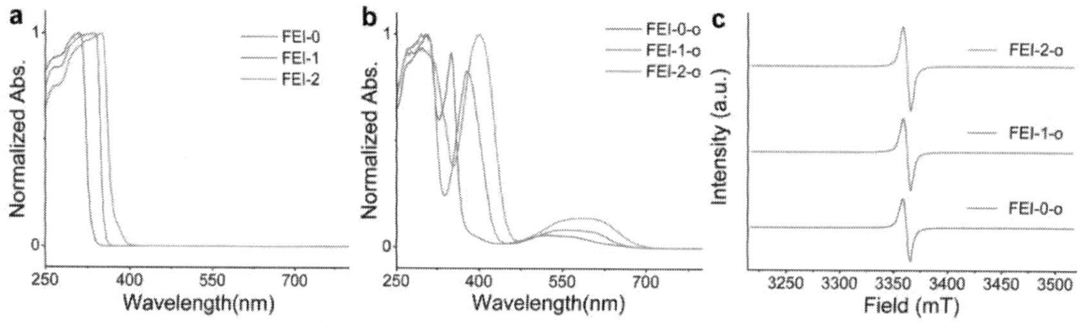


**Figure 2.** (a) UV-vis absorption spectra for FEI-0, FEI-1, and FEI-2; (b) UV-vis absorption spectra for FEI-0-o, FEI-1-o, and FEI-2-o; (c) ESR spectra for FEI-0-o, FEI-1-o, and FEI-2-o.

First, we conducted UV-vis absorption spectroscopy tests on FEI-0, FEI-1, FEI-2, FEI-0-o, FEI-1-o, and FEI-2-o in solution(dichloromethane). All concentrations are set at approximately $10^{-5}$ mol/L. The normalized absorption spectra are displayed in Figure 2a. The results reveal distinct absorption ranges for each molecule: FEI-0 primarily absorbs between 250 nm and 300 nm, FEI-1 between 230 nm and 350 nm, and FEI-2 between 300 nm and 350 nm (Figure 2a). As the number of phenyl rings increases, a redshift is observed in the absorption spectra of the molecules before oxidation.

Furthermore, when comparing the pre-oxidation (Figure 2a) and post-oxidation samples (Figure 2b), a second absorption peak appears in the spectra of the oxidized molecules: FEI-0-o exhibits this peak between 330 nm and 360 nm, FEI-1-o between 340 nm and 400 nm, and FEI-2-o between 400 nm and 450 nm. The intensity of the second peak is lower than the first in FEI-0-o and FEI-1-o but slightly stronger in FEI-2-o (Figure 2b). Notably, both before and after oxidation, the absorption spectra of the molecules show a significant redshift with an increase in the number of phenyl rings. After oxidation, distinct radical-associated absorption peaks are evident, with FEI-0-o displaying a wide absorption range from 450 nm to 550 nm, FEI-1-o from 500 nm to 600 nm, and FEI-2-o from 550 nm to 650 nm. These peaks all exhibit a redshift correlated with the increasing number of phenyl rings (Figure 2b).

Subsequently, we conducted ESR tests on the radical molecules, the results of which are shown in Figure 2c. The signals from the radical molecules were clear and displayed the characteristic feature of a single radical having only one signal. Based on the ESR test results, we were able to confirm that FEI-0-o, FEI-1-o and FEI-2-o had been successfully oxidized.

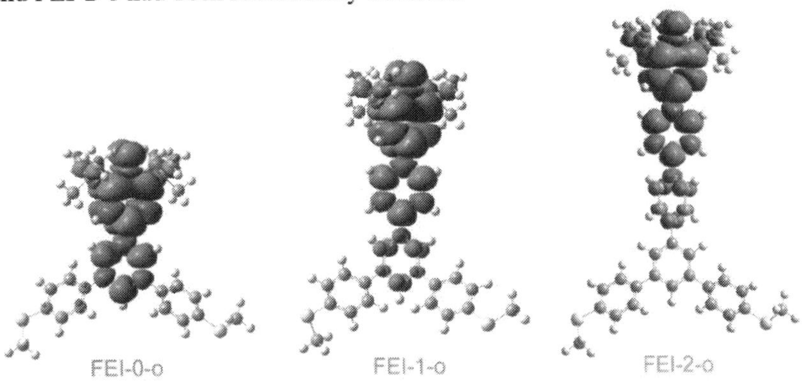


**Figure 3.** Spin density diagrams for FEI-0-o, FEI-1-o, and FEI-2-o obtained from DFT calculations.

**Table 1.** Summary of DFT Calculations for HOMO, LUMO, and $\Delta E_g$ Results for FEI-0, FEI-1, FEI-2, FEI-0-o, FEI-1-o, and FEI-2-o.

| Molecules | HOMO/eV | LUMO/eV | $\Delta E_g$/eV |
|---|---|---|---|
| FEI-0 | -5.40 | -0.78 | -4.62 |
| FEI-0-o | -5.43 | -1.10 | -4.33 |
| FEI-1 | -5.36 | -0.93 | -4.43 |
| FEI-1-o | -5.33 | -1.24 | -4.09 |
| FEI-2 | -5.32 | -1.12 | -4.20 |
| FEI-2-o | -5.31 | -1.39 | -3.91 |

Upon confirming the presence of radicals, we conducted DFT calculations on the series of molecules, with the results shown in Figure 3 and Table 1. It was observed that as the phenyl ring chain increases, the probability cloud of unpaired electrons delocalizing into the conductive pathway significantly decreases (Figure 3). For instance, a substantial amount of unpaired electrons in FEI-0-o delocalize to the central phenyl ring of the conductive pathway, whereas only a few do so in FEI-1-o, and almost none in FEI-2-o after an extension of two phenyl rings (Figure 3).

The HOMO-LUMO energy gaps of FEI-0-o, FEI-1-o, and FEI-2-o are found to be lower than those of their unoxidized counterparts, FEI-0, FEI-1, and FEI-2 (Table 1). This is primarily because, although the oxidation process increases the HOMO levels, the LUMO levels rise even more significantly, thereby reducing the energy difference between them. A comparison of the HOMO-LUMO energy gaps of FEI-0, FEI-1, and FEI-2 shows a gradual decrease in the HOMO levels with the addition of phenyl rings, while the LUMO levels increase, leading to a continuous decrease in the HOMO-LUMO energy gap. This trend is also observed in the oxidized samples, FEI-0-o, FEI-1-o, and FEI-2-o, confirming that both oxidation and the addition of phenyl rings contribute to the reduction of the HOMO-LUMO energy gap. Thus, these processes concurrently affect the electronic properties of the organic radical molecules (Table 1).

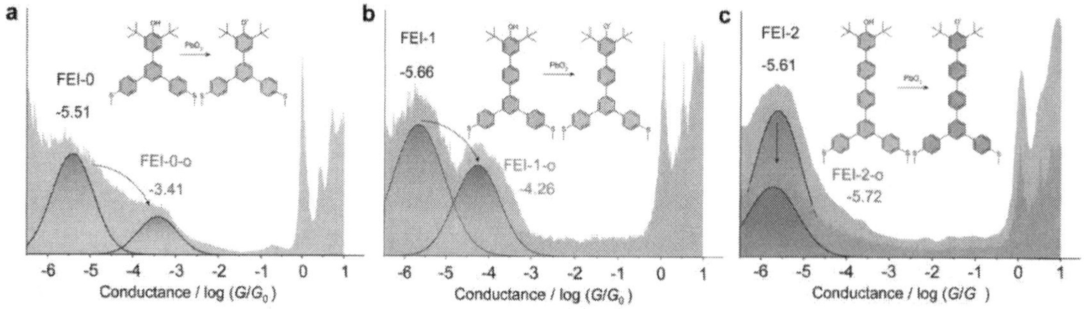


**Figure 4.** (a) 1D logarithmic conductance histograms for FEI-0 (gray) and FEI-0-o (blue); (b) 1D logarithmic conductance histograms for FEI-1 (gray) and FEI-1-o (purple); (c) 1D logarithmic conductance histograms for FEI-2 (gray) and FEI-2-o (red).

To conduct single-molecule conductance tests, the samples were dissolved in 1,2,4-trichlorobenzene (TCB) to prepare a solution with a concentration of $1x10^{-4}$ mol/L. Conductance tests were performed on molecules using STM-BJ setup, with results shown in Figure 4. Prior to oxidation, the molecules do not contain unpaired electrons, hence no electron delocalization occurs, and thus they do not influence the conductive pathway. The conductance pathways for FEI-0, FEI-1, and FEI-2 consist of bis(methylthio)triphenylene core, resulting in pre-oxidation conductance values around $10^{-5.5}G_0$ to $10^{-5.6}G_0$, which are within the expected error range (Figure 4a-c).

Post-oxidation STM-BJ tests were conducted for the three molecules FEI-0-o, FEI-1-o, and FEI-2-o. The conductance for FEI-0-o was $10^{-3.41}G_0$, which is an increase of 2.1 orders of magnitude compared to its pre-oxidation value of $10^{-5.51}G_0$ (Figure 4a). Notably, for FEI-0-o, a conductance peak at $10^{-5.51}G_0$ was observed, indicating incomplete oxidation (Figure 4a). For FEI-1, the conductance was measured at $10^{-4.26}G_0$, which represents an increase of 1.4 orders of magnitude from the pre-oxidation value of $10^{-5.66}G_0$ (Figure 4b). In contrast, the conductance for FEI-2-o (Figure 4c) showed virtually no change from its pre-oxidation level, thus confirming our DFT calculations that suggest a significant delocalization of electron spins to the central phenyl ring in FEI-0-o, affecting the conductive pathway (Figure 3). In FEI-1-o, the delocalization of electrons to the conductive pathway was significantly reduced, resulting in a smaller increase in conductance compared to FEI-0-o. For FEI-2-o, the probability of spin distribution on the central phenyl ring is nearly zero, thereby having almost no impact on the conductive pathway, which is why there is no significant change in conductance relative to the pre-oxidation levels of FEI-2 (Figure 3 and Figure 4c).

## 3. Conclusion

The integration of STM-BJ testing with experimental and theoretical analysis provided profound insights into the conductive behavior of organic radical-based single-molecule junctions. The phenoxyl radicals exhibited distinct electron delocalization patterns, which significantly influenced their conductance properties depending on the number of phenyl rings incorporated into the molecular structure. The DFT results were consistent with experimental findings, confirming that the molecular modifications led to a decrease in the HOMO-LUMO energy gap, which correlates with increased conductivity in certain configurations. This study not only confirms the pivotal role of molecular architecture in influencing the electronic properties of radical-based devices but also paves the way for future explorations into the design of more efficient molecular electronic components.

**Acknowledgments**
We gratefully acknowledge the financial support from the National Natural ScienceFoundation of China National Natural Science Foundation of China (52203134, 22375029).

**References**

[1] Alkadi, H., A review on free radicals and antioxidants. Infectious Disorders-Drug Targets (Formerly Current Drug Targets-Infectious Disorders) 2020, 20 (1), 16-26.
[2] Porter, N. A.; Caldwell, S. E.; Mills, K. A., Mechanisms of free radical oxidation of unsaturated lipids. *Lipids* 1995, **30** (4), 277-290.
[3] Spruell, J. M., Molecular recognition and switching via radical dimerization. *Pure Appl. Chem.* 2010, **82** (12), 2281-2294.
[4] Moad, G.; Solomon, D. H., The chemistry of radical polymerization. *Elsevier*: 2005.
[5] Modarelli, D. A.; Lahti, P. M.; George, C., Solid-state photochemical generation of a very stable phenoxyl-phenoxyl radical pair. *J. Am. Chem. Soc.* 1991, **113** (16), 6329-6330.
[6] Chaudhuri, P.; Hess, M.; Müller, J.; Hildenbrand, K.; Bill, E.; Weyhermüller, T.; Wieghardt, K., Aerobic oxidation of primary alcohols (including methanol) by Copper (II) and Zinc (II) phenoxyl radical catalysts. *J. Am. Chem. Soc.* 1999, **121** (41), 9599-9610.
[7] Platz, J.; Nielsen, O.; Wallington, T.; Ball, J.; Hurley, M.; Straccia, A.; Schneider, W.; Sehested, J., Atmospheric chemistry of the phenoxy radical, $C_6H_5O$ (•): UV spectrum and kinetics of its reaction with NO, $NO_2$, and $O_2$. *J. Phys. Chem. A* 1998, **102** (41), 7964-7974.
[8] Huang, M.; Han, Y.; Xiang, W.; Zhong, D.; Wang, C.; Zhou, T.; Wu, X.; Mao, J., In situ-formed phenoxyl radical on the CuO surface triggers efficient persulfate activation for phenol degradation. *Environ. Sci. Technol.* 2021, **55** (22), 15361-15370.
[9] Mukai, K.; Nishiguchi, H.; Ishizu, K.; Deguchi, Y.; Takaki, H., Solvent effects in the electron spin resonance spectra of some phenoxyl, nitroxide, and anilino radicals. *Bull. Chem. Soc. Jpn.* 1967, **40** (12), 2731-2739.
[10] Lazdins, D.; Karplus, M., Spin delocalization in the allyl and ethyl radicals. *J. Chem. Phys.* 1966, **44** (4), 1600-1611.
[11] Zhang, N.; Samanta, S. R.; Rosen, B. M.; Percec, V., Single electron transfer in radical ion and radical-mediated organic, materials and polymer synthesis. *Chem. Rev.* 2014, **114** (11), 5848-5958.
[12] Herges, R.; Geuenich, D., Delocalization of electrons in molecules. *J. Phys. Chem. A* 2001, **105** (13), 3214-3220.
[13] Poater, J.; Solà, M.; Duran, M.; Fradera, X., The calculation of electron localization and delocalization indices at the Hartree–Fock, density functional and post-Hartree–Fock levels of theory. *Theor. Chem. Acc.* 2002, **107**, 362-371.
[14] Heath, J. R., Molecular electronics. *Annu. Rev. Mater. Res.* 2009, 39 (1), 1-23.
[15] Li, Z.; Borguet, E., Determining charge transport pathways through single porphyrin molecules using scanning tunneling microscopy break junctions. *J. Am. Chem. Soc.* 2012, **134** (1), 63-66.
[16] Gehring, P.; Thijssen, J. M.; van der Zant, H. S., Single-molecule quantum-transport phenomena in break junctions. *Nat. Rev. Phys.* 2019, **1** (6), 381-396.

# Numerical simulation of aluminum deoxidization in the RH process

**Xinyu Zhao***

Dept. of Technology, Zenith Steel Group Co., Ltd., Changzhou, Jiangsu 213011 China

* Corresponding author: zhaoxinyu@zt.net.cn

**Abstract:** Metal aluminum is usually used to deoxidize as a deoxidant in RH reactor for producing ultra-low carbon Al-killed steel. Studying the process of aluminum deoxidization is of significance for forecasting the distribution of aluminum and optimization of deoxidization duration. An inhomogeneous two-phase model is used to simulate the RH circulating process. A multicomponent equation is used to simulate the diffuse process of elements, aluminum and oxygen. The control of the source term of the component equation is used to simulate the reaction process. Then the numerical model of aluminum deoxidization for producing ultra-low carbon Al-killed steel is established. The following results are concluded: the results of the simulation are accorded with the results of actual production. So the model is authentic and veracious. When the flux of lift gas is 96 $Nm^3/h$, the mass fraction of oxygen is exhausted to balance concentration about 200 s after aluminum is thrown into liquid steel. Then the reaction stops. The element aluminum is uniform about 500 s after aluminum is thrown into liquid steel. The maximum value of the mass fraction of aluminum decreases from 0.14% to 0.12% and the time of deoxidization decreases from 179.2 s to 149.4 s while the flux of lift gas increases from 72 $Nm^3/h$ to 120 $Nm^3/h$.

## 1. Introduction

The production route of ultra-low carbon aluminum killed steel by a domestic steel plant is BOF-RH-CC. RH stands for Vacuum Circulation Degassing Refining Method, which maintains a certain level of dissolved oxygen content in the molten steel after steelmaking in the converter, and decarburizes it through vacuum circulation in RH. After decarburization, there is still a certain amount of dissolved oxygen in the molten steel. For ultra-low carbon steel, if this oxygen is not removed, Fe FeO or FeO FeS low melting point eutectic crystals will appear at the grain boundaries when the steel solidifies. At this time, the steel is prone to thermal brittleness during rolling. Meanwhile, excessive dissolved oxygen content in steel can lead to an increase in the content of inclusions in the molten steel during subsequent processes, thereby affecting the purity of the final product. The commonly used deoxidation method currently is to add a deoxidizer - metallic aluminum - to the molten steel. Adding a certain amount of aluminum metal to the molten steel can achieve both deoxidation and alloying. For aluminum deoxidation, there is a reaction equation 2 [Al]+3 [O]=Al2O3. From the Gibbs free energy of the reaction, it can be obtained that at a steel temperature of 1873 K and a [% Al] of 0.01, the dissolved oxygen [% O] content in the steel is 0.0004. From this, it can be seen that Al is an extremely strong reducing agent, so using metallic aluminum for deoxidation can achieve lower dissolved oxygen content in a short period of time [1].

Studying the deoxidation process of RH aluminum is of great significance for predicting the distribution of metallic aluminum and optimizing deoxidation time, and can provide initial data for further research on the polymerization and removal of deoxidation products in molten steel. However, RH smelting is a black box process, and it is very difficult to directly observe and study an RH refining process that coexists with high temperature and chemical reactions. Therefore, numerical simulation methods have become an important means of studying the deoxidation process of RH smelting aluminum. At present, there is no reported mathematical model for the RH aluminum deoxidation process. In this paper, a non-homogeneous two-phase flow model is used to simulate the RH flow process, and a multi-component equation is used to simulate the diffusion process of aluminum and oxygen elements in molten steel. The source term of the component mass fraction is controlled to simulate the chemical reaction process. A mathematical model for the RH aluminum deoxidation process of ultra-low carbon aluminum killed steel is established, and the model is validated through actual production data. The influence of process parameters on the deoxidation process is studied by changing the operating conditions, providing theoretical data support for on-site operations.

## 2. Numerical model of reaction of aluminum and oxygen

### 2.1. Assumption

Due to the complexity of the RH decarburization process, establishing a mathematical model requires the following assumptions for the transport process [2][3]: (1) The molten steel is considered as a continuous incompressible Newtonian flow; (2) Bubbles are discrete phase particles of fixed size; (3) The fluctuation of the liquid level in the vacuum chamber is ignored; (4) The slag layer is ignored. The following assumptions are made about the chemical reaction process [4]: (1) Only chemical reactions between Al and O occur in the system, and the product is only $Al_2O_3$; (2) Al and O dissolve in the molten steel in the form of ions; (3) The reaction between surface slag and steel slag is ignored.

### 2.2. The equation of the chemical reaction

A non-homogeneous two-phase flow model is established to simulate the flow process of RH. RH flow is mainly achieved by blowing argon gas into the molten steel and using the density difference generated to promote the flow of the molten steel in the ladle and vacuum chamber. The fluid in this model is composed of molten steel and argon gas. The flow control model is shown in references [5]-[7]:

The equation of chemical reaction in RH is built as:

$$\frac{\partial \rho[\%Al]}{\partial t}+\nabla(\rho U_j[\%Al])-\nabla(\rho\Gamma_{Aleff}\nabla[\%Al])=-\rho\min([\%Al],\frac{[\%O]}{n}) \tag{1}$$

$$\frac{\partial \rho[\%O]}{\partial t}+\nabla(\rho U_j[\%O])-\nabla(\rho\Gamma_{Oeff}\nabla[\%O])=-\rho\min([\%O],[\%Al]*n) \tag{2}$$

$$\frac{\partial \rho[\%Al_2O_3]}{\partial t}+\nabla(\rho U_j[\%Al_2O_3])-\nabla(\rho\Gamma_{Al_2O_3eff}\nabla[\%Al_2O_3])=4(1+n)\min([\%Al],\frac{[\%O]}{n}) \tag{3}$$

where $[\%Al]$, $[\%O]$, $[\%Al_2O_3]$ are mass fractions of aluminum, oxygen, and aluminum oxide, $\Gamma_{Aleff}$, $\Gamma_{Oeff}$, and $\Gamma_{Al_2O_3eff}$ are effective diffusion coefficients of aluminum, oxygen, and aluminum oxide respectively, and n is the mass consumption ratio coefficient of aluminum and oxygen elements in the chemical reaction process.

### 2.3. Parameters of experiment

Based on the analysis of actual production data from a domestic steel plant, a deoxygenation mathematical model is established. The simulation parameters and their values are shown in Table 1.

**Table 1.** Parameters and values of simulation.

| Parameter | Value | Parameter | Value |
|---|---|---|---|
| Gas flow rate, $Nm^3/h$ | 72, 96, 120 | Mass fraction of oxygen in liquid steel, mass% | 0.0658 |
| Weight of liquid steel, t | 200 | Weight of aluminum, kg | 336 |

## 3. Results and discuss

### *3.1. Simulation result and verification*

The mass fraction of dissolved oxygen after decarburization of molten steel is 0.06579%, the amount of aluminum added is 336 kg, the gas flow rate is increased by 96 $Nm^3/h$, and the time of adding metallic aluminum is set to time 0. After the simulation is completed, the real-time mass fractions of aluminum and oxygen elements in the molten steel are compared with the actual measured values, and the results are shown in Figure 1. From the graph, it can be seen that with the addition of metallic aluminum, the mass fraction of aluminum element in the molten steel increases rapidly and reaches its peak. Then, as the chemical reaction progresses, the aluminum element is gradually consumed due to the chemical reaction, and the mass fraction of the aluminum element in the molten steel gradually decreases. When the mass fraction of dissolved oxygen in the molten steel is below 0.0004%, the aluminum oxygen reaction reaches equilibrium, the chemical reaction is in equilibrium, and the aluminum element remains at a relatively stable level after a certain equilibrium fluctuation. At this point, comparing the equilibrium mass fraction of aluminum element in the simulation results with the measured values in the molten steel, it is found that the simulation results are highly consistent with the measured values. At time 0, the mass fraction of dissolved oxygen in the molten steel is 0.06579%. When aluminum metal is added, the dissolved oxygen in the molten steel is rapidly consumed by chemical reactions, but its consumption rate gradually decreases with the decrease of dissolved oxygen mass fraction. When the mass fraction of dissolved oxygen in the molten steel is lower than 0.0004%, the chemical reaction reaches an equilibrium state. After the equilibrium fluctuation, the mass fraction of dissolved oxygen in the molten steel reaches an equilibrium value. At this point, comparing the mass fraction of dissolved oxygen in the simulation results with the measured values in the molten steel, it is found that the simulation results are highly consistent with the measured values. From this, it can be seen that the mathematical model of RH aluminum deoxidation is real and reliable, and can be used to predict the aluminum deoxidation process of ultra-low carbon aluminum killed steel RH smelting.

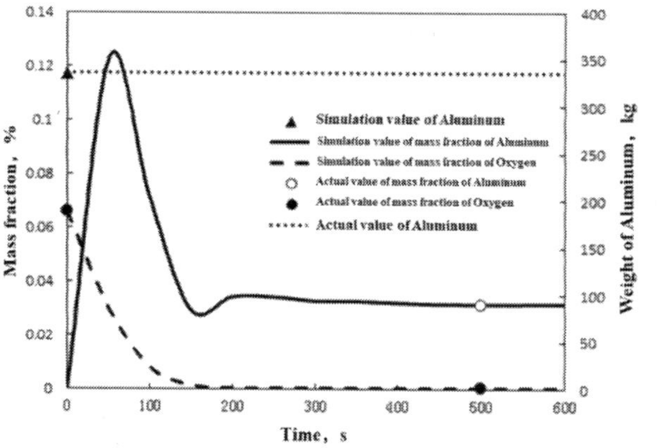


**Figure 1.** Curve of comparison between simulation and measured results.

The cloud map of the mass fraction distribution of dissolved oxygen in the molten steel at different times after aluminum addition is shown in Figure 2. At the 3-second time of aluminum addition, the

Journal of Physics: Conference Series **2961** (2025) 012019 doi:10.1088/1742-6596/2961/1/012019

aluminum element is enriched near the free liquid surface in the vacuum chamber, and the dissolved oxygen at this position is first consumed. As the aluminum element diffuses in the molten steel under cyclic driving, a chemical reaction occurs between the aluminum and oxygen elements. In the area where the aluminum and oxygen elements coexist, the oxygen element is consumed to the equilibrium concentration. At the 200 s time after aluminum addition, the metallic aluminum is basically transported to the entire container, and the chemical reaction is basically stopped. The average mass fraction of dissolved oxygen drops to 0.0004% of the equilibrium concentration.

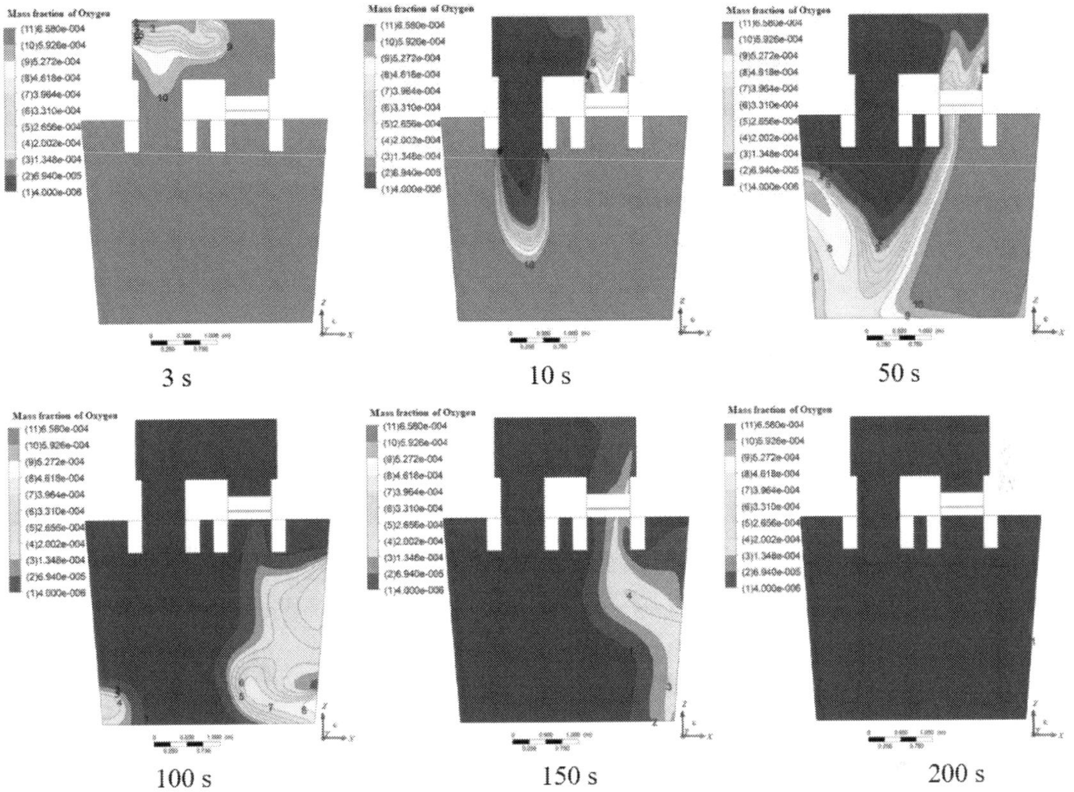


**Figure 2.** Contour of mass fraction of dissolved oxygen at various moments.
a) 3 s b) 10 s c) 50 s d) 100 s e) 150s f) 200 s

### *3.2. Effect on the mass fraction of gas flow rate*

When the mass fraction of dissolved oxygen after decarburization of molten steel is 0.06579% and the amount of aluminum added is 336 kg, the gas flow rates are changed to 72 $Nm^3/h$, 96 $Nm^3/h$, and 120 $Nm^3/h$, respectively, to study the effect of different gas flow rates on the aluminum deoxidation process. The comparison curve of the average mass fraction of aluminum element in the molten steel under different gas flow rates is shown in Figure 3. It can be seen from the figure that at time 0, aluminum is added to the molten steel, and the mass fraction of the aluminum element in the molten steel is assumed to be zero. With the addition of aluminum element, the average mass fraction of aluminum element in the molten steel begins to increase rapidly and reaches its maximum value. From the comparison of different gas flow rates, as the gas flow rate increases, the peak value of the aluminum element gradually decreases. When the gas flow rate increases from 72 $Nm^3/h$ to 120 $Nm^3/h$, the maximum value of the average mass fraction of metal aluminum element in the molten steel decreases from 0.14% to 0.12%. This is because the diffusion of the aluminum element in the molten steel accelerates with the increase of gas flow rate. The completion time of the aluminum oxygen reaction decreases, the consumption of aluminum element accelerates, and the maximum mass fraction of aluminum element in the molten steel decreases. Due to the fixed initial dissolved oxygen content and aluminum addition in this operating condition, although the diffusion time of aluminum element is different under different gas flow rates, the average mass fraction of aluminum element in the steel liquid at equilibrium is the same, which is around 0.0318%.

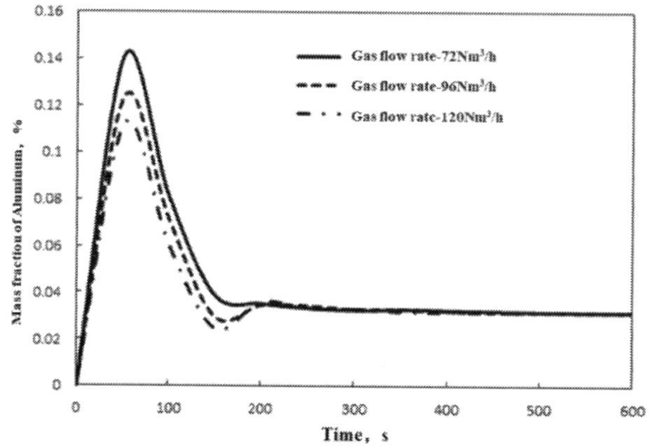


**Figure 3.** Curve of comparison of mass fraction of aluminum at various flux of lift gas.

The time when the average mass fraction of dissolved oxygen in the molten steel reaches the equilibrium value of 0.0004% is defined as the deoxidation time. The deoxidation time under different gas flow rates is taken out, as shown in Figure 4. It can be seen from the figure that when the gas flow rate increases from 72 $Nm^3/h$ to 120 $Nm^3/h$, the deoxidation time decreases from 179.2 s to 149.4 s. It can be seen that as the gas flow rate increases, the deoxygenation time decreases and the deoxygenation efficiency improves.

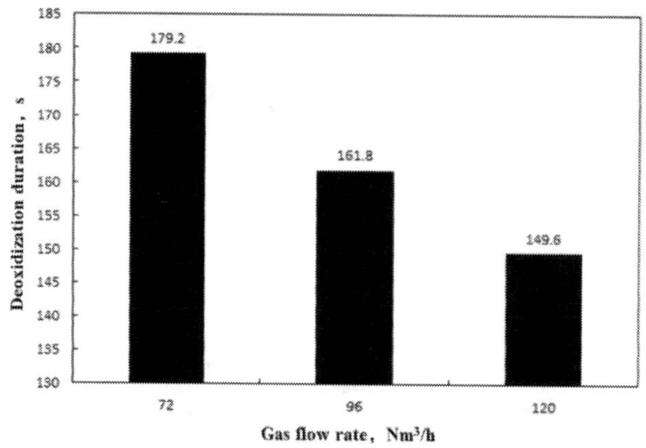


**Figure 4.** Curve of comparison of deoxidization time at various flux of lift gas.

## 4. Conclusions

A mathematical model for the RH deoxidation process of ultra-low carbon aluminum killed steel is established, validating the model through actual production data, and studying its effects on the deoxidation process and aluminum diffusion process by changing the gas flow rate. The following conclusions are drawn:

(1) Using a non-homogeneous two-phase flow model to simulate the RH flow process, a multi-component equation is used to simulate the diffusion process of aluminum and oxygen elements in molten steel, and the source term of the component mass fraction is controlled to simulate the chemical reaction process. A mathematical model of the ultra-low carbon aluminum killed steel RH

aluminum deoxidation process is established, and the model is validated through actual production data. The results show that the simulated data is in good agreement with the actual production data, and the model is authentic and reliable.

(2) According to the cloud map analysis of aluminum and oxygen elements during the simulation process, the aluminum deoxidation process is as follows: metal aluminum is added from the vacuum chamber, and under the action of flow, the metal aluminum first enters the ladle through the downcomer with the molten steel, and then diffuses inside the Baosteel. After about 200 seconds of aluminum addition, the oxygen element in the molten steel is consumed to 0.0004% of the equilibrium concentration, and the chemical reaction stops. The average mass fraction of the aluminum element does not change, but it still diffuses uniformly with the flow in the molten steel. After about 500 seconds of aluminum addition, the aluminum element in the molten steel is basically in a uniform state. At this time, the maximum mass fraction of aluminum element is 0.0360%, and the minimum value is 0.0312%.

(3) Increasing the gas flow rate accelerates the diffusion of aluminum elements in the molten steel, reduces the completion time of the aluminum oxygen reaction, accelerates the consumption of aluminum elements, and reduces the maximum average mass fraction of aluminum elements in the molten steel. When the gas flow rate is increased from 72 $Nm^3/h$ to 120 $Nm^3/h$, the maximum average mass fraction of aluminum elements in the molten steel decreases from 0.14% to 0.12%. At the same time, under other unchanged operating conditions, changing the gas flow rate without changing the average mass fraction of aluminum element in the molten steel at equilibrium state is around 0.0318%.

(4) As the gas flow rate increases, the deoxygenation time decreases and the deoxygenation efficiency improves. When the gas flow rate is increased from 72 $Nm^3/h$ to 120 $Nm^3/h$, the deoxygenation time decreases from 179.2 s to 149.4 s

**References**

[1] Hanjie G. Metallurgical physical chemistry course. Beijing: Metallurgical Industry Press, 2004
[2] Kensuke T, Iori S, Yoshihiko H, Numerical analysis for the interaction of fluid and sphere penetrating into a liquid bath. ISIJ Inter. 2023, 63(2): 244.
[3] Liu C. Investigation on multiphase fluid flow, mixing, and decarburization during RH vacuum refining process. University of Science and Technology Beijing, 2021,
[4] Kwon Y J, Zhang J, Lee H G. A CFD-based nucleation-growth-removal model for inclusion behavior in a gas-agitated ladle during molten steel deoxidation. ISIJ Int. 2008, 48(7): 891
[5] Jiongming Z, Liang L, Xinyu Z, et al. A mathematical model for decarburization process in RH refining process. ISIJ International, 2014, 54(7):1560
[6] Sha J, Kaijun N, Alberto N. Multiphase modeling of steel-slag mass transfer through Distorted Interface in Bottom-stirred Ladle, ISIJ International, 2024, 64(1):52
[7] Wang J, Zhang L, Ren Y. Advances in the thermodynamics and kinetics of the steelmaking process and their application in numerical simulation. Chinese Journal of Engineering, 2024, 46(11): 1960

# Bimetallic MOF doped with reduced graphene oxide enables rapid detection of chloramphenicol

**Youheng Pan[1], Tianyi Zhang[1] and Lanlan Li[1, *]**

[1]College of Mechanical and Electrical Engineering, Henan Agricultural University, Zhengzhou, 450002, China

[*]Corresponding author: lanlan.li@henau.edu.cn

**Abstract:** A broad-spectrum antibiotic called chloramphenicol (CAP) stops the creation of bacterial proteins, which stops bacteria from growing and reproducing. When chloramphenicol is misused, it can cause major side effects, secondary infections, bacterial resistance, delays in effective treatment, and higher treatment costs. As a result, creating an electrochemical sensor that is quick and sensitive is crucial. The electrocatalytic activity of CAP determination was enhanced in this work by producing reduced graphene oxide doped bimetallic organic framework (Fe/Al-MOF/rGO) nanocomposites in a single pot, which had more active sites and better electrical conductivity. The constructed sensor has a linear range of 20 - 1600 μM and a detection line (LOD) of 1.5 nM (S/N=3). Furthermore, the sensor's recovery rate for sample detection is good enough.

## 1. Introduction

A common broad-spectrum antibiotic in clinical settings is chloramphenicol (CAP)[3]. With the increasing awareness that chloramphenicol may cause severe adverse reactions such as aplastic anemia, many countries have gradually restricted or banned the use of chloramphenicol. As a result, developing a sensitive, quick, and accurate approach for detecting traces of CAP is crucial. Electrochemical detection stands out because it has the advantages of dependability, fast reaction, high sensitivity, easy operation, and low cost. Traditional detection methods have numerous steps, high cost, lengthy time, and other drawbacks, making it impossible to perform rapid detection.

A novel class of crystalline porous materials is known as metal-organic framework (MOF) materials. Coordination interactions between organic ligands and metal ions or clusters result in the formation of MOFs. Monomeric MOFs feature uniform porosity, a high specific surface area, low density, and strong thermal stability. Shu et al. reported a Cu-MOF for the detection of Staphylococcus aureus[5]. In contrast, binuclear MOFs contain different REDOX states of metals as active reaction sites, typically exhibiting a larger specific surface area, higher void volume, and more carbon matrix. Mukundan synthesized a NiCo-MOF hydrogel material for the detection of urea in human tears[4]. However, most MOFs have poor electrical conductivity, limiting their direct application as electrocatalysts. In this case, the electrical conductivity of the MOF material is usually improved by doping phosphide, transition metal sulfides, carbon-based materials, or metal nanoparticles. Graphene is a hexagonal lattice of carbon atoms, which has high electrical conductivity, high strength, lightweight, and excellent thermal conductivity. It is widely used in electronics, energy storage, and composite materials.

In this study, Fe/Al-MOF/rGO composites were successfully prepared and characterized by various analyses and spectra. The sensing properties of CAP were studied using the electrochemical method.

The results show that the material has good selectivity, repeatability, stability, and satisfactory recovery rate in practical samples.

## 2. Experimental

### *2.1. Synthesis of rGO*

GO (5mg) was dissolved in 100 mL of deionized water (DW) and then underwent ultrasound for 3 hours. To reduce the suspension, 35μL of ethylenediamine was added, followed by 5 hours of reflux in a water bath at 90°C[1]. After washing and drying, 22.5mg of rGO was obtained.

### *2.2. Synthesis of Fe/Al-MOF*

$Fecl_3.6H_20$ (0.3784g) and $NH_2BDC$ (0.2540g) were dissolved in DMF (35mL), underwent ultrasound for 30min, and were labeled as liquid A. Al $(NO_3)_3.9H_20$ (0.5224g) and 2,6-naphthalene dicarboxylic acid (0.2594g) were dissolved in DMF (30mL), underwent ultrasound for 30min, and labeled as liquid B. Mix liquids A and B thoroughly. Then, they were transferred to a Teflon liner and heated at 125°C for 24h. After cooling to ambient temperature, the products were washed with DMF, DW, and ethanol, and then vacuum-dried at 60°C for 24h.

### *2.3. Synthesis of Fe/Al-MOF/rGO*

The preparation method of Fe/Al-MOF/rGO is similar to that of Fe/Al-MOF. rGO (0.294g) was mixed with liquids A and B, and the mixture was transferred to a Teflon liner after ultrasound and heated at 130°C for 24 h. After cooling to ambient temperature, we wash and centrifuge with DMF, DW, and ethanol 3 times, and dry at 60°C for 24h.

### *2.4. Modification of electrode*

The polished GCE was ultrasonically cleaned with ethanol and distilled water before being treated further. The Fe/Al-MOF (5mg) was equally dissolved in DW (5mL), and the mixture (5μL) was applied to the GCE, which was subsequently labeled as Fe/Al-MOF/GCE after natural air drying for 30 minutes. Using the same approach, Fe/Al-MOF/rGO/GCE were prepared.

## 3. Results and discussion

### *3.1. Characterization of Fe/Al-MOF/rGO nanocomposites*

As shown in Figure 1 (a), the spindle shape of some Fe-MOF can be observed, while the binuclear MOF shows irregular particle shape, and Fe/Al-MOF shows disordered agglomerated particles, which increases the surface area and is conducive to the adhesion of the active site. As shown in Figure 1 (b), the prepared Fe/Al-MOF/rGO presented wrinkled nanosheets due to the doping of rGO, and Fe/Al-MOF was uniformly distributed on the surface of rGO. XPS testing of Fe/Al-MOF/rGO nanocomposites reveals the elements contained (Figure 1c). The high-resolution fe2p spectra of Fe-MOF have two peaks, fe2p3/2 and fe2p1/2(Figure 1d). The Al 2p peak confirms the successful synthesis of Al-MOF (Figure 2e).

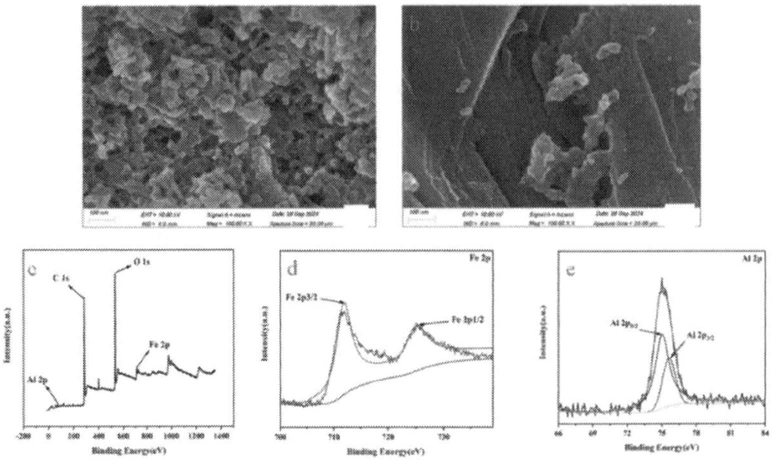


**Figure 1**. SEM images of Fe/Al-MOF(a) and Fe/Al-MOF/rGO (b). XPS testing of Fe/Al-MOF/rGO(c), Fe 2p(d) and Al 2p(e).

### *3.2. Electrochemical determination of CAP*

Figure 2a shows the cyclic voltammetry (CV) of GCE modified with different materials in 0.1M PBS solution (pH=7). Due to the excellent electrical conductivity of rGO, Fe/Al-MOF/rGO/GCE has the largest response compared with Fe/Al-MOF/GCE. The relationship between analyte concentration and current response is an essential factor in the manufacture of electrochemical sensors. Differential Pulse Voltammetry (DPV) technology is a commonly used analytical method in electrochemical testing, which is mainly used to measure and analyze the reaction of the electrode surface and the concentration of electrochemically active substances. Figure 2b shows DPV measurements when 50 - 1550μM CAP solution was successfully added to 0.1M PBS (pH=7) solution. Figure 2 (b, c) With the increase of CAP concentration, the current also increases, and the peak current is positively correlated with the reactant concentration. The linear range of CAP detection is 20 -1600μM, and LOD is 1.5nM (S/N=3).

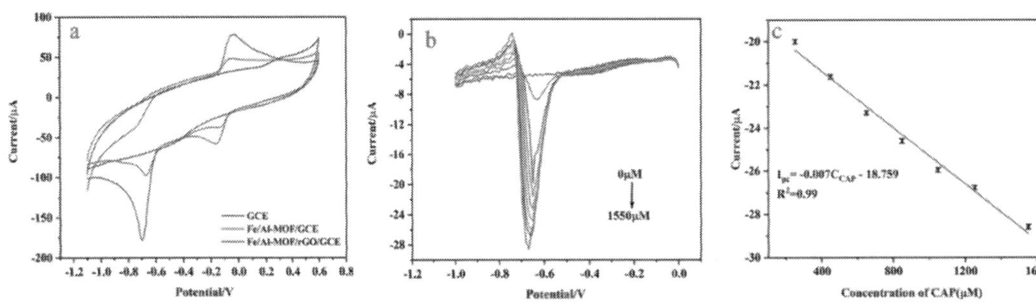


**Figure 2.** (a) CVs of GCE, Fe/Al-MOF and Fe/Al-MOF/rGO. (b、c) DPV curves and corresponding linear fitting curves of CAP detection.

### *3.3. Selectivity, reproducibility and stability*

The stability, repeatability, and anti-interference of electrochemical sensors are very important for practical sample detection. Therefore, we tested the detection effect of one electrode for 14 consecutive days, and the relative error (RSD) was 2.5%, indicating that the sensor had good stability (Figure 3a). In addition, five electrodes were selected to study the repeatability of the sensor (Figure 3b), whose RSD was 1.59%, indicating that the sensor had good repeatability [2]. A 0.1M PBS solution containing 100μM CAP (pH=7) was configured, and $Na^+$, $K^+$, $Ca^{2+}$, $CO_3^{2-}$, $Cl^-$, $H_2PO_4^-$, $SO_4^{2-}$, $NO_2^-$ and tetracycline (TC) with 30 times the concentration of interfering ions were added. When the relative signal changes within 10%, the effect of interfering ions can be ignored (Figure 3c).

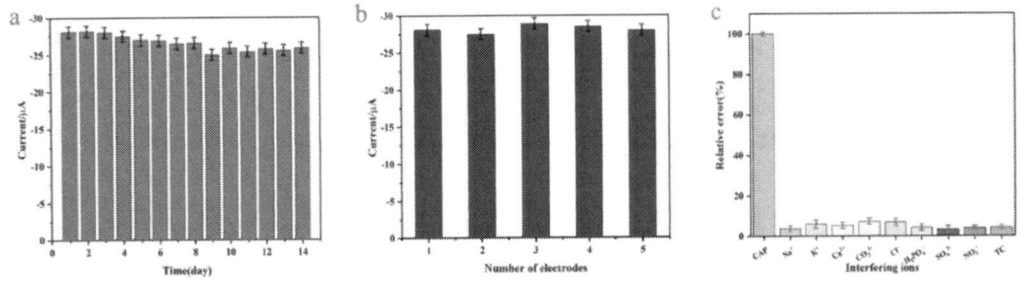


**Figure 3.** (a) The stability for two consecutive weeks of detection. (b) 5 different electrodes were used to test CAP. (c) The selectivity of CAP detection with different interfering ions.

### *3.4. Real sample analysis*

Real sample tests were conducted to demonstrate the practical applicability of the Fe/Al-MOF/rGO/GCE. A certain volume of lake water was taken and filtered to remove impurities, then mixed with a PBS solution. Different concentrations of CAP solution were added, and the current intensity increased with the concentration. The results are shown in Table 1. The Fe/Al-MOF/rGO electrochemical sensor exhibited good recovery rates, indicating its effectiveness for detecting CAP in real water samples.

**Table 1.** Detection of CAP in actual water samples.

| Sample number | Actual sample (μM) | Fe/Al-MOF/rGO/GCE(μM) | Recovery (%) |
|---|---|---|---|
| 1 | 50 | 47.16 | 94.3 |
| 2 | 80 | 77.89 | 97.4 |
| 3 | 120 | 118.45 | 98.7 |

## 4. Conclusion

In this study, we prepared a binuclear metal-organic framework electrochemical sensor doped with reduced graphene oxide for the detection of chloramphenicol in water. Graphene has good electrical conductivity, the metal-organic frame has a larger surface area, and the prepared sensor has both advantages, improving the detection performance. The linear range is 20 - 1600uM, the detection limit is 1.5nM, and the sensitivity is high. In addition, the method is also applied to the actual sample detection with a good recovery rate. In conclusion, the modified MOF has better performance and is more conducive to environmental monitoring of chloramphenicol.

**Acknowledgments**
The Top Talent Program of Henan Agricultural University.

**References**

[1]Ali M, Alotaibi B M, Alrowaily A W, Alyousef H A, Al-Sehemi A G, Henaish A M A and Ahmad K (2024) Construction of efficient SrMnO3@rGO 3 @rGO nanohybrid as an electrode material for improved electrocapacitive performance. Journal of Energy Storage, 97: 112901.
[2]Elanthamilan E and Wang S-F (2024) Flower-like 3D SnS decorated on nickel metal-organic framework for electrochemical detection of dimetridazole in food samples. Food Chemistry, 452: 139575.
[3]He G, Zhao S, Yang C, Chen L, Liu Y, Hu Q and Yang Y (2024) Point-of-care monitoring of milk quality by rapid Immunofluorescence with mechanical deformation of the hydrogel microspheres. Sensors and Actuators B-Chemical, 417: 136160.

Journal of Physics: Conference Series **2961** (2025) 012020 doi:10.1088/1742-6596/2961/1/012020

[4]Mukundan G and Badhulika S (2024) Nickel-cobalt metal-organic frameworks based flexible hydrogel as a wearable contact lens for electrochemical sensing of urea in tear samples. Microchimica Acta, 191: 252.

[5]Wang S, Deng W, Yang L, Tan Y, Xie Q and Yao S (2017) Copper-Based Metal Organic Framework Nanoparticles with Peroxidase-Like Activity for Sensitive Colorimetric Detection of Staphylococcus aureus. Acs Applied Materials & Interfaces, 9 (29): 24440-24445.

# Sterilization of escherichia coli in cotton fabric by supercritical carbon dioxide fluid

**Jun Yan[1,*], Yitong Chen[1], Huimin Zhang[1,*], Hong Li[1], Yongping Liao[1] , Bing Du[1] and Dawei Gao[2]**

[1] School of Textile and Material Engineering, Dalian Polytechnic University, Dalian Liaoning, 116034, China.
[2]School of Textile and Clothing, Yancheng Institute of Technology, Yancheng, Jiangsu, 22400, China.

[*]Corresponding author's e-mail: yanjun@dlpu.edu.cn
[*]Corresponding author's e-mail: zhanghm@dlpu.edu.cn

**Abstract:** Sterilizing clothing is essential due to the health risks posed by bacteria, which thrive in fabrics. Traditional sterilization methods have limitations, making supercritical carbon dioxide ($scCO_2$) an innovative alternative. This technology effectively inactivates microorganisms at lower temperatures and moderate pressures without generating waste. Using first-order inactivation kinetics, results show that increased temperature and pressure lead to higher E. coli mortality rates. The stages of the deactivation process can be divided into two stages: a rapid phase lasting 1 to 15 minutes, followed by a slower phase after 15 minutes. Performance tests reveal that $scCO_2$-treated cotton fibers exhibit greater surface roughness, fracture strength, and elongation, though their structure becomes denser and permeability decreases. Under optimal conditions of 50°C, 12 MPa pressure, and 25 minutes, the inactivation logarithm reaches 6.7 orders of magnitude, suggesting a promising new method for fabric sterilization.

## 1. Introduction

Textiles play a crucial role in our daily lives, with cotton fabrics being the most prevalent [1]. Cotton fibers primarily consist of cellulose, which provides an ideal environment for microorganisms, allowing bacteria to thrive on the fabric [2]. This poses a significant risk to human health. Common sterilization methods for textiles include dry heat [3], steam [4], ultraviolet [5], and ethylene oxide [6] sterilization. However, these conventional techniques often come with drawbacks such as toxicity, lengthy processing times, and high costs. There is a need for eco-friendly, and resultful disinfection technologies in the textile industry.

Supercritical carbon dioxide ($scCO_2$) has emerged as a promising solution [7]. It is a high-density fluid that combines the diffusion properties of both gases and liquids, and it has been used in the research of sterilization process [8]. Previous research by Mehmet Orhan et al. [9] demonstrated the effectiveness of $scCO_2$ in inactivating microbes on polyester fabrics. They synthesized acrylamide and treated the polyester at 120°C and 30 MPa for 6 hours, achieving significant antibacterial properties. Additionally, Aslanidou et al. [10] investigated $scCO_2$'s effectiveness on silk fabrics, finding that at 40°C and 15 MPa, with the addition of water and 5% $Ca(OH)_2$, they achieved a 97% inactivation rate. These studies illustrate that $scCO_2$ can serve as an effective alternative for achieving complete bacterial spore inactivation on both polyester and silk textiles.

## 2. Experiments

### *2.1. Preparation before operation*
A nutrient medium was prepared using yeast powder, NaCl, and tryptone in ultrapure water, adjusted to pH 7.4, and solidified with 15% agar. After culturing E. coli for 12-14 hours, the bacterial solution was diluted and plated. Cotton fabrics were sterilized in an autoclave, dried, and inoculated with the bacterial solution.

### *2.2. scCO$_2$ sterilization treatment*
The reactor was disinfected with alcohol before loading the bacterial cotton samples. Experimental parameters included pressures from 8 to 12 MPa and temperatures from 30 to 50°C, with reaction times of 5-30 minutes. After treatment, samples were analyzed for bacterial counts compared to a control group.

## 3. Results and analysis

### *3.1. The influence of pressure on sterilization*
The bactericidal effect of supercritical carbon dioxide (scCO2) on E. coli in cotton fabrics was studied at pressures from 7 MPa to 14 MPa, at a temperature of 30°C for 10 minutes. Results, shown in Figure 1, indicated that at pressures from 7 to 14 MPa at 30°C for 10 minutes, increased pressure significantly improved sterilization rates, showing a transition from a slow to a fast inactivation phase. The rapid sterilization between 10 and 12 MPa was due to enhanced mass transfer and $CO_2$ solubility.

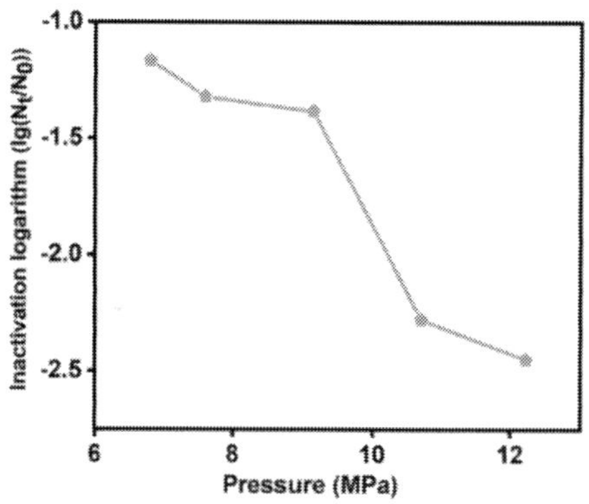


**Figure 1.** Pressure on the inactivation logarithm at 30°C and 10 min.

### *3.2. The influence of temperature on sterilization*
Temperature affects the mass transfer rate of $CO_2$ and the dissolution characteristics of $CO_2$. Therefore, the effect of scCO$_2$ pressure at 12 MPa, temperature ranging from 30°C to 50°C, and reaction time at 10 min on the bactericidal effect of cotton fabric E. coli was studied. Figure 2 shows that, at 12 MPa, temperatures from 30°C to 50°C also enhanced bactericidal effects, with the most significant inactivation occurring as temperature increased, suggesting a reduction in E. coli tolerance.

IOP Publishing
doi:10.1088/1742-6596/2961/1/012021

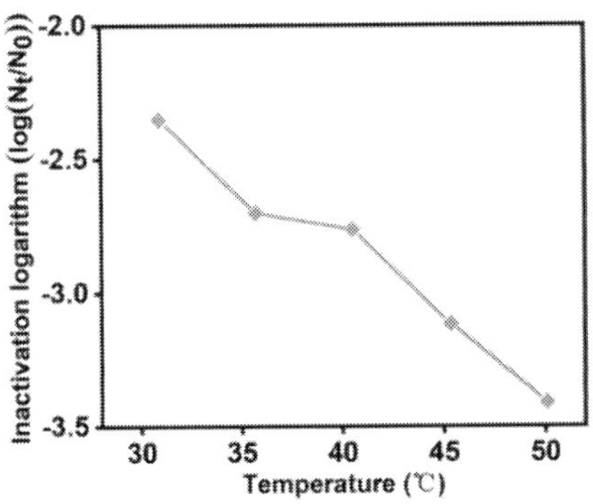


**Figure 2.** The temperature on the inactivation logarithm at 12 MPa and 10 min.

*3.3. Kinetic analysis of E. coli in $scCO_2$-Treated cotton fabrics*

*3.3.1. Survival logarithm and treatment time at 7 MPa.* Figure 3 shows the change in the survival logarithm of *E. coli* with different temperatures at subcritical $CO_2$ conditions and 7 MPa. With the increase of time, the survival logarithm of *E. coli* was decreased. The sterilization process at different temperatures could be divided into two stages. At the pressure of 7 MPa, when the temperature was increased from 30°C to 50°C, the sterilization rates in the first stage increased from 0.2945 $min^{-1}$ to 0.5100 $min^{-1}$, and the D value decreased from 7.819 min to 4.515 min. In the second stage, the sterilization rates were increased from 0.0838 $min^{-1}$ to 0.2211 $min^{-1}$, and the D value decreased from 27.482 min to 10.415 min. Under the condition of 7 MPa and 50°C, $scCO_2$ can reduce the sterilization of *E. coli* by 4.88 orders of magnitude.

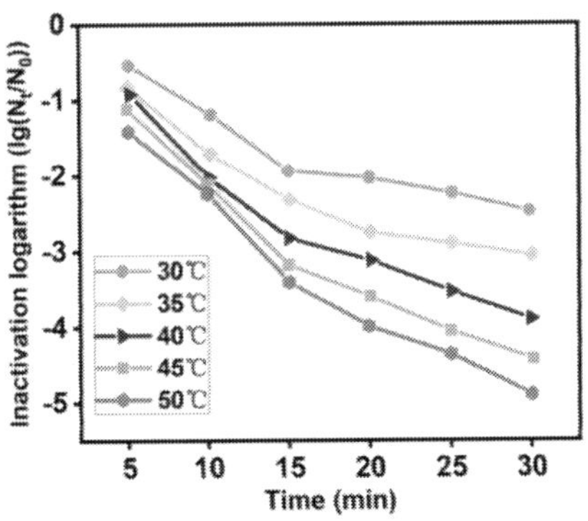


**Figure 3.** Inactivation logarithm of *E. coli* with different temperatures at 7 MPa.

*3.3.2. Relationship between E. coli survival logarithm and treatment time at 12 MPa.* Figure 4 depicts the survival logarithm of E. coli at 12 MPa. The logarithmic survival declined significantly over time. The temperature increased from 30°C to 50°C, the first-stage sterilization rate increased from 0.444 $min^{-1}$ to 0.759 $min^{-1}$, while the D value decreased from 5.178 min to 3.030 min. In the second stage, the sterilization rate rose from 0.0992 $min^{-1}$ to 0.244 $min^{-1}$, and the D value decreased from 23.212 min to 9.40 min. Under 12 MPa and 50°C, $scCO_2$ reduced E. coli levels by 6.69 logarithmic levels, which is 1.81 levels more than at 7 MPa. The heat destroyed the bacterial cell structure and denatured enzymes, leading to bacterial death.

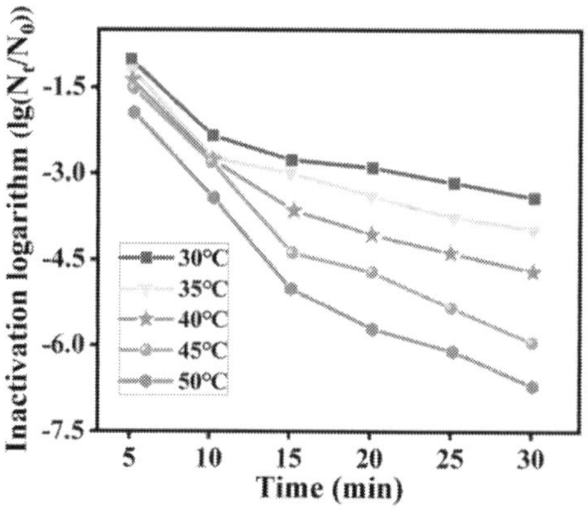


**Figure 4.** Inactivation logarithm of *E. coli* with different temperatures at 12 MPa.

### *3.4. Fabric testing*

*3.4.1 Microstructure of fabric.* Figure 5 shows the images of untreated and $scCO_2$-treated cotton fibers. Untreated fibers had a smooth surface with some granules and shallow grooves. Post-treatment, the fibers exhibited increased surface roughness with more vertical grooves, small bumps, and folds. This alteration was due to increased pressure precipitating zwitterions on the fiber surface and enhanced molecular movement from elevated temperatures.

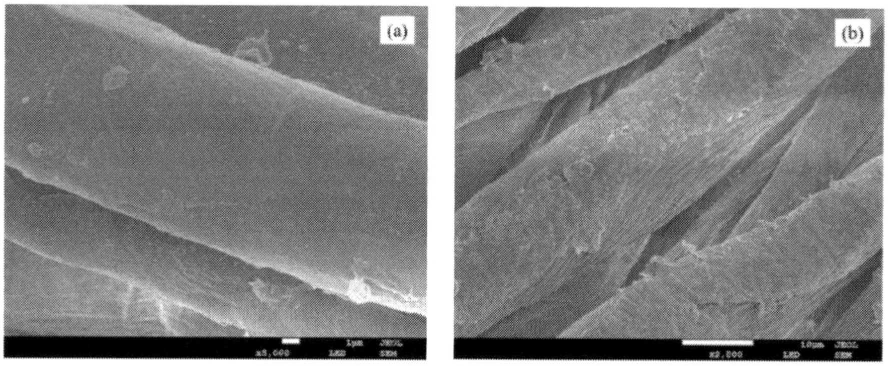


(a) Untreated fabrics (b) Treated fabrics

**Figure 5.** SEM image of cotton fibers before and after $scCO_2$ fluid sterilization treated.

*3.4.2 FTIR analysis.* Figure 6 presents the infrared spectra of cotton fabrics before and after $scCO_2$ treatment. Characteristic absorption peaks of cotton were observed, including C-O-C stretching (1106 $cm^{-1}$), C=O stretching from lignin (1627 $cm^{-1}$), -CH stretching (2889 $cm^{-1}$), and -OH stretching (3800–400 $cm^{-1}$). While some molecular-level changes occurred, they were minimal and did not significantly impact the cotton fabric's performance. This indicates that $scCO_2$ treatment effectively sterilized without compromising the material's quality.

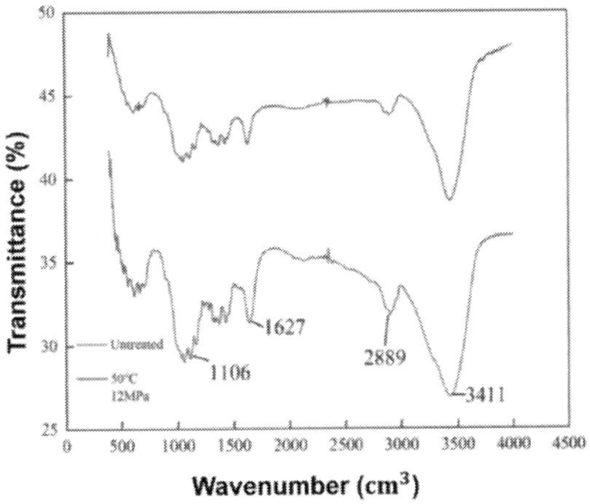


**Figure 6.** FTIR spectra of cotton fabric specimens before and after $scCO_2$ fluid sterilization treatment.

## 4. Conclusion

In conclusion, this study highlighted the potential of supercritical carbon dioxide ($scCO_2$) as an effective sterilization method for E. coli on cotton fabric samples. The investigation systematically examined the impact of various treatment conditions, specifically temperature, pressure, and time, on the bactericidal efficacy of $scCO_2$. The results clearly indicated that the optimal sterilization conditions were achieved at a temperature of 50°C, a pressure of 12 MPa, and a treatment duration of 25 minutes, which resulted in an impressive 6.7-log inactivation of E. coli. The enhanced bactericidal effect observed at higher pressures and temperatures can be attributed to improved solubilization of $CO_2$, allowing for deeper penetration into the bacterial cell walls. This facilitates a more effective disruption of cellular integrity, ultimately leading to greater inactivation rates. Furthermore, the established bactericidal kinetics within the ranges of 30-50°C, 7-12 MPa, and 5-30 min underscore the robustness of $scCO_2$ as a sterilizing agent. Specifically, the study demonstrated a significant increase in the death rate of E. coli, from 0.2945 $min^{-1}$ to 0.795 $min^{-1}$, along with a notable decrease in the D value from 7.819 min to 3.03 min, corresponding to the elevated temperature and pressure conditions. These findings not only reinforce the viability of $scCO_2$ as an alternative to traditional sterilization methods but also pave the way for further research into its application across various materials and pathogens. The ability to effectively sterilize textiles without the use of harsh chemicals presents a compelling opportunity for the development of safer and more sustainable sterilization practices in the textile industry and beyond. Future studies should explore the scalability of this process, its impact on fabric integrity, and its effectiveness against a broader spectrum of microbial contaminants. Overall, the promising results of this research suggest that $scCO_2$ could play a significant role in advancing sterilization technology in both industrial and clinical settings.

## References

[1] Dan Z, Guolin Z, Chuang Z, Yuhe W, and Zhu L. (2019) Preparation and characterization of wear-resistant superhydrophobic cotton fabrics. Progress in Organic Coatings. 134:226-33.

[2] Li L, Zhou D, Zhu Y, Zhang T, Zhang Z, and Jing L. (2024) Multifunctional antibacterial cotton fabrics based on a triazole-halamine-diethyl phosphite. Industrial Crops and Products. Dec 15;222:119571.

[3] Oh C, Araud E, Puthussery JV, Bai H, Clark G G, Wang L, Verma V and Nguyen TH. (2020) Dry heat as a decontamination method for N95 respirator reuse. Environmental Science & Technology Letters. 7(9):677-82.

[4] Bosman A, van Kemenda D, van Wezel R A, Buijs J G, Bulkmans D A, van Rosmalen-de Widt A M, Zwinkels W S, Steegh P, Mulders M, Plutz C and Waegemans M. (2023) Improving parametric load release for steam sterilization. The Journal of Hospital Infection. S0195-6701.

[5] Liu Y, Wu X, Liu R, Chen Y., Fu J, and Ou H. (2022) Modifications of ultraviolet irradiation and chlorination on microplastics: Effect of sterilization pattern. Science of the Total Environment, 812, 152541.
[6] Horakova J, Mikes P, Saman A, Jencova V, Klapstova A, Svarcova T, Ackermann M, Novotny V, Suchy T and Lukas D. (2018) The effect of ethylene oxide sterilization on electrospun vascular grafts made from biodegradable polyesters. Materials Science and Engineering: C. 92:132-42.
[7] Abou Elmaaty T, Abd El-Aziz E. (2018) Supercritical carbon dioxide as a green media in textile dyeing: a review. Textile Research Journal. 88(10):1184-212.
[8] Perrut, Michel. Sterilization and virus inactivation by supercritical fluids (a review). (2012) The Journal of Supercritical Fluids. 66: 359-371.
[9] Mehmet O, Fatma D, Hasan B K and Vincent N. (2020) Supercritical Carbon Dioxide Application Using Hydantoin Acrylamide for Biocidal Functionalization of Polyester. The Journal of Supercritical Fluids. 10(4):77-83.
[10] Aslanidou D, Tsioptsias C, Panayiotou C. (2013) A novel approach for textile cleaning based on supercritical $CO_2$ and Pickering emulsions. The Journal of Supercritical Fluids. 76: 83-93.

Journal of Physics: Conference Series **2961** (2025) 012022 doi:10.1088/1742-6596/2961/1/012022

# Quantitative analysis of slurry cement powder injection based on microwave plasma torch

**Zongjun Zhu** [1]**, Haoze Wei** [2]**, Dengjie Yu** [2]**, Wei Jin** [2] **and Bingwen Yu** [*2]

[1] Huzhou University, Zhejiang 31300, China
[2] Huzhou Research Institute, Zhejiang University, Zhejiang 310058, China

[*]Corresponding author's e-mail: yubingwen007@126.com

**Abstract**: The elements of Mg, Al, Ca, and Si in cement were quantitatively analyzed by slurry powder injection microwave plasma torch emission spectrometry (MPT-OES). Through the comparative test of powders with different particle sizes, 40 μm powder was selected to configure 0.05% m/v slurry, and 0.01% m/v glycerol + 1% v/v HCl was selected as a dispersant. Based on the principle of the Babington nebulizer, a slurry nebulizer with a good atomization effect was designed and fabricated to meet the experimental conditions, and the atomization efficiency was 3.28%. The slurry was stabilized by ultrasonic shaking for 20 min before injection, and the MPT light source was introduced through the nebulization chamber through a peristaltic pump (0.2 mL/min). The oxygen shielding gas was 1.2 L/min, the argon maintenance gas was 0.6 L/min, and the argon carrier gas was 0.2 L/min. The mass concentrations of Mg, Al, Ca, and Si were 2.85~10.92, 17.96~34.76, 192.11~211.00, and 9.38~54.81 μg/mL, respectively, and the intensity in the concentration range was linear. The linear correlation coefficients were greater than 0.997, and the relative standard deviations were 1.14%-4.56% (n=50). The results show that this method has a good detection effect under different powers, and has a good application prospect in the online detection of solid powders.

## 1. Introduction

In cement production, the control of ingredient quality is crucial for product quality, energy saving, and carbon reduction. In the laboratory using inductively coupled plasma (ICP) to detect the elemental composition of solid powders [3, 4, 5, 12, 13, 14], it is usually used to dissolve the solid with an acidic or alkaline solvent, and then atomize the solution, and the sample is passed into the ICP in the form of an aerosol for detection, which is complex and time-consuming. In addition, the additional components may introduce contamination and cause inaccurate test results. The slurry method has been successfully used in the analysis of Atomic Absorption Spectroscopy (AAS) in graphite furnaces [1-2], in which solid powders are evenly dispersed in solution to prepare a stable suspension with good fluidity, atomized into aerosols by a nebulizer, filtered in a swirl nebulization chamber, and finally passed into a detection instrument for analysis. The slurry method saves a lot of time in sample preparation, is simple to operate, reduces the possibility of contamination, and has the advantages of solution injection and direct solid injection, but it is limited by the single-element detection ability of AAS and cannot meet the needs of use. Therefore, slurry powder injection coupled with ICP has a good application prospect, Table 1 lists the instances that are successfully applied to the ICP.

**Table 1.** Examples of ICP detection by slurry injection.

| Method | Ingredients | Element | Particle size | Effect |
|---|---|---|---|---|
| ICP-AES | Chia seeds [3] | Pb Cd Se Hg | < 10 μm | RSD 0.7-3.9% |
| ICP-AES | Soil [4] | Li Be V Cr Zn | < 15 μm | RSD 0.9-14.2% |
| ICP-OES | $AL_2O_3$ [5] | Fe Si Ti Cr | < 20 μm | RSD 0.7-6.5% |
| ICP-AES | Cement [12] | Si Fe Mg Al Ca | < 10 μm | RSD 0.16-3.07% |
| ICP-OES | Cement [13] | Si Fe Mg Al Ca | < 38 μm | 95% confidence |
| ICP-AES | Bauxite [14] | Fe Ca Si Mg Ti | < 4 μm | RSD 0.34-5.5% |

In the process of cement production, it is difficult to detect the quality of cement at each stage at the sample preparation site due to the working principle and the limiting factors such as volume size. In 1985, Professor Jin Qinhan's research group proposed and successfully developed a new microwave plasma excitation light source - microwave plasma torch (MPT) [6], and later cooperated with Professor Hieftje of the United States to study and improve the performance of the light source [7-10]. MPT's overall volume is small, with the structure of three concentric metal tubes, similar to the structure of an ICP device, and the plasma formed is an inverted funnel structure, which is conducive to the introduction of samples. It has a strong resistance to the introduction of solutions and molecular gases. The purpose of this paper is to establish a simple and stable microwave plasma torch atomic emission spectroscopy (MPT-OES) method for the quantitative analysis of fine cement powders and to evaluate the feasibility of the method.

## 2. Experimental methods

### *2.1. Slurry nebulizer*

A suitable injection system is the focus of the analysis, where the nebulizer is required to have good nebulization results. The traditional concentric atomizer adopts a capillary structure, which is easy to cause blockage when atomizing the slurry, so the concentric atomizer is not suitable for the slurry atomization method. Therefore, the slurry atomizer is designed and manufactured to not clog the nebulizer. Based on the principle of the Babington nebulizer [14], a "V" tank slurry atomizer was fabricated. The atomization process is simulated by Fluent simulation software, and the nebulizer principle [11] is shown in the figure 1: the main components of the nebulizer are a gas pipe and the nebulizer housing. The gas and liquid flow to the outlet through an internal channel in the housing. The liquid outlet is a small hole with an inner diameter of 1 mm, the small hole is located on a section of the "V" groove, and the liquid flows down the V-shaped groove through the 0.2 mm hole of the gas outlet to be atomized into aerosols. The front section of the nebulizer is cylindrical with an outer diameter of 6 mm, which is suitable for the existing nebulization chamber in the laboratory and can be directly connected to the nebulizer chamber for use, without the need for additional connection devices, which avoids the loss of aerosols in the transmission process and reduces the experimental error.

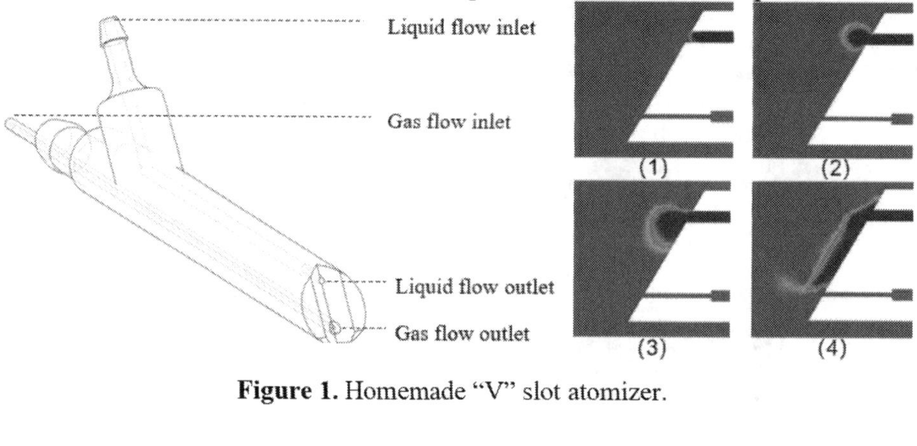


**Figure 1.** Homemade "V" slot atomizer.

### *2.2. Main instruments and parameters*

**Table 2.** Experimental equipment and operating conditions.

| Microwave source | 2, 450 MHz, 300-1, 000 W |
|---|---|
| Magnetron | Panasonic 2M259G-04ULN |
| MPT | Lab-made |
| Gas mass flowmeter | Sevenstar, CS200 |
| Spectrometer | ESA 4000, 200-870 nm |
| Ultrasonic cleaner | Kunshan Ultrasonic Company |
| Argon, Oxygen | Purity of 99.999% |
| Observation mode | End-view |
| Microwave power | 650-900 W |
| Argon carrier gas | 0.2 L/min |
| Argon support gas | 0.6 L/min |
| Oxygen shielding gas | 1.2 L/min |
| Spectrometer integral time | 200 ms |
| Spectrometer amplification | 2, 900 |
| Peristaltic pumps | 0.2 mL/min |

### *2.3. Slurry configuration*

The P.O42.5 cement powder was dried at 110°C in the drying oven for 2 h and then taken out, and the cement powder with a particle size of 40, 50, and 60 μm was screened out by using 250 mesh, 300 mesh, 350 mesh, and 400 mesh screens, and the four powders screened out were continued to be used in the drying box for later use. Using an analytical balance, we weigh 0.05 g of sample powder into a 100 mL dry volumetric flask, add about 50 mL of pure water, shake well, add an appropriate amount of dispersant (0.01 g glycerol + 1 mL HCl), and dilute it with pure water to the reticle. After the configuration, four 0.05% m/v sample slurries were placed in an ultrasonic cleaner for 20 min and then introduced into the MPT light source by a peristaltic pump through the self-made V-groove nebulizer and atomization chamber. Table 3 shows the percentage concentrations of each element in the sample cement.

**Table 3.** Percentage concentration of each element in the sample cement (%).

| Simple | Mg | Si | Al | Ca |
|---|---|---|---|---|
| 1 | 3.64 | 23.49 | 6.79 | 53.79 |
| 2 | 0.95 | 4.02 | 13.14 | 59.08 |
| 3 | 2.57 | 15.76 | 9.31 | 93.15 |
| 4 | 2.78 | 17.24 | 8.83 | 82.08 |

## 3. Results and discussion

### *3.1. Particle size selection*

As can be seen from Table 1, the particle size of most detection schemes is less than 10 μm, and excessive particle size will lead to serious precipitation of slurry, which will affect the detection effect. To test the effect of cement at 40-60 μm particle size, the concentration X of the sample is taken as the abscissa, and the intensity value Y is used as the ordinate for linear regression, taking the regression equation measured by the analysis line of Mg element 280.269 as an example. It can be seen from Figure 2 that with the increase of particle size, the $R^2$ value gradually decreases, the relative standard deviation (RSD) gradually increases, and the detection effect is good and stable at 40 μm, so the 40 μm particle size powder is selected for the experiment.

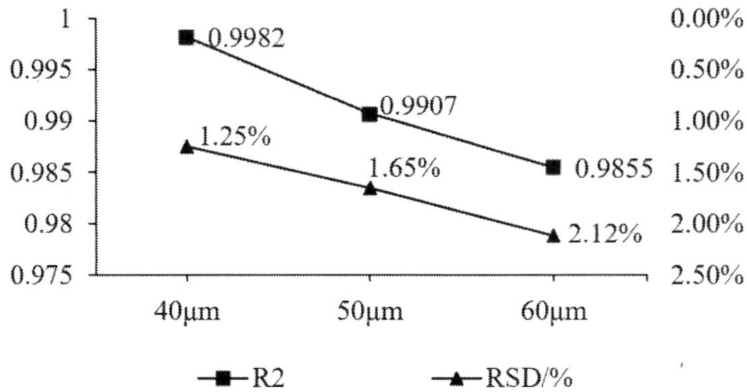


**Figure 2.** Detection effect of three particle sizes of cement.

### *3.2. Atomization efficiency*

The atomization efficiency refers to the percentage of the mass that can reach the plasma after the solution is atomized, which is an important parameter to evaluate the performance of the atomizer. In the test, the atomization efficiency of the nebulizer was measured by the differential subtraction method, and the measurement steps were as follows: the weight of the injection solution ($m_1$, $m_2$), waste liquid ($m_3$, $m_4$), nebulizer and nebulization chamber ($m_5$, $m_6$), and peristaltic pump and pump tubing ($m_7$, $m_8$) before and after the test were measured. The $\eta$ of atomization efficiency can be derived from Equation (1):

$$\eta = \frac{(m_1 - m_2) - (m_3 - m_4) - (m_5 - m_6) - (m_7 - m_8)}{(m_1 - m_2)} \times 100\% \quad (1)$$

In the test process of the nebulizer, we maintain a certain test time to test the atomization efficiency of atomizers, and the test conditions and atomization efficiency are shown in the following table 4:

**Table 4.** The atomization efficiency of each atomizer.

| Test conditions | Nebulizer type | Efficiency |
|---|---|---|
| Carrier gas flow: 0.6 L/min | Concentric nebulizers | 3.93% |
| Peristaltic pump speed: 1 L/min<br>Sample: Deionized water | High salt atomizer | 3.04% |
| Atomization time: 15 min | Homemade V-groove nebulizer | 3.28% |

### *3.3. Standard working curves*

Under the experimental instruments and working conditions in Table 2, the sample slurry configured in 2.3 was tested, 50 sets of data were collected for each sample, and the relative standard deviation of each element was calculated. Table 5 shows the test results. The relative standard deviations (RSD) of Mg, Al, Ca, and Si were 1.49%, 3.52%, 1.63%, and 2.39%, respectively, and the linear correlation coefficients ($R^2$) were greater than 0.997, indicating that the powder injection method of MPT-OES combined with slurry had high stability and accuracy.

**Table 5.** Operating curve equation and detection limit.

| Element | Linear | $R^2$ | RSD/% | Limit of detection /($mg \cdot L^{-1}$) |
|---|---|---|---|---|
| Mg | Y=709.91X−151.65 | 0.9997 | 1.49% | $7.75e^{-4}$ |
| Al | Y=170.37X−1, 577.1 | 0.9995 | 3.52% | $2.23e^{-4}$ |
| Ca | Y=143.2X−18, 904.1 | 0.9990 | 1.63% | $4.14e^{-4}$ |
| Si | Y=51.748X+4, 902.5 | 0.9970 | 2.39% | $3.61e^{-4}$ |

### 3.4. *Method feasibility analysis*

Table 6 compared with the experimental results of cement sample preparation using the melting method in ICP [15-16], the MPT slurry method greatly shortens the sample preparation time, has excellent detection performance without introducing any flux, and avoids the loss of element content.

**Table 6.** Comparison of the effects of different detection methods.

| | $R^2$ | | | RSD | | | Detection accuracy | | |
|---|---|---|---|---|---|---|---|---|---|
| | ICP-MS | ICP-AES | MTP-OES | ICP-MS | ICP-AES | MTP-OES | ICP-MS | ICP-AES | MTP-OES |
| Mg | 0.999 | 0.999 | 0.999 | 5.17% | 0.87% | 1.49% | 92.4% | 99.1% | 99.7% |
| Al | 0.999 | 0.999 | 0.999 | 5.81% | 1.04% | 3.52% | 92.9% | 99.5% | 99.9% |
| Ca | 0.999 | 0.999 | 0.999 | 5.30% | 1.24% | 1.63% | 97.8% | 99.5% | 99.9% |

MPT-OES can detect cement powders under different power and gain conditions. Figure 3 takes Mg as an example, and the coefficient of determination ($R^2$) and the relative standard deviation (RSD) under the three different power conditions of 600 W, 750 W, and 900 W have good detection effects, and the method has good stability and practicability.

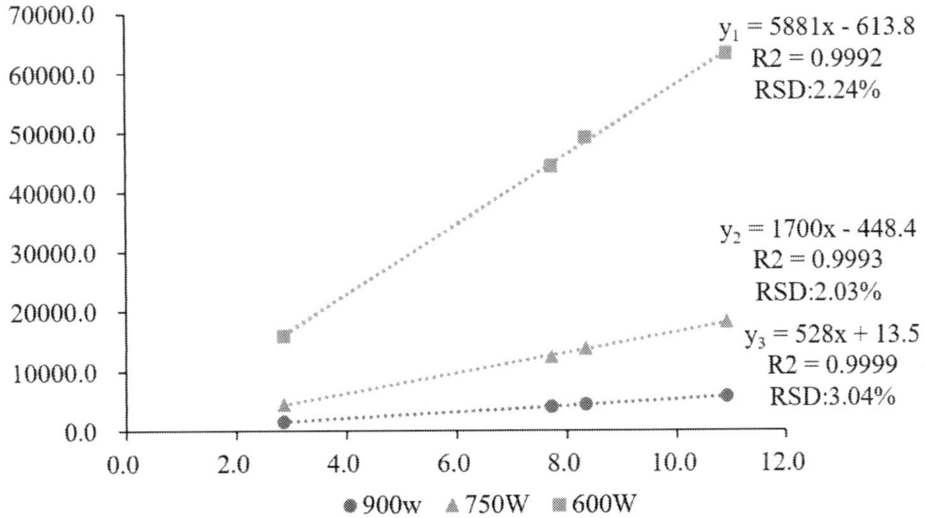


**Figure 3.** Detection effect at different power levels.

## 4. Conclusion

It is feasible to analyze cement powder by slurry powder injection coupled with MPT-OES, and the device is small in size and has strong method applicability, so it has great potential and application prospects in powder online detection. According to the special physical properties of cement slurry, based on the principle of Babington atomizer, an atomizer with a good atomization effect was designed and manufactured. The slurry method avoids the complex sample preparation process, omits the grinding treatment of the powder in the process of preparing the stable slurry, and has good precision and accuracy when the particle size of the cement powder is 40 μm. Compared with other test cases, MPT-OES can be detected in the case of larger powder particle size and has good detection results at different powers.

Journal of Physics: Conference Series **2961** (2025) 012022 doi:10.1088/1742-6596/2961/1/012022

**References**

[1] Y. N. Zong (2023). Suspension injection-graphite furnace atomic absorption spectrometry for the determination of lead in solid beverages. Guangzhou Chemical, 51 (13): 150-152.
[2] G. L. Chen (2022). Suspension injection-graphite furnace atomic absorption spectrometry for the determination of trace amounts of silver in soil. Chemical Research and Applications, 34 (07): 1, 680-1, 684.
[3] X. Liu and X. Shen (2023). Suspension emulsion injection-Ike-Mack method for the determination of lead, cadmium, arsenic, and chromium in chia seeds. Food industry, 44 (05): 324-327.
[4] L. J. Zhang and T. B. Xu (2019). Ultrafine Solids Suspension Injection-Inductively Coupled Plasma Mass Spectrometry Determination in Soiltrace Elements. Rock and mine testing, 38 (02): 147-153.
[5] Z. Y. Zhao (2017). Determination by ICP-OES assay by direct injection of suspension Fe, Ti, Si, and Cr in high-purity alumina. Chemical analysis and metrology, 25 (01): 55-57.
[6] D. J. Yu (2024). Study on a modified spectral standardization method and quantitative analysis of cement based on microwave plasma torch. Journal of Analytical Atomic Spectroscopy, 7 (38): 1, 402-1, 411.
[7] B. Yu, W. Jin, Y. Ying, H. Yu, C. Xu, and Q. Jin (2016). High-speed photographic analysis of microwave plasma torch source behavior. J. Anal. At. Spectrom, 31 (3): 759-766.
[8] Q. Jin, F. Wang, C. Zhu, and G. M. Hieftje (1990). Atomic emission detector for gas chromatography and supercritical fluid chromatography. J. Anal. At. Spectrom, 5 (6): 487-494.
[9] W. Yang, H. Zhang, A. Yu, and Q. Jin (2000). Microwave plasma torch analytical atomic spectrometry. Microchem. J., 66 (1): 147-170.
[10] Q. Jin, C. Zhu, and G. M. Hieftje (1991). A microwave plasma torch assembly for atomic emission spectrometry. Spectrochimica Acta Part B: Atomic Spectroscopy, 46 (3): 417-430.
[11] X. C. Liu (2004). Application of suspension injection technique in ICP-AES. Beijing General Research Institute of Nonferrous Metals, Beijing.
[12] Cıntia S. Silva (2002). Analysis of cement slurries by inductively coupled plasma optical emission spectrometry with axial viewing. Spectrochimica Acta Part B, 57 (1): 29-33.
[13] X. C. Liu and Y. Liu (2004). Suspension injections were injected by ICP-AES for the determination of Fe, Mg, Si, Ca, and Ti in bauxite. Analytical laboratory, 2004, 23 (3): 34-36.
[14] LES EBDON, MICHAEL FOULKES, and KAREN SUTTON (1997). Slurry Nebulization in Plasmas. Journal of Analytical Atomic Spectrometry, 12 (2): 213-229.
[15] C. X. Zhang (2018). Rapid determination of chemical composition in cement by ICP-MS. CEMENT, 2018 (06): 49-51.
[16] D. D. Zhang (2014). The ICP-AES method simultaneously determined K, Ca, Na, Mg, Al, Fe, and Ti in cement. CEMENT, 2014 (10): 53-54.

Journal of Physics: Conference Series **2961** (2025) 012023 doi:10.1088/1742-6596/2961/1/012023

# Identifying the concentration of the ethanol by utilizing surface plasmon resonance technology

**Jiaping Yang[1], Guofu Wang[2], Hongqi Li[3], Xiaobo Jia[4], Jin Zhou[5] and Yansheng Liu***

[1,2,3,4,5,*]Guangxi University of Science and Technology, No.2, Wenchang Road, Liuzhou City, 545000, Guangxi, China.

*Corresponding author's e-mail: yansheng.liu@gxust.edu.cn

**Abstract:** Ethanol is an important raw material in the chemical, pharmaceutical, and food industries. Due to the volatile nature of ethanol products, it is essential to monitor the concentration of ethanol in these products to prevent degradation caused by evaporation. This paper presents a sensor chip for detecting ethanol concentration. An aqueous solution is used as the ethanol solvent, and a surface plasmon resonance (SPR) sensor serves as the receiver. When ethanol interacts with the surface of the SPR sensor, the refractive index is modulated and then detected by the SPR sensor. Based on this principle, the sensor was fabricated and tested. Real-time experiments of identifying ethanol solutions with concentrations ranging from 0.5% to 4% demonstrated that the system exhibits good repeatability and durability. Furthermore, the device offers high detection sensitivity and low cost, making it suitable for widespread application in industrial production.

## 1. Introduction

Ethanol is a crucial raw material in various industries, including food and medicine. In the pharmaceutical sector, ethanol serves as an effective solvent for organic substances and is commonly used in dissolving agents. Many liquid medications contain a small amount of ethanol. Additionally, in the food industry, ethanol is utilized as both a colorant and flavoring agent to cater to the growing demand for unique sensory experiences [1]. Various methods are commonly used to measure ethanol content, including liquid chromatography-mass spectrometry (LC-MS) [2], colorimetric methods [3], and electrochemical sensing methods. Although these technologies can also detect ethanol, most of them have difficulty achieving real-time monitoring. Surface plasmon resonance (SPR) sensors address the limitations of traditional sensing technologies, offering high accuracy, resistance to electromagnetic interference, compact size, and low complexity [4]. In addition, the SPR can combine sensors and optical communication devices, which can achieve array sensing [5]. SPR offers good sensitivity, small sensing units, excellent accuracy, and label-free detection. In the field of real-time SPR biosensors, real-time data analysis and processing are essential. It is necessary to process large volumes of data within a short time frame to enable the comparison of two data sets and fully achieve the purpose of real-time detection. In this study, we developed a differential phase SPR system capable of real-time detection for measuring ethanol concentration. By using a WFT-based approach, we collected data every 10 seconds and performed a comparative analysis with the raw data to calculate the offset every 10 seconds, allowing for a more intuitive observation of the real-time changes in ethanol concentration in the solution. The sensitivity was tested using ethanol concentrations ranging from 0.5% to 4%. Based on experimental comparisons, the detection limit (LOD) of the SPR system was determined to be 1%. The consequences indicate that stable monitoring of ethanol concentration in the solution can be achieved through the real-time monitoring system, demonstrating strong practical application value [6].

## 2. Experimental Section

### *2.1. Materials*

Materials: ethanol was purchased from Beijing InnoChem Science & Technology Co., Ltd. The water used was purified water.

### *2.2. Apparatus*

Through this research, we independently developed a SPR system that can be used for differential phase measurement. The main components of the system are a light source with a wavelength of 500-800nm, a collimator (I1 and I2) for concentrating light, a polarizer (P1 and P2), a birefringence crystal (BC) for generating two waves, a lens (FL) for focusing, a slit, a sensing chip, and an injection pump, as shown in Figure 1.

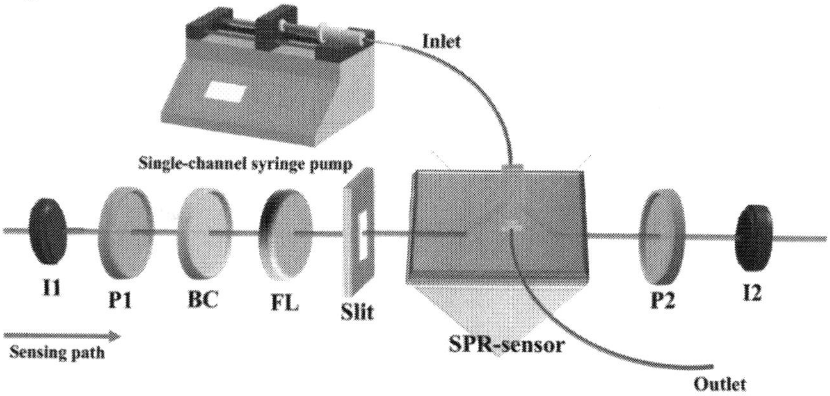


**Figure 1.** SPR system schematic diagram.

### *2.3. Production of SPR chip*

First, we used BK7 glass as the carrying medium of the gold film and cleaned the BK7 glass with water for 15 min using an ultrasonic instrument. This operation was repeated three times. Then, we put the cleaned BK7 glass into the thermal evaporation machine and set the evaporation of 50nm gold film so that a layer of 50nm gold film is deposited on the BK7 glass. The PDMS microflow channel is then covered with a gold film, and the chip is finished.

## 3. Results and discussion

The SPR sensor is highly sensitive and can be easily affected by various factors such as light source stability, environmental temperature, and humidity. Ideally, the entire operation should be carried out in the dark to avoid interference from other light sources, as well as to minimize the vibration of the platform. During the experiment, we observed that environmental temperature had a significant impact on the SPR sensor. To address this issue, we installed an air conditioner in the laboratory and set the temperature to 26°C. During the experiment, the sensor surface must remain free of air bubbles. Additionally, small bubbles may form during the detection process, causing significant fluctuations that can affect the accuracy of the experimental results. If this occurs, we need to restart the experimental procedure to ensure reliable and accurate results.

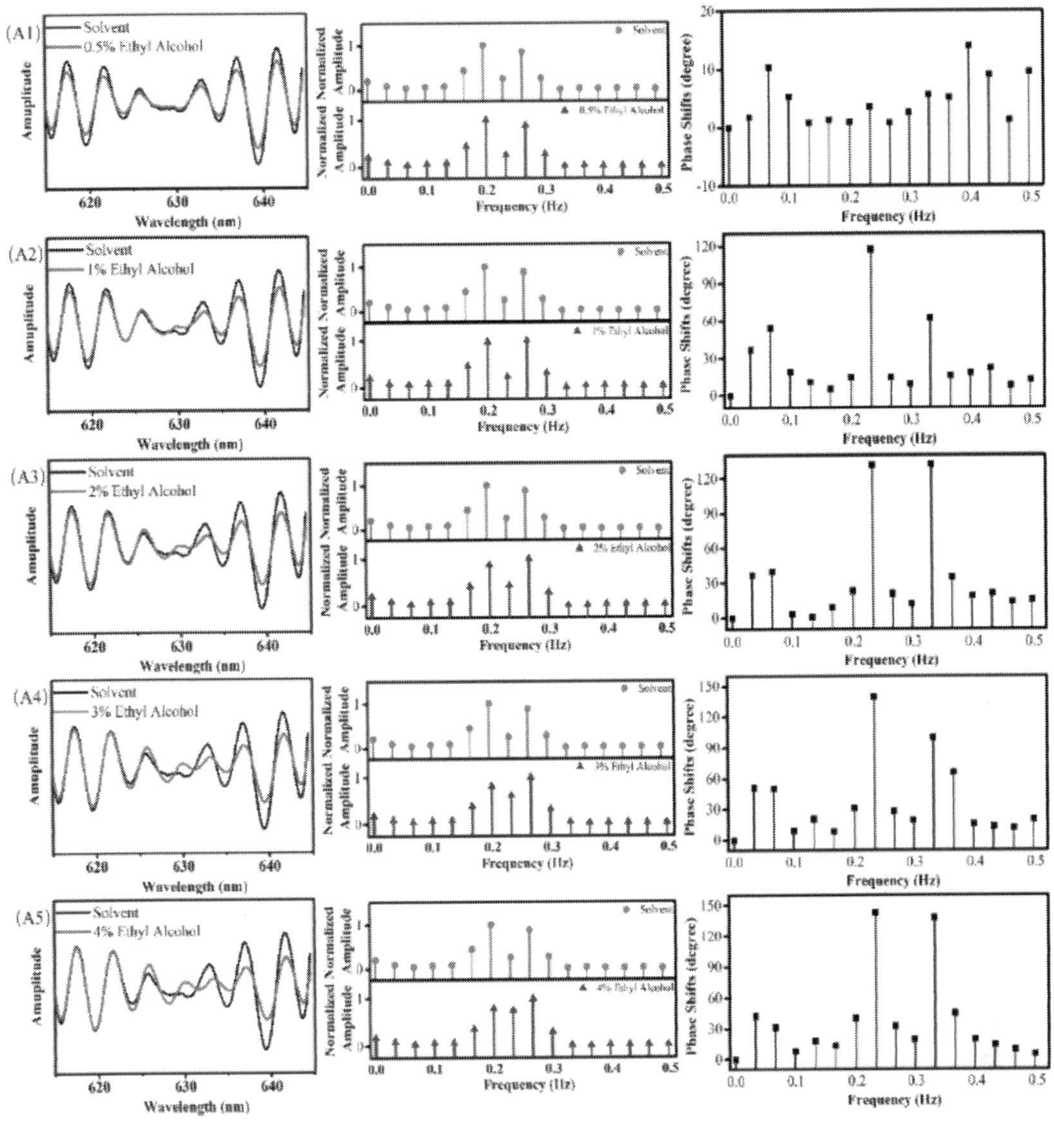


**Figure 2.** (A1) to (A5) represent different concentrations of ethanol.

Figure 2 shows the phase shift between ethanol and water at different concentrations. The calculated phase shits were plotted in Figure 3, which were calculated from the spectral interferograms. By calculation, the phase shift for ethanol with concentrations of 0.5%, 1%, 2%, 3%, and 4% were 3.04°, 33.63°, 53.03°, 67.01°, and 85.7°, respectively. It can be observed from the experimental results that within the error range, the minimum detection limit of the surface plasmon resonance system for sodium chloride is 1%.

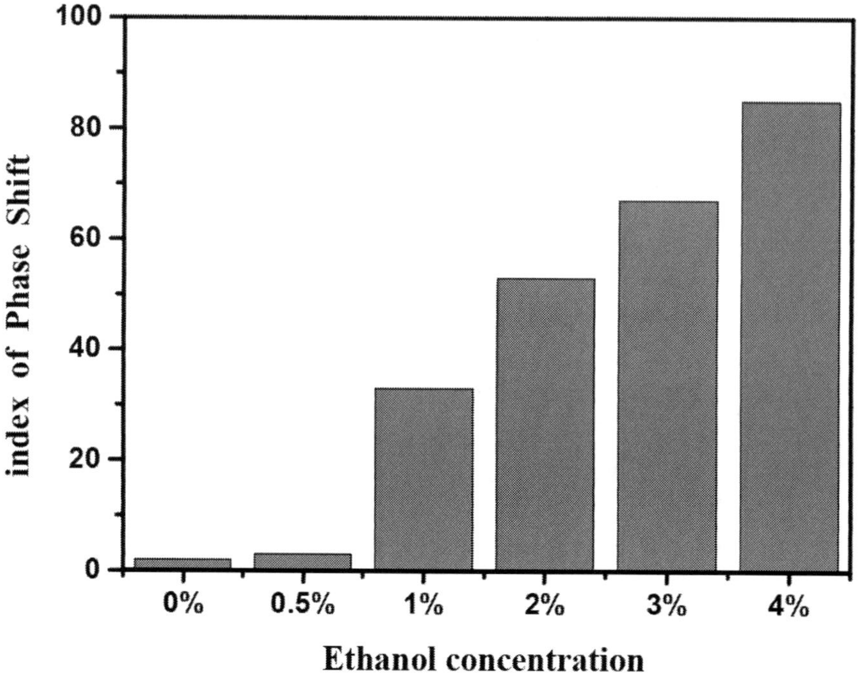


**Figure 3.** Phase shift diagram of different ethanol concentrations.

## 4. Conclusion

An SPR real-time detection system based on the WFT method has been successfully proposed and validated. By combining the Wavelet Fourier Transform (WFT) method with SPR technology, we calculated the frequency domain results extracted from the SPR spectral interferograms and used the WFT method to calculate the phase shift changes, which allowed us to determine the specific variation. We used ethanol as the analyte in this system and tested ethanol solutions with concentrations ranging from 0.5% to 4%. The experimental results indicated that the detection limit of this system for ethanol was 1%. The findings of this study provide a new method for detecting ethanol concentration, and due to the incorporation of real-time detection technology in SPR, this system has practical application value in terms of both sensitivity and stability. Beyond the application of ethanol detection using SPR technology, this method can also be extended to ion detection, molecular detection, protein detection, and DNA detection, showing potential in specific areas of environmental monitoring, food safety, etc. The real-time detection system based on WFT proposed in this study can have many applications.

## Acknowledgments

This work has been supported by the National Key Research and Development Program of China (No. 2022YFE0134600), Guangxi Science and Technology Project (No. AD22035215), Guangxi Natural Science Foundation Project (No. 2021GXNSFBA196049)

## References

[1] MS Neo, S. Gupta, T.M. Khan, MJPJ Gupta, Quantification of Ethanol Content in Traditional Herbal Cough Syrups, 9(2017) 821-7.
[2] S.J. Hird, B.P.-Y. Lau, R. Schuhmacher, R.J.T.i.A.C. Krska, Liquid chromatography-mass spectrometry for the determination of chemical contaminants in food, 59(2014) 59-72.

[3] S. Sumbhate, S.P. Nayak, D. Goupale, A. Tiwari, R.S. Jadon, Colorimetric Method for the Estimation of Ethanol in Alcoholic-Drinks, 2012.
[4] E. Fontana, H.D. Dulman, D.E. Doggett, R.H.J.I.I. Pantell, P. Measurement Technology Conference Sensing, Networking. IMTC Proceedings, Surface plasmon resonance on a single mode fiber, 1(1997) 611-6 vol.1.
[5] T. Liu, Fibre Optic Sensors for Coal Mine Hazard Detection, in: G.-D. Peng (Ed.) Handbook of Optical Fibers, Springer Singapore, Singapore, 2018, pp. 1-27.
[6] T.K. Sayğı, E.G. Pınar, I. Taştekil, P.Ö. Sarıca, G. Topçu, E.A. Turkoglu, Elucidation of the Binding Interaction between β-sitosterol and Lysozyme using Molecular Docking, Molecular Dynamics and Surface Plasmon Resonance Analysis. %J Chemistry & biodiversity, (2024) e202401470.

Journal of Physics: Conference Series **2961** (2025) 012024 doi:10.1088/1742-6596/2961/1/012024

# Mixed adsorption characteristics of magnetic calcium silicate for Pb(II), Cd(II), and Cr(III) in solution

**Jianfeng Han[1] and Xiaoping Chen[1, *]**

[1]School of Energy and Environment, Southeast University, Nanjing, Jiangsu, 211189, China

[*]Corresponding author's e-mail: xpchen@seu.edu.cn

**Abstract:** Adsorption methods have been widely employed for heavy metal wastewater treatment, yet traditional adsorbents pose challenges in terms of recovery. Magnetic adsorbent materials have garnered attention due to their ease of recovery and reuse. However, studies on their simultaneous adsorption of multiple heavy metal ions are limited. In this study, a magnetic mesoporous calcium silicate adsorbent ($Fe_3O_4$@MCS) was synthesized via an improved solvothermal and chemical precipitation method and applied to treat simulated wastewater containing Pb, Cd, and Cr ions. The adsorption capability of $Fe_3O_4$@MCS was assessed in a series of batch studies under a range of circumstances, including adsorbent dosage, pH, temperature, oscillation intensity, contact time, and initial metal concentration. The results demonstrated that $Fe_3O_4$@MCS exhibited excellent adsorption capacities for Pb, Cd, and Cr. Enhancements in adsorbent dosage, pH, temperature, and oscillation intensity improved the adsorption performance. After five cycles, the removal percentages of $Pb^{2+}$, $Cd^{2+}$, and $Cr^{3+}$ decreased by 21.48%, 13.56%, and 23.54%, respectively. The adsorption mechanisms were found to include surface precipitation, surface complexation, physical adsorption, and ion exchange.

## 1. Introduction

With the continuous progress of science and technology in human society, many harmful heavy metals are released into the environment. Heavy metals, including Pb, Cd, and Cr, are hazardous, persistent, and bioaccumulative in water bodies, and heavy metal pollution poses a serious threat to the environment and public health [1-3]. These contaminants are commonly discharged from industrial activities, including mining, electroplating, and battery manufacturing, necessitating efficient and effective removal methods [4]. Adsorption is one of the many treatment technologies that has shown promise because of its ease of use, high efficiency, and capacity to manage low quantities of contaminants [5].

For the removal of heavy metals, traditional adsorbents, including zeolites and activated carbon, have been utilized extensively. However, these materials often suffer from limitations related to recovery and regeneration, which hinder their practical application on a large scale. Because of the convenience of separation and reusability, magnetic adsorbent materials have drawn a lot of attention as an option for these issues. A magnetic organic nanosorbent (PAN-Mel@$Fe_3O_4$) was designed by Dogari et al. to capture lead and cadmium from aqueous solutions. The maximal adsorption capabilities for $Pb^{2+}$ and $Cd^{2+}$ under ideal circumstances were 416.66 mg/g and 454.54 mg/g, respectively. The adsorption performance of PAN-Mel@$Fe_3O_4$ did not decrease after four adsorption cycles [6]. In addition, magnetic organic polymer composite (CMNC) was synthesized for the removal of nickel and cobalt from water. It showed high adsorption performance (30.03 mg/g for $Ni^{2+}$ and 53.19 mg/g for $Co^{2+}$) under optimum conditions [7].

Numerous magnetic adsorbents have been shown in earlier research to be efficient at extracting individual metal ions from aqueous solutions [8]. Nevertheless, the coexistence of multiple heavy metals may result in competitive adsorption, affecting both the total adsorption capacity and the selectivity. Therefore, it is crucial to understand the mixed adsorption characteristics of magnetic adsorbents in multi-metal systems to develop optimized treatment strategies.

This study aims to investigate the mixed adsorption characteristics of magnetic calcium silicate ($Fe_3O_4$@MCS) for Pb, Cd, and Cr ions in solution. With high specific surface area and strong magnetism, $Fe_3O_4$@MCS is a novel kind of adsorbent material that brings together the benefits of magnetic nanoparticles and silicon-based materials. By synthesizing $Fe_3O_4$@MCS using an improved solvothermal and chemical precipitation method, we seek to evaluate its adsorption performance under various conditions, including adsorbent dosage, pH, temperature, oscillation intensity, contact time, and initial heavy metal concentration. Additionally, we aim to elucidate the underlying adsorption mechanisms and assess the reusability of the adsorbent through multiple adsorption-desorption cycles. Understanding the adsorption behavior of $Fe_3O_4$@MCS in multi-metal systems will provide valuable insights into the design and application of efficient adsorbents for the treatment of complex wastewater, contributing to environmental sustainability and public health protection.

## 2. Experimental section

### *2.1. Synthesis of $Fe_3O_4$@MCS*

First, 1.08 g of ferric chloride hexahydrate ($FeCl_3 \cdot 6H_2O$) and 2.88 g of sodium acetate ($Na_2SiO_3$) were dissolved in 40 mL of 90% (v/v) ethylene glycol aqueous solution. After mechanically stirring the mixture for 30 minutes, it was transferred to a hydrothermal reactor. The reactor was then placed in an electric thermostatic drying oven at 200°C for 8 h. The reactor was given time to spontaneously cool to ambient temperature before the black product was separated. $Fe_3O_4$ magnetic microspheres were produced by repeatedly washing the product in deionized water and anhydrous ethanol, respectively, and then drying it.

A blend of 0.50 g of the aforementioned synthesized $Fe_3O_4$ microspheres and 0.8526 g of sodium metasilicate nonahydrate ($Na_2SiO_3 \cdot 9H_2O$) was combined with 50 mL of deionized water and mechanically stirred to achieve a homogeneous suspension, designated as Suspension I. Separately, 0.3330 g of anhydrous calcium chloride ($CaCl_2$) was thoroughly dissolved in 5 mL of deionized water to create Solution II. This Solution II was then gradually introduced into Suspension I at a rate of 2.5 mL/h, with continuous mechanical stirring of Suspension I during the entire injection process. After the completion of the injection, the mixture was continuously stirred at room temperature for an additional 3 hours. Finally, the gray product was separated from the suspension, washed with ultrapure water and anhydrous ethanol, and then dried to yield the magnetic mesoporous calcium silicate heavy metal adsorbent ($Fe_3O_4$@MCS).

### *2.2. Adsorption experiments*

In order to investigate the adsorption properties of $Fe_3O_4$@MCS on wastewater containing both $Pb^{2+}$, $Cd^{2+}$, and $Cr^{3+}$, some batch experiments were carried out. All adsorption experiments were conducted with a minimum of three replicates, and erroneous data were excluded. The experimental results were reported as the mean of multiple measurements.

(1) The experimental procedure for investigating the effect of $Fe_3O_4$@MCS dosage on adsorption is as follows: Using analytical grade lead nitrate, cadmium nitrate, and chromium nitrate nonahydrate, a 250 mL solution containing $Pb^{2+}$, $Cd^{2+}$, and $Cr^{3+}$ ions at concentrations of 50 mg/L each was prepared. The solution pH was raised to 5.5 and then equally divided into five portions. Different doses of $Fe_3O_4$@MCS were added to each solution separately. After allowing for static adsorption for 6 hours,

solid-liquid separation was performed using a magnet. The supernatant was carefully extracted with a pipette and analyzed for $Pb^{2+}$, $Cd^{2+}$, and $Cr^{3+}$ concentrations using ICP-OES.

(2) The experimental procedure for investigating the effect of initial solution pH on adsorption is as follows: The preparation steps for the simulated solution are the same as described in (1), with a fixed $Fe_3O_4$@MCS dosage of 20 mg. The initial pH of the solution was varied, and after allowing for static adsorption for 6 hours, the concentrations of $Pb^{2+}$, $Cd^{2+}$, and $Cr^{3+}$ in the supernatant were measured.

(3) The experimental procedure for investigating the effect of adsorption temperature on adsorption is as follows: The preparation steps for the simulated solution and the supernatant detection steps are the same as described in (1). The $Fe_3O_4$@MCS dosage was fixed at 20 mg, and the solution pH was changed to 5.5. During the adsorption process, the solution temperatures were adjusted to 20°C, 40°C, 60°C, and 80°C, respectively. The adsorption was allowed to proceed for 6 hours under these conditions.

(4) The experimental procedure for investigating the effect of oscillation intensity on adsorption is as follows: The preparation steps for the simulated solution and the supernatant detection steps are the same as described in (1). The $Fe_3O_4$@MCS dosage was fixed at 20 mg, and the solution pH was changed to 5.5. The beakers were kept at room temperature in a water bath shaker, with oscillation rates set to 0, 50, 100, 150, and 200 rpm. The adsorption process was carried out for 6 hours under these conditions.

(5) The experimental procedure for investigating the effect of interaction time on adsorption is as follows. A 100 mL solution containing $Pb^{2+}$, $Cd^{2+}$, and $Cr^{3+}$, each at an initial concentration of 50 mg/L, was prepared. Then, 40 mg of $Fe_3O_4$@MCS was added to the solution and the solution pH was changed to 5.5. Finally, at different time intervals, the supernatant was extracted, and the concentration of each heavy metal ion was measured.

### *2.3. Adsorption-desorption cyclic regeneration experiment*

A 50 mL solution containing $Pb^{2+}$, $Cd^{2+}$, and $Cr^{3+}$ is prepared at a concentration of 50 mg/L each, and the pH of the solution is adjusted to 5.5. Subsequently, 20 mg of $Fe_3O_4$@MCS is added to the solution, and it is allowed to be absorbed at ambient temperature for 6 hours. After the adsorption process, solid-liquid separation is performed using a magnet, and the supernatant is extracted with a pipette to measure the concentrations of $Pb^{2+}$, $Cd^{2+}$, and $Cr^{3+}$ in the supernatant. The corresponding adsorption amounts and removal rates are calculated. The $Fe_3O_4$@MCS used in the adsorption experiments is collected with the $Pb^{2+}$, $Cd^{2+}$, and $Cr^{3+}$ solutions and immersed in a 1 mol/L NaOH solution for desorption experiments. Then, solid-liquid separation is performed, and the adsorbent is washed several times with deionized water. The adsorbent is dried at 60°C in an electric thermostatic drying oven until it reaches a consistent weight. The dried adsorbent is used for adsorption experiments with $Pb^{2+}$, $Cd^{2+}$, and $Cr^{3+}$ solutions (50 mL, 50 mg/L each). This process is repeated five times to investigate the adsorption regeneration performance of $Fe_3O_4$@MCS.

## 3. Results and discussion

### *3.1. Influential factors for adsorption*

*3.1.1 $Fe_3O_4$@MCS dosage.* Figure 1 illustrates how the dosage of $Fe_3O_4$@MCS affects the remaining $Pb^{2+}$, $Cd^{2+}$, and $Cr^{3+}$ concentrations in the solution. The total residual concentration of heavy metal ions in the solution (SUM) steadily dropped as the dosage of $Fe_3O_4$@MCS increased, with an overall linear trend from 150 mg/L to 30 mg/L (at a $Fe_3O_4$@MCS dosage of 772 mg/L). The mass concentration of $Cr^{3+}$ decreased at the fastest rate with increasing $Fe_3O_4$@MCS dosage, followed by $Pb^{2+}$, and finally $Cd^{2+}$. Interestingly, the increase in $Fe_3O_4$@MCS dosage did not lead to a decrease in the residual concentration of $Cd^{2+}$, which remained relatively stable. In Figure 1, SUM1 represents the total residual concentration of heavy metal ions in the solution after treatment with $Fe_3O_4$@MCS for a single heavy metal. The adsorption sites on the magnetic mesoporous calcium silicate are limited, and when multiple heavy metal ions are present simultaneously in the solution, they compete for these adsorption sites. Additionally, different heavy metal ions may interact with each other in the solution, forming complexes or co-precipitates. These phenomena can modify the effective concentration of each ion, thereby influencing the adsorption process and resulting in a reduction in the total adsorption capacity.

*3.1.2 pH value.* The impact of the solution's initial pH on the adsorption capacity is depicted in Figure 2. At lower initial pH values, $Fe_3O_4$@MCS has a comparatively low ability to adsorb the three heavy metal ions. When the pH rose, the residual concentrations of $Pb^{2+}$, $Cd^{2+}$, and $Cr^{3+}$ decreased rapidly, reaching their lowest points near neutral pH, where the removal rates of all three ions reached 100%. Beyond this pH range, the residual concentrations remained unchanged.

*3.1.3 Solution temperature.* The impact of the adsorption temperature on the adsorption capacity is depicted in Figure 3. As the temperature rose, the residual mass concentrations of heavy metal ions in the solution first declined and subsequently increased. At an adsorption temperature of 40°C, the residual mass concentrations of $Pb^{2+}$ and $Cr^{3+}$ reached their minimum values of 8.46 mg/L and 11.41 mg/L. $Pb^{2+}$ and $Cr^{3+}$ were removed at rates of 83.08% and 77.18%, respectively. For $Cd^{2+}$, the minimum residual mass concentration was 24.14 mg/L at an adsorption temperature of 60°C, with a removal rate of 51.72%. This trend suggests that the residual concentrations of the three heavy metal ions decrease initially and then increase with rising temperature.

Below 40°C, increasing the temperature promotes the exposure of active sites on the $Fe_3O_4$@MCS surface or alters its surface chemistry, enhancing its affinity for heavy metal ions. Additionally, higher temperatures increase the thermal motion of ions, facilitating the access of heavy metal ions to the active sites on the adsorbent, thereby improving adsorption efficiency. However, above 40°C, the interaction between the heavy metal ions and the adsorbent may weaken, resulting in the desorption of already adsorbed ions from the adsorbent surface, shifting the adsorption-desorption equilibrium towards desorption. Furthermore, heavy metal hydroxides may form precipitates, whose solubility typically increases with temperature, potentially leading to re-dissolution at higher temperatures and consequently reducing adsorption efficiency.

*3.1.4 Oscillation intensity.* The effect of varying oscillation rates on the residual mass concentrations of $Pb^{2+}$, $Cd^{2+}$, and $Cr^{3+}$ in solution during adsorption experiments is shown in Figure 4. Mass transfer effects can be improved, and the diffusion layer thickness can be decreased by increasing the oscillation rate, thereby increasing convection and diffusion. This enables heavy metal ions to approach the surface of $Fe_3O_4$@MCS more quickly and frequently. Additionally, higher oscillation rates can disrupt local agglomerates of $Fe_3O_4$@MCS, exposing more active adsorption sites to the solution and thereby increasing the effective utilization of these sites. Consequently, the adsorption efficiency of $Fe_3O_4$@MCS for $Pb^{2+}$, $Cd^{2+}$, and $Cr^{3+}$ is improved.

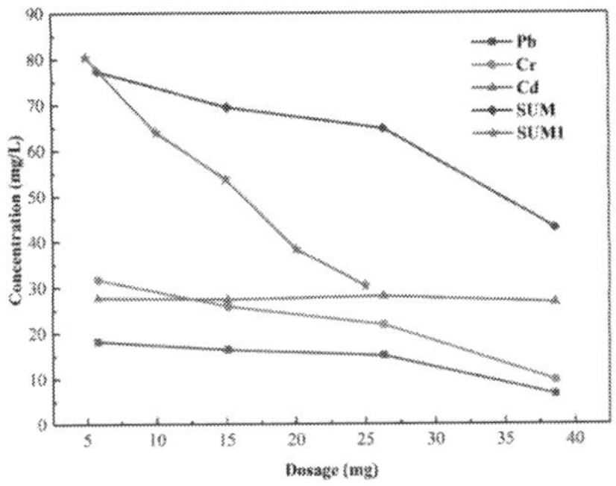


**Figure 1.** The impact of $Fe_3O_4$@MCS dosage on the adsorption.

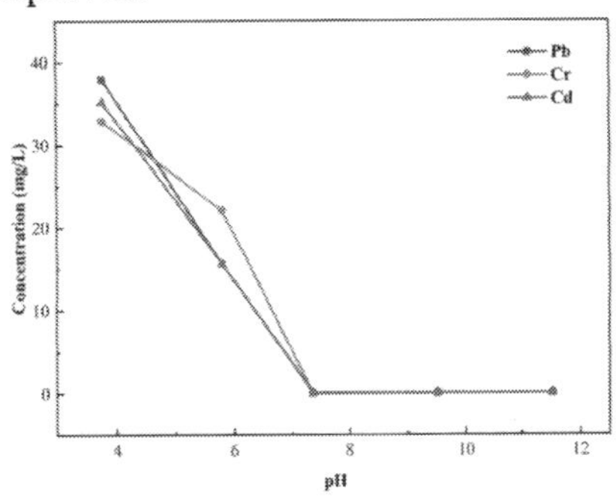


**Figure 2.** The impact of pH on the adsorption.

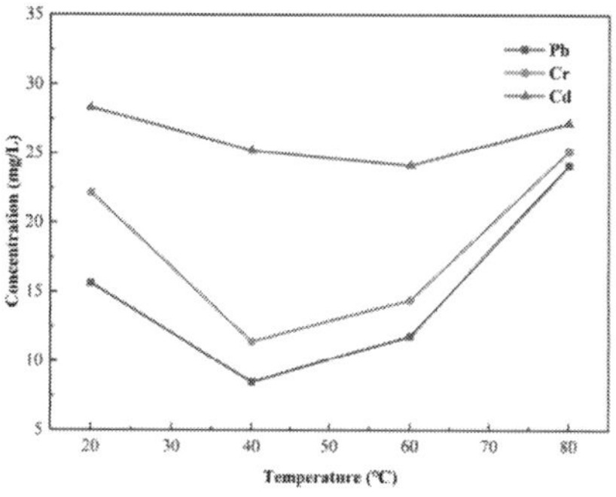


**Figure 3.** The impact of temperature on the adsorption.

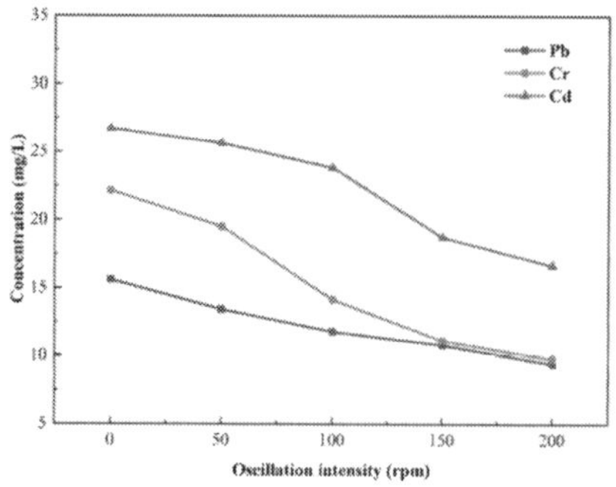


**Figure 4.** The impact of oscillation rate on the adsorption.

*3.1.5 Contact time.* Figure 5 displays the $Pb^{2+}$, $Cd^{2+}$, and $Cr^{3+}$ adsorption quantities ($q_t$) as a function of adsorption time ($t$) during competitive adsorption. The adsorption rates of the three heavy metal ions are initially rapid, then gradually slow down until reaching equilibrium. This trend can be attributed to the abundance of available adsorption sites on the $Fe_3O_4$@MCS surface during the initial fast adsorption phase, which is primarily controlled by mass transfer. Heavy metal ions can swiftly migrate to the surface of $Fe_3O_4$@MCS and be adsorbed. However, as adsorption time extends, the adsorption sites on the $Fe_3O_4$@MCS surface become quickly occupied, resulting in a drop in the amount of heavy metal ions and a corresponding reduction in their diffusion capability. These combined effects lead to a gradual slowdown in the adsorption rate until equilibrium is reached.

### *3.2. Adsorbent regeneration performance*

The recyclability of $Fe_3O_4$@MCS in 1 mol/L NaOH solution after competitive adsorption is illustrated in Figure 6. The adsorption efficiency shows a declining trend with increasing cycles. During the first two cycles, the adsorption efficiency decreases rapidly and then gradually stabilizes after three cycles. After five cycles of reuse, the removal percentages of $Pb^{2+}$, $Cd^{2+}$, and $Cr^{3+}$ decrease by 21.48%, 13.56%, and 23.54%, respectively.

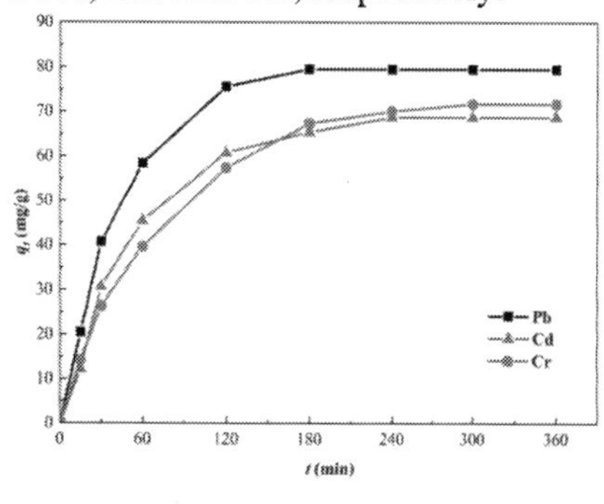


**Figure 5.** Adsorption kinetic curve.

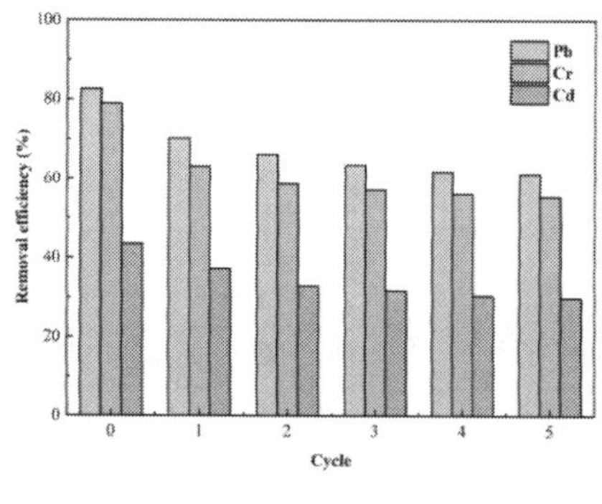


**Figure 6.** The adsorption efficiency of $Fe_3O_4$@MCS over five consecutive desorption-adsorption cycles.

### *3.3. Adsorption mechanisms*

Figure 7 presents the XPS full spectrum stacking plot of magnetic mesoporous calcium silicate after the adsorption of heavy metal ions, along with high-resolution scans for Pb, Cd, and Cr. As shown in Figure 7a, the survey spectrum of $Fe_3O_4$@MCS-Pb/Cd/Cr displays peaks corresponding to Pb 4f, Cd 3d, and Cr 2p, indicating that $Fe_3O_4$@MCS is capable of adsorbing all three heavy metal ions simultaneously on its surface. However, the peak intensities for all three metals are lower compared to their individual adsorption, suggesting that the adsorption amounts of $Pb^{2+}$, $Cd^{2+}$, and $Cr^{3+}$ are reduced during competitive adsorption relative to their individual adsorption scenarios.

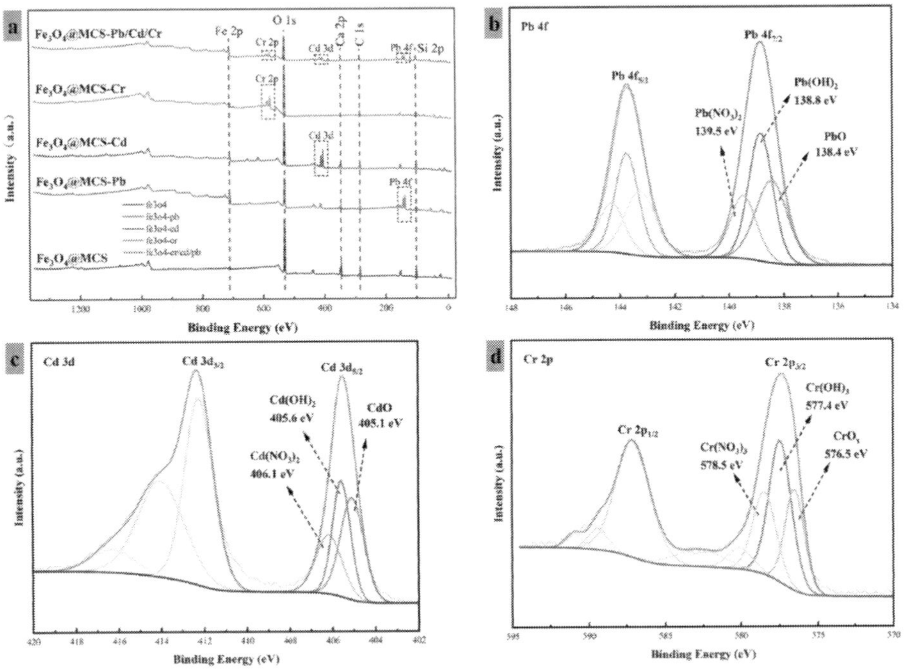


**Figure 7.** XPS spectrum of $Fe_3O_4$@MCS after competitive adsorption.

To investigate the forms in which Pb, Cd, and Cr exist on the surface of $Fe_3O_4$@MCS, XPS analysis was performed on $Fe_3O_4$@MCS-Pb/Cd/Cr after heavy metal ion adsorption. Figure 7a shows the binding energy spectrum of Pb 4f electrons, with two peaks corresponding to the spin-orbit components Pb $4f_{7/2}$ and Pb $4f_{5/2}$. Deconvolution of the Pb $4f_{7/2}$ peak yields three peaks at binding energies of 138.4 eV, 138.8 eV, and 139.5 eV, corresponding to PbO, $Pb(OH)_2$, and $Pb(NO_3)_2$, respectively. This indicates that $Pb^{2+}$ forms complexes with hydroxyl groups or Si-O on the $Fe_3O_4$@MCS surface, while some $Pb^{2+}$ forms hydroxides with $OH^-$ and is removed from the solution. The peak at 139.5 eV suggests that $Pb(NO_3)_2$ is likely due to the physical adsorption of $Pb^{2+}$ by $Fe_3O_4$@MCS.

Similarly, the deconvolution of the Cd $3d_{5/2}$ peak in Figure 7c shows peaks at binding energies of 405.1 eV, 405.6 eV, and 406.1 eV, corresponding to CdO, $Cd(OH)_2$, and $Cd(NO_3)_2$, respectively. In Figure 7d, the binding energies of $CrO_x$, $Cr(OH)_3$, and $Cr(NO_3)_3$ are shown by the peaks at 576.5 eV, 577.4 eV, and 578.5 eV, respectively, indicating that the removal mechanisms of $Cd^{2+}$ and $Cr^{3+}$ include physical adsorption, surface chemical precipitation, and surface complexation. The removal mechanisms of $Pb^{2+}$, $Cd^{2+}$, and $Cr^{3+}$ are similar, involving competition for active adsorption sites, which reduces their individual adsorption capacities compared to single-ion adsorption scenarios.

## 4. Conclusions

In summary, $Fe_3O_4$@MCS was synthesized by improved solvothermal and chemical precipitation methods, and the adsorption effect of $Fe_3O_4$@MCS could be effectively improved by increasing the adsorbent dosage, increasing the solution pH and appropriately elevating the adsorption temperature and supplemented with oscillation. When $Fe_3O_4$@MCS is used to treat a simulated solution containing $Pb^{2+}$, $Cd^{2+}$, and $Cr^{3+}$ each at an initial concentration of 50 mg/L, the optimal adsorption operating conditions are a pH of 7.2, an adsorption temperature of 40°C, and an oscillation intensity of 200 r/min. The adsorption of Pb, Cd, and Cr by $Fe_3O_4$@MCS can reach equilibrium within 6 hours, and it maintains a high adsorption efficiency even following five cycles of adsorption-desorption. The removal mechanisms of Pb, Cd, and Cr from $Fe_3O_4$@MCS included surface precipitation reaction, surface complexation, physical adsorption, and ion exchange. These results collectively confirm that the magnetic calcium silicate adsorbent demonstrates considerable adsorption capabilities in multi-metal systems, positioning it as a heavy metal adsorbent with broad industrial applications.

**Acknowledgments**
This work is part of the interim results of the National Key Research and Development Program of China (Grant No. 2020YFC1908700).

**References**
[1] Li S, Wang L, Lu P, Li Y, Li Y, Wang Y, Yao L and Qiu D. (2025) Nanoconfined polyethyleneimine in mesoporous MCM-41 silica for heavy metal ions removal. Sep. Purif. Technol., 353: 128421.
[2] Sun H, Wang L, Wang X, Dong Y, Xin J, Ren J, An J and Jing R. (2024) Construction of recyclable magnetic PEI-modified GO/$Fe_3O_4$ adsorbents for high-efficient removal of Cr(VI) ions from aqueous solutions. Solid State Sci., 154: 107611.
[3] Yazdi F, Sepehrian M and Anbia M. (2024) Selective removal of trace Pb(II) from water using tin dioxide nanoparticle-modified porous materials: A comprehensive study on batch, and fixed-bed column adsorption. J. Water Process Eng., 64: 105532.
[4] Saravanan P, Saravanan V, Rajeshkannan R, Arnica G, Rajasimman M, Baskar G and Pugazhendhi A. (2024) Comprehensive review on toxic heavy metals in the aquatic system: sources, identification, treatment strategies, and health risk assessment. Environ. Res., 258: 119440.
[5] Villafranca J C, Berton P, Ferguson M, Clausen R, Arancibia-Miranda N and Martinis E M. (2024) Aluminosilicates-based nanosorbents for heavy metal removal - A review. J. Hazard. Mater., 474: 134552.
[6] Dogari H, Salehi M M, Hassanzadeh-Afruzi F, Saeidirad M and Maleki A. (2023) Magnetic polyacrylonitrile-melamine nanoadsorbent (PAN-Mel@$Fe_3O_4$) for effective adsorption of Cd (II) and Pb (II) from aquatic area. Mater. Sci. Eng. B, 298: 116871.
[7] Sharifi M J, Nouralishahi A and Hallajisani A. (2023) $Fe_3O_4$-chitosan nanocomposite as a magnetic biosorbent for removal of nickel and cobalt heavy metals from polluted water. Int. J. Biol. Macromol., 248: 125984.
[8] Usman M, Ahmed A, Yu B, Rafiq M, Ji Z, Shen Y and Cong H. (2024) Installation of synergetic binding sites in β-Cyclodextrin-Bipyridine ionic liquid based magnetic sorbent for simultaneous removal of anionic PFAS and Cr (VI) in water matrix. Sep. Purif. Technol., 335: 126190.

# Effects of the chain extenders on the properties of waterborne polyurethanes

**Xiue Ren*, Yiqiang Huang, Jiacheng Yu, Zenan Zheng, Tingbin Fan and Sihong Wang**

Zhuhai College of Science and Technology, Zhuhai 519000, Guangdong, China

*Corresponding author's e-mail: renxiue19@zcst.edu.cn

**Abstract**: This study investigated the effect of different chain extenders on the properties and degradability of waterborne polyurethanes (WPUs). Three types of waterborne polyurethanes were prepared by polyethylene glycol (PEG) and isophorone diisocyanate (IPDI) with the addition of isosorbide (IS), 1,4-butanediol (BDO), and imidazolidinyl urea (IU) as chain extender, respectively. The experimental results showed that the waterborne polyurethanes with imidazolidinylurea (IU) as the chain extender performed better in terms of mechanical properties and degradability, and displayed good hydrophilicity and solubility. These findings are of great significance for the design and application of polyurethane materials with environmental friendliness.

## 1. Introduction

As global environmental consciousness grows, demand for eco-friendly, biodegradable polymers like polyurethane (PU) is rising. PU, used in packaging, furniture, and construction, is a crucial polymer material [1, 2]. Polyurethane (PU) materials are synthesized from polyol and isocyanate, forming multiblock copolymers with repeating urethane units (-NHCOO-) in their molecular chains [3, 4]. Polyurethane (PU) materials' versatility stems from their core components, polyols, isocyanates, chain extenders, and cross-linkers. Their varied combinations enable PU to tailor properties for various applications. The performance of Waterborne Polyurethane (WPU) is significantly influenced by the types of polyols and chain extenders, the ratio of hard to soft segments, and chain extension methods. By scientifically adjusting and optimizing WPU's structure, its performance can be greatly improved.

Most studies on polyurethanes (PUs) have concentrated on how chain extenders affect their mechanical properties and chemical stability. There has been less focus on their degradability, particularly for waterborne PUs and their environmental impact. Moreover, the use of new chain extenders like imidazolidinyl urea (IU) in waterborne PUs remains understudied, offering a promising research avenue.

Isosorbide (IS) is a biobased bicyclic diol with two cis-tetrahydrofuran rings and two hydroxyl groups [5]. IS can be prepared through the sequential dehydration of sorbitol and is widely used as a plasticizer, monomer, polymer, or in biomedical applications [6]. The introduction of isosorbide (IS) into copolymers can reduce crystallinity, elevate the glass transition temperature of the copolymers, and enhance their ultimate mechanical properties [7]. IS, an abundant dihydroxyhexanol, is biodegradable, chemically stable, and thermally stable, making it suitable for reinforcing long-chain polyesters [8]. In addition, due to its harmless nature, it can be used as a substitute for Bisphenol A [9]. The hydroxyl groups in their molecular structure exhibit high reactivity, effectively reacting with isocyanate groups to enhance the cross-linking degree and molecular weight of polyurethane chains. 1,4-Butanediol (1,4-BDO) is a key non-natural diol that has a wide range of applications in many fields such as pharmaceuticals, chemicals, textiles, and light industries [10]. 1,4-Butanediol (BDO) is a commonly used linear diol with a straight-chain structure that facilitates the formation of long-chain linear block copolymers. The introduction of BDO typically increases the flexibility and ductility of polyurethane materials for application scenarios that require a higher modulus of elasticity. Imidazolidinyl urea (IU), on the other hand, is a functional chain extender containing an imidazole ring, and its special chemical structure imparts unique properties to polyurethane materials. In addition, the presence of an imidazole ring may also promote the biodegradation process of polyurethane materials, which is in line with the current trend of green chemistry.

Waterborne polyurethane, recognized as an environmentally friendly polymer with water as its dispersion medium, has attracted considerable attention in a broad spectrum of industrial applications owing to its sustainability and low toxicity. Waterborne polyurethanes were developed to minimize the use of volatile organic compounds (VOCs). This paper innovatively investigates how three chain extenders, IS, BDO, and IU affect the properties and degradability of waterborne polyurethanes made from polyethylene glycol (PEG) and isophorone diisocyanate (IPDI). Notably, the study found that using imidazolidinyl urea (IU) as a chain extender enhanced the mechanical properties and degradability of the polyurethanes, offering valuable data for material development in this sector.

## 2. Experimental section

### *2.1. Materials*

Polyethylene glycol (PEG, Mn2000) was purchased from McLean, Shanghai; isophorone diisocyanate (IPDI) was purchased from McLean, Shanghai; bismuth neodecanoate (OB) was purchased from McLean, Shanghai; isosorbide (IS) was purchased from McLean, Shanghai; 1,4-butanediol (BDO) was purchased from McLean, Shanghai; imidazolidinylurea (IU) was purchased from McLean, Shanghai; hydroxyethyl acrylate (HEA) was purchased from McLean, Shanghai; acetone was purchased from Guangdong Yudan Bali Technology Co., Ltd.; camphor quinone (CQ) was purchased from McLean Shanghai.

### *2.2. Preparation of WPU*

The aqueous polyurethane (WPU) preparation process begins with drying 50.2 g of PEG2000 under vacuum at 80°C for 8 hours. Next, 12.33 ml of IPDI and bismuth neoaurate catalyst are added and stirred at 70°C for 1 hour. The reaction is then expanded with 2 g each of isosorbide (IS), 1,4-butanediol (BDO), and imidazolidinylurea (IU), stirred for 2 more hours at 70°C. Afterward, 5 ml of HEA is added dropwise and stirred for 3 hours at 70°C to form a prepolymer. Ultrapure water is added under 1, 500 rpm stirring for 10 minutes, followed by vacuum distillation to remove acetone, yielding photocurable WPU-IS, WPU-BDO, and WPU-IU.

### *2.3. Preparation of polyurethane film*

Photoinitiator camphor quinone (CQ) was dissolved in anhydrous ethanol and mixed with aqueous polyurethane to form a homogeneous solution. This solution was then poured into polytetrafluoroethylene molds, cured under 460 nm light for 5 minutes, air-dried for 24 hours at room temperature, and finally heated in a blast oven at 30°C for 48 hours to produce the polyurethane film.

### *2.4. Measurement and characterization*

*2.4.1 Fourier transform infrared spectroscopy (FTIR).* Three samples with chain extenders, namely WPU-IS, WPU-BDO, and WPU-IU, were subjected to infrared testing using the coating method. The dried samples were dissolved in tetrahydrofuran, dropped onto potassium bromide salt plates, and dried to form thin films. Infrared spectroscopy settings included 20 scans, a resolution of 4 $cm^{-1}$, and a scanning range of 400-4, 000 cm$^{-1}$. For testing of cured PU films, ATR infrared was used with 15 scans, a resolution of 4 cm$^{-1}$, and the same scanning range.

*2.4.2 Scanning electron microscope (SEM).* WPU-IS, WPU-BDO, and WPU-IU underwent brittle fracturing using liquid nitrogen. The cross sections were then sprayed with gold for preparation. Micro-morphology was examined by using scanning electron microscopy.

*2.4.3 Water contact angle.* 5 μL of ultrapure water was dropped on the surface of dry films with different ratios of WPU-IS, WPU-BDO, and WPU-IU, and the contact angles of the surfaces of the dry films of WPU-IS, WPU-BDO, and WPU-IU on the water were tested by a contact angle meter, which was then used to study the wettability of the surface of the polyurethane films on water.

*2.4.4 Rate of swelling.* The dried WPU-IS, WPU-BDO, and WPU-IU films with constant weight $m_0$ were immersed in ultrapure water at room temperature. After a time interval of t, the films were removed, and excess water was removed by using filter paper before weighing to obtain $m_t$. The films were then re-immersed until they reached a constant weight due to water absorption. The water absorption swelling ratio S of the PU films was calculated by using the formula.

$$S = (m_t - m_0) \div m_0$$

*2.4.5 Mechanical properties.* Specimens of WPU-IS, WPU-BDO, and WPU-IU dry films were prepared in a dumbbell shape. Testing was conducted by using a universal tensile testing machine with a tensile speed set at 50 mm/min to calculate the tensile strength and elongation at break.

*2.4.6 Bond strength test.* The PEG film was cut into 1 cm×10 cm size. After removal, the 1 cm×10 cm size polyurethane sample was evenly coated on the film and cured by a UV lamp, and the shear strength was measured by a universal tensile tester after bonding, and the tensile speed was set to 20 mm/min.

*2.4.7 Degradability tests.* We take a certain amount of cured WPU-IS, WPU-BDO, and WPU-IU dry film in a beaker, take a certain amount of PH7.4 buffer solution, pour it into the beaker that did not cover the sample, respectively, remove once, wipe clean, and put in the blower oven at 28°C drying for 24 h, weighing, calculating the quality of the loss, and then putting back to the beaker again.

## 3. Results and analysis

### *3.1. Infrared spectral analysis*

FTIR analysis was conducted on the samples of WPU-IS, WPU-BDO, and WPU-IU. The results are shown in Figure 1.Before curing, the infrared spectra of these samples exhibited high similarity. The characteristic peaks observed included the stretching vibration peaks for $-CH_2$ and $-CH_3$ at 2, 962 cm$^{-1}$ and 2, 868 cm$^{-1}$ respectively, as well as the stretching and in-plane bending vibration peaks for N-H at 3, 379 cm$^{-1}$ and 1, 541 cm$^{-1}$. The C=O stretching vibration peak of the urethane group was located at 1, 714 cm$^{-1}$. Additionally, the asymmetric and symmetric bending vibration peaks for C-N at 1, 454 cm$^{-1}$ and 1, 354 cm$^{-1}$ confirmed the formation of urethane bonds. The peak at 1, 097 cm$^{-1}$ represented the stretching vibration of C-O-C, while the peak at 1, 651 cm$^{-1}$ indicated the presence of C=C bonds, proving successful HEA capping. For WPU-IS, a stronger peak at 1, 103 cm$^{-1}$ suggested the introduction of isosorbide's ether bond. In contrast, the enhanced peak at 1, 722 cm$^{-1}$ in WPU-IU originated from the carbonyl group of imidazolinyl urea, indicating its successful incorporation into the polymer structure.

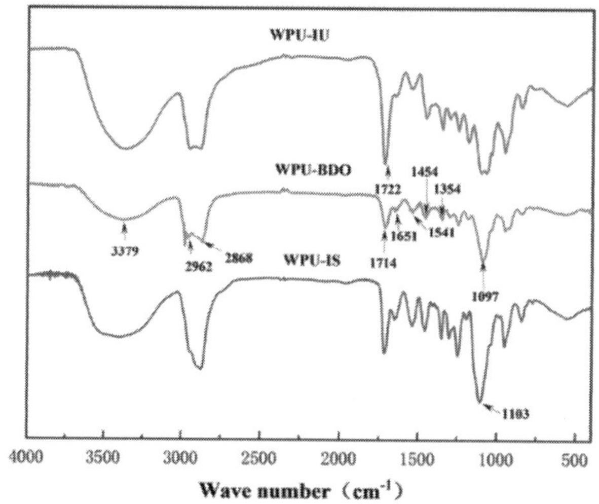


**Figure 1.** WPU infrared spectrogram before curing.

### *3.2. Scanning electron microscopy analysis*

Images of polyurethane (WPU) films observed under scanning electron microscope (SEM) reveal the cross-section characteristics of the films with three different soft and hard segment contents. The results are shown in Figure 2, where the WPU-BDO film exhibits a smoother cross-section due to its higher soft segment content, which is attributed to the formation of a continuous phase composed of soft segments after fracture. In contrast, the WPU-IS and WPU-IU films exhibit a microphase-separated scale-like structure in their cross sections due to their increased hard segment content, where the brighter regions are formed by hard segments and the darker regions are continuous phases composed of soft segments. This structural difference reflects the different distribution of soft and hard segments within the material, which in turn affects the physical properties of the films.

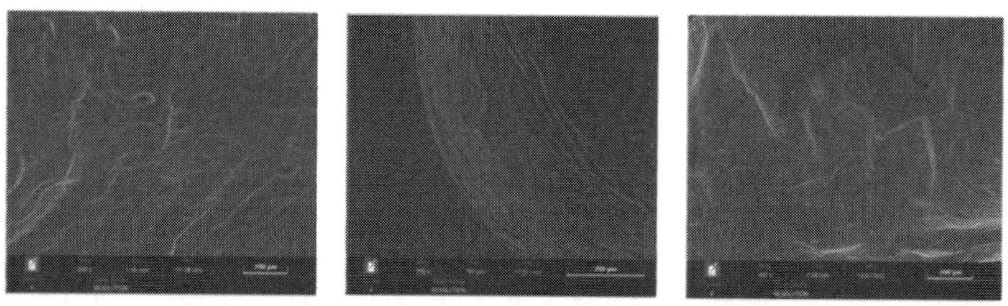

**Figure 2.** SEM plots of WPU-IS, WPU-BDO, and WPU-IU.

### *3.3. Water contact angle analysis*

The water contact angle test of the WPU dry film was performed, and the results are shown in Figure 3, which shows that the water contact angles of polyurethane dry films with different chain extenders can be observed to be different. When imidazolidinyl urea (IU) is the chain extender, due to the ester group in its structure, the cross-linking network formed during light curing is denser, and the hydrophobicity is better than that of the polyurethane dry film formed by IS, and BDO chain extender, so the water contact angle on the surface of the WPU-IU film becomes larger.

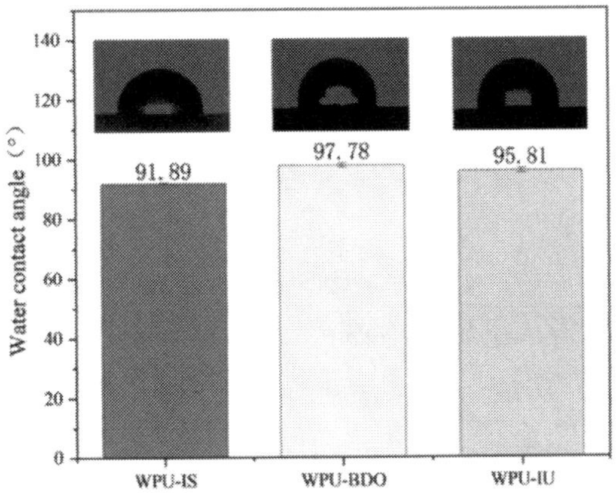


**Figure 3.** The water contact angle of WPU dry film.

### *3.4. Water absorption*

To assess the impact of chain extenders on polyurethane (PU) film properties, we conducted a swelling rate test.The results are showed in Figure 4. The WPU-IS film reached a 332.6% swelling rate at equilibrium after 11 hours of water absorption. The WPU-BDO film achieved a 476.1% rate after 10 hours, and the WPU-IU film reached a higher 642.7% rate at the same time. The higher swelling rate of WPU-IU is due to its increased polar groups, enhancing hydrophilicity. WPU-BDO, using 1,4-butanediol, likely formed a more regular network, resulting in a lower water absorption rate than WPU-IU but higher than WPU-IS. These insights are crucial for optimizing PU material design.

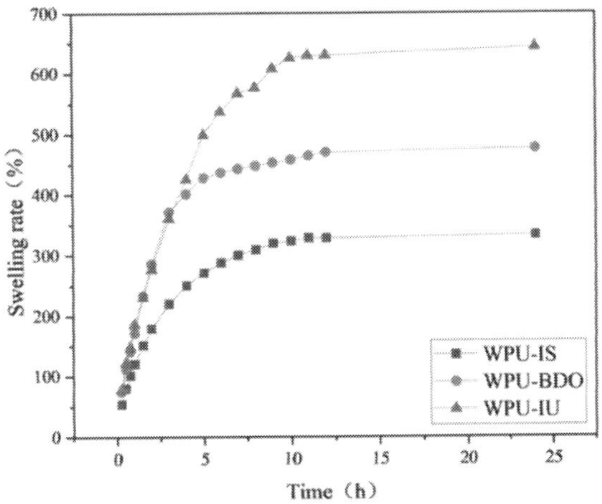


**Figure 4.** The dissolution rate of WPU films with time.

### *3.5. Mechanical properties*

This study explores the impact of chain extenders on the mechanical properties of polyurethane (PU) films. The results are showed in Figure 5. WPU-IS films, with low cross-linking, exhibit low tensile strength (0.37 MPa) but high elongation (271%). WPU-IU films perform better, with a tensile strength of 3.49 MPa and elongation of 950.1%. WPU-BDO films also show high tensile strength (3.58 MPa) and elongation (648.6%). The introduction of an IU chain extender is crucial, as it introduces multiple hydrogen bonds, enhancing intermolecular interactions and chain entanglement, thus improving mechanical strength. The reversible non-covalent bonds (e.g., hydrogen bonds) in polymers dynamically break and reform under stress, facilitating chain stretching and sliding during deformation, endowing the material with good elasticity.

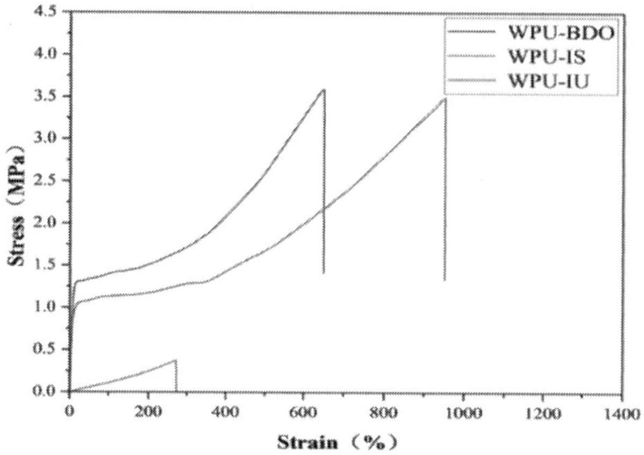


**Figure 5.** Stress-strain curve of WPU film.

### *3.6. Adhesion strength analysis*

The bonding test results of WPU-IS, WPU-BDO, and WPU-IU are shown in Figure 6. It can be seen that WPU-IU has the best bonding ability. This is because imidazolidinylurea contains several polar groups, such as amino groups, which can form strong hydrogen bonding interactions with the surface of the PEG film, so the polyurethanes with imidazolidinylurea as the chain extender show better adhesion.

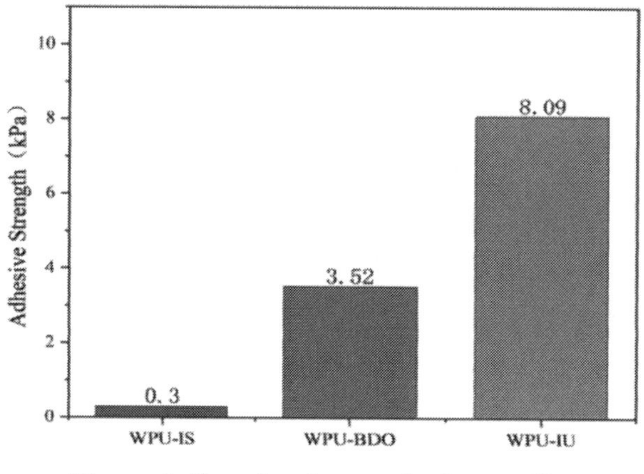


**Figure 6.** Shear bond strength of WPU films.

### *3.7. In vitro biodegradability*

The enzymatic hydrolysis of polyurethane (PU) heavily depends on breaking the urethane bond. Lipase first attacks and hydrolyzes the ester bond in PU, and then hydrolyzes the amide bond, resulting in the breakdown of the polymer chain into low molecular weight compounds. These PU fragments are further degraded. Figure 7 displays the mass loss curves of three waterborne polyurethane (WPU) membranes at 28°C with 0.01 M PBS buffer and lipase. The results indicate that all WPUs degraded over time, with the fastest rate in the first 50 minutes. WPU-BDO showed the least degradation at 10%, followed by WPU-IS, while WPU-IU had the highest at 28%. This could be due to the stability of 1,4-butanediol (BDO), a linear aliphatic diol with a stable molecular structure, making WPU-BDO more resistant to degradation. In contrast, imidazolidinyl urea (IU), used as a chain extender, contains multiple nitrogen and oxygen atoms that may enhance degradation, especially under microbial or hydrolytic conditions. The high hydrophilicity of IU increases water sensitivity and degradation rate for WPU-IU. Conversely, the hydrophobicity of BDO and isosorbide (IS) reduces water uptake and degradation in their respective WPUs.

IOP Publishing
doi:10.1088/1742-6596/2961/1/012025

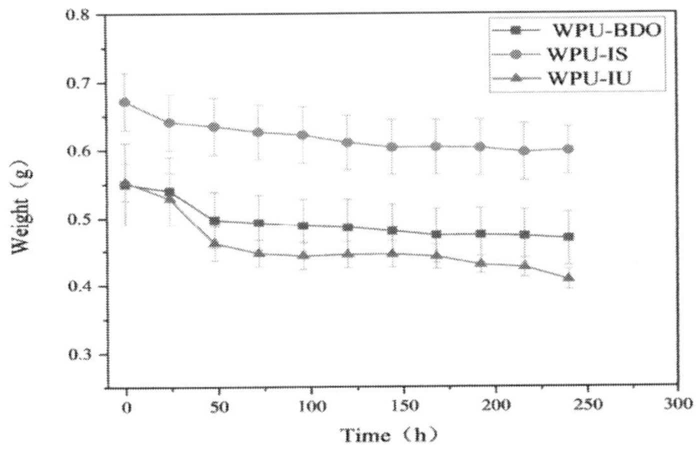


**Figure 7.** Degradability curve of WPU film.

## 4. Conclusion

This study explored the impact of various chain extenders, isosorbide (IS), 1,4-butanediol (BDO), and imidazolidinylurea (IU) on the properties and degradability of waterborne polyurethanes (WPUs) synthesized from polyethylene glycol (PEG) and isophorone diisocyanate (IPDI). The results indicated that WPU with IU as the chain extender had superior mechanical properties and degradability, along with desirable hydrophilicity and solubility. This finding suggests that selecting the right chain extenders can enhance WPU performance and environmental sustainability, offering new insights for developing eco-friendly materials. Future research could investigate other extenders to further improve WPU properties.

## Acknowledgments

This research was financially supported by the Guangdong Provincial Engineering Technology Research Center of the Green Polyurethane Adhesives, Zhuhai College of Science and Technology (Grant No. 2021GCZX010), the Key Construction Academic Subject for Scientific Research Quality Improvement of Guangdong Province (No. 2021ZDJS139), and the Key Research Platforms and Projects of Universities in Guangdong Province (No. 2022ZDZX3039).

## References

[1] Cornille A., R. Auvergne, O. Figovsky, B. Boutevin, and S. Caillol (2017). A perspective approach to sustainable routes for non-isocyanate polyurethanes. European Polymer Journal 87: 535-552.
[2] Engels H. W., H. G. Pirkl, R. Albers, R. W. Albach, J. Krause, A. Hoffmann, H. Casselmann, and J. Dormish (2013). Polyurethanes: Versatile Materials and Sustainable Problem Solvers for Today's Challenges. Angewandte Chemie International Edition, 52 (36): 9, 422-9, 441.
[3] Jutrzenka Trzebiatowska P., A. Santamaria Echart, T. Calvo Correas, A. Eceiza, and J. Datta (2018). The changes in crosslink density of polyurethanes synthesized using the recycled component. Chemical structure and mechanical properties investigations. Progress in Organic Coatings, 115: 41-48.
[4] Fu K., C. Lu, Y. Liu, H. Zhang, B. Zhang, H. Zhang, F. Zhou, Q. Zhang, and B. Zhu (2021). Mechanically robust, self-healing superhydrophobic anti-icing coatings based on a novel fluorinated polyurethane synthesized by a two-step thiol click reaction. Chemical Engineering Journal, 404: 127, 110.
[5] Song D., J. He, Q. Xie, X. Lv, Y. Liu, and L. Zheng (2024). Nonisocyanate Polyurethanes Originated from Biobased Isosorbide with Good Combined Properties. ACS Applied Polymer Materials, 6 (14): 8, 133-8, 141.

Journal of Physics: Conference Series **2961** (2025) 012025 doi:10.1088/1742-6596/2961/1/012025

[6] Saxon D. J., A. M. Luke, H. Sajjad, W. B. Tolman, and T. M. Reineke (2020). Next-generation polymers: Isosorbide as a renewable alternative. Progress in Polymer Science, 101: 101, 196.
[7] Pan X., Y. Tian, J. Li, Q. Tan, and J. Ren (2022). Bio-based polyurethane reactive hot-melt adhesives derived from isosorbide-based polyester polyols with different carbon chain lengths. Chemical Engineering Science, 264: 118, 152.
[8] de Andrade, Â. G. L. L., and S. H. Pezzin (2022). Production of "Green" Thermoplastic Polyurethanes from Decamethylene Sebacate-co-Isosorbide Copolymers. Macromolecular Symposia 406 (1): 2, 200, 051.
[9] Hong S. M., H. J. Kwon, and C. W. Lee (2023). Synthesizing Polyurethane Using Isosorbide in Primary Alcohol Form and Its Biocompatibility Properties, 15 (2): 418.
[10] Chocarro Wrona C., J. de Vicente, C. Antich, G. Jiménez, D. Martínez Moreno, E. Carrillo, E. Montañez, P. Gálvez Martín, M. Perán, E. López Ruiz, and J. A. Marchal (2021). Validation of the 1,4-butanediol thermoplastic polyurethane as a novel material for 3D bioprinting applications. Bioengineering & Translational Medicine, 6 (1): e10, 192.

# Research on the synthesis and properties of EPEG high water reducing polycarboxylate superplasticizer at room temperature

**Xianxian Yang**

KZJ New Materials Group Guangdong Co., Ltd., Foshan, Guangdong, 528000, China

Corresponding author's e-mail: 1849462918@qq.com

**Abstract:** A highly water reducing polycarboxylate superplasticizer was prepared at room temperature using ethylene glycol monovinyl polyethylene glycol ether (EPEG, content of 60%), acrylic acid (AA), hydroxyethyl acrylate (HEA), and multiple esters as the main synthetic materials, under the action of initiators and based on the principle of the free radical polymerization reaction. The influence of different synthetic factors on the dispersibility of polycarboxylate superplasticizers was studied through orthogonal experimental analysis and single-factor adjustment. The results showed that the optimal synthesis process was as follows: the acid ether ratio was 2.2, and the amounts of reducing agent, hydrogen peroxide, and mercaptoethanol were 0.07%, 0.55%, and 0.24% of the amount of EPEG, respectively. The synthesis temperature was controlled within the range of 25°C to 35°C. At this time, the synthesized polycarboxylate superplasticizer had a better water reduction rate and slump retention performance.

## 1. Introduction

In recent years, the construction of large-scale infrastructure such as highways, intercity railways, subways, tunnels, bridges, and cross-sea bridges in China has entered a period of rapid development. The requirements for polycarboxylate water reducers in practical engineering applications have become increasingly stringent. It has become extremely necessary and urgent to develop polycarboxylate water reducers with different characteristics to meet different construction needs [1-4].

This article synthesized EPEG high water reducing polycarboxylate superplasticizer by free radical solution copolymerization of polyether monomer EPEG and acrylic acid at room temperature, and studied the effect of different synthesis conditions on the dispersibility of the superplasticizer.

## 2. Experiment

### *2.1. Synthetic raw materials*

The raw materials required for the synthesis experiment are as follows: ethylene glycol monovinyl polyethylene glycol ether (EPEG, 60%), acrylic acid (AA), hydrogen peroxide ($H_2O_2$, 30%), ferrous sulfate heptahydrate ($FeSO_4 \cdot 7H_2O$), unsaturated phosphoric acid, hydroxyethyl acrylate (HEA), polyester, mercaptoethanol (BME), sodium hydroxide (NaOH, 32%), reducing agent, all of which are industrial grade; Standard tap water.

### *2.2. Testing raw materials*

The testing raw materials include: Cement, Guangdong Jinyang Cement P•042.5; Washed sand, with a mud block content of 0.1% and a fineness modulus of 3.6%; Mechanized sand with a clay content of 0.2% and a fineness modulus of 3.4; Crushed stone, with a nominal particle size of 10-20 mm and continuous grading; Fly ash, Grade II; Mineral powder, S95 grade; Water, tap water that meets the testing requirements; Additive, commercially available polycarboxylate superplasticizer (PCE-W), with a solid content of 40%.

### *2.3. Synthesis process design*

We weigh a certain amount of polyether monomer EPEG and water and add them into a four-necked flask equipped with a stirrer, temperature control device, and peristaltic pump. We start stirring to evenly mix the macromonomer with water, then add a certain amount of acrylic acid and unsaturated phosphoric acid, adjust the temperature of the water bath to the reaction temperature, and sequentially add a certain amount of ferrous sulfate heptahydrate ($FeSO_4 \cdot 7H_2O$) and $H_2O_2$. After stirring for 5 minutes, we start synchronously adding material A (dissolving acrylic acid, hydroxyethyl acrylate, and polyester in water) and material B (dissolving BME and reducing agent in water). After the dropwise addition is complete, we continue to react at a constant temperature for 1 h and finally add a certain amount of NaOH and water to prepare a 50% solid content polycarboxylate chelating agent, which is named PCE-2.

### *2.4. Testing and characterization*

(1) Fluidity of cement paste: It is tested according to GB/T 8077-2012 "Test Method for Homogeneity of Concrete Admixtures", with a polycarboxylate superplasticizer content of 0.18% and a water cement ratio of 0.29.

(2) Performance of concrete mixture: According to GB/T 50080-2016 "Test Method for Performance of Ordinary Concrete Mixture" and GB/T 50081-2019 "Standard Test Method for Physical and Mechanical Properties of Concrete", the performance of concrete mixture and hardened concrete are determined. The test concrete mix ratio is shown in Table 1.

**Table 1.** Concrete mix ratio. Unit: kg/m$^3$.

| Strength grade | C | F | K | S1 | S2 | G | W | Admixtures/% |
|---|---|---|---|---|---|---|---|---|
| C30 | 180 | 70 | 80 | 580 | 300 | 980 | 160 | 2.2 |

(3) FTIR characterization: The Perkin Elmer Spectrum 100 Fourier transform infrared spectrometer was used to fully dry the synthesized polycarboxylate chelating agent, and its infrared spectrum was detected by KBr pellet pressing. The characteristic peaks of the infrared spectrum in the absorption wave range of 400 cm-1 to 4000 cm-1 were tested.

## 3. Experimental results and discussion

### *3.1. Orthogonal result analysis*

By fixing the dosage of EPEG and other small materials, AA dosage, reducing agent dosage, $H_2O_2$ dosage, and BME dosage (all calculated as percentages of EPEG dosage) were used as orthogonal experimental design factors to conduct a three-level orthogonal experiment to study the effect of synthetic raw material ratio on the dispersibility and slump retention performance of the synthesized slump protectant. The level of orthogonal experimental factors is shown in Table 2, and the results of orthogonal analysis are shown in Tables 3 and 4.

**Table 2.** Level table of orthogonal experiment factors.

| Orthogonal design | AA dosage/% (A) | Quantity of reducing agent/% (B) | $H_2O_2$ dosage/% (C) | BME dosage/% (D) |
|---|---|---|---|---|
| Level 1 | 2.0 | 0.07 | 0.5 | 0.22 |
| Level 2 | 3.0 | 0.09 | 0.6 | 0.24 |
| Level 3 | 1.0 | 0.05 | 0.4 | 0.20 |

**Table 3.** Results of orthogonal experiment.

| Number | A | B | C | D | Fluidity of cement paste/mm | Initial expansion degree of concrete/mm | Concrete collapse difference/mm |
|---|---|---|---|---|---|---|---|
| $1^{\#}$ | 1 | 1 | 1 | 1 | 249 | 590 | 140 |
| $2^{\#}$ | 1 | 2 | 2 | 2 | 268 | 585 | 160 |
| $3^{\#}$ | 1 | 3 | 3 | 3 | 269 | 590 | 170 |
| $4^{\#}$ | 2 | 1 | 2 | 3 | 280 | 610 | 180 |
| $5^{\#}$ | 2 | 2 | 3 | 1 | 290 | 600 | 235 |
| $6^{\#}$ | 2 | 3 | 1 | 2 | 305 | 600 | 180 |
| $7^{\#}$ | 3 | 1 | 3 | 2 | 153 | 540 | 70 |
| $8^{\#}$ | 3 | 2 | 1 | 3 | 210 | 520 | 105 |
| $9^{\#}$ | 3 | 3 | 2 | 1 | 184 | 490 | 85 |

**Table 4.** Range analysis of orthogonal experiment.

| Factor | | A | B | C | D |
|---|---|---|---|---|---|
| Fluidity of cement paste | $K_1$ | 262.00 | 214.00 | 254.70 | 241.00 |
| | $K_2$ | 291.67 | 256.00 | 244.00 | 228.70 |
| | $K_3$ | 169.00 | 252.70 | 224.00 | 253.00 |
| | R | 122.67 | 42.00 | 30.67 | 24.33 |
| Initial expansion degree of concrete | $K_1$ | 588.33 | 570.00 | 570.00 | 560.00 |
| | $K_2$ | 603.33 | 568.30 | 561.67 | 565.00 |
| | $K_3$ | 506.67 | 560.00 | 566.70 | 573.30 |
| | R | 96.67 | 10.00 | 8.33 | 13.33 |
| Concrete collapse difference | $K_1$ | 156.67 | 130.00 | 141.7 | 153.3 |
| | $K_2$ | 198.33 | 166.7 | 141.67 | 136.67 |
| | $K_3$ | 86.67 | 145.0 | 158.3 | 151.7 |
| | R | 111.67 | 36.67 | 16.67 | 16.67 |

From Table 4, it can be seen that the degree of influence of various factors on the flowability of cement slurry, from high to low, is as follows: A>B>C>D; The order of influence on the initial expansion of concrete from high to low is: A>D>B>C; The degree of influence on concrete collapse, from high to low, is as follows: A>B>C=D. Taking into account the flowability of the slurry, the initial expansion of the concrete, and the difference in concrete collapse, $7^{\#}$ had the best collapse retention performance among the 9 orthogonal experiments. Through orthogonal analysis, the optimal horizontal process formula was determined as A3B1C3D2, with AA, reducing agent, $H_2O_2$, and BME dosages of 1.0%, 0.07%, 0.4%, and 0.24%, respectively. The polycarboxylate chelating agent synthesized under condition $7^{\#}$ is named PCE-1.

*3.2. The influence of acid ether ratio on the dispersibility of slump protectants*

The above orthogonal experimental results show that the concrete with the best slump retention performance synthesized by orthogonal $7^{\#}$ ratio has poor initial dispersibility. Therefore, this summary improves the dispersibility of the slump retention agent by adjusting the acid ether ratio [n (AA): n (EPEG)] based on $7^{\#}$. The effects of acid ether ratios [n (AA): n (EPEG)] of 1.6, 1.9, 2.2, 2.5, and 2.8 on the dispersibility of synthesized polycarboxylate chelating agents were investigated with a fixed $H_2O_2$ dosage of 0.5%, a reducing agent dosage of 0.07%, and a BME dosage of 0.20%. The test results are shown in Figure 1.

From Figure 1, it can be seen that as the acid ether ratio gradually increases from 1.6 to 2.8, the flowability of both 0 h and 1 h cement slurries shows a gradually increasing trend. However, when the acid ether ratio exceeds 2.2, the growth of 1 h cement slurry flowability begins to slow down, and the

Journal of Physics: Conference Series **2961** (2025) 012026
doi:10.1088/1742-6596/2961/1/012026

dispersibility of cement also weakens accordingly. When the acid ether ratio is 2.2, the flowability of the 0 h cement slurry is good, and the loss of flowability is minimal at 1 h. At this time, the synthesized polycarboxylate superplasticizer exhibits good dispersibility and dispersion retention. When the ratio of acid ether is small, the number of carboxyl groups in the main chain of the synthesized polycarboxylate chelating agent molecules is small, resulting in poor dispersibility of cement; Continuing to increase the acid ether ratio, the number of carboxyl groups in the polycarboxylate superplasticizer gradually increases, and its adsorption capacity also improves, thereby exerting a dispersing effect on the surface of cement particles. When the acid-ether ratio is high, the carboxyl content in the main chain of the polycarboxylate chelating agent molecule is too high, which weakens the steric hindrance effect. Therefore, although the resulting polymer has strong dispersibility, its dispersibility retention is weak[5-6].

### *3.3. The effect of $H_2O_2$ dosage on the dispersibility of slump protectants*

Under the condition of maintaining an acid ether ratio of 2.2 and keeping other experimental conditions unchanged, the influence of $H_2O_2$ at 0.40%, 0.45%, 0.50%, 0.55%, and 0.60% of the total mass of EPEG on the dispersibility of synthesized polycarboxylate slump protectants was investigated. The test results are shown in Figure 2.

From Figure 2, it can be seen that with the continuous increase of $H_2O_2$ dosage, the flowability of the 0 h and 1 h cement slurries mixed with synthetic polycarboxylate superplasticizer shows a trend of first increasing and then decreasing. When the dosage of $H_2O_2$ is 0.55%, the initial flowability of the synthesized polycarboxylate superplasticizer reaches its maximum value, and the loss of flowability is minimal after 1 h, indicating that the superplasticizer at this time is in the best state in terms of dispersibility and dispersion retention performance. When the amount of $H_2O_2$ exceeds the optimal value, an excessive amount of free radicals are generated, resulting in a rapid reaction rate and incomplete polymerization, which reduces the adsorption capacity of cement and deteriorates its dispersibility. When the dosage of $H_2O_2$ is lower than the optimal value, the molecular weight of the synthesized polymer is too large, which can also lead to a decrease in the dispersibility of cement[7].

### *3.4. The influence of reaction temperature on the dispersibility of slump protectants*

The reaction temperature plays a crucial role in the synthesis and performance of polycarboxylate chelating agents, and needs to be strictly controlled within an appropriate range to ensure high-quality products are obtained. This study investigated the effect of synthesis reaction temperatures of 15°C, 20°C, 25°C, 30°C, 35°C, and 40°C on the dispersibility of synthesized polycarboxylate superplasticizer, while maintaining an acid ether ratio of 2.2, $H_2O_2$ dosage of 0.55%, and other experimental conditions unchanged. The test results are shown in Figure 3.

From Figure 3, it can be seen that as the reaction temperature increases from 15°C to 40°C, the 0 h and 1 h flowability of the synthesized polycarboxylate superplasticizer in cement slurry show a trend of first increasing and then gradually decreasing. When the reaction temperature is at 15°C, the flowability of the cement slurry after 0 h is 225 mm, and the flowability loss after 1 h reaches 29 mm. At this time, the initial dispersibility and dispersibility retention of the synthesized slump inhibitor are poor. When the reaction temperature rises to 25°C, the flowability of the cement slurry reaches 240 mm after 0 h, and the flowability only loses 8 mm after 1 h. At this time, the slump protector exhibits excellent dispersion and slump retention effects in the cement slurry. Continuing to increase the temperature, the flowability of the cement slurry tends to flatten until the temperature reaches 40°C, at which point the flowability of the slurry sharply decreases. This indicates that the optimal synthesis temperature for polycarboxylate chelating agents is within the range of 25°C to 35°C, which is room temperature conditions. When the reaction temperature is below 25°C, the reaction rate slows down, the conversion rate of macromonomers decreases, and the dispersibility of the slurry decreases. If the reaction temperature is higher than 35°C, the reaction rate will be too fast, which is highly likely to cause uncontrolled phenomena such as explosion and polymerization.

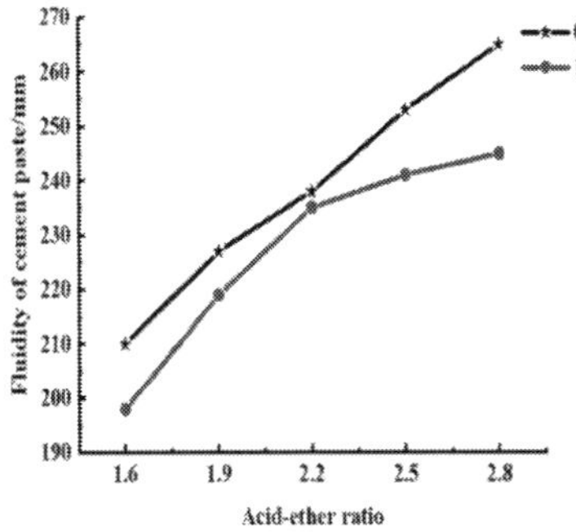


**Figure 1.** Effect of acid ether ratio on the dispersibility of polycarboxylate

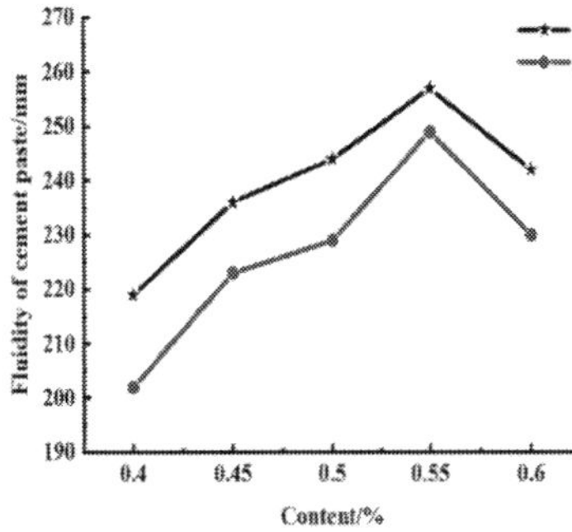


**Figure 2.** The effect of $H_2O_2$ dosage on the dispersibility of polycarboxylate chelating agents

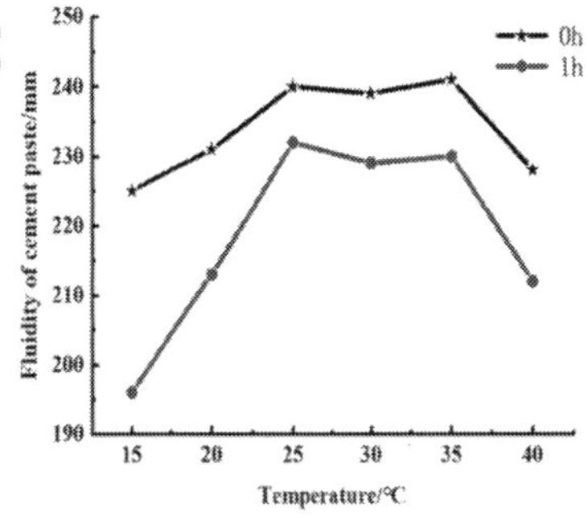


**Figure 3.** Effect of reaction temperature on the dispersibility of polycarboxylate chelating agents

*3.5. Comparison of polycarboxylate superplasticizer slurry and concrete performance*

The flowability of cement paste and concrete properties were tested using the synthesized polycarboxylate superplasticizers PCE-1 and PCE-2, as well as the commercially available polycarboxylate superplasticizer PCE-W. The results are shown in Table 5.

**Table 5.** Performance test results.

| Sample number | Fluidity of cement paste/mm | Slumps/mm | | | Extensibility/mm | | |
|---|---|---|---|---|---|---|---|
| | | 0 h | 2h | 3h | 0 h | 2h | 3h |
| PCE-1 | 185 | 230 | 220 | 190 | 560 | 480 | 360 |
| PCE-2 | 218 | 230 | 225 | 220 | 600 | 580 | 500 |
| PCE-W | 214 | 230 | 220 | 210 | 590 | 540 | 400 |

From Table 5, it can be seen that under the same solid content, comparing PCE-1, PCE-2, and PCE-W, PCE-2 has better dispersibility and dispersion retention than PCE-1; PCE-2 has comparable initial dispersibility to PCE-W, but its slump retention performance is significantly better than PCE-W, indicating that the homemade polycarboxylate slump inhibitor PCE-2 has excellent water reduction and slump retention effects.

*3.6. FTIR*

Infrared spectroscopy analysis was performed on the synthesized polycarboxylate chelating agent PCE-2, and the results are shown in Figure 4.

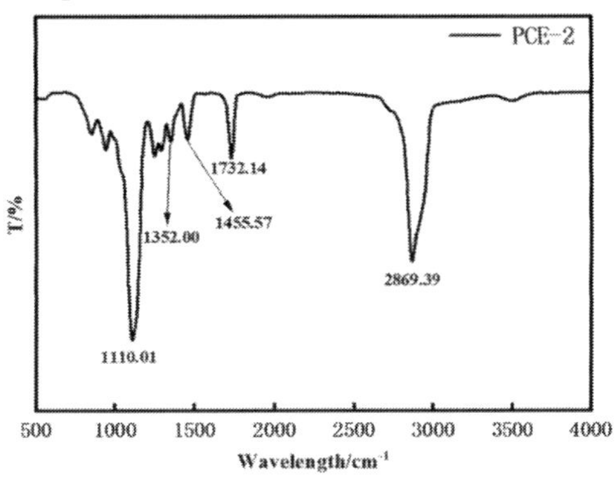


**Figure 4.** Infrared spectrum.

From Figure 3, it can be seen that the synthesized polycarboxylate chelating agent exhibits a symmetric stretching vibration characteristic peak of ether group (C-O-C) at 1110.01 $cm^{-1}$; C-H symmetric bending vibrations of methyl groups appear at 1352.00 $cm^{-1}$ and 1455.57 $cm^{-1}$; A C=O stretching vibration peak of carboxylic acid appears at 1732.14 $cm^{-1}$; A stretching vibration peak of methyl ($-CH_3$) appears at 2869.39 $cm^{-1}$, and the absorption peak is relatively small, indicating a high conversion rate of the reaction. In summary, the functional groups contained in the molecular structure of the synthesized polycarboxylate chelating agent are basically consistent with the expected molecular structure design.

## 4. Conclusion

(1) The optimal synthesis process for polycarboxylate superplasticizer was determined by studying the effects of different synthesis factors. The acid ether ratio was 2.2, and the amounts of reducing agent, hydrogen peroxide, and mercaptoethanol were 0.07%, 0.55%, and 0.24% of the amount of EPEG used, respectively. The optimal synthesis temperature range was between 25°C and 35°C, i.e., at room temperature.

(2) The net slurry flowability and concrete performance tests were conducted on PCE-1, PCE-2, and PCE-W. The results showed that PCE-2 had better dispersibility and dispersion retention than PCE-1 and PCE-W.

(3) Through infrared spectroscopy analysis, it was shown that the molecular structure of the synthesized polycarboxylate slump protector is basically consistent with the expected design.

## References

[1] Zhongquan Wu. Development of water-reducing and slump-retaining composite polycarboxylate superplasticizer and its application in concrete [D]. South China University of Technology, 2019. DOI: 10.27151/d. cnki. ghnlu. 2019.004425.

[2] Huiqun Li. Research and preparation of slow-release polycarboxylate superplasticizer [D]. Beijing University of Technology, 2010.

[3] Tuanhua Wu. Discussion on the influence of concrete raw materials on the application performance of polycarboxylate superplasticizer [J]. Construction Materials & Decoration, 2020, (02): 43-44.

[4] Wenjie Liao, Heng Wang and Changtai Ruan. Preparation and performance study of water-reducing and slump-retaining composite polycarboxylate superplasticizer [J]. Guangzhou Chemical Industry, 2021, 49(22): 51-54.

[5] Fang Y , Chen Z , Zhang X F ,et al.Effect of Ether-Typed Polycarboxylate Superplasticizer on the Adsorption, Rheology and Concrete Properties[J].Key Engineering Materials, 2022, 905:320 - 324.DOI:10.4028/www.scientific.net/KEM.905.320.

[6] Lin Z , Zhang X , Chen Z ,et al.Study on adsorption, rheology and hydration behaviours of Polycarboxylate(PCE) superplasticizer synthesized by different acid to ether ratio[J].Journal of Physics: Conference Series, 2021, 2133.DOI:10.1088/1742-6596/2133/1/012007.

[7] Liang Tan, Wenhai Yan, Hong Yang, et al. Study on the influence of redox initiation system on the performance of polycarboxylate superplasticizer [J]. Modern Chemical Research, 2021, (18): 28-29.

# First-Principles study of Hydrogen, Carbon, Nitrogen, and Oxygen atoms adsorption on $PuO_2$(110) surfaces

**Huilong Yu[1,2], Laiyang Li[1,2], Shuaipeng Wang[1], Ruizhi Qiu[1] and Suolong Yang[1,*]**

[1] Insititute of Materials, China Academy of Engineering Physics, P.O.Box 9071-12, No.9, Huafengxincun, Jiangyou City, Sichuan Province, 621907, PR China

[2] The authors contributed equally

* Corresponding author's e-mail: yuhuilong2002@126.com

**Abstract:** The investigation into the adsorption behavior of atoms of hydrogen, carbon, nitrogen, and oxygen located on the (110) face of plutonium dioxide ($PuO_2$) utilizing the projector augmented wave (PAW) method, coupled with density functional theory (DFT) incorporating on-site Coulombic interaction corrections (GGA+U), reveals a spectrum of adsorption tendencies. Within this framework, a nuanced interplay emerges, showcasing both weak and robust binding affinities with the $PuO_2$(110) surface. Detailed scrutiny of the charge density differences (CDD) and partial density of states (PDOS) at strategic adsorption locales unveils distinct mechanisms governing these interactions. Specifically, electrostatic forces emerge as the driving factor behind the weaker associations, while the more substantial adhesions are characterized by the presence of covalent bonding. This comprehensive analysis not only delineates the diversity in adsorption behaviors but also underscores the multifaceted nature of surface interactions on $PuO_2$(110), providing valuable insights into the underlying physics governing such phenomena.

## 1. Introduction

The exploration of plutonium oxides is critical owing to their significance in radioactive energy source processes and radioactive waste management, alongside their distinct chemical and physical characteristics. In dry air, the behavior of metallic plutonium is similar to that of other active metals, forming a plutonium dioxide layer ($PuO_2$), adorned with a slender coating of plutonium sesquioxide ($Pu_2O_3$) formed between the plutonium surface and the $PuO_2$ layer[1][2][3][4]. The plutonium dioxide layer could potentially come into contact with the air in fission reactors and various additional uses. Therefore, understanding this adsorptive interaction process of H, C, N, and O atoms on $PuO_2$ is essential for comprehending the subsequent corrosion and oxidation of this metal in the air[2][5].

Exploring the atomic and electronic structural characteristics of H, C, N, and O atoms adsorption on plutonium dioxide surfaces through experimental means is exceptionally challenging. This difficulty primarily stems from stringent international regulations and the radio-toxicity associated with plutonium atoms, which have substantially limited experimental investigations into plutonium and its compounds. Conversely, atomic simulations employing computational methods are not subject to these restrictions, offering a viable alternative for investigating such materials.

In studying the electronic characteristics of $PuO_2$, standard DFT employing either the LDA or the GGA fails to account for the significant on-site Coulombic repulsion among the 5f valence electrons. As a result, it incorrectly classifies $PuO_2$ as a ferromagnetic (FM) conductor[6], rather than the experimentally verified antiferromagnetic (AFM) Mott insulator[7]. Thankfully, the LDA methods and the GGA+U methods were formulated to rectify these computational inaccuracies in the analysis of compounds of actinides[8][9][10]. The successful adjustment of conventional DFT through LDA/GGA+U has been extensively validated in research on $UO_2$[8][9][10] and $PuO_2$[11][12][13][14]. Nevertheless, scant DFT investigations on the $PuO_2$ (110), (100), and (111) surfaces have been documented in literary works[15][16].

In our investigation, we employ the DFT+U approach to analyze the adsorption of the surface of $PuO_2$(110), which is covered with atoms of hydrogen, carbon, nitrogen, and oxygen, concentrating on identifying stable adsorption configurations and characterizing the adsorption mechanism. This includes determining whether the adsorption is chemisorption or physisorption, assessing any changes in the surface geometry caused by the adsorbates, and examining the outcome of the adsorbates on the valence electron arrangement of the surface.

## 2. Methodology

In this article, VASP software was the tool employed for all the calculations founded on DFT[17]. We use the PAW method to characterize the interplay of charged particles and electrons[18]. In our calculations, we exclusively employed the method of PBE[19]. Our cutoff energy was set to be greater than 450 eV, and our Kpoint is designated as 4×4×1. And Hellmann-Feynman force tolerance of 0.01 eV/Å. Given the challenge of accurately representing the 5f orbitals in $PuO_2$ employing standard DFT techniques[20][21], And opt to utilize the U value in the DFT plus U method[11][12][13][14]. In this calculation, our U value was set to 4 eV, consistent with commonly used values in the literature[11][12].

After our calculations, the optimized cell parameter value is 5.464 Å, which closely matches the experimental value (a=5.398 Å)[22]. A supercell with dimensions of 10.927 Å×7.727 Å×20.795 Å was employed to model the surface. In the surface model we constructed, there are a total of 48 atoms, as illustrated in Figure 1.

The equation used to determine the adsorption energy of surface molecules is:

$$E_{ads-atom/PuO_2(110)} = E_{atom/PuO_2(110)} - E_{PuO_2(110)} - E_{atom} \quad (1)$$

The stable configurations of adsorbates and their adsorption energies were calculated for each surface configuration. Additionally, the differential charge analysis was determined to provide insights into the bonding characteristics of atoms of hydrogen, carbon, nitrogen, and oxygen that are adsorbed onto the surface .

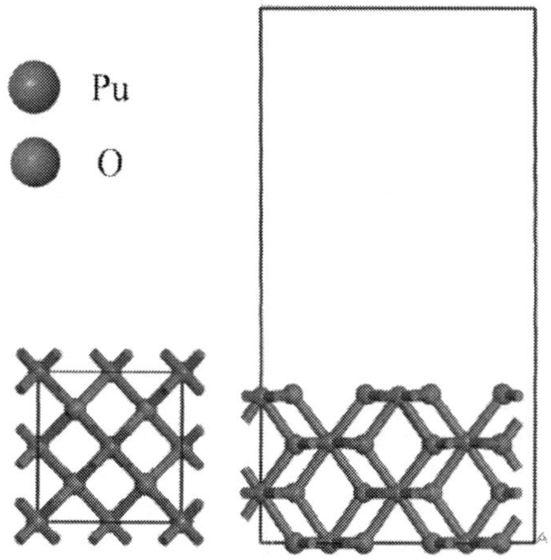


**Figure 1.** Structure of plutonium dioxide in 3D, including side and top views.

## 3. Results

### *3.1. Surface adsorption.*

In Figure 2, we clearly demonstrate the five adsorption sites on the surface of $PuO_2(110)$. Table 1 displays the adsorption energies of various adsorption structures.

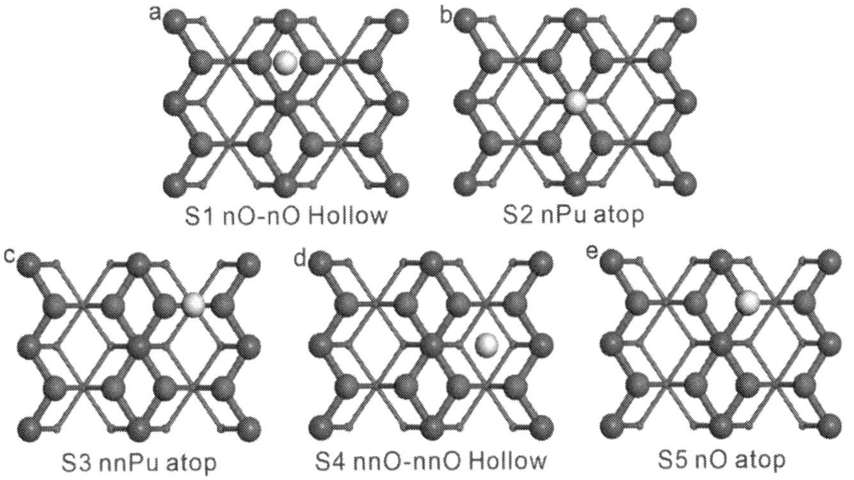


**Figure 2.** Displaying different adsorption sites of adsorbed atoms on the surface.

**Table 1.** Adsorption energy data.

| | Site | | | | |
|---|---|---|---|---|---|
| | S1 | S2 | S3 | S4 | S5 |
| H-Eads (eV) | -1.70 | -1.68 | -1.59 | -1.62 | -5.15 |
| C-Eads (eV) | -7.92 | -1.90 | -8.58 | -9.00 | — |
| O-Eads (eV) | -3.50 | -4.41 | -2.65 | -3.00 | -4.75 |
| N-Eads (eV) | -2.48 | -3.43 | -3.16 | -6.51 | -6.18 |

The most stable form of $\Delta_{Eads}$ of a hydrogen atom on this surface is -5.15 electron volts, corresponding to the structure H-S5(Figure 3a). In this configuration, the hydroxyl distance is 0.99 Å, which aligns well with the literature findings[23]. Additionally, hydrogen atoms also undergo physical adsorption on the surface, as illustrated by the H-S1 structure in Figure 3b.

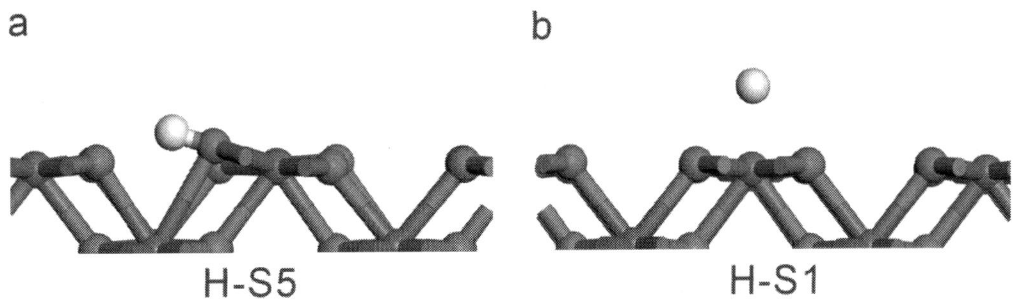


**Figure 3.** Two typical adsorption configurations of H atom.

The computational results indicate that the C-S4 adsorption configuration is the strongest adsorption configuration, with an impressive adsorption energy of -9.00 electron volts. This stable adsorption configuration is illustrated in Figure 4a. The lengths of chemical bonds of C-O extend from 1.2 angstroms to a maximum of 1.3 angstroms. Moreover, there is also physical adsorption of the C atom on the surface, as illustrated in Figure 4b.

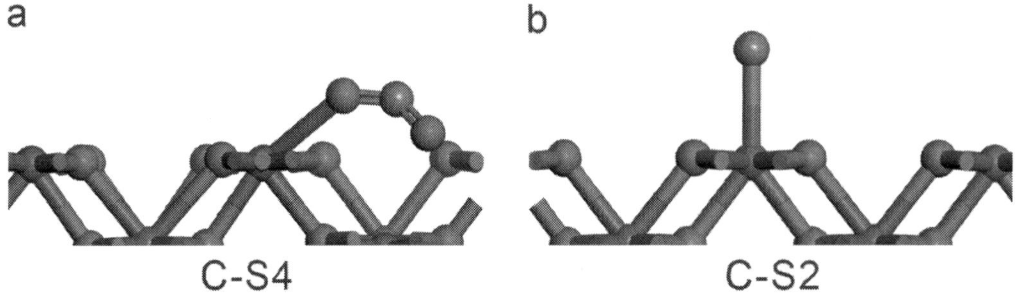


**Figure 4.** Two typical adsorption configurations of C atom.

The most intense adsorption, characterized by an adsorption energy of -4.75 eV, takes place at the surface spot nO-atop in the O-S5 top layer arrangement, positioned above an O atom from the first atomic layer. Figure **5a** depicts a side view of the O-S5 configuration, demonstrating that the O atom tends to move towards the neighboring oxygen atom between two surface Pu atoms (nPu). An evident attraction among the oxygen atom in contact with the surface Pu atom is observed, with Pu–O bond lengths measuring 2.588 Å. As shown in Figure 5b, We achieved highly comparable adsorption outcomes for the O-S2 configuration as observed in O-S5 after structural relaxations. In the O-S2 configuration, the O atom adsorbs at the nPu-atop site, flanked by four surface oxygen atoms (nO), having a separation of 1.881 angstroms between $PuO_2$ and O atom. This type is characterized as weak adsorption since the oxygen atom undergoes minimal physical changes upon adsorption.

After structural relaxations, we have obtained the following outcomes. The structure of N-S4 is shown in Figure 6a. N-S4 is the most stable configuration, and it has fewer energy advantages than N-S5. This N atom adsorbs at the nO-nO hollow point. For configuration N-S1(Figure.6b), N-S2, and N-S3, the N atom adsorbs on the first layer surface with the adsorption energies from -2.48 to -3.43 eV. This type of adsorption is characterized as weak adsorption, as the N atom undergoes minimal physical changes upon adsorption.

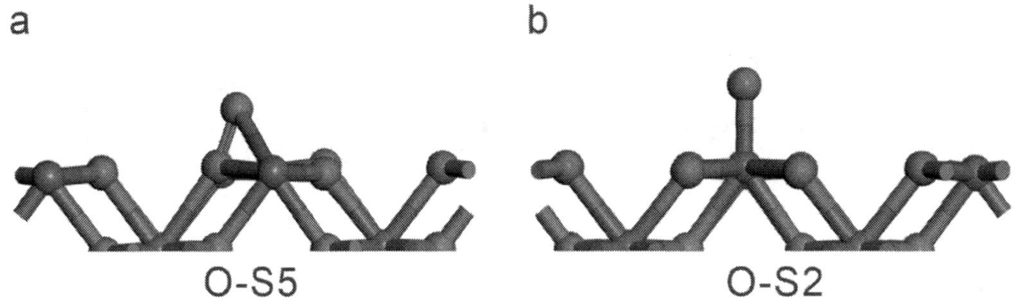


**Figure 5.** Two typical adsorption configurations of O atom.

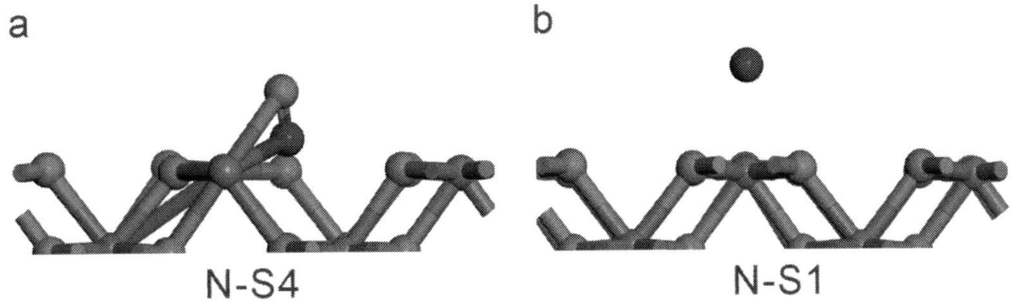


**Figure 6.** Two typical adsorption configurations of N atom.

Overall, there is a combination of robust chemisorption and mild physisorption evident across the adsorption arrangements of Atoms of hydrogen, carbon, nitrogen, and oxygen on the (110) face of $PuO_2$, as evidenced by $\Delta E_{ads}$. This is not the same as the stoichiometric $CeO_2(110)$ surface, which possesses chemical characteristics similar to plutonium dioxide, showcasing primarily feeble adsorption modes.

*3.2. Electron configuration*

To unveil the underlying mechanisms concerning the relationship between H, C, N, or O atoms and the $PuO_2$ (110) surfaces, thorough studies of the electron configuration of the aforementioned adsorptive conditions were conducted. Initially, to grasp the redistribution of charge density during the surface attachment process, the $\Delta\rho(r)$ was constructed for specific adsorption configurations of interest. $\Delta\rho(r)$ is determined using the following formula:

$$\Delta\rho(r) = \rho_{atom+PuO_2(110)}(r) - \rho_{PuO_2(110)}(r) - \rho_{atom}(r) \tag{2}$$

where $\rho_{atom+PuO_2(110)}(r)$, $\rho_{PuO_2(110)}(r)$, and $\rho_{atom}(r)$ represent the overall charge density data after structural optimization, the isolated adsorbate, and the isolated substrate, respectively. When calculating $\rho_{PuO_2(110)}(r)$ and $\rho_{atom}(r)$, the positions of the surface and atoms remained fixed. In this instance, we have calculated charge density differences for two distinct adsorption modes of H, C, N, or O atoms on the $PuO_2(110)$ surface, respectively, as portrayed in Figure 7-10. This unit for the differential charge density(DCD) map is $e/Å^3$.

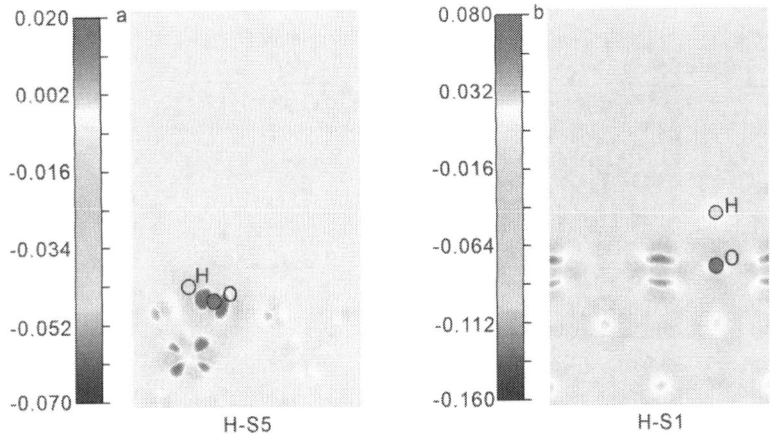


**Figure 7.** The DCD maps after surface adsorption of H atoms.

Journal of Physics: Conference Series **2961** (2025) 012027 doi:10.1088/1742-6596/2961/1/012027

The DCD map in Figure 7a reveals the electronic association between the hydrogen atom and the oxygen atom on the surface, indicating a complex charge redistribution during adsorption, distinctly demonstrating the covalent bond formation between the hydrogen atom and the adjacent oxygen on the surface. Furthermore, the complex surface charge redistribution phenomenon extends to the adjacent layer.

In Figure 7b, the weak adsorption mode (H-S1) is contrasted with the strong adsorption mode (H-S5), showing notable differences. In the adsorption configuration of H-S1, the differential charge density map exhibits high symmetry with respect to the H atom, indicating nearly equivalent H-O bonds. In this configuration, no covalent bonds exist, and weak interactions may be caused by the van der Waals attraction between the H atom and oxide substrate.

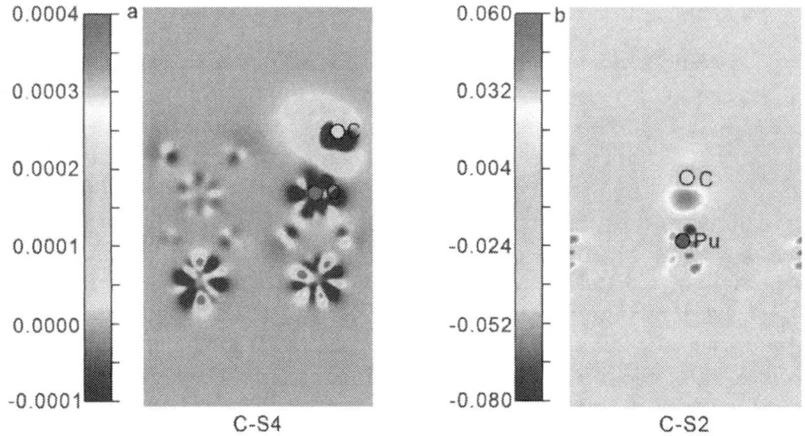


**Figure 8.** The DCD maps after surface adsorption involving carbon atoms.

The DCD map in Figure 8a illustrates the charge variation upon strong adsorption of a C atom on the surface. However, because of the absence of notable electron localization between the C atom and O atom, we speculate that the stability of adsorption may originate from the deformation of the substrate surface. Upon adsorptive attachment of the C atom to the $PuO_2$ (110) surface, two O atoms protrude slightly above the flat surface.

As the DCD map shown in Figure 8b, it is evident that there is minimal charge accumulation between the carbon atom interacting with the surface Pu atom, leading to the inference that there is no covalent bond between Pu and C. The surface has a low affinity for the C atom, which is attributed to van der Waals interactions.

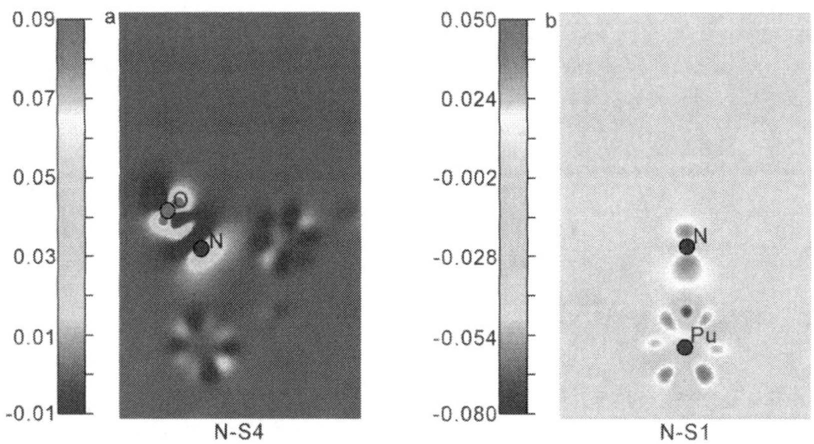


**Figure 9.** The DCD maps after surface adsorption of N atoms.

In the structure depicted in Figure 9a, the N atom embeds into the interior of the substrate during the adsorption process, displacing one O atom to the surface position. The differential charge density map indicates minimal electron localization between N and O atoms. The strong adsorption is attributed to the rearrangement of the surface structure.

Figure 9b presents the weak adsorption mode of the N atom on the surface, where no complex charge redistribution occurs during adsorption, and there is minimal impact on the charge of the substrate's second layer. It is evident that there is no covalent bond between the N atom and Pu atom, with weak interactions partially attributed to vdW interactions between the N atom and the substrate.

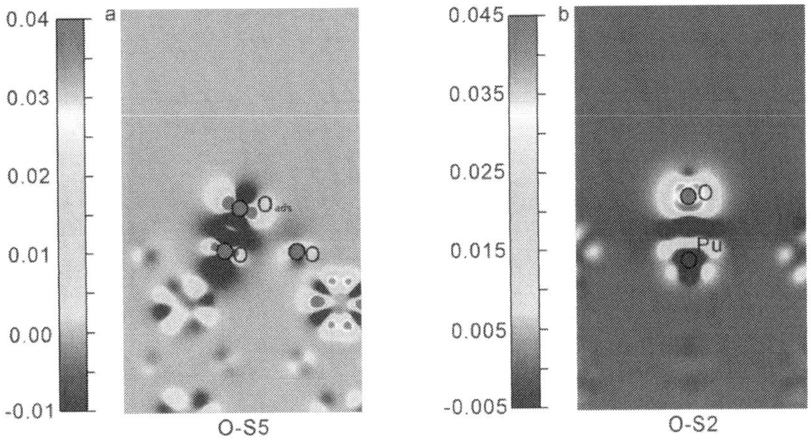


**Figure 10.** The DCD maps after surface adsorption of O atom.

The DCD map in Figure 10a reveals no significant charge accumulation among the adsorbed O atom and the surface O atoms, suggesting that the strong adsorption may be attributed to interactions with the surface Pu atom.

In Figure 10b, the weak adsorption mode (O-S2) depicted is analogous to the interactions of Hydrogen, Carbon, and Nitrogen with the substrate. Minimal charge polarization was observed between the oxygen atom and the plutonium surface atom, suggesting the absence of a covalent bond between O and Pu. The O atom is evidently adsorbed above the substrate primarily through van der Waals forces. Next, we calculated the density of states (DOS) concerning the adsorbed atom/$PuO_2$(110) system, and these results are plotted in Figure 11. To enhance clarity in the illustration, the shift of the Fermi level is now towards 0 eV. Figures a, c, e, and g collectively represent the DOS diagrams for stable adsorption of H, C, O, and N atoms on the $PuO_2$(110) surface, respectively. At the same time, Figures b, d, f, and h collectively depict the DOS diagrams for weak adsorption of H, C, O, and N atoms on the $PuO_2$(110) surface, respectively. In regions of denser band structures, there are more states within a given energy range, resulting in higher DOS. Conversely, in regions of sparser band structures, the density of states is lower. From the figures, it is evident that the influence of adsorbed atoms on the surface density of states primarily varies with the different atomic species. For a given atomic species, whether it is bound to the surface via chemisorption or weak van der Waals interactions, the impact on the DOS has a tendency to be consistent.

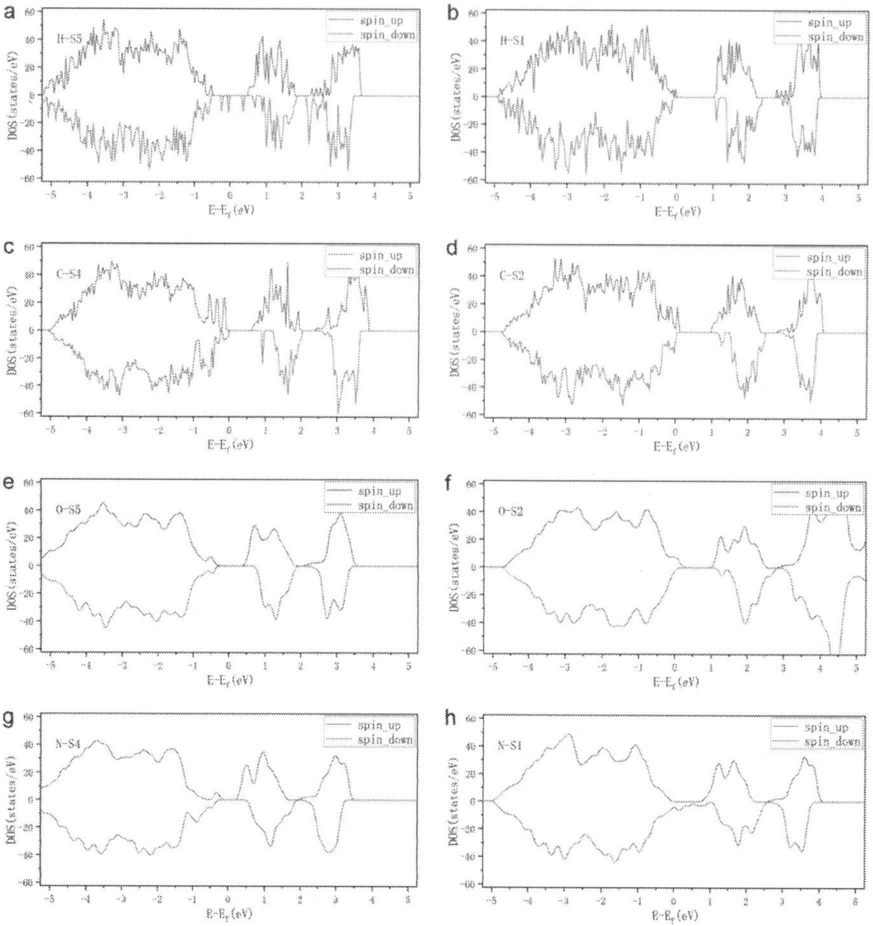


**Figure 11.** The density of states diagram for H, C, O, and N atoms.

When the relationship between the adsorbed atoms and the base material produces significant charge movement or level adjustments, the DOS diagram may exhibit pronounced irregularities. In such cases, the presence of adsorbed atoms significantly alters the electronic structure of the substrate, leading to localized density enhancements or suppressions within the DOS diagram, rendering it irregular and non-smooth, as shown in Figure 11a, b, c, d.

Conversely, when the structural and symmetry differences between the adsorbed atoms and the substrate are minimal, and charge transfer and level adjustments do not cause significant local density changes, the DOS diagram may appear smoother. O or N atoms may interact more uniformly with the $PuO_2(110)$ surface, with minimal charge transfer and level adjustments upon surface adsorption. In such instances, the influence of the adsorbed atoms on the substrate is relatively minor, resulting in a smoother DOS diagram, as depicted in Figure 11e, f, g, h.

## 4. Conclusions

The relationship between H, C, N, or O atoms accompanied by the $PuO_2(110)$ surface was investigated using the PAW method using the DFT approach employing the GGA approximate calculation. It was discovered that various adsorption configurations of H, C, N, or O atoms are possible on the $PuO_2(110)$ surface, with both weak and strong adsorption configurations observed. The weak interactions are characterized by electrostatic interactions, whereas the strong interactions are marked by the development of covalent bonds.

This investigation focused solely on the stoichiometric, flawless $PuO_2(110)$ surface. The presence of imperfections like step edges, reconfigurations, and vacancies could potentially make it possible for atoms to penetrate the $PuO_2(110)$ surface, should these imperfections exist. More research within this domain has recently been conducted.

**Acknowledgments**
Financial backing for this research was provided by CAEP, as indicated by Grant Number YZJJZL2024112.

**References**
[1] D.H. Templeton and C.H. Dauben (1952) University of California Radiation Laboratory Report No.UCRL-1886.
[2] Zhang, L. Sun, B. Zhang, Q.L. Liu, H.F and Song, H.F (2023) Theoretical insights into the hydroxyl-promoted H2 releasing reaction after H2O splitting on Pu-oxide surfaces. J. Nucl. Mater. 585, 154642.
[3] Zhang, L. Sun, B. Zhang, Q. Liu, H. Liu, K and Song, H (2020) First-Principles Study of the Hydrogen Resistance Mechanism of PuO2. ACS Omega 5, 7211.
[4] J.M. Haschke, T.H. Allen and L.A. Morales (2000) Surface and corrosion chemistry of plutonium Los Alamos Sci. 26, 252.
[5] Zhang, L. Sun, B. Zhang, Q. Liu, H. Liu, K and Song, H (2020) First-principles study of hydrogen behavior in α-Pu2O3 Comput. Mater. Sci. 179, 109688.
[6] C. E. Boettger and A. K. Ray (2022) Fully relativistic density functional calculations on hydroxylated actinide oxide surfaces Int. J. Quantum Chem. 90, 1470 -1477.
[7] C. McNeilly (1964) The electrical properties of plutonium oxides J. Nucl. Mater. 11, 53-58.
[8] J L Chen, Peter B and Nikolas K (2023) DFT + U Simulation of the X-ray Absorption Near-Edge Structure of Bulk UO2 and PuO2 J. Phys. Chem. C 127, 17994-18000
[9] Samuel M, Adam R, Joshua S.T., James D, Joseph M.F, Stephen C.P., David J.C., Marco M (2020) Phys. The energetics of carbonated PuO 2 surfaces affects nanoparticle morphology: a DFT+ U study Chem. Chem. Phys. 22, 7728
[10] J L Chen, Nikolas K (2022) DFT + U Study of Uranium Dioxide and Plutonium Dioxide with Occupation Matrix Control J. Phys. Chem. C 126, 11426-11435
[11] L Zhang, B Sun, Q L Zhang, H F Liu, K Z Liu and H F Song (2021) The adsorption thermodynamics of H2O and CO2 on PuO2 (1 1 1) surface: A comparative study based on DFT+ U-D3 Appl. Surf. Sci. 537, 147882.
[12] X Y Han and Nikolas K (2023) Effect of point defects on water adsorption on the ThO2 {111} surface: A first-principles computational study J. Nuclear Matt. 575, 154240.
[13] Gerald J and Francois B (2011) Thermodynamic stability of PuO surfaces: Influence of electronic correlations Physical Review B. 84, 195469.
[14] H Wang and Kenji K (2012) LDA+ U study of Pu and PuO2 on ground state with spin–orbital coupling Journal of Alloys and Compounds 533, 53-57.
[15] X. Wu and A.K. Ray (2001) A density functional study of plutonium dioxide Eur. Phys. J. B. 19, 345-351.
[16] X. Wu and A.K. Ray (2022) Density-functional study of water adsorption on the PuO2(110) surface Phys. Rev. B. 65, 085403.
[17] G. Kresse and Hafner (1993) Ab initio molecular dynamics for open-shell transition metals Phys. Rev. B. 48, 13115.
[18] G. Kresse and D. Joubert (1999) From ultrasoft pseudopotentials to the projector augmented-wave method Phys. Rev. B. 59, 1758.
[19] J. P. Perdew, K. Burke and M. Ernzerhof (1996) Generalized Gradient Approximation Made Simple Phys. Rev. Lett. 77, 3865-3868.
[20] C. McNeily (1964) The electrical properties of plutonium oxides J. Nucl. Mater. 11, 53-58.

[21] C.E. Boettger and A.K. Ray (2022) Fully relativistic density functional calculations on hydroxylated actinide oxidesurfaces Int. J. Quantum Chem. 90, 1470-1477.
[22] J.M. Haschke, T.H. Allen and L.A. Morales (2000) Reaction of plutonium dioxide with water: formation and properties of PuO2+x Science 287, 287.
[23] M.M. Islam, M. Calatayud and G. Pacchioni (2011) Hydrogen adsorption and diffusion on the anatase TiO2 (101) surface: a first-principles investigation J. Phys. Chem. C. 115, 680914.

Journal of Physics: Conference Series **2961** (2025) 012028 doi:10.1088/1742-6596/2961/1/012028

# Evolution and preparation of graphene oxide and deep learning image characterization

**Ziyun Zhang**[1,*]

[1]Fuzhou University, Fujian, Fuzhou, 350000, China

[*]Corresponding author's e-mail: 1694105819@qq.com

**Abstract:** Graphene-polymer nanocomposites have garnered significant scientific attention because of their excellent electron transport properties, exceptional mechanical strength, large surface area, size-selective ability, and high Young's modulus. When effectively organized at both the structural and interfacial levels, these graphene-based polymer nanocomposites demonstrate remarkable structural integrity and multifunctional capabilities. Graphene oxide (GO) stands out as one of the most promising materials within the graphene family. This review provides an overview of the synthetic process and characterization of GO and GO-based polymer nanocomposites, including GO-cement and GO-hydrogels. The dispersion of GO in composites is analyzed using techniques such as Raman spectroscopy, atomic force microscopy, scanning electron microscopy, and others. Despite some challenges, GO-based polymer nanocomposites exhibit superior performance in certain aspects and demonstrate wide potential applications.

## 1. Introduction

Polymer nanocomposites incorporating dispersed graphene have been utilized for improved mechanical, thermal, electrical, size-selective and gas barrier properties [1]. The interfacial contact between graphene and different matrices markedly enhances the uniform distribution of graphene throughout the polymer matrix. It significantly leads to enhancement in the overall performances of nanocomposites and diverse applications of polymer nanocomposites.

Graphene is a planar network of carbon atoms arranged in a hexagonal pattern with $sp^2$ hybridization, as illustrated in Figure 1. It is considered the fundamental building block of all other graphitic carbon allotropes with varying dimensionalities [2]. For example, graphite consists of graphene layers stacked with an interlayer distance of 0.342 nm. Fullerenes, a 0-dimensional carbon allotrope, can be conceptualized as sections of graphene sheets wrapped into a spherical shape. Carbon nanotubes and nanoribbons, both one-dimensional carbon allotropes, are formed by rolling or slicing graphene sheets. The initial attempt to produce graphene nanosheets in the 1970s was not perfect. However, the exfoliation of free-standing single-layer graphene was first achieved by using micromechanical cleavage in 2004.

Single-layer graphene, exhibiting Young's modulus of 1 TPa and ultimate strength of 130 GPa, is recognized as the strongest material tested to date. It also boasts an impressive thermal conductivity of 5000 W/(m·K), aligning with the highest values ever recorded. Furthermore, single-layer graphene exhibits exceptionally high electrical conductivity, reaching up to 6000 S/cm, and unlike carbon nanotubes, its electrical properties are not influenced by chirality. These incredible characteristics, coupled with an extraordinarily high surface area (theoretical limit: 2630 $m^2/g$) and gas impermeability, underscore graphene's significant potential to enhance the mechanical, electrical, thermal, and gas barrier properties of graphene-based polymer nanocomposites.

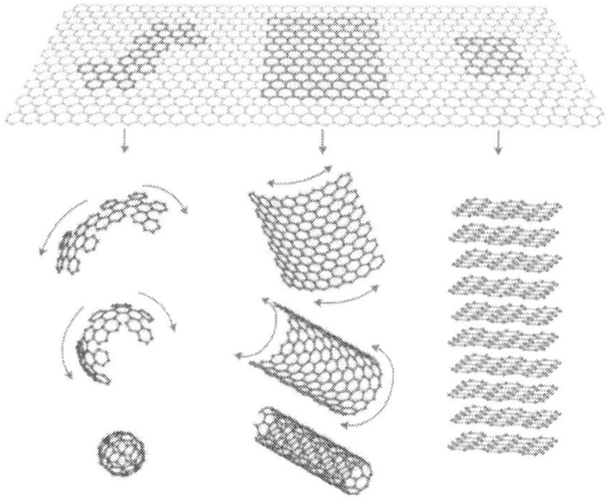

**Figure 1.** Graphene serves as a fundamental two-dimensional material for various carbon structures. It can be encapsulated into zero-dimensional buckyballs, rolled into one-dimensional nanotubes, or layered into three-dimensional graphite [2].

## 2. Graphene Oxide

Graphene oxide, derived from graphene flakes, is pivotal in material science owing to its twofold functionality. Primarily, GO serves as a precursor for reduced graphene oxide (r-GO). Additionally, GO possesses distinctive and noteworthy physical and chemical characteristics. In the field of polymer nanotechnology, the importance of GO cannot be overstated. This is primarily due to its exceptional ability to remain stably dispersed in a wide array of solvents, enhancing its versatility and making it an indispensable component in various applications. The structural makeup of GO is particularly noteworthy. Graphene oxide contains multiple oxygen-containing functional groups. These include epoxy and hydroxyl groups located on the basal plane, as well as carboxyl groups found at the edges, as illustrated in Figure 2. These groups are instrumental in imparting GO with its hydrophilic nature and its chemical reactivity, which are critical for its applications. The synthesis of GO involves a chemical process that can yield different chemical compositions. Examples of these compositions include $C_8O_{3.5-4.3}H_{2.5-2.9}$, $C_8O_{3.78-5.05}H_{2.9-4.4}$, $C_8O_{2.54}H_{3.91}$, and $C_8O_{4.61}H_{6.70}$. These differences arise from the varying amounts and types of oxygen-bearing functional groups that are incorporated during the oxidation process. The ability to manipulate these compositions allows scientists and engineers to customize the properties of GO for specific needs. For instance, GO can be tailored for use in composites, where its reinforcing properties can significantly enhance material strength. In coatings, GO's dispersion stability and reactivity can improve the performance and durability of the coating. Furthermore, in the realm of electronic devices, GO's unique properties can be exploited to develop advanced electronic components. This adaptability makes GO a highly versatile material with a broad range of potential applications.

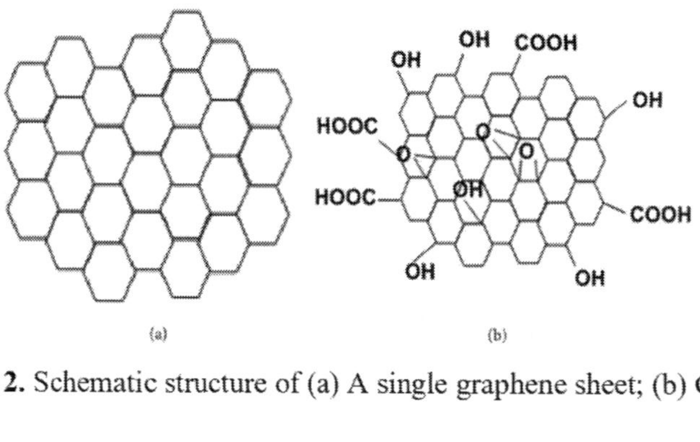


**Figure 2.** Schematic structure of (a) A single graphene sheet; (b) GO [3].

### *2.1. X-ray diffraction method*

X-ray diffraction analysis was utilized to examine the structural properties of graphene as well as graphene oxide, as shown in Figure 3. The diffraction patterns reveal a pronounced peak for graphite crystals, observed at an angle of 26.5° (2θ). In contrast, GO exhibits a peak at a significantly lower angle, around 10° (2θ). This shift in peak position is indicative of a substantial increase in the interlayer distance, which expands from 0.341 nm in pristine graphite to 0.958 nm in GO. The intrinsic stability of graphite can be attributed to its robust structural properties, which present considerable resistance to chemical reactions. The tightly packed layers in graphite form strong interactions, making it a chemically inert material under standard conditions. However, the oxidation process fundamentally alters this structure. During oxidation, the incorporation of oxygen-bearing functional groups expands the interlayer distance, which reduces the van der Waals forces binding the graphene layers. This increase in interlayer distance is a critical factor in the dispersibility of GO. With the interactions between the graphene layers significantly weakened, GO can be more readily dispersed in aqueous solutions. This expansion of the interlayer spacing is a pivotal property for various applications, particularly in creating homogeneous GO suspensions for further chemical modifications and integration into composite materials. The ability of GO to form stable aqueous dispersions opens numerous possibilities in fields such as materials science, biomedicine, and environmental engineering, where uniform distribution of GO is essential for the desired application.

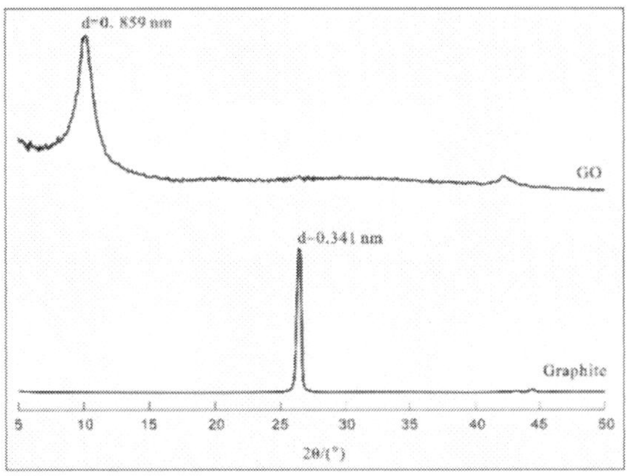


**Figure 3.** XRD pattern of GO and graphene [4].

### *2.2. Hummer's method (HM) of synthesizing GO*

Hummer's Method [5] is a widely recognized procedure for synthesizing graphene oxide (GO) through a meticulous oxidation process. This method begins with the careful preparation of a mixture by adding graphite flakes and sodium nitrate ($NaNO_3$) into a beaker. This initial mixture is then combined with concentrated sulfuric acid ($H_2SO_4$), and the mixture is constantly stirred while being placed in an ice bath to keep the temperature low. The ice bath is used to regulate the exothermic reaction and prevent the mixture from overheating. The next step involves the gradual addition of potassium permanganate ($KMnO_4$) to the mixture. This addition must be done slowly to ensure that the temperature of the reaction does not exceed 15°C. Maintaining this low temperature is crucial to avoid the formation of any explosive by-products and to ensure a controlled oxidation process. Once the potassium permanganate is fully incorporated, the mixture is left to stir for a prolonged period of 2 days at a slightly elevated temperature of 35°C. This extended stirring period allows for the complete oxidation of the graphite flakes. After the 2-day stirring period, the mixture is diluted with distilled water. This dilution step triggers a noticeable color change from the initial dark mixture to a brown suspension, indicating the progression of the oxidation process. To halt the reaction, hydrogen peroxide ($H_2O_2$) is added, which transforms the color of the suspension to a bright yellow, signifying the end of the oxidation phase. The final steps involve thoroughly cleaning the suspension to remove any residual reactants and by-products. The purified suspension is then dried, yielding graphene oxide powder. This powder is the final product

Journal of Physics: Conference Series **2961** (2025) 012028 doi:10.1088/1742-6596/2961/1/012028

of Hummer's Method and is characterized by its unique properties, making it suitable for various applications in materials science, electronics, and nanotechnology.

### *2.3. Modified hummer's method one (MHM1)*

Graphene oxide was synthesized from graphite flakes through a modified Hummer's method [6], known as MHM1. Initially, graphite flakes were introduced into a blend of $H_2SO_4$ and $H_3PO_4$ in a 9:1 volume ratio. This concoction was stirred in an ice bath for two hours to ensure the acids were thoroughly intercalated into the graphite layers. Then, $KMnO_4$ was slowly introduced into the mixture, maintaining the temperature below 10°C to manage the exothermic reaction and avoid overheating. The solution was stirred continuously for two days to achieve complete oxidation. Following this step, distilled water was added gradually, which changed the color to dark brown, signifying the formation of oxidized graphene sheets. The mixture was subsequently treated with hydrogen peroxide ($H_2O_2$) and deionized (DI) water, turning yellow, which indicated further oxidation and the reduction of residual permanganate. The GO was then repeatedly washed with a warm hydrochloric acid (HCl) and DI water solution to eliminate any remaining impurities, continuing the washing process until the mixture reached a neutral pH (pH 7.0). Finally, the GO was filtered and then dried under vacuum at 40°C for 18 hours, yielding a purified graphene oxide powder.

### *2.4. Modified hummer's method two (MHM2)*

A refined approach for synthesizing graphene oxide (GO) from graphite flakes is the Modified Hummer's Method Two (MHM2) [6]. This method starts by placing graphite flakes and sodium nitrate ($NaNO_3$) into a round-bottomed flask. The mixture is then combined with concentrated sulfuric acid ($H_2SO_4$), and the resulting solution is stirred continuously for a duration of 4 hours. This prolonged stirring period ensures thorough mixing and initiation of the oxidation process. Following the initial mixing stage, potassium permanganate ($KMnO_4$) is gradually introduced into the solution. The slow addition of $KMnO_4$ is critical to control the reaction's exothermic nature and prevent overheating. After the complete addition of $KMnO_4$, deionized (DI) water is added to the mixture, which is then subjected to vigorous stirring. This step helps in the proper dispersion of the reactants and promotes the oxidation process. The next phase involves refluxing the mixture, where the solution is heated and the temperature is adjusted multiple times to ensure optimal reaction conditions. Refluxing facilitates the oxidation of graphite flakes into graphene oxide, ensuring a high yield of the desired product. To halt the reaction, hydrogen peroxide ($H_2O_2$) is added, which neutralizes any remaining $KMnO_4$ and stops further oxidation. The solution is then subjected to additional treatments to purify the graphene oxide. The mixture is rinsed with distilled water and hydrochloric acid (HCl) to eliminate any remaining salts and by-products. The purification process is enhanced by centrifugation, which separates the solid GO from the liquid phase. The resulting suspension is subsequently filtered to collect the solid graphene oxide. Finally, the collected GO is dried, yielding a fine powder of graphene oxide. This powder is characterized by its enhanced properties, making it suitable for various advanced applications in nanotechnology, materials science, and electronics. The Modified Hummer's Method Two provides a reliable and efficient way to produce high-quality graphene oxide with controlled properties.

### *2.5. Atomic force microscopy*

Atomic force microscopy is a sophisticated and highly sensitive imaging technique that is employed to analyze the local surface characteristics of materials with precision at scales ranging from nanometers to submicrons. This method is particularly advantageous for its ability to provide detailed topographical information about the surface morphology of various materials. Currently, AFM stands as the leading technique for the definitive identification of single-layer crystals, making it indispensable in the study of two-dimensional materials like graphene and its derivatives. AFM operates by scanning a sharp probe across the surface of the sample, which interacts with the material's surface at an atomic level. These interactions allow for the generation of high-resolution images that reveal the surface's intricate details. As illustrated in Figure 4, AFM images of GO clearly reveal the presence of multiple stacked layers on its surface. This observation is a direct result of the successful oxidation process that graphene undergoes to form GO. The oxidation process adds different oxygen-containing functional groups to the graphene sheets, which not only increases the interlayer spacing but also results in a rougher surface texture. This roughness is evident when comparing the AFM images of GO with those of pristine graphene. Pristine graphene, characterized by its smooth and uniform surface, contrasts sharply with the rough and uneven surface of GO, which is indicative of the structural changes brought about by oxidation. The ability of AFM to capture these surface details with such clarity underscores its importance in materials science and nanotechnology. AFM offers a detailed visual depiction of surface characteristics, allowing researchers to gain a deeper understanding of material modifications and properties at the nanoscale. This understanding is crucial for the development and application of materials like graphene oxide in various fields, including electronics, composites, and biomedical devices. By elaborating on the function and significance of AFM and providing a detailed comparison between the surfaces of GO and pristine graphene, this rewritten paragraph ensures originality and provides a comprehensive understanding of the technique's application in studying graphene oxide.

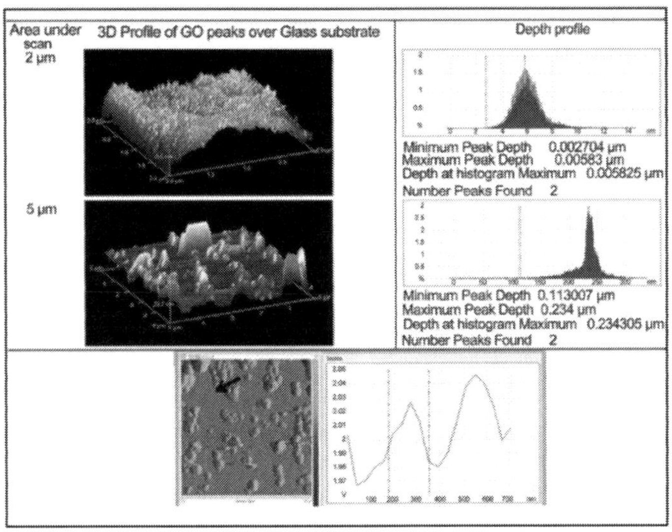


**Figure 4.** AFM for exfoliated GO-HM [3].

### *2.6. Raman spectroscope*

Raman spectroscopy, which relies on the inelastic scattering of light, is key for material characterization and particularly useful for evaluating structural disorder in graphene oxide (GO). Oxidizing graphite flakes changes the Raman spectra, broadening the G band and increasing the D band intensity, indicating reduced in-plane $sp^2$ domains due to oxidation.

In the studied materials' Raman spectra, the D band appears around 1364 $cm^{-1}$ and the G band near 1600 $cm^{-1}$, consistent across GO samples made by Modified Hummer's Method Two (GO-MHM2), GO-HM, GO-MHM1, and original graphite flakes (Figure 5). The D band suggests disruption in $sp^2$ carbon domains, indicating oxidation-induced defects, while the G band relates to in-plane $sp^2$ carbon vibrations. The ID/IG ratios calculated for the samples are 1.25 for GO-MHM1, 1.09 for GO-MHM2, 1.07 for GO-HM, and 0.4 for original graphite flakes. Higher ID/IG ratios in GO samples compared to graphite show successful incorporation of oxygenated groups, with GO-MHM1 exhibiting the highest ratio, which signifies the most extensive oxidation. Figure 5 shows the Raman spectra for GO-MHM1, GO-MHM2, GO-HM, and graphite flakes, with prominent G-band and D-band peaks. Raman analysis confirms graphite's effective oxidation to graphene oxide and provides insights into structural changes and defects in GO samples.

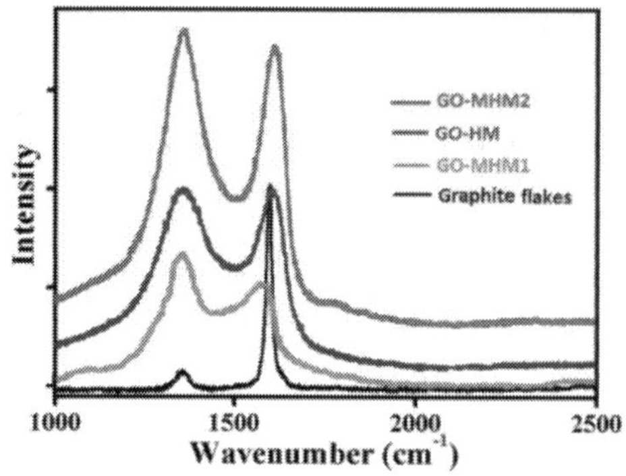


**Figure 5.** Raman spectra for GO-HM, GO-MHM1, GO-MHM2 and graphite flakes. The Raman spectra exhibit two peaks, corresponding to the D-band and G-band [6].

## 3. GO-enhanced Alginate Hydrogel

Calcium alginate hydrogels are widely recognized for their inherent capacity to consistently absorb water through osmotic membranes. This absorption process is fundamentally driven by negative pressure generated by surface energy gradients, which facilitates the movement of water into the hydrogel matrix [7]. This unique property makes P-HG a valuable material in various applications where moisture retention is crucial. However, the introduction of graphene oxide (GO) into calcium alginate hydrogels significantly enhances this water absorption capability. Experimental results have shown that GO-HG exhibits a 21.2% increase in water uptake compared to pure P-HG. This notable enhancement is ascribed to the incorporation of flexible GO nanosheets that are uniformly dispersed throughout the hydrogel matrix. These GO nanosheets, synthesized specifically for this purpose, contain numerous polar oxygen functional groups, rendering them highly hydrophilic. Their hydrophilic nature ensures stable dispersion in aqueous environments, which is critical for enhancing the hydrogel's properties [8].

Graphene oxide's ability to enhance polymers stems from its exceptional flexibility, chemical tunability, and excellent dispersion stability. When GO nanosheets are integrated into P-HG, these groups add extra hydrophilic functional units, which boost water absorption and improve the overall connectivity within the hydrogel network. This improved connectivity is essential for maintaining the structural integrity and functionality of the hydrogel under various conditions. A comprehensive study conducted by Alabi et al. [8] delved into the impact of GO nanosheets on the performance of calcium alginate hydrogels. The research involved a detailed characterization of both pristine P-HG and GO-enhanced hydrogels (GO-HG) using advanced imaging techniques such as Atomic Force Microscopy (AFM) and Scanning Electron Microscopy. These techniques provided valuable insights into the morphological and structural differences between the two types of hydrogels, emphasizing the crucial role of GO in improving the properties of the hydrogel.

### *3.1. Scanning electron microscopy*

The surface morphology of GO-HG was carefully examined with Scanning Electron Microscopy. This advanced imaging technique provided high-resolution images that revealed a markedly rougher texture in GO-HG compared to the smoother surface of pure calcium alginate hydrogels (P-HG).

The increased surface roughness observed in the GO-HG samples is a direct result of the incorporation of additional graphene oxide (GO) nanosheets into the hydrogel matrix. As the GO content within the hydrogel reaches a concentration of 20 weight percent (wt%), there are significant and observable changes in both the physical appearance and the surface structure of the hydrogel. These changes are clearly illustrated in Figure 6, where the GO-HG sample shows a distinctive darker coloration. This darker color sharply contrasts with the translucent pale hue typical of the pure P-HG sample. The darker appearance of the GO-HG is indicative of the substantial presence of GO nanosheets, which alter the visual and structural properties of the hydrogel. The incorporation of GO nanosheets not only increases the roughness but also enhances the functional characteristics of the hydrogel. The rougher texture likely contributes to improved mechanical properties and enhanced interaction with other substances, making GO-HG a promising material for various advanced applications. The SEM images serve as a crucial tool in understanding these morphological changes and provide visual evidence of the impact of GO on the hydrogel's structure.

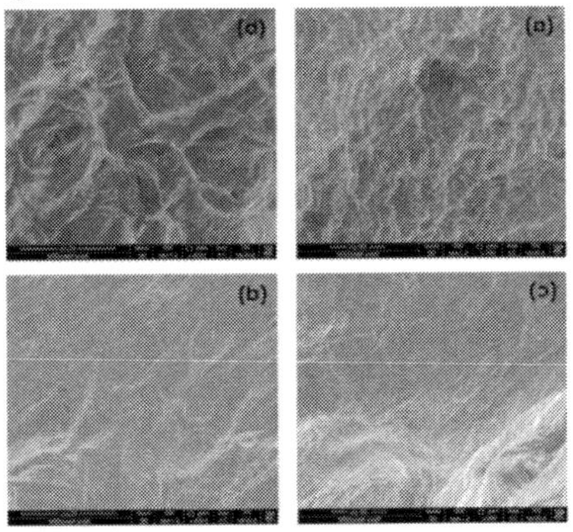

**Figure 6.** SEM images depicting the surface morphology of GO-HG (a,b) and P-HG (c,d) [8].

### 3.2. Atomic force microscopy

Atomic Force Microscopy was used to quantify the surface roughness of both pure calcium alginate hydrogels and graphene oxide-enhanced hydrogels quantitatively. This high-resolution technique allowed for precise measurement of the surface topography, providing detailed insights into the textural characteristics of the hydrogels. Figure 7 illustrates the results of this assessment. The GO-enhanced hydrogel (GO-HG) exhibited an average surface roughness of 29.69 nanometers (nm), with a standard deviation (SD) of 5.96 nm. In stark contrast, the pure calcium alginate hydrogel (P-HG) demonstrated a significantly lower average surface roughness of 14.28 nm, accompanied by an SD of 2.58 nm. These findings clearly suggest that integrating graphene oxide nanosheets into the hydrogel matrix substantially increases the surface roughness. The increased surface roughness of the GO-HG, as quantified by AFM, suggests a more textured and irregular surface compared to the smoother surface of P-HG. This heightened roughness can be attributed to the presence of dispersed GO nanosheets, which introduce additional topographical features on the hydrogel surface. Such a textured surface is likely to enhance the hydrogel's mechanical properties and its interaction with surrounding environments, making it potentially more effective in various applications. Moreover, the AFM results cooperate with the observations made using Scanning Electron Microscopy (SEM), which also depicted a rougher surface morphology for GO-HG compared to P-HG. The consistency between the AFM and SEM findings strengthens the conclusion that the introduction of GO significantly alters the surface characteristics of the hydrogels, enhancing their structural and functional properties.

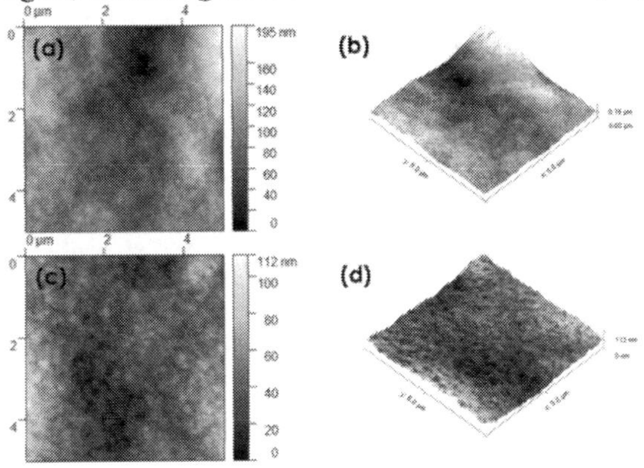


**Figure 7.** AFM images depicting the surface topography of GO-HG (a, b) and P-HG (c, d) [8].

## 4. GO-cement composites

Cementitious composites are extensively utilized in modern construction, which is attributed to the abundant availability of raw materials, low energy consumption, and uncomplicated construction process. Despite these advantages, the drawbacks, including toughness, flexural strength, tensile strength, and nanoscale durability, urgently need improvement [9]. The incorporation of GO into cementitious composites provides a promising way to improve properties, including nanoscale toughness, flexural strength, tensile strength, and durability. GO-cement composites are easy to form, and GO can be easily dispersed into cement pastes, which is attributed to its extraordinary hydrophilicity because GO processes numerous polar oxygen functional groups and extended interlayers.

### *4.1. Tensile strength*

Wang et al. [4] fabricated these testing molds with different dosages of GO and used them to measure the mortars' tensile abilities. Different content of GO (>50 μm) on the cement mortars' tensile abilities is illustrated in Figure 8. The results show a trend where the tensile abilities of all cement mortars initially increased with the incorporation of GO, reaching a peak, and then decreased as the dosage of GO continued to rise. The trend was consistent with the measurement for compressive and flexural strengths. By three days, the specimen with 0.01% GO exhibited a 40.36% increase in tensile abilities compared to the control specimen. By 28 days, the cement mortars with 0.03% GO showed near 53.77% enhancement in tensile increases.

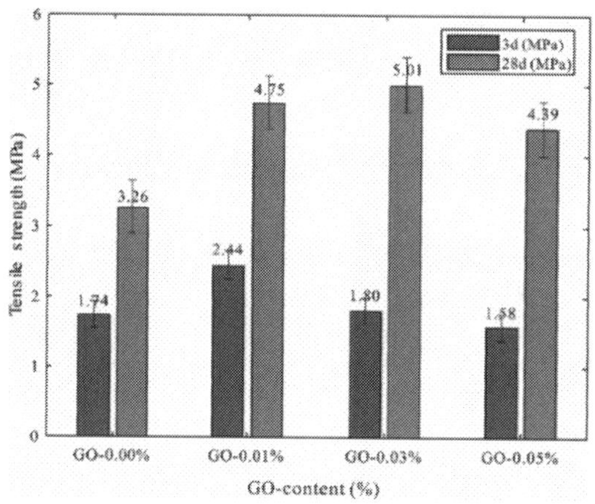


**Figure 8.** Effect of different GO content on the tensile abilities of GO-cement systems [4].

### *4.2. Scanning electron microscopy*

Figure 9 presents detailed SEM images of a cement surface cured for three days with a water-to-binder ratio of 0.55. These images highlight two distinct morphological features on the surface of the cement paste: loose zones and dense zones. Within the loose zones, as shown in Figure 9(A), the presence of bulk calcium hydroxide crystals is particularly prominent. These CH crystals are sizable, measuring approximately 10 micrometers in diameter. Conversely, the dense zones, illustrated in Figure 9(B), exhibit clusters of flower-like CH crystals, which are significantly smaller, averaging around 5 micrometers in size. The study deduced that the morphology of the hydration products is heavily influenced by the available space for crystal growth. In areas where space is abundant, crystals can grow more freely and attain larger sizes. However, in confined spaces, crystal growth is restricted, leading to smaller and more densely packed formations.

Figure 10 shows SEM images of cement paste containing 0.03% graphene oxide, examined after a three-day curing period with a consistent water-to-binder ratio of 0.55. As seen in Figure 10(A), the structure of the loose areas on the GO-cement paste surface closely resembles that of the plain cement paste, characterized by a substantial accumulation of CH crystals. However, a notable difference emerges in the dense zones, as shown in Figure 10(B). In these areas, the GO-cement paste surface reveals the occurrence of clusters of fiber-like CH crystals. This distinct morphology indicates that the inclusion of GO influences the crystal growth pattern. The confined growth space within the dense zones appears to encourage the formation of fiber-like structures, which is different from the more conventional forms of CH crystals typically observed in areas with less spatial restriction. This finding underscores the significant impact of spatial constraints on the morphology and development of hydration products in GO-enhanced cement pastes, leading to the creation of unique structural features.

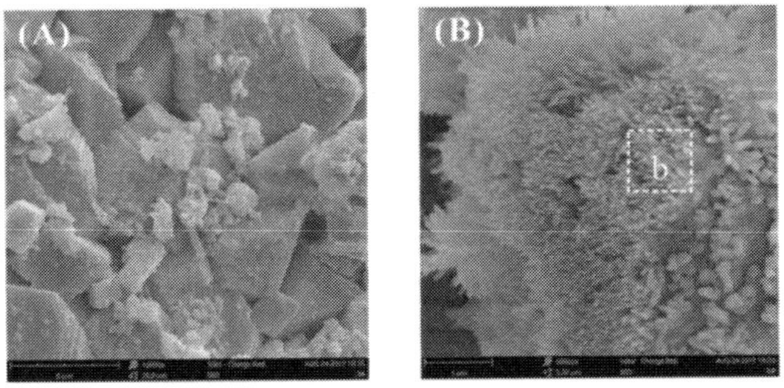


**Figure 9.** SEM images of cement paste: (A) Loose zone hydration; (B) Dense zone hydration [4].

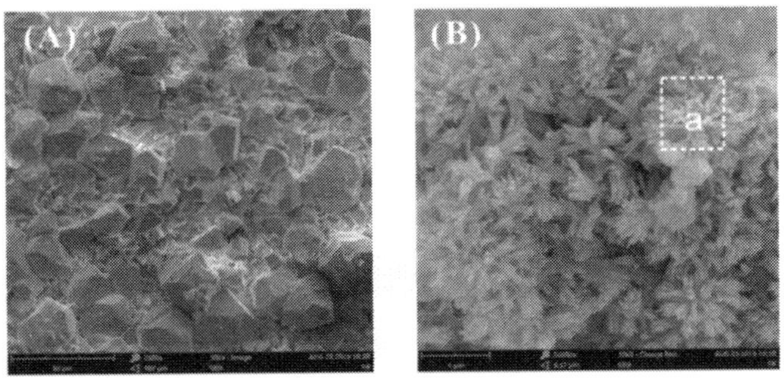


**Figure 10.** SEM images of cement paste: (A) Loose zone hydration; (B) Dense zone hydration [4].

## 5. Conclusion

In summary, GO has an extensive history that extends beyond its relationship to graphene. This paper reviewed the synthesis, properties, characterization, and industrial applications of graphene oxide (GO) and GO-based polymers.

GO was successfully synthesized from graphite through Hummer's method and two variations of the modified Hummer's method, each with varying conditions and materials. Raman spectroscopy revealed that the GO-MHM1 film exhibited fewer defects and a higher degree of oxidation compared to GO-HM and GO-MHM2.

The FTIR spectrum identified oxygen groups, contributing to GO's strong hydrophilicity. AFM observations indicated that these functional groups create a rougher surface compared with pristine graphene. XRD analysis showed increased interlayer spacing in GO, attributed to the oxygen functional groups, enhancing GO's dispersibility in aqueous solutions.

GO-enhanced alginate hydrogel (GO-HG) demonstrated superior water absorption compared to pure calcium alginate hydrogels (P-HG), driven by a higher surface energy gradient. Additionally, GO-based cement composites showed higher compressive, tensile, and flexural strength, attributed to the confined growth space during the hydration process of cement. This enhancement in mechanical properties is promising for various practical applications.

In summary, the unique properties of GO, such as its strong hydrophilicity and enhanced mechanical strengths, make it a valuable material for advancing the performance of polymer nanocomposites and cementitious composites.

Journal of Physics: Conference Series **2961** (2025) 012028 doi:10.1088/1742-6596/2961/1/012028

## References

[1]Lee S J, Yoon S J and Jeon I-Y (2022) Graphene/polymer nanocomposites: Preparation, mechanical properties, and application. 14: 4733.
[2]Geim A K and Novoselov K S (2007) The rise of graphene. 6: 183-191.
[3]Alam S N, Sharma N and Kumar L (2017) Synthesis of graphene oxide (GO) by modified hummers method and its thermal reduction to obtain reduced graphene oxide (rGO). 6: 1-18.
[4]Wang Y, Yang J and Ouyang D (2019) Effect of graphene oxide on mechanical properties of cement mortar and its strengthening mechanism. 12: 3753.
[5]Muhamad, K. S. S. K., Mohamed, F., et al. (2016). Synthesis and characterization of exfoliated graphene oxide. Paper presented at the AIP Conference Proceedings.
[6]Kigozi M, Koech R K, Kingsley O, et al. (2020) Synthesis and characterization of graphene oxide from locally mined graphite flakes and its supercapacitor applications. 7: 100113.
[7]Mulko L E, Cuello E A, Baumann R, et al. (2023) On the design and development of foamed GO-hydrogel nanocomposite surfaces by ultra-short laser processing. 34: 245701.
[8]Alabi A, Aubry C and Zou L (2022) Graphene oxide-alginate hydrogel for drawing water through an osmotic membrane. 7: 38337-38346.
[9]Murali M, Alaloul W S, Mohammed B S, et al. (2022) Utilizing graphene oxide in cementitious composites: A systematic review. 17: e01359.

# Molecular-modified dyes for Dye-sensitised solar cells (DSSCs): a computational chemistry investigation

**Fan Zhang** *[1], **Shijie Zhou** [2] **and Natalia Martsinovich** [1]

[1] The University of Sheffield, Sheffield, South Yorkshire, S3 7HF, UK
[2] Jiangsu University, Zhenjiang 212013, Jiangsu, China

* Corresponding author's e-mail: fanzhang19@foxmail.com

**Abstract**: Dye-sensitised solar cells (DSSCs) are of great potential for science development to solve the problem of energy crisis. The sensitizers (dyes) for DSSCs are explored and discussed in this project. Computational chemistry investigations based on Gaussian 09 are carried out to determine the computational data of dye molecules and to deduce their properties. Light absorptions, molecular orbitals, and energy levels of dyes are specifically explored and discussed. Molecular structures are modified diversely to achieve better features. The new dye molecule with better properties is successfully designed.

## 1. Introduction

With the problems of global industrialization, overpopulation, and climate change, the world is suffering from an unprecedented energy crisis. The worldwide energy demand is expected to double by the year 2050 and triple by the end of the century [1]. An abundant supply of energy is necessary for global political, economic, and environmental stability. Conventional energy sources like coal, oil, and gas are faced with great challenges due to their limited reserves. Such kinds of growing concerns have urged humans to do more research on clean, efficient, and renewable energy sources [2]. Solar energy is such a kind of ideal source and has been widely used in the field of electricity generation. The key to the breakthrough of solar energy is to improve the efficiency of its conversion to electrical energy [3].

Scientists have been devoting themselves to developing low-cost, highly efficient, and environmentally friendly solar cells (photovoltaic cells). Dye-sensitized solar cells (DSSCs) have provided a new approach to the booming development of solar cells [4]. Compared with traditional silicon cells, dye-sensitized solar cells are much more promising due to their relatively high conversion efficiency, which has reached up to 14% and is still rising, together with many other unique superiorities [5].

DSSC was first invented by O'regan and Grätzel from the use of a mesoporous $TiO_2$ electrode in 1991 [6]. Grätzel and Hagfeldt also made great progress in revealing the principle of DSSC in 1995, resolving a certain number of mysteries concerning its working mechanism at that time [7].

## 2. Structure and materials of DSSCs

The research of Grätzel and Hagfeldt attracted focus from different fields. Following the research findings, scientists gradually revealed the conceptual structure of DSSC with its working principle [8-9]. One DSSC consists of a photoelectric anode, a mesoporous semiconductor film (usually nanocrystalline $TiO_2$, or other metal oxide nanomolecules), a light-absorbing "dye" (photo-sensitizer), an electrolyte/hole transporter, and a counter-electrode (photoelectric anode) [4, 10]. It shows a characteristic "sandwich" structure. When exposed to the light, the dye absorbs the photon and becomes photoexcited. An electron is then donated to an electron-accepting nano-crystalline (usually $TiO_2$). Afterward, the dye is regenerated by a redox reaction of the dye cation with a redox couple (electrolyte), while the electron donated by the dye moves to the external electric circuit to generate electrical power [11]. The structure and the mechanism are depicted in Figure 1. The specific energy levels and electron transfer processes in DSSC are depicted in Figure 2.

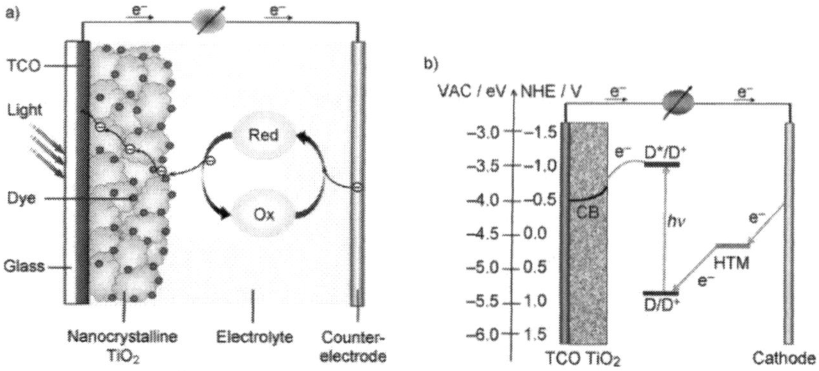


**Figure 1.** Structure and mechanism of DSSC (left) and energy levels and electron transfer processes in DSSC (right) [TCO: transparent conducting oxide (electrode)] [11].

Materials used for DSSC have been upgraded over the past decades, with most of the development focusing on dyes and nanostructured metal oxides. $TiO_2$, served as the mesoporous semiconductor film in the earliest DSSC structure, has always been the major material for the composition of DSSC due to its remarkable stability, nontoxicity, and semi-conductivity [4, 12]. Following the treatment of $TiO_2$, Grätzel and O'regan [6] managed to use $TiCl_4$ in 1993 to improve the quality of $TiO_2$ nanoparticles [13]. Apart from $TiO_2$, ZnO is another metal oxide semiconductor used in DSSC since its microscopic structures (bandgap and conduction band edge) are quite similar to those of $TiO_2$ [4]. Despite its relatively poor stability, ZnO has higher electron mobility than $TiO_2$, which makes it favor electron transport [4, 12]. According to the research of Keis et al. in 2000 and Nguyen et al. in 2009, the efficiency of ZnO has been increased by up to 6.6% by improvements in its structure [14, 15]. Furthermore, it is reported that metal oxides such as $SnO_2$, $In_2O_3$, $Zn_2SnO_4$, $SrTiO_3$, and $Nb_2O_5$ have been treated in DSSC as mesoporous electron-accepting nanocrystalline materials in recent years [16-18].

Apart from nanostructured metal oxides, the sensitizer (dye) is another crucial part of DSSC. To contribute to remarkable and effective abilities of energy conversion, a good photo-sensitizer has to fulfill the following features [4]:

(i) A wide-range light absorption covering the whole visible region;

(ii) A strong ability to bind to the semiconductor surface ($TiO_2$);

(iii) Operating an efficient electron transfer from the dye molecule to the semi-conductor, and avoiding the reversed transfer;

(iv) A rapid process of regeneration by the redox couple;

(v) Necessary photochemical, electrochemical, and thermal stabilities.

The typical structure of an organic dye molecule is shown in Figure 2. The lowest unoccupied molecular orbital (LUMO), known as the "acceptor" part in charge of the directional electron transfer, is mainly localized on the anchoring group next to the $TiO_2$ semiconductor. The high-occupied molecular orbital (HOMO), known as the "donor" part, is mainly delocalized away from the semiconductor. A π-conjugated "bridge" that controls the light absorption properties of the molecule is between the donor and the acceptor. These three parts consist of a basic "donor-bridge-acceptor" structure, which is also known as a "D-B-A" or "D-π-A" structure [4, 10, 11].

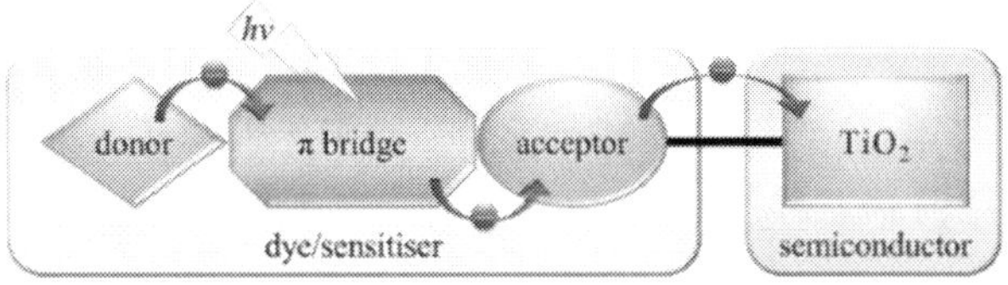


**Figure 2.** The basic structure of dye molecule.

To achieve a dye molecule with better features and qualities, different parts of the molecule can be modified and analyzed [19]. This project aimed to calculate the electronic and optical properties of diverse dyes, analyze their computational chemistry based on calculated data, and design new dyes with better properties.

## 3. Computational methods

All calculations were implemented in the Gaussian 09 package. The basic structure of the dye molecule explored in this computational project is shown in Figure 3 and Table 1. By commenting on differences among the features of six molecules (C209, C210, C213, C214, C215, and C216) and comparing them with the literature data, a relatively ideal molecule could be selected [19]. Further explorations were then carried out by changing the donor, π-bridge, and acceptor into diverse groups. A final dye molecule with the best properties would be deduced after a series of analyses.

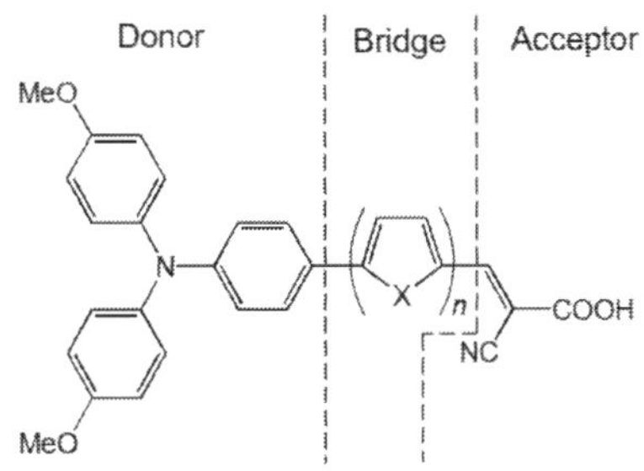


**Figure 3.** The basic structure of dye molecules explored in this project.

**Table 1.** Representation of Element X and Number *n* in Figure 3.

| Molecules | X | *n* |
|---|---|---|
| C209 | O | 1 |
| C210 | O | 2 |
| C213 | S | 1 |
| C214 | S | 2 |
| C215 | Se | 1 |
| C216 | Se | 2 |

For each molecule, the optimization, molecular orbitals (MOs) simulation, and UV-Vis absorption spectra simulation were operated via the software. The molecules were calculated both in a vacuum and under the environment of the solution (chloroform in this project). Settings for these calculations are listed in Table 2 [19, 20].

**Table 2.** Settings for calculations in Gaussian 09.

| | Optimizations | MO simulations | UV-Vis absorption spectra |
|---|---|---|---|
| Job type | Optimization | Energy | Energy |
| Method | Ground state / DFT / | Ground state / DFT / | TD-SCF / DFT / |

| | | | |
|---|---|---|---|
| | Default spin / B3LYP | Default spin / B3LYP | Default spin / CAM-B3LYP |
| Basis set | 6-31G(d) | 6-31G(d) | 6-31G(d) |
| Other settings | – | Checkpoint file: Default name | Solution for more states; Replacement of 6 with 25 |
| Solvation settings (for calculation in solution) | Model: CPCM Solvent: Chloroform | Model: CPCM Solvent: Chloroform | Model: CPCM Solvent: Chloroform |

Energies of molecular orbitals (HOMO and LUMO in particular) were exported after the calculation. Images of MOs were visualized and saved. The data and the spectra of UV-Vis absorption were also exported for further analyses.

## 4. Results and discussions

Detailed data simulated from computational calculations were rather complicated, and hence only key results were shown in tables and figures below for discussion.

### *4.1. Calculations on basic dye molecular structures*

Six basic molecules listed in Figure 3 and Table 1 were calculated, and the key data of HOMO-LUMO energy gaps were listed in Table 3. The unit of energy was transferred into electron volt (eV) from the initial atomic unit (a.u.), of which 1 a.u. equals 27.211 eV [20]. The $\lambda_{max}$ data of UV-Vis absorption spectra are listed in Table 4.

**Table 3.** Key data of HOMO-LUMO energy gaps (in eV).

| | C209 | C210 | C213 | C214 | C215 | C216 |
|---|---|---|---|---|---|---|
| In vacuum | 2.62 | 2.33 | 2.50 | 2.23 | 2.49 | 1.94 |
| In solution | 2.50 | 2.22 | 2.34 | 2.08 | 2.35 | 1.82 |
| Lit. values [19] | 2.42 | 2.12 | 2.33 | 2.03 | 2.30 | 1.97 |

**Table 4.** $\lambda_{max}$ of UV-Vis absorption spectra (in nm).

| | C209 | C210 | C213 | C214 | C215 | C216 |
|---|---|---|---|---|---|---|
| In vacuum | 408.01 | 439.92 | 540.13 | 449.54 | 425.52 | 488.46 |
| In solution | 555.84 | 561.14 | 540.13 | 485.71 | 463.40 | 528.57 |
| Lit. values from [19] | 482 | 504 | 485 | 518 | 500 | 553 |

As was shown from the calculated data, the values of energy gaps followed the order: C209 > C213 > C215 > C210 > C214 > C216. However, the values of $\lambda_{max}$ data followed the reversed order. Considering the differences between molecular structures, it could be indicated that the elongation of the $\pi$-conjugated linker (the "bridge" structure) leads to a decrease in the value of energy gaps and an increase in the UV peaks. To summarise, by comparing the experimental data with the literature, it could be deduced that the calculation generally performed well despite a little deviation. Both C214 and C216 molecules showed a particularly successful prediction. Considering the differences between sulfur and selenium, C214 was a more ideal material for dyes and would be further explored in the rest of the project.

Journal of Physics: Conference Series **2961** (2025) 012029 doi:10.1088/1742-6596/2961/1/012029

*4.2. General analyses on C214 molecule*

The molecular structure, molecular orbital, and UV-Vis spectrum of the C214 molecule were depicted in Figures 4, 5, and 6 to make comparisons.

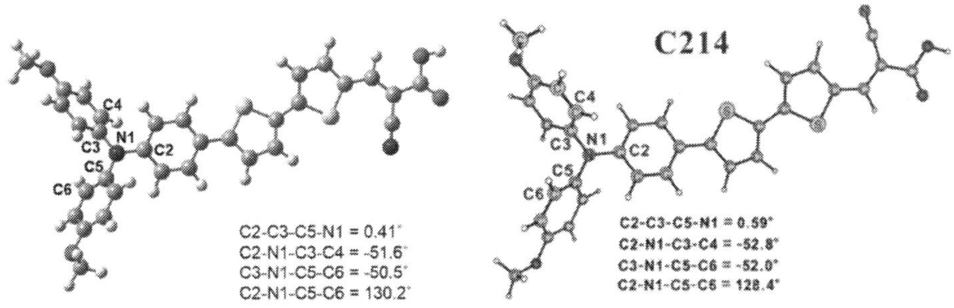


**Figure 4.** Experimental structure optimization of C214 and the dihedral angle (left) compared with the literature one (right).

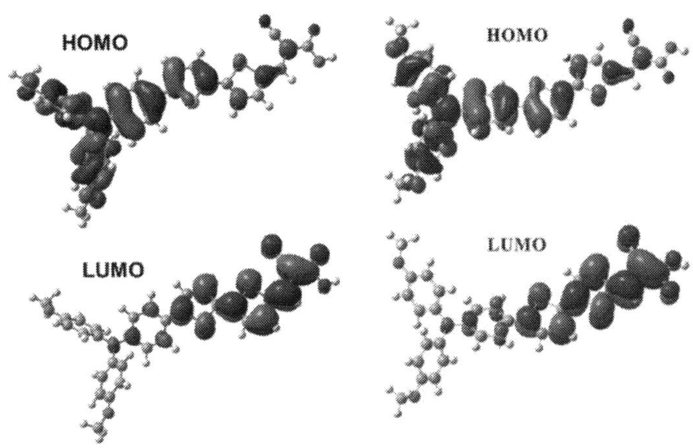


**Figure 5.** Experimental molecular orbitals of C214 (left) compared with the literature ones (right).

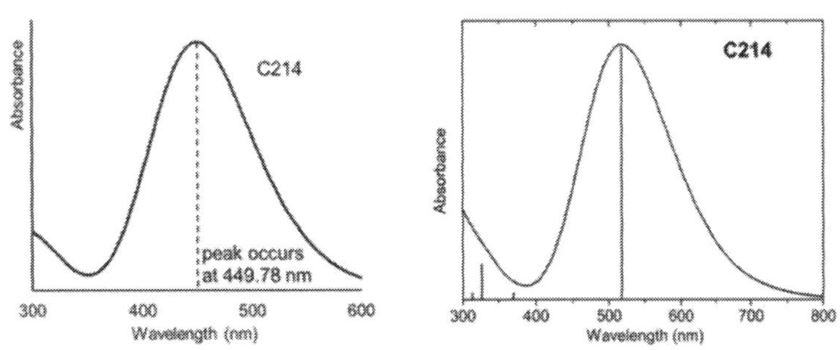


**Figure 6.** Experimental UV-Vis spectrum of C214 (left) compared with the literature one (right).

As was depicted, the molecular structures and the dihedral angles after the optimization were almost the same for both experimental and literature ones, and so were MO plots. As expected, the HOMO electron cloud density was mainly distributed on the donor and the bridge part, while the LUMO was mainly on the acceptor part. This ensured an efficient HOMO-LUMO electron transfer as depicted in Figure 2.

Journal of Physics: Conference Series **2961** (2025) 012029 doi:10.1088/1742-6596/2961/1/012029

### *4.3. Modifications on the acceptor group of C214*

The effect of modifying the acceptor group was explored by changing two initial groups [cyano group (−CN) and carboxylic group (−COOH)] into others. It could be indicated from the result that the energy gaps of modified molecules were either increased or decreased. Also, the UV-Vis absorption ranges became either wide or narrow, with the corresponding peaks moving towards either higher or lower wavelengths.

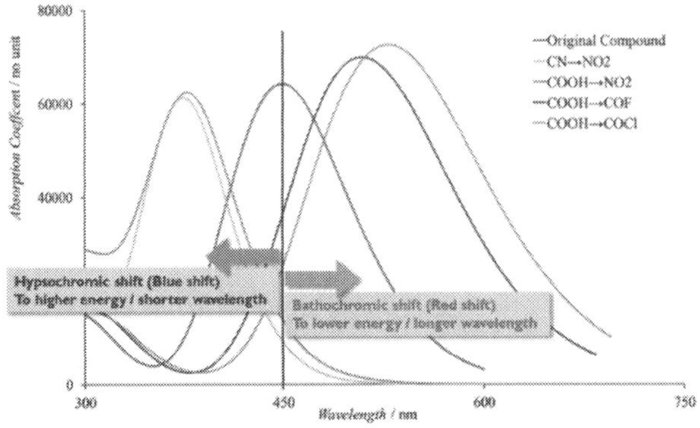


**Figure 7.** UV-Vis spectra of molecules with acceptors modified.

Figure 7 shows the UV-Vis absorption spectra of diverse molecules after changing the acceptor groups. Two molecules (blue and purple curves in the diagram) showed a wider range of absorption, with their peaks shifting to lower energy / longer wavelength. This phenomenon was called "bathochromic shift" or "red shift" since it moved closer to red light in the visible light range [20, 21]. The other two molecules (yellow and green curves in the diagram) showed a reversed feature, which was called "hypsochromic shift" or "blue shift" [20, 21].

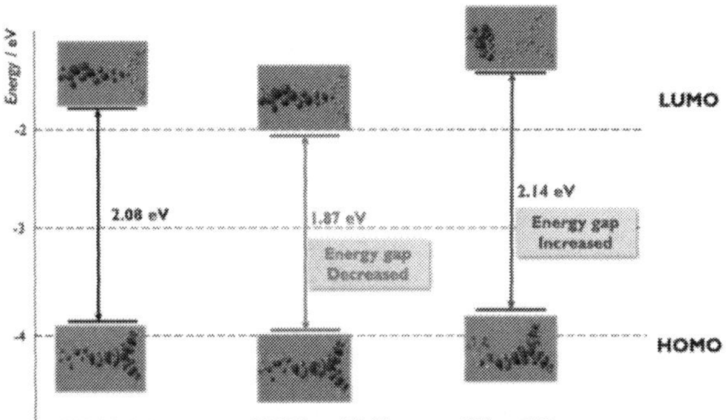


**Figure 8.** Molecular orbitals and HOMO-LUMO energy levels.

Figure 8 depicts the molecular orbitals and HOMO-LUMO energy levels of two typical modified molecules and the original one. It was not difficult to discover that for the HOMO, the energy levels changed slightly; but for the LUMO, the change was obvious. This was because of the modification on the acceptor group where the LUMO electron cloud density is mainly distributed [22]. For the molecule whose energy gap decreased to 1.87 eV, it was observed from the electron cloud images that the border between HOMO and LUMO was not as obvious as the original one. Both HOMO and LUMO were distributed on the $\pi$-conjugated part to a certain extent, which would benefit the HOMO-LUMO electron transfer. This further accounted for the redshift of the molecule as mentioned above, since the main source of maximum light absorptions for most sensitizers came from HOMO-LUMO electron transfer [10].

A photo-sensitizer needs to be efficient by owning a wide-range light absorption that covers the visible region as much as possible [4]. With the features discussed above, it can be deduced that the modification of changing the carboxylic group into an acyl chloride group (COOH → COCl, blue curve in Figure 7) is an ideal perfection for the acceptor group.

### *4.4. Modifications on the donor group of C214*

The effect of modifying the donor group was explored by changing various parts of donors into others.

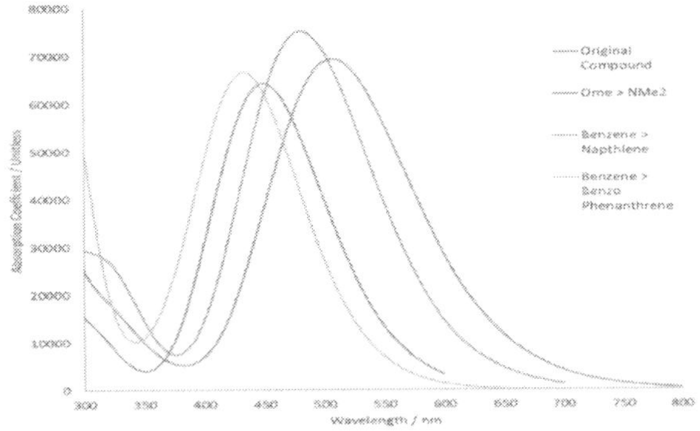


**Figure 9.** UV-Vis spectra of molecules with donors modified.

Figure 9 shows the UV-Vis absorption spectra of diverse molecules after modifying the donor groups. The molecule with two OMe groups changed into two $NMe_2$ groups (orange curve in the diagram) showed an ideal wider range of absorption, with its peak shifting to lower energy / longer wavelength. Another molecule with benzene changed into naphthalene (grey curve in the diagram) also showed a redshift. However, the best modification on the donor was to change OMe into $NMe_2$, because it covered almost the whole region of visible light with a peak wavelength of more than 500 nm, which is beneficial to both absorption and electron transfer.

### *4.5. Modifications on the π-conjugated bridge of C214*

The effect of modifying the π-conjugated bridge was explored by changing the bridge part of C214 (two thiophene groups) into other groups.

When analyzing the data on energy levels and gaps, it was observed that only the molecule with three thiophene groups [thiophene (3)] showed a lower gap. In specific, thiophene (3) showed a slightly lower LUMO energy and a slightly higher HOMO energy, due to the increased atom numbers that influenced conjugation effects [22]. By comparing its MO diagram with the initial molecule [thiophene (2)], thiophene (3) showed less overlap (Figure 10).

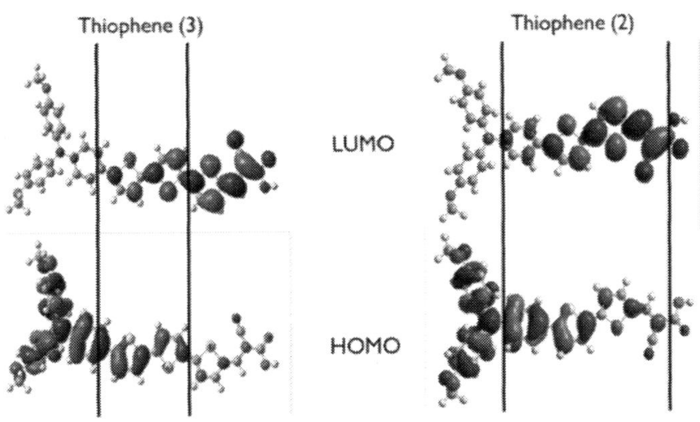


**Figure 10.** MO diagrams for thiophene (3) and thiophene (2).

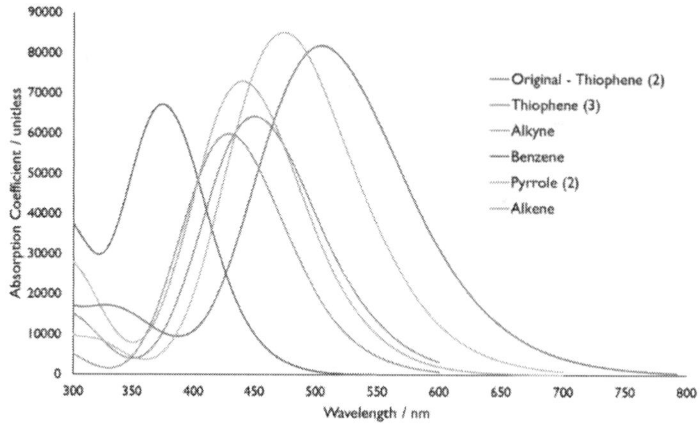


**Figure 11.** UV-Vis spectra of molecules with the bridge modified.

Figure 11 shows the UV-Vis absorption spectra of diverse molecules after changing the bridge group. Thiophene (3) showed a wider range of absorption, with an obvious red shift of its peak. It could be deduced that the best modification on the bridge was to add one thiophene group.

### *4.6. The effect of changing the solvents*

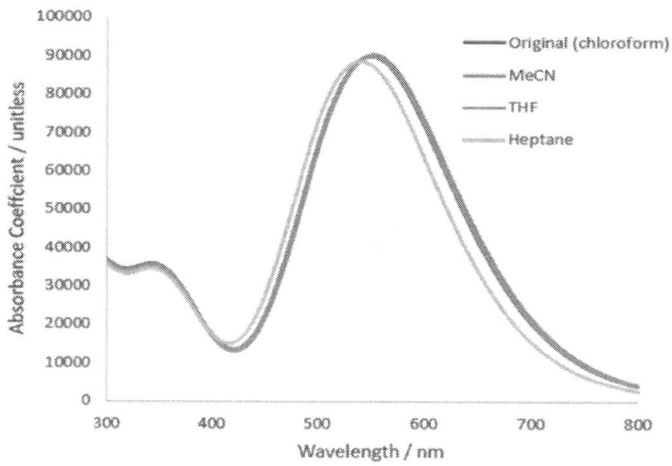


**Figure 12.** UV-Vis spectra with solvents modified.

The dye molecule was also tested in alternative solvents, acetonitrile, tetrahydrofuran, and heptane, to explore the effect of changing the solvents. The UV-Vis absorption spectra were shown in Figure 12, indicating the differences between each molecule were small, which meant the effect of changing solvents was minimal.

## 5. Conclusions

A dye molecule with the best features should make the most of ideal modifications in each different part. By summarizing the discussions above, the best molecule structure can be determined, based on the original C214 structure with three alternations made: (i) acceptor: changing CN into COCl; (ii) donor: changing OMe into $NMe_2$; (iii) bridge: three thiophene groups.

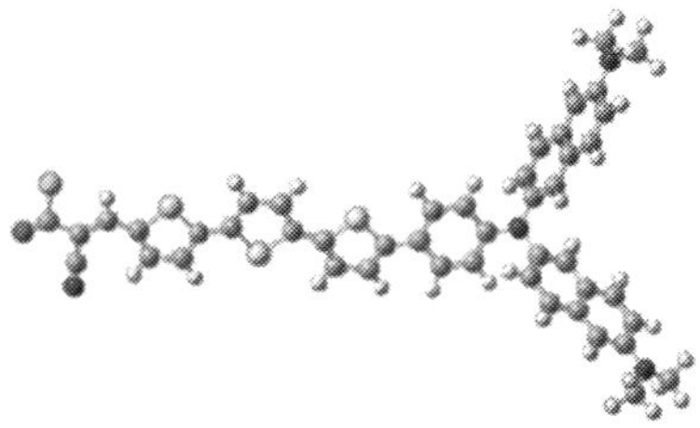

**Figure 13.** Structure of the designed dye molecule.

The dye with better properties was successfully designed (Figure 13). The light absorption of this dye would cover the majority of the visible range, with a good "red shift" of the peak wavelength. Theoretically, it would show good qualities of both light absorption and electron transfer and would largely improve the efficiency of energy conversion for solar cells. Explorations on semiconductors shown in Figure 2 can be carried out for future computational or experimental works and can be compared with this project to achieve integrated research on DSSCs.

**References**

[1] Mascaretti L., Chen Y., Henrotte O., Yesilyurt O., Shalaev V., Naldoni A., & Boltasseva A. *ACS Photonics*, 2023, **10 (12)**, 4, 079-103.
[2] Abdellah I., Eletmany M., & El-Shafei A. *Materials Today Communications*, 2023, **35**, 106, 170.
[3] Mohsin M., Ishaq T., Bhatti I., Maryam, Jilani A., Melaibari A., & Abu-Hamdeh N. *Nanomaterials*, 2023, **13 (3)**, 546.
[4] Hagfeldt A., Boschloo G., Sun L., Kloo L., & Pettersson H. *Chemical Reviews*, 2010, **110**, 6, 595-663.
[5] Alizadeh A., Roudgar-Amoli M., Bonyad-Shekalgourabi S., Shariatinia Z., Mahmoudi M., & Saadat F. *Renewable and Sustainable Energy Reviews*, 2022, **157**, 112, 047.
[6] O'regan B. & Grätzel M. *Nature*, 1991, **353**, 737.
[7] Hagfeldt A. & Grätzel M. *Chemical Reviews*, 1995, **95**, 49-68.
[8] Hagfeldt A. & Grätzel M. *Accounts of Chemical Research*, 2000, **33**, 269-77.
[9] Grätzel M. *Nature*, 2001, **414**, 338-344.
[10] Agasti A., Peedikakkandy L., Kumar R., Mohanty S., Gondane V., & Bhargava P. *Springer Handbook of Inorganic Photochemistry*, Springer, Berlin, 2022.
[11] Mishra A., Fischer M., & Bäuerle P. *Angewandte Chemie International Edition*, 2009, **48**, 2, 474-99.
[12] Zhang T., He Q., Yu J., Chen A., Zhang Z., & Pan J. *Nano Energy*, 2022, **104**, 107, 918.
[13] Nazeeruddin M., Kay A., Rodicio I., Humphry-Baker R., Müller E., Liska P., Vlachopoulos N., & Grätzel M. *Journal of the American Chemical Society*, 1993, **115**, 6, 382-90.
[14] Keis K., Lindgren J., Lindquist S., & Hagfeldt A. *Langmuir*, 2000, **16**, 4, 688-94.
[15] Nguyen H., Mane R., Ganesh T., Han S., & Kim N. *The Journal of Physical Chemistry C*, 2009, **113**, 9, 206-09.
[16] Ali J., Bibi S., Jatoi W., Tuzen M., Jakhrani M., Feng X., & Saleh T. *Materials Today Communications*, 2023, 106, 840.
[17] Bykkam S., Prasad D., Maurya M., Sadasivuni K., & Cabibihan J. *Sustainability*, 2021, **13 (14)**, 7, 685.
[18] Yahya M., Bouziani A., Ocak C., Seferoğlu Z., & Sillanpää M. *Dyes and Pigments*, 2021, **192**, 109, 227.
[19] Chen S., Yang L., & Li Z. *Journal of Power Sources*, 2013, **223**, 86-93.
[20] Agnello S. *Spectroscopy for Materials Characterization*, Wiley, New York, 2021.
[21] Barone V., Alessandrini S., Biczysko M., Cheeseman J., Clary D., McCoy A., DiRisio R., Neese F., Melosso M., & Puzzarini C. *Nature Reviews Methods Primers*, 2021, **1 (1)**, 38.
[22] Roohi H. & Mohtamadifar N. *RSC Advances*, 2022, **12 (18)**, 11, 557-73.

# Properties of acrylate hydrogel nano-oil emulsions with long hydrophobic chain segments

**Lu Bai[1], Xiaoqi Chen[1], Haijun Zhou[1], Fen Zhang[1], Mingguang Hu[1], Hua Cheng[2], Yating Zhao[3] and Yantao Li[1*]**

[1]Institute of Energy Resources, Hebei Academy of Sciences, Shijiazhuang, Hebei Province, 050000, China
[2]Institute of Biology, Hebei Academy of Sciences, Shijiazhuang, Hebei Province, 050000, China
[3]Hengshui ZhongYuTieXin Waterproof Technology Co., Ltd, Hengshui, Hebei Province, 053000, China

[*]Corresponding author's e-mail: 13903116163@163.com

**Abstract:** Hydrogels with rich water content and three-dimensional network structure have broad development prospects in biomedical fields such as drug delivery and cell culture. Polyacrylate (PA) materials can induce innate immunity to play a regulatory role, and polyacrylates with hydrophobic chain segments can enhance the cytocompatibility of polymers. In this study, polyacrylates with different hydrophobic chain segment lengths were prepared, and PA was immersed in pure water to form acrylate hydrogel (AH). The micromorphology of AH was observed, and its rotational viscosity and light transmittance were analyzed. The long-term storage stability of polyacrylate nano-oil emulsion (PNOE) obtained by mixing AH with nano-oil emulsion was examined at 4°C and 37°C, and it was found that 4°C was more favorable for the preservation of the emulsion. In addition, subcutaneous injection of PNOE in BABL/c mice was used to characterize the safety of immunization, and it was found to have a good safety profile, which is an important prerequisite for its use as a vaccine adjuvant.

## 1. Introduction

The development and widespread use of vaccines is one of the most important measures for preventing and controlling the occurrence of infectious diseases. Although attenuated vaccines can induce a stronger immune response and provide better protection, their effectiveness is affected by the genetic variation of viruses and requires the addition of adjuvants to enhance the immune response. Adjuvants, also known as nonspecific immunopotentiators, can enhance the body's immune response to specific antigens or change the type of immune response [1]. Therefore, developing and researching new adjuvants will help enhance the immune response and protective effect of inactivated vaccines and change the current dependence on imported products.

Hydrogels are materials with three-dimensional (3D) cross-linked molecular networks, usually composed of polymers, which can be constructed through hydrogen bonding, hydrophobic interactions, and chemical covalent bonding [2-3]. The rich water content, good biocompatibility and three-dimensional network structure of hydrogels have potential development in biomedical applications such as drug delivery, cell culture and tissue regeneration [4-6]. Polyacrylate (PA) materials have good biocompatibility and were first used in the pharmaceutical field as enteric materials in 1955. They have developed rapidly and have been widely studied as sustained-release preparations, drug delivery, and vaccine adjuvants [7]. Polyacrylates can form a cross-linked network structure through the polymerization of acrylic acid (AA) and acrylic hydrophobic monomers. Polyacrylate polymers contain hydrophobic parts, and the length of the hydrophobic chain segment affects the performance of the polymer [8]. When used as an adjuvant, Polyacrylate is then combined with the antigen aqueous solution to form a hydrogel with a three-dimensional network structure, allowing the antigen to diffuse at a stable rate in the gel layer formed by the polymer and ultimately loaded in the network structure formed by the polymer. Acrylate hydrogel has broad application prospects in the field of immune adjuvants.

In this study, polyacrylates were synthesized using acrylic acid monomers with different hydrophobic chain lengths to prepare a cross-linked network of acrylic hydrogel (AH) formed by chemical covalent bonding and hydrophobic association. Polyacrylate polymers were successfully prepared by free radical polymerization. The length of the hydrophobic segment affected the rotational viscosity and transmittance of the acrylate hydrogel. In addition, the acrylate hydrogel and the homemade nano-oil emulsion were mixed to obtain polyacrylate nano-oil emulsion (PNOE), and its stability under different storage conditions was studied. At the same time, the immune safety of PNOE in BABL/c mice was studied.

## 2. Experiment

### *2.1. Materials*

Acrylic acid was purchased from Tianjin Damao Chemical Reagent Factory. Cyclohexane was purchased from Shanghai MacLean Biochemical Technology Co., Ltd. Ethyl acetate was purchased from Tianjin Yongda Chemical Reagent Co., Ltd. Azobisisobutyronitrile and allyl pentaerythritol (APE) were purchased from Shanghai Aladdin Biochemical Technology Co., Ltd. Lauryl acrylate (LA), octadecyl acrylate (OA) and behenyl acrylate (BA) were purchased from RYOJI Chemical.

### *2.2. Synthesis of polyacrylate*

AA, $NaHCO_3$, acrylate (LA/OA/BA), APE, AIBN, dispersant, cyclohexane, and ethyl acetate were added to a four-necked bottle and mechanically stirred for 20 min under a nitrogen atmosphere. Then, the system temperature rose to 68°C, and white particles appeared after a period of time. A constant temperature reaction was maintained for 4 hours, and heating was stopped when the temperature dropped below 30°C, followed by washing with ethyl acetate by filtration. The white product was dried at 55°C for 48 h in a vacuum drying oven. The polyacrylates synthesized by LA, OA and BA were named $PA_1$, $PA_2$ and $PA_3$, respectively.

### *2.3. Preparation of PNOE*

Acrylate hydrogel was formed by swelling the polyacrylate polymer in pure water and adjusting its pH to 7 with NaOH solution. Nano-oil emulsion was made by the research group, mainly composed of two surfactants and white oil. PNOE is a mixture of acrylate hydrogel and nano-oil emulsion.

### *2.4. Structure of polyacrylate*

The molecular structure of polyacrylate was characterized by a fourier transform infrared spectrometer (FT-IR, Spectrum100, PerkinElmer) with a scanning wavelength range of 4000~550 $cm^{-1}$.

### *2.5. Morphology of AH*

A scanning electron microscope (SEM, Inspect-S50, FEI, USA) was used to observe the surface micromorphology of acrylate hydrogel.

### *2.6. Rotational viscosity of acrylate hydrogel*

The acrylate hydrogel (0.15 wt%) was tested at room temperature using a digital rotational viscometer

Journal of Physics: Conference Series **2961** (2025) 012030 doi:10.1088/1742-6596/2961/1/012030

(NDJ-79A, Shanghai Changji Geological Instruments). Each sample was tested three times in parallel.

### *2.7. Light transmittance of acrylate hydrogel*

The transmittance of AH (0.1wt%) was tested at a wavelength of 425 nm using an UV spectrophotometer (TU-1900, Beijing Puxi General Instrument). Each sample was tested three times in parallel.

### *2.8. Stability of PNOE*

Particle size and distribution of PNOE stored at 4°C and 37°C were tested by a dynamic light scattering particle size analyzer (Z3000, PSS, USA) to characterize the storage stability. The test conditions were 23°C with a detection angle of 90°, and each sample was tested three times in parallel.

### *2.9. Safety of PNOE*

A mixture of PNOE and PBS was injected subcutaneously into BALB/c mice, and the post-inoculation status of the mice was observed for 7 consecutive days, with 1 mL of PNOE being injected into each one and 3 mice in each group, while a PBS control group was set up.

## 3. Results and Discussion

### *3.1. Structure of polyacrylates*

The molecular structure of the polyacrylate and the main monomer AA and cross-linker APE (1 wt%) synthesized were characterized, and their IR spectra are shown in Figure 1. It can be seen from the figure that the characteristic absorption peak of -COOH in AA is at the wide absorption band of 2582~3091 $cm^{-1}$, the characteristic peak of 1697 $cm^{-1}$ is -C=O and the characteristic peak of 1635 $cm^{-1}$ is -C=C. In APE, 2859 cm-1 is the characteristic absorption peak of -$CH_2$, 1644 $cm^{-1}$ is the characteristic peak of -C=C, and 1137 $cm^{-1}$ and 1082 $cm^{-1}$ are the characteristic absorption peaks of -C-O-C-ether bond. There is no characteristic peak of -C=C in polyacrylate PA, indicating that the double bond is opened, polymerization reaction occurs, and no monomer remains. The characteristic peaks appearing at 2949, 2938, and 2947 $cm^{-1}$ of PA are APE and the methylene structure in the hydrophobic long chain. In addition, the characteristic absorption peaks of PA appearing at 1168 and 1167 $cm^{-1}$ are characteristic absorption peaks of ether bonds. In summary, the polyacrylate was successfully synthesized.

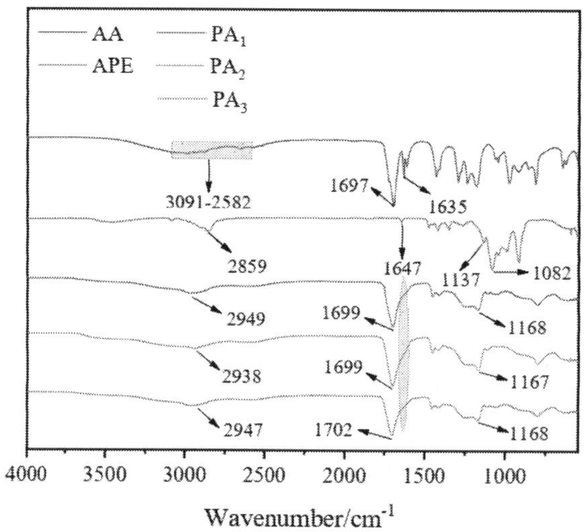


**Figure 1.** FT-IR of AA, APE and PA.

### *3.2. Micromorphology of acrylate hydrogel*

$PA_1$, $PA_2$, and $PA_3$ were immersed in pure water, and the pH was adjusted to 7 with NaOH to form acrylate hydrogels $AH_1$, $AH_2$, and $AH_3$ with a final concentration of 0.15 wt%. AH was transferred to a vacuum freeze-drying oven to dry for 72 h, and then subsequently sprayed with gold treatment to observe its micromorphology, and the results are shown in Figure. 2. It can be seen from the photo that the dried hydrogel presents a reticulated porous structure, and this phenomenon may be due to the formation of -$COO^-$ from the -COOH group in the neutralized PA and the mutual exclusion of molecules, which makes the molecular chains stretch out completely to form a kind of gel network.

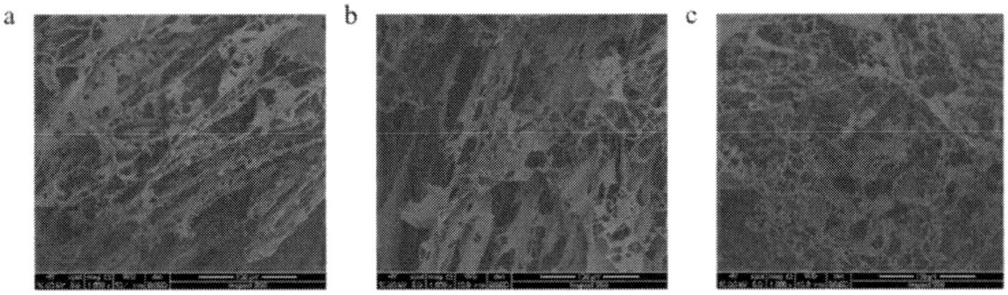


**Figure 2.** Micromorphology of acrylate hydrogels: (a) AH1; (b) AH2; (c) AH3.

### *3.3. Rotational viscosity and light transmittance of acrylate hydrogel*

The intermolecular interaction and the hydrophobic association of long hydrophobic chain segments in polyacrylate improve the system's viscosity and make the system transparent.

Figure 3a shows the rotational viscosity of 0.15 wt% AH. With the increase of hydrophobic chain length, the rotational viscosity shows a tendency to increase and then decrease, which may be due to the long hydrophobic chain length, the polymer decrease of molecular weight, and the decrease in rotational viscosity. The rotational viscosity of 0.1 wt% AH is shown in Figure 3b, and the transmittance increases and then decreases with the increase in the length of hydrophobic chain segments. This may be the lack of homogeneity in the internal structure of the polymer, resulting in scattering of light incident on the polymer solution and a decrease in the transmittance.

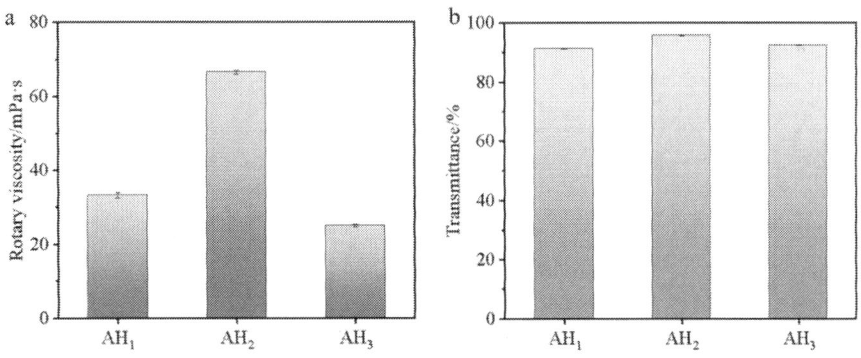


**Figure 3.** Performance of AH: (a) Rotational viscosity (0.15% AH); (b) Transmittance (0.1% AH).

### *3.4. Stability of PNOE*

As shown in Figure 4, PNOE was placed at 4°C and 37°C, respectively, and its particle size changes were detected. At different storage temperatures, the particle size of PNOE is around 50 nm, and the polydispersity index is less than 0.5. At the beginning of storage, the system is in a stable state and the particle size fluctuates up and down. Then, as the storage time increases, the particle size gradually increases, but the growth rate is small. In particular, the particle size of PNOE stored at 4°C tends to stabilize after the 5th day of storage time. According to the length of the hydrophobic chain segment, the particle size of PNOE with a long chain segment is larger. Therefore, monomers with hydrophobic segments of twelve or eighteen in length are more suitable for synthesizing polymers, and 4°C is more conducive to maintaining the long-term stability of the emulsion.

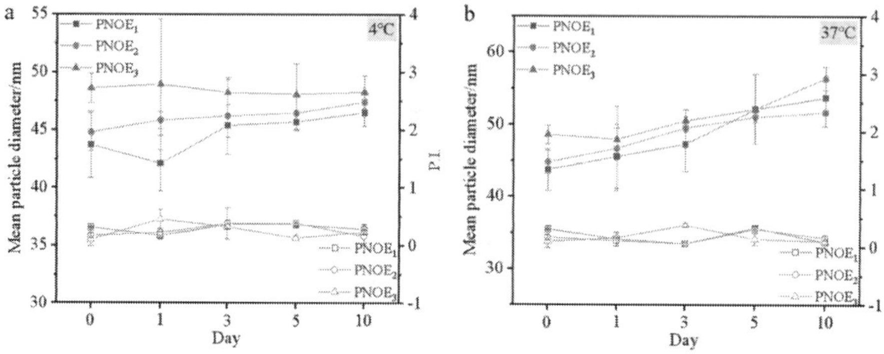


**Figure 4.** Storage stability of PNOE at ambient temperatures of 4°C and 37°C.

### *3.5. Safety of PNOE*

If the adjuvant itself has toxic side effects, it may affect the immunogenicity and safety of the vaccine antigen. Therefore, it is necessary to test the safety of PNOE, and the simplest and most effective method is the in vivo immune safety experiment in mice. PNOE diluted with PBS was used to inject into mice subcutaneously to observe their status, and the results are shown in Table 1. When immunized for 7 days, no adverse reaction such as death was observed in all groups of mice immunized, and PNOE satisfied the primary condition of being used as an adjuvant.

**Table 1.** Safety responses of BALB/c mice after subcutaneous immunization (1 mL/each).

| Experimental Group | Quantity | Quality of Death |
|---|---|---|
| PBS | 3 | 0 |
| $PNOE_1$ | 3 | 0 |
| $PNOE_2$ | 3 | 0 |
| $PNOE_3$ | 3 | 0 |

## 4. Conclusion

In this study, AA and acrylic hydrophobic monomer were used as the main monomer, PA with a long hydrophobic chain segment was synthesized by free radical polymerization of double bonds, and acrylic ester was obtained by immersion in pure water. It was found that when the hydrophobic chain segment was too long, the rotation viscosity and light transmittances decreased. In addition, the PNOE had good safety when subcutaneously immunized with 1 mL in BALB/c mice and had good storage stability at 4°C. The preparation of polyacrylates with hydrophobic long chains in this study provides a technological reserve for vaccine adjuvants.

## Acknowledgement

This work was financially supported by the Basic Research Expenses System Pilot Project of the Hebei Academy of Sciences (2024PF01).

## Reference

[1] Zhao T, Cai Y, Jiang Y, He X, Wei Y, Yu Y and Tian X. (2023) Vaccine adjuvants: mechanisms and platforms, Signal transduction and targeted therapy, 8(1), 283.

[2] Martin N and Youssef G. (2018) Dynamic properties of hydrogels and fiber-reinforced hydrogels, Journal of the mechanical behavior of biomedical materials, 85, 194-200.

[3] Osaki M, Sekine T, Yamaguchi H, Takashima Y and Harada A. (2021) Material adhesion through direct covalent bond formation assisted by noncovalent interactions, ACS Applied Polymer Materials, 3(4), 2189-2196.
[4] Lou J and Mooney D. (2022) Chemical strategies to engineer hydrogels for cell culture, Nature Reviews Chemistry, 6(10), 726-744.
[5] Xie C, Xu J, Wang X, Jiang S, Zheng Y, Liu Z, Jia Z, Jia Z and Lu X. (2024) Smart Hydrogels for Tissue Regeneration, Macromolecular Bioscience, 24(3), 2300339.
[6] Dreiss C. (2020) Hydrogel design strategies for drug delivery, Current opinion in colloid & interface science 48, 1-17.
[7] Parker R, Deville S, Dupuis L, Bertrand F and Aucouturier J. (2009) Adjuvant formulation for veterinary vaccines: Montanide™ Gel safety profile, Procedia in vaccinology, 1(1), 140-147.
[8] Kreuter J. (1985) Poly (alkyl acrylate) nanoparticles, Methods in enzymology, 112, 129-138.

# Application of multiple regression modeling and BP neural network in the analysis of prepared C4 olefins

**Kaiyao Huang** *

Xiangtan University Xiangtan, China

*Corresponding author's e-mail: 1095074649@qq.com

**Abstract:** Ethanol serves as a crucial raw material for synthesizing C4 olefins, making the investigation of catalytic synthesis process conditions highly valuable. This study conducted a correlation analysis on experimental data from the catalytic coupling of ethanol into C4 olefins under varying temperatures and catalyst conditions, determining the Pearson correlation coefficient between ethanol conversion, C4 olefin selectivity, and temperature. The analysis revealed a significant relationship between C4 olefin selectivity and temperature. Following this, gray correlation analysis was employed to pinpoint the experimental conditions with the highest correlation coefficients. Using six independent variables, including catalyst combinations and temperature, alongside dependent variables such as ethanol conversion rate and C4 olefin selectivity, a multiple regression model was established. Results indicated that temperature significantly influences the dependent variables, with $Co/SiO_2$ dosage and Co loading being the most pivotal factors affecting ethanol conversion. Specifically, while $Co/SiO_2$ and Co loading have minimal impact on ethanol conversion, HAP dosage and temperature are positively correlated. Ethanol dosage is negatively correlated. Additionally, C4 olefin selectivity is less influenced by the amounts of $Co/SiO_2$ and ethanol. Finally, a neural network model was developed linking five influencing factors to C4 olefin yield. Bayesian regularization was used to train the model, optimizing the experimental conditions for maximizing C4 olefin yield.

## 1. Introduction

C4 olefins are primarily utilized in the manufacturing of polyolefins and synthetic rubbers, while cyclic olefins are commonly presented in plant essential oils, many of which serve as fragrances. Ethanol serves as the principal raw material in the production and synthesis of C4 olefins [1]. Throughout the preparation process, various chemical reactions occur, with adjustments in catalyst configurations (such as Co loading, $Co/SiO_2$ and HAP loading ratios, and ethanol concentration) and temperature influencing the reaction pathways. This in turn affects the selectivity and yield of C4 olefins [2, 3]. Consequently, exploring the optimal synthesis conditions and appropriate catalyst selection for C4 olefin reactions is crucial. This paper examines the relationships among ethanol conversion, C4 olefin selectivity, and temperature, as well as the effects of different catalyst combinations and temperature variations on ethanol conversion and C4 olefin selectivity. Additionally, a neural network model was developed to determine the optimal catalyst combinations and temperatures to maximize C4 olefin yield under consistent experimental conditions.

Journal of Physics: Conference Series **2961** (2025) 012031 doi:10.1088/1742-6596/2961/1/012031

## 2. Study on the dependence of C4 olefin preparation on temperature

The data under different catalyst compositions were first pre-processed and visualized. Some of the visualization results are shown in Figure 1.

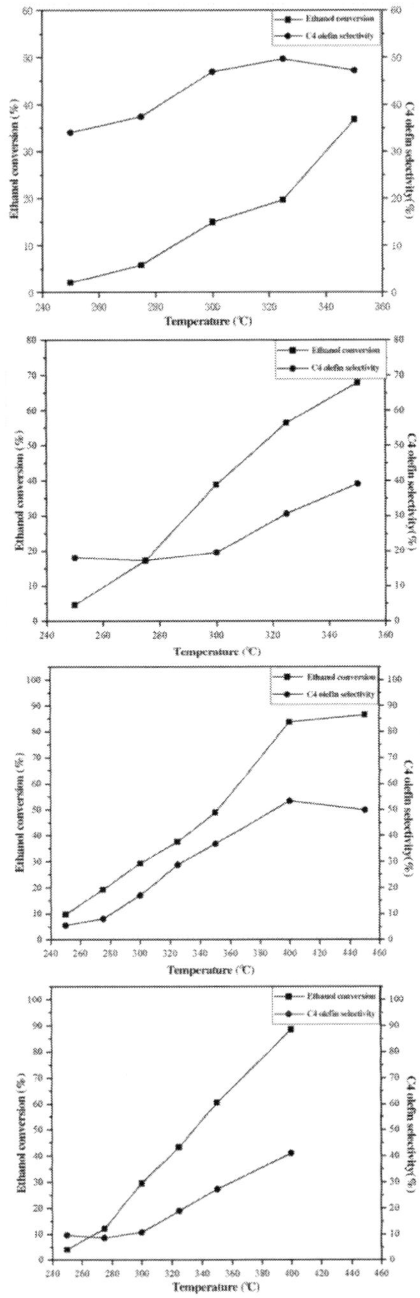


**Figure 1.** Group differences in ethanol conversion and C4 olefin selectivity.

As can be seen from the figure above, under different reaction conditions, both ethanol yield and C4 olefin selectivity have a certain correlation with the reaction temperature. Their correlations are discussed further below.

Journal of Physics: Conference Series **2961** (2025) 012031
doi:10.1088/1742-6596/2961/1/012031

The dependence among ethanol conversion, C4 alkene selectivity, and reaction temp for every catalyst combination is discussed by calculating Pearson related factors [4].

$a_i$ represents different temperatures with the same catalyst combination, $b_i$ represents the corresponding ethanol conversion and $c_i$ represents the corresponding C4 olefin selectivity.

Pearson correlation coefficient for ethanol conversion is:

$$r = \frac{n\sum_{i=1}^{n} a_i b_i - \sum_{i=1}^{n} a_i \sum_{i=1}^{n} b_i}{\sqrt{n\sum_{i=1}^{n} a_i^{\,2} - (\sum_{i=1}^{n} a_i)^2}\sqrt{n\sum_{i=1}^{n} b_i^{\,2} - (\sum_{i=1}^{n} b_i)^2}} \tag{1}$$

Pearson correlation coefficients for C4 olefin selectivity are:

$$r = \frac{n\sum_{i=1}^{n} a_i c_i - \sum_{i=1}^{n} a_i \sum_{i=1}^{n} c_i}{\sqrt{n\sum_{i=1}^{n} a_i^{\,2} - (\sum_{i=1}^{n} a_i)^2}\sqrt{n\sum_{i=1}^{n} c_i^{\,2} - (\sum_{i=1}^{n} c_i)^2}} \tag{2}$$

The results of the calculations are shown in Table 1.

**Table 1.** Relationship coefficients among ethanol conversion, c4 olefin selectivity, and temperature.

| Catalyst portfolio | Ethanol conversion rate (%) | C4 Olefin Selectivity (%) | Catalyst portfolio | Ethanol conversion rate (%) | C4 Olefin Selectivity (%) |
|---|---|---|---|---|---|
| A1 | 0.9655 | 0.8871 | A12 | 0.9631 | 0.9832 |
| A2 | 0.9950 | 0.9143 | A13 | 0.9363 | 0.9883 |
| A3 | 0.9820 | 0.9554 | A14 | 0.9639 | 0.9592 |
| A4 | 0.9975 | 0.9578 | B1 | 0.9621 | 0.9858 |
| A5 | 0.9345 | 0.9696 | B2 | 0.9293 | 0.9848 |
| A6 | 0.9836 | 0.8854 | B3 | 0.8901 | 0.9710 |
| A7 | 0.9994 | 0.9682 | B4 | 0.8993 | 0.8950 |
| A8 | 0.9771 | 0.9916 | B5 | 0.9124 | 0.9776 |
| A9 | 0.9210 | 0.9974 | B6 | 0.9399 | 0.9821 |
| A10 | 0.9229 | 0.8615 | B7 | 0.9361 | 0.9944 |
| A11 | 0.9031 | 0.9890 | | | |

Table 1 shows that among the 21 different catalyst compositions, the Pearson correlation coefficients of 19 groups were greater than 0.9, especially in the catalyst composition of the A7 group. The Pearson correlation coefficient reached 0.9994, so it can be assumed that the data of the ethanol conversion rate and the C4 olefin selectivity have a strong correlation under different catalyst compositions. Hence, the relationship between ethanol conversion rate and C4 olefin selectivity can be modeled using a nonlinear regression approach, with a portion of the fitting results displayed in Figure 2.

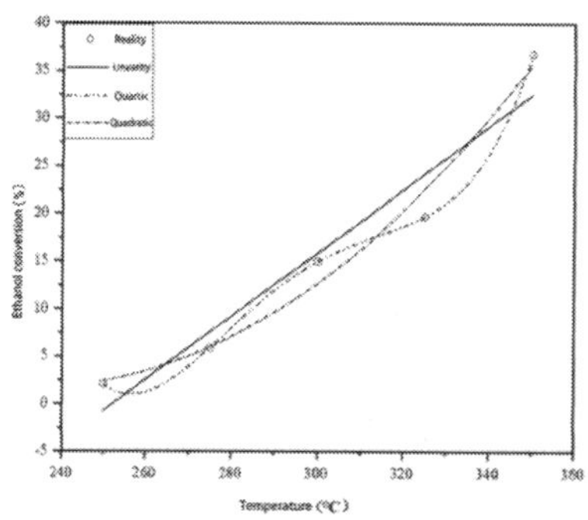

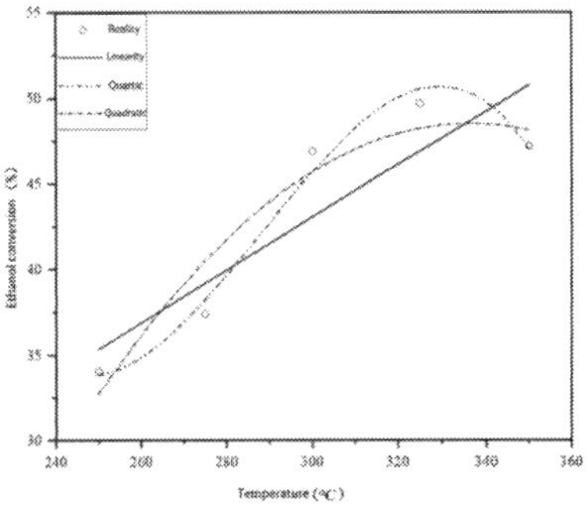


**Figure 2.** Comparison of ethanol conversion and C4 olefin selectivity fitted images (partial).

In order to better fit the function image, multiple fitting models were tried and the optimal fitting function was selected by comparing the correlation coefficients of the fitting functions.

In the fitting graph of the ethanol conversion rate, it can be seen that for the linear fitting, the error is larger and the fitting accuracy is not high, while the quadratic polynomial fitting results are more accurate and the fitted curves are in good agreement with the actual points. In a better choice of the fitting method, the correlation test is carried out below for the different fitting methods. The results of the test are shown in Table 2.

**Table 2.** Correlation coefficients for different fits (partial).

| Fitting method | R2 | Fitting method | R2 |
|---|---|---|---|
| Biometrics | 0.932 | Biometrics | 0.786 |
| Quadratic | 0.980 | Quadratic | 0.916 |
| Quartic | 1.000 | Quartic | 0.984 |

Based on Table 2, the correlation coefficient $R^2$ of the fitted values exceeds 98% when optimal values for ethanol conversion rate and C4 olefin selectivity are selected. This indicates a significant correlation between ethanol conversion, the selectivity of C4 olefins, and temperature.

## 3. Investigate the impact of different conditions on the preparation of C4 olefins

### *3.1. Grey correlation analysis*

Firstly, grey relevance level analysis was carried out to rank each evaluation object based on the magnitude of the grey-weighted relevance level and to determine the relevance order of the evaluation objects [5]. Different combinations of catalyst compositions and experimental conditions were given according to the data. There were m evaluation objects (m=114) representing different experimental conditions, and n evaluation indexes (n=2) for ethanol conversion and C4 alkene selectivity, respectively. The weights of ethanol conversion and C4 olefin selectivity in the two evaluation criteria were 0.5 and 0.5, respectively.

Substituting the grey correlation coefficient formula can get the grey correlation coefficients of 114 evaluation objects for ethanol converting rate and C4 alkene selection. Then we multiply them by their corresponding weight ratios to get the composite scores of all the evaluation objects. Their homogeneity scores can be obtained from their grey correlation analyses.

$$\xi_i(k) = \frac{\underset{s}{min}\underset{t}{min}|x_0(t) - x_s(t)| + \rho\underset{s}{max}\underset{t}{max}|x_0(t) - x_s(t)|}{|x_0(k) - x_i(k)| + \rho\underset{s}{max}\underset{t}{max}|x_0(t) - x_s(t)|} \tag{3}$$

Journal of Physics: Conference Series **2961** (2025) 012031 doi:10.1088/1742-6596/2961/1/012031

After calculating and normalizing the grey correlation coefficients for 114 different experimental conditions, the highest correlation coefficient was found for catalyst combination A3 (comprising 200 mg of 1 wt% $Co/SiO_2$, 200 mg HAP, with an ethanol flow rate of 0.9 ml/min) at 400°C. This condition yielded higher ethanol conversion and increased C4 olefin selectivity. A selection of the grey correlation coefficients is presented in Table 3.

**Table 3.** Table of gray correlation coefficients (partial data).

| Reaction condition | The gray correlation coefficient for ethanol conversion | C4 olefin selectivity gray correlation coefficient | Gray weighted correlation coefficient |
|---|---|---|---|
| A1-250 | 0.3548 | 1.0000 | 0.0095 |
| A1-275 | 0.3509 | 1.0000 | 0.0112 |
| A1-300 | 0.3333 | 1.0000 | 0.0156 |
| A1-325 | 0.3382 | 1.0000 | 0.0172 |
| A1-350 | 0.4760 | 1.0000 | 0.0200 |
| A2-250 | 0.5499 | 1.0000 | 0.0058 |
| A2-275 | 0.8021 | 1.0000 | 0.0080 |
| A2-300 | 1.0000 | 0.6523 | 0.0130 |
| A2-325 | 1.0000 | 0.6070 | 0.0194 |
| A2-350 | 1.0000 | 0.6044 | 0.0239 |

*3.2. Regression modeling*

The data analysis software was utilized to examine the regression coefficients of ethanol conversion and C4 olefin selectivity, resulting in the coefficients presented in Table 4.

**Table 4.** Table of regression coefficients.

| | Mould | Ratio | Standard error | Lower confidence interval 95% | The confidence interval limit is 95% |
|---|---|---|---|---|---|
| Y1 | Constant | -48.732 | 5.112 | -58.866 | -38.598 |
| | x1 | 0.003 | 0.050 | -0.097 | 0.103 |
| | x2 | -3.164 | 0.635 | -4.422 | -1.905 |
| | x3 | 0.086 | 0.049 | -0.010 | 0.183 |
| | x4 | 2.673 | 1.460 | -0.221 | 5.568 |
| | x5 | 0.181 | 0.013 | 0.155 | 0.208 |
| Y2 | Constant | -80.521 | 7.232 | -94.856 | -66.186 |
| | x1 | -0.034 | 0.071 | -0.175 | 0.107 |
| | x2 | 0.133 | 0.898 | -1.647 | 1.914 |
| | x3 | 0.141 | 0.069 | 0.004 | 0.278 |
| | x4 | -8.766 | 2.066 | -12.861 | -4.672 |
| | x5 | 0.333 | 0.019 | 0.296 | 0.371 |

The regression equation from the above table can be obtained from the ethanol conversion rate:

$$y_1 = -0.0337x_1 + 0.133435x_2 + 0.14124x_3 - 8.76642x_4 + 0.333198x_5 - 80.5211 \quad (4)$$

Also according to the above table, it can be seen that the confidence intervals of $x_1$ and $x_2$ contain zeros, indicating that the amount of $Co/SiO_2$ and the loading of Co do not have a significant effect on the conversion of ethanol. So, these two variables, the amount of $Co/SiO_2$ and the amount of loading of Co were removed when solving the regression equation.

The same C4 olefin selective multiple linear regression equation is:

$$y_2 = 0.002989x_1 - 3.16389x_2 + 0.086447x_3 + 2.67345x_4 + 0.181223x_5 - 48.732 \quad (5)$$

Since the confidence intervals at $x_1$ and $x_4$ contain zeros, indicating that these two independent variables, $Co/SiO_2$ dosage, and ethanol dosage rate, do not exert a considerable influence on the selectivity of C4 olefins. They were removed from the solution of the regression equation. Therefore, the accuracy of the multiple linear regression model was further improved by examining the confidence intervals of the respective variables and thus removing some of them.

The multiple linear regression is performed again after removing some of the variables. The regression equation can be found as follows.

$$y_1 = 0.109867x_3 - 8.68085x_4 + 0.333209x_5 - 80.7574 \quad (6)$$

Since the confidence intervals for all coefficients do not include zero, a significance test was performed on the regression equation. The F value was obtained by referencing the relevant data, indicating the overall significance of the regression model. This indicates that the predictor variables hold statistical significance in accounting for the variability in the response variable, thereby affirming the reliability of the regression model.

$$F = 142.2, F_{0.05}(3, 112) = 2.686 \quad (7)$$

Since 142.2 > 2.686, the equation passes the test.

The regression equation for C4 olefin selectivity is also obtained as follows.

$$y_2 = -2.90762x_2 + 0.082637x_3 + 0.180272x_5 - 44.178 \quad (8)$$

Significance test was calculated for the regression equation.

$$F = 85.6, F_{0.05}(3, 112) = 2.686 \quad (9)$$

Since 85.6 > 2.686, the equation passes the test.

The multivariate linear model results indicate that temperature has a significant impact on both ethanol conversion and C4 olefin selectivity. Specifically, for ethanol conversion, the influence of $Co/SiO_2$ dosage and Co loading was minimal and lacked explanatory power. In contrast, HAP dosage and temperature showed a positive correlation with ethanol conversion, whereas ethanol dosage exhibited a negative correlation. Similarly, for C4 olefin selectivity, the dosages of $Co/SiO_2$ and ethanol did not have a notable effect on selectivity.

## 4. BP neural network modeling

To achieve the highest C4 olefin yields, it is important to identify the right catalyst combination and temperature. It is necessary to maximize the C4 alkene yield based on different catalyst combinations and temperatures. Firstly, it is necessary to consider the effect of different combinations of catalysts and temperatures on the C4 olefin yield. To solve this problem, this paper chooses to use an artificial neural network model that can be approximated with arbitrary accuracy for modeling and solving [6], which is closer to the actual data and has higher reliability.

### *4.1. Modeling*

In order to establish five C4 olefins recovery indicators and a neural network model of C4 olefins recovery, the input data of the model are $Co/SiO_2$ usage, Co loading, HAP usage, ethanol concentration, and temperature. The output data are C4 olefins recovery. The neural network model is shown in Figure 3. The output data are C4 olefins production rate. The neural network model is shown in Figure 3.

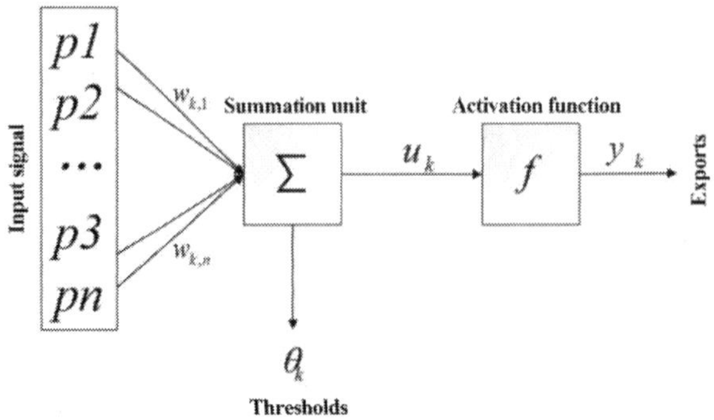


**Figure 3.** Artificial neural network architecture.

The structure models a network of connections analogous to biological neurons (synapses), where each link's strength is represented by a weight. Positive weights signify activation, while negative weights indicate inhibition. This setup includes a summation unit that computes a weighted sum (linear combination) of the input signals and a nonlinear activation function that serves as a nonlinear mapper, constraining the neuron's output within a specified range (typically limited to (0, 1) or (-1, 1)) by setting a threshold $\theta_k$ (or bias $b_k = -\theta_k$). The mathematical equation is expressed as follows.

$$u_k = \sum_{j=1}^{p} w_{kj} p_j, v_k = u_k - \theta_k, y_k = f(v_k) \tag{10}$$

where $p_1, p_2, \ldots, p_j$ denote the input signals; $w_{k1}, w_{k2}, \ldots, w_{kn}$ denote the weights of neuron $k$; $u_k$ is the result of the linear combination and $\theta_k$ is the threshold.

Then the activation function is:

$$f(v) = \frac{2}{1 + e^{-2v}} - 1 \tag{11}$$

$y_k$ is the output of neuron $k$.

### *4.2. Validation of BP neural network models*

Initially, the study examines five independent variables: the dosage of $Co/SiO_2$, the loading of Co, the dosage of HAP, the dosage of ethanol, and the temperature—which were input into the MATLAB neural network toolbox as predictors. These data were divided into three subsets: training set (70%), validation set (15%), and test set (15%), with the target being the dependent variable, C4 olefin yield. After multiple iterations and testing, a configuration of 10 hidden neurons was chosen. The model was trained using Bayesian regularization.

After completing the training of the model, the parameters need to be tested.

Firstly, the training effect of the model is tested, followed by the gradient and mean error of the training data. The test results are shown in Figure 4 and Figure 5.

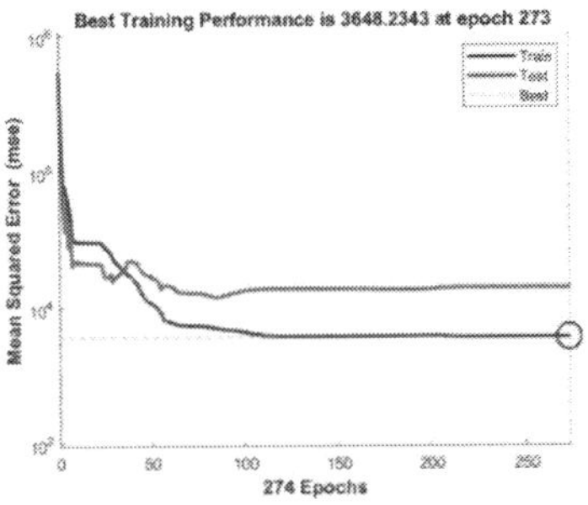


**Figure 4.** The 273rd generation of optimal training results.

From Figure 4 and Figure 5, it can be seen that the model reaches the optimization after the 273rd training. This generation model is used as the training model. Meanwhile, from the figure, the gradient and mean error of the training data tend to stabilize with the increase of the number of generations during the continuous training process, so the training model is stable.

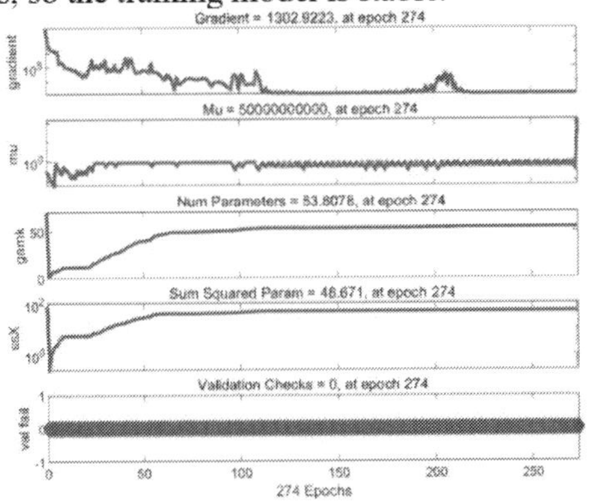


**Figure 5.** Gradient and mean error plots for training data.

Secondly, the historical residuals of the training data are examined. The results are shown in Figure 6, which shows that the training data are roughly normally distributed, and therefore satisfy the need for iteration of the number of generations.

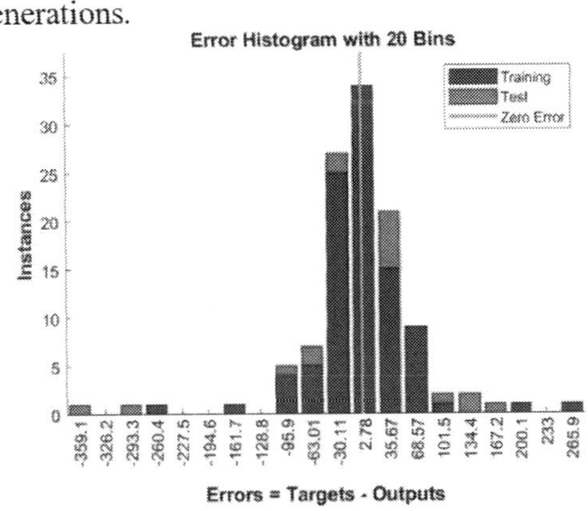


**Figure 6.** Historical residuals for training data.

Finally, we need to examine the R of the training set, test set, and validation set. Figure 7 illustrates that the R of the three sets of data is very close to 1. This suggests that the alignment of the three data sets is highly satisfactory.

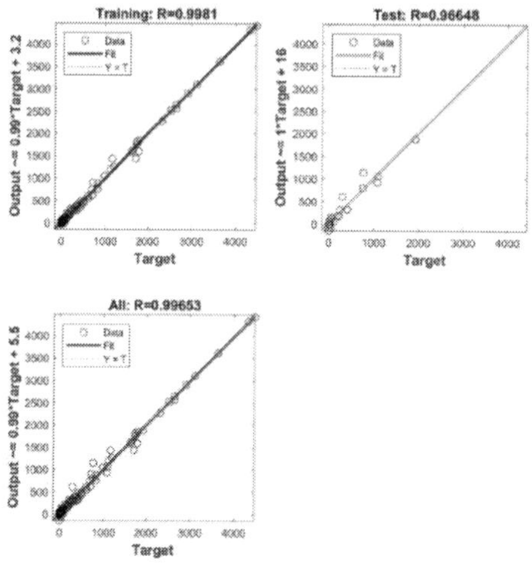


**Figure 7.** BP neural network test results.

### *4.3. Solving BP neural network models*

Various catalyst combinations and temperatures were integrated using a BP (Backpropagation) neural network model. The trained model was then employed to systematically evaluate these combinations, leading to the identification of the optimal conditions. The final results, showcasing the optimal solution, are presented in Table 5.

**Table 5.** Optimal reaction conditions.

| | $Co/SiO_2$ dosage | Coloading | HAPdosage | Ethanol concentration | Temp | C4 Olefin yield |
|---|---|---|---|---|---|---|
| Temp>350°C | 200 | 5 | 200 | 2.1 | 450 | 67.01% |
| Temp<350°C | 200 | 2 | 200 | 2.1 | 350 | 28.42% |

The BP neural network analysis revealed that the highest yield of C4 olefin, 67.01%, was achieved at 450°C using a catalyst combination of 200 mg 5 wt% $Co/SiO_2$, 200 mg HAP, and an ethanol flow rate of 2.1 ml/min. Additionally, a yield of 28.42% was observed at 350°C with a catalyst composition of 200 mg 2 wt% $Co/SiO_2$, 200 mg HAP, and an ethanol concentration of 2.1 ml/min. When the temperature was restricted to 350°C with the same catalyst combination (200 mg 2 wt% $Co/SiO_2$ and 200 mg HAP, with an ethanol flow rate of 2.1 ml/min), the maximum yield of C4 olefin remained at 28.42%.

## 5. Conclusion

Grey association analysis for C4 olefin selection decisions can be used for a wide range of uncertainties and their correlations, combining quantitative and qualitative organically. In this way, an otherwise complicated decision problem becomes straightforward and easy to calculate. The correlation analysis was carried out. The better-fitting functions were selected. The R2 fitted by each function was very close to 1. The fitting effect was close to reality. At the same time, the artificial neural network model, which can be approximated with arbitrary accuracy, is chosen for modeling and solving, which is closer to the actual data and more reliable.

**References**

[1] Lv Shao-Pei. Preparation of butanol and C_4 olefins by ethanol coupling [D]. Dalian University of Technology, 2018.

[2] Xia Wei, Wang Junguo, Qian Chen, et al. Comprehensive experiments on the preparation of low carbon olefins from ethanol catalyzed by Zr/ZSM-5 molecular sieves [J]. Experimental Technology and Management, 2021, 38(08): 149-153.DOI: 10.16791/j.cnki.sjg.2021.08.032.

[3] TIAN Yuan, SAN Yuhua, SHI Jun et al. Influence of catalyst preparation methods on the synthesis of 1, 3-butadiene from ethanol [J]. modern chemical, 2016, 36(07): 104-108. DOI: 10.16606/j.cnki.issn0253-4320.2016.07.026.

[4] Cheng Juanjuan. Empirical research on the relationship between research and teaching in universities—an analysis based on Pearson's correlation coefficient [J]. Science and Technology in Chinese Universities, 2022(10): 46-52.DOI: 10.16209/j.cnki.cust.2022.10.016.

[5] ZHANG Yana, HU Daohua. Research on Influential Factors of Tourism Economic Development in Shanxi Province Based on Gray Correlation Analysis [J]. Journal of Hunan University of Technology, 2023, 37(05): 78-85.

[6] Xia Liang. Intelligent algorithm optimization of BP neural network in wave height prediction [D]. Tianjin University, China2020.DOI: 10.27356/d.cnki.gtjdu.2020.002365.

Journal of Physics: Conference Series **2961** (2025) 012032 doi:10.1088/1742-6596/2961/1/012032

# Bactericidal properties of magnesium hydroxide nanodispersions and intelligent feedback platforms

**Chang Cao** [1] **and Linan Dun** [2, 3 *]

[1]School of Stomatology, Southern Medical University, Guangzhou 510515, China
[2]College of Materials Science and Engineering, Northeastern University, Shenyang 110004, Liaoning, China
[3]Hoffmann Institute of Advanced Materials, Shenzhen Polytechnic, Shenzhen 518055, Guangdong, China

[*]Corresponding author's e-mail: 2110204@stu.neu.edu.cn

**Abstract**: In this study, 3-(Trimethoxysilyl) propyl methacrylate was used as a modifier for surface modification of magnesium hydroxide to prepare magnesium hydroxide nanodispersions. Among them, the magnesium in magnesium hydroxide comes from the brine of the Charkhan Salt Lake. Magnesium hydroxide nanodispersions are well dispersed in water, and the UV transmittance of aqueous nanodispersions with a powder mass ratio of 10 wt% can reach more than 60%. Aqueous-phase magnesium hydroxide nanodispersions were used for the preparation of a bactericidal solution, which had a good scavenging effect on Escherichia coli, providing a new idea for the application of magnesium hydroxide in the field of bactericides.

## 1. Introduction

Given the increasing worldwide attention to health and cleanliness, the innovation of novel materials with antimicrobial properties has emerged as a significant research priority [1, 2]. Employing conventional antimicrobial substances, including heavy metal ions and organic compounds, can partially suppress microbial growth. Nevertheless, such agents could potentially exert negative impacts on both environmental and human health [3, 4]. In light of these considerations, the search for safe, efficient, and environmentally friendly antimicrobial materials has become a significant focus within the field of materials science [5, 6].

Magnesium hydroxide ($Mg(OH)_2$) is emerging as a promising candidate for antimicrobial applications due to its biocompatibility, minimal toxicity, and outstanding thermal stability. In recent times, the emergence of nanotechnology has paved the way for innovative applications of $Mg(OH)_2$. $Mg(OH)_2$ nanodispersions demonstrate superior antimicrobial efficacy over traditional materials, a result of their unique physicochemical properties such as an elevated specific surface area, enhanced surface reactivity, and improved solubility.

Research by Marta M Alves et al. suggests that meticulously designed rosehip-derived magnesium nanoparticles could serve as an eco-friendly, non-toxic source of polyphenols for various antibacterial applications within the biomedical sector. According to Lu et al., $Fe_3O_4$@$Mg(OH)_2$ nanoparticles, synthesized from waste bischofite, show promise as an ozonation catalyst for antibiotics, enhancing water purification and promoting the reuse of bischofite. In the study by Dhaouadi et al., $Mg(OH)_2$ nanorods were synthesized via a simple hydrothermal method employing hexadecyltrimethylammonium bromide (CTAB) as the surfactant. Furthermore, magnesium oxide

Journal of Physics: Conference Series **2961** (2025) 012032 doi:10.1088/1742-6596/2961/1/012032

(MgO) nanorods were obtained through the calcination of $Mg(OH)_2$ nanorods at a temperature of 700°C. Using the disk diffusion technique, it was found that these $Mg(OH)_2$ nanorods exhibited enhanced antibacterial effects against a range of tested bacterial strains. Qu et al. have developed a composite nanogel, named $Mg(OH)_2$NPs/CS hydrogel, which consists of $Mg(OH)_2$ nanoparticles ($Mg(OH)_2$NPs), chitosan (CS), sodium alginate, and polyacrylamide. This hydrogel exhibits superior mechanical properties, excellent physiological compatibility, and enhanced antimicrobial capabilities across a wide range of bacteria. Notably, the incorporation of $Mg(OH)_2$NPs enhances the hydrogel's rigidity and tensile properties without compromising its ductility. Furthermore, the $Mg(OH)_2$NPs/CS hydrogel has demonstrated a significant enhancement in the rate of wound healing in a rat model of dermal lesions, showing superior performance compared to both sterile gauze and multi-sentry bandages. Significantly, the components used in the hydrogel are cost-effective and easily obtainable, indicating that the $Mg(OH)_2$NPs/CS hydrogel represents a more feasible and budget-friendly choice. These results imply that the $Mg(OH)_2$NPs/CS hydrogel has the potential to serve as a proficient antibacterial wound care material, possibly contributing to the advancement of innovative antimicrobial substances.

However, dispersion of magnesium hydroxide in the aqueous phase has not been extensively reported. Therefore, in this paper, aqueous phase $Mg(OH)_2$ nanodispersions were prepared by using 3-(Trimethoxysilyl)propyl methacrylate as a modifier. Concurrently, the aqueous suspensions of $Mg(OH)_2$ nanoparticles exhibited remarkable bactericidal activity against Escherichia coli (E. coli), offering novel insights into the application potential of $Mg(OH)_2$. As opposed to prior techniques, the synthesis of magnesium hydroxide in this study was prepared by electrochemical deposition using the brine from Salt Lake as the raw material. This approach ensured the comprehensive utilization of the brine from the Charkhan Salt Lake and significantly enhanced the utilization value of magnesium resources in the West. Furthermore, the convenient electrical energy in Qinghai offers a promising avenue for the comprehensive utilization of Salt Lake magnesium resources.

## 2. Experiments and methods

### *2.1. Chemicals*

Ethanol ($C_2H_5OH$), sodium hydroxide (NaOH), 3-(Trimethoxysilyl) propyl methacrylate (KH-570), menthol, glycerol, xylitol, peptone, yeast extract, and agar were procured from Shanghai McLean Reagent Co. Commercially available mouthwash are purchased from local supermarkets. It was not ensured that all chemicals were purified prior to use.

### *2.2. Preparation of $Mg(OH)_2$ nanodispersions*

$Mg(OH)_2$ nanodispersions were synthesized by utilizing a combination of direct precipitation and wet surface modification methods. As described in Reference, $Mg(OH)_2$ nanodispersions were prepared by dispersing it in ethanol to create a suspension. Subsequently, 10% by weight of $Mg(OH)_2$ of KH-570 was introduced into the suspension, followed by heating to 70°C and maintaining stirring for 4 hours. After cooling to room temperature, vacuum filtration was performed, and washed, and the filter cake was transferred directly to deionized water for ultrasonic dispersion to obtain transparent aqueous-phase $Mg(OH)_2$ nanodispersions. The detailed experimental procedure is depicted in Figure 1.

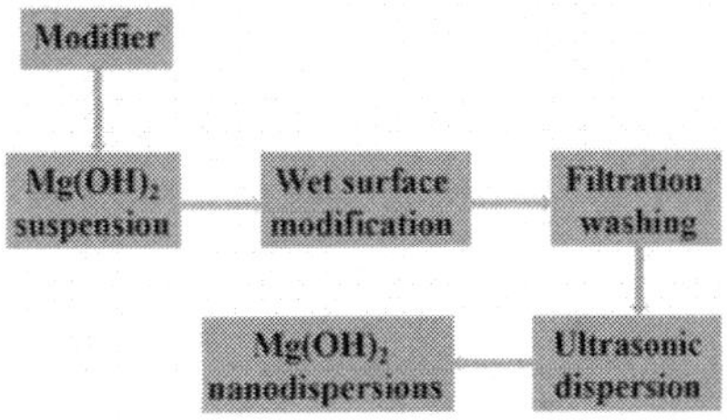


**Figure 1.** Experimental procedure for the preparation of $Mg(OH)_2$ nanodispersions.

*2.3. Characterization*

The unique functional groups of $Mg(OH)_2$ nanodispersions were characterized by using a PE Frontier FTIR spectrometer over the wavenumber range of 400 to 4, 000 $cm^{-1}$, with KBr as the substrate. The surface structure of Escherichia coli (E. coli) was observed with a scanning electron microscope (SEM) operating at 10 kV on a JEOL/JSM-IT800 instrument (SHL, Japan). The transmittance of the $Mg(OH)_2$ nanodispersions was measured by using a UV-vis-NIR spectrometer (solidspec-3700, Shimadzu, Japan) across the wavelength range of 300 to 800 nm. X-ray diffraction patterns were obtained with an X-ray diffractometer (Rigaku Co., Japan) scanning from 5° to 80°. For the preparation of samples, a 2% suspension was made by dissolving 2 g of $Mg(OH)_2$ nanodispersion in 100 mL of ultrapure water. The particle size distribution was analyzed by using a laser particle size analyzer (Mastersizer 3000, Malvern Instruments Co., Ltd., UK) under conditions of a 35% obscuration rate and a scattering intensity between 10-20%.

## 3. Results and discussions

*3.1. Synthesis of $Mg(OH)_2$ nanodispersions*

Figure 2 displays SEM photographs and elemental distributions of $Mg(OH)_2$ nanodispersions. The surface morphology of the $Mg(OH)_2$ nanodispersions presents a lamellar structure with a particle size of approximately 100 nm, as well as good dispersion, with minimal agglomeration of particles. This indicates that the addition of KH-570 not only improves the dispersion of $Mg(OH)_2$ nanoparticles in $H_2O$ but also effectively inhibits the growth of the particles so that transparent and stable aqueous-phase $Mg(OH)_2$ nanodispersions can be prepared. The elemental composition analysis indicated that carbon, oxygen, magnesium, and silicon were evenly dispersed on the nanoparticle surfaces.

**Figure 2.** SEM photographs of $Mg(OH)_2$ nanodispersions and elemental distribution of C, O, Mg, and Si.

Figure 3 (a) displays the XRD patterns for $Mg(OH)_2$ and its nanodispersions. It is observable from Figure 3 (a) that the diffraction peaks of both samples align with the standard pattern (JCPDS No.7-239), confirming that the synthesized $Mg(OH)_2$ nanodispersions possess a hexagonal crystalline structure. Moreover, the treatment of $Mg(OH)_2$ with KH-570 did not compromise the integrity of its crystalline framework. Notably, the absence of extraneous peaks in the XRD patterns suggests a high level of purity in the samples. Additionally, the (001) plane of the $Mg(OH)_2$ nanodispersions shows a higher degree of polarity compared to the (101) plane of pristine $Mg(OH)_2$, signifying improved dispersibility. This is corroborated by SEM imaging. As depicted in Figure 3 (c), to further delineate the particle size distribution of the $Mg(OH)_2$ nanodispersions, the nanoscale powder was analyzed for particle dimensions, which substantiated the SEM findings and determined an average particle diameter of 125 nm.

Figure 3 (b) presents the FTIR spectra for $Mg(OH)_2$, KH-570, and $Mg(OH)_2$ nanodispersions. In the FTIR spectrum (a) of $Mg(OH)_2$, the intense symmetric absorption band at 3, 700 $cm^{-1}$ is associated with the O-H stretching vibration within the $Mg(OH)_2$ lattice, while the distinctive peak at 440 $cm^{-1}$ represents the Mg-O stretching vibration. In the FTIR spectrum (b) of KH-570, the characteristic absorption bands at 2, 918 $cm^{-1}$ and 2, 890 $cm^{-1}$ are due to the symmetric and asymmetric C-H stretching vibrations of KH-570, respectively, and the peak at 1, 092 $cm^{-1}$ corresponds to the Si-O stretching vibration. In comparison to spectra (a) and (b), the FTIR spectrum (c) of the $Mg(OH)_2$ nanodispersions exhibits characteristic absorption bands for both O-H and Mg-O from $Mg(OH)_2$, as well as bands for C-H stretching vibrations in the range of 2, 980-2, 800 $cm^{-1}$. Additionally, the Si-O absorption bands at 1, 024 $cm^{-1}$ show a noticeable shift relative to those in the spectrum (b). This suggests that KH-570 has been effectively immobilized onto the $Mg(OH)_2$ surface through a chemisorption process.

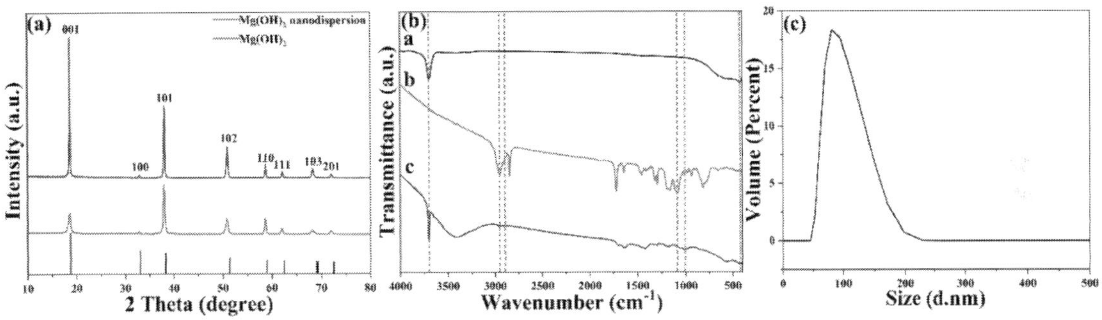


**Figure 3.** The XRD spectra (a), FTIR spectra (b), and particle size distribution diagram (c) of $Mg(OH)_2$ nanodispersions.

### *3.2. Dispersion properties of $Mg(OH)_2$ nanodispersions*

To study the dispersive behavior of $Mg(OH)_2$ nanodispersions in $H_2O$, $Mg(OH)_2$ nanodispersions with different solid contents were dispersed in $H_2O$. Figure 4 illustrates the physical diagrams of aqueous $Mg(OH)_2$ nanodispersions with different $Mg(OH)_2$ contents. As depicted in Figure 4, the $Mg(OH)_2$ nanodispersions exhibited a slight reduction in transparency with rising $Mg(OH)_2$ concentrations, they retained superior transparency relative to unmodified $Mg(OH)_2$, a clear Tyndall effect could be observed, and they could be left to stand for several months without settling.

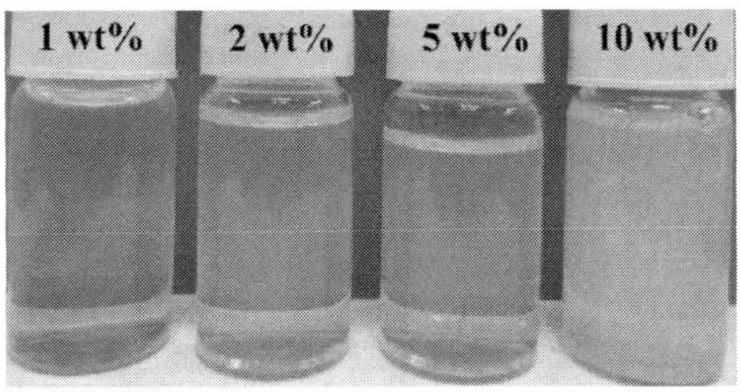


**Figure 4.** Pictures of aqueous phase $Mg(OH)_2$ nanodispersions with different solid contents.

### *3.3. Bactericidal activity of $Mg(OH)_2$ nanoscale suspensions against E. coli*

Figure 5 illustrates the viability of E. coli with a concentration of 10 mg/mL $Mg(OH)_2$ nanodispersions over various time intervals. In the absence of $Mg(OH)_2$ nanodispersions, it is apparent that E. coli grows rapidly, and the number of colonies of E. coli after 2 h and 4 h decreases significantly, which is discernible to the naked eye. Subsequently, after 8 h of incubation in the culture solution containing $Mg(OH)_2$ nanodispersions, there were no surviving E. coli. It is apparent that the number of E. coli in the culture solution containing $Mg(OH)_2$ nanodispersions gradually decreased with time, i.e., it is shown that the $Mg(OH)_2$ nanodispersions have a bactericidal effect on E. coli. The viability of E. coli diminishes over time in the presence of $Mg(OH)_2$ nanodispersions.

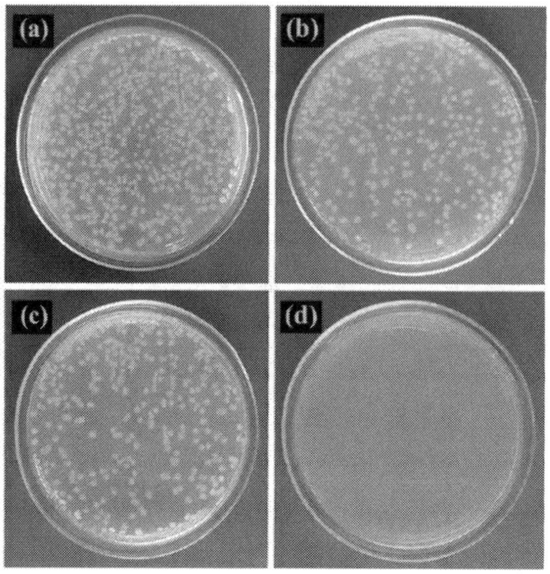


**Figure 5.** Antibacterial effect of nano $Mg(OH)_2$ dispersions against Escherichia coli at different times, (a) 0 h, (b) 2 h, (c) 4 h, and (d) 8 h.

To get direct proof of the bactericidal activity of $Mg(OH)_2$ nanodispersions against E. coli, experiments were conducted to evaluate the changes in E. coli morphology post-exposure to nanoparticles, as observed through SEM imaging. Figure 6 illustrates the distinct morphological changes in E. coli, both pre-and post-exposure to a 2 mg/mL concentration of $Mg(OH)_2$ nanodispersions for 2 hours. Figure 5 (a) presents the SEM image of the E. coli cells in their typical, healthy state prior to the application of nanodispersions. As shown in Figure 6, the E. coli cell surface was smooth and full, with regular morphology and uniform refractive index. The individual cells were intact and structurally dense, and there was no obvious breakage on the surface. However, in the E. coli cells treated with $Mg(OH)_2$ nanodispersions, as shown in Figure (b), the cells have been ruptured and their surfaces are severely broken. Further analysis of the two image sets underscores the bactericidal impact of $Mg(OH)_2$ on E. coli.

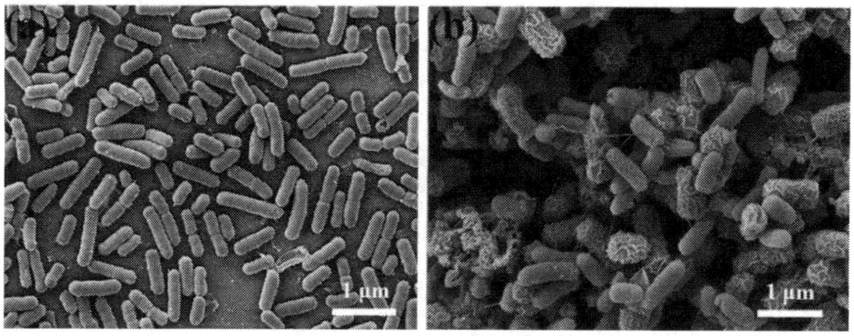


**Figure 6.** SEM images of E. coli: untreated (a) and post-treatment (b) with 2 mg/mL $Mg(OH)_2$ nanodispersions over 2 hours.

*3.4. Design of an experiential feedback system for bactericidal materials in $Mg(OH)_2$ nanodispersions*
To collect users' experiences and feedback, we have established an online feedback platform, which is linked to the online feedback platform by printing QR codes on the packaging of magnesium hydroxide sterilization products. Users access the online feedback platform by scanning the QR code to record and share their experiences. Researchers analyze the feedback data and use it to improve product formulations and marketing strategies. The platform promises to do the following:

*3.4.1 Incentives of user feedback.* (1) We offer incentives such as coupons, points, or small gifts to encourage users to submit feedback.
(2) We ensure that the feedback process is quick and not cumbersome to increase user engagement.

*3.4.2 Data collection and analysis.* (1) We design effective data collection methods to ensure that valuable user feedback can be collected.
(2) We use data analysis tools to identify common problems, user preferences, and potential market opportunities.

*3.4.3 Protection of user privacy.* (1) We inform users of how their data are collected and used, and obtain their consent.
(2) We ensure the security of information and compliance with relevant data protection legislation.

*3.4.4 Response to user feedback.* (1) We have a rapid response mechanism to respond to and process user feedback promptly.
(2) We regularly update users on product improvements and market strategy adjustments.

*3.4.5 Continuous optimization.* (1) We continuously optimize the product and user experience based on the feedback collected.
(2) We regularly review and update the functionality and user interface of the online feedback platform.

*3.4.6 Transparency and honesty.* We maintain transparency in communicating with users and honestly demonstrate the product's strengths and areas for improvement.

*3.4.7 Multi-channel integration.* We consider combining online feedback platforms with other market research tools and user feedback channels to gain more comprehensive market insights.

## 4. Conclusion
In this research, $Mg(OH)_2$ nanodispersions were synthesized by utilizing electrochemical deposition and wet surface modification techniques. The experimental data revealed that the nanomaterials possessed bactericidal activity against E. coli. Its low toxicity and high biocompatibility make it potentially applicable in the medical and food industries. Future studies will focus on optimizing the preparation process and exploring its antibacterial mechanism in depth. Additionally, it offers novel insights into the full exploitation of magnesium reserves found in the salt lakes of Western China.

**Acknowledgments**
This research was funded by the Priority Basic Research Projects as part of the Basic Enhancement Programme (Grant number: 2022-173ZD-226).

**Reference**
[1] Duan S., Wu R. N., Xiong Y. H., et al. "Multifunctional Antimicrobial Materials: From Rational Design to Biomedical Applications". Progress in Materials Science, Vol. 125, 100, 887, 2023. https://doi.org/10.1016/j.pmatsci.2021.100887.
[2] Kirthika S. K., Goel. G., Matthews A., et al. "Review of the Untapped Potentials of Antimicrobial

Materials in the Construction Sector". Progress in Materials Science, Vol. 133, 101, 065, 2023. https://doi.org/10.1016/j.pmatsci.2022.101065.
[3] Xiao D., Ding W., Zhang J. B., et al. "Fabrication of a Versatile Lignin-based Nano-trap for Heavy Metal Ion Capture and Bacterial Inhibition". Chemical Engineering Journal, Vol. 358, No.15, pp. 310-320, 2019. https://doi.org/10.1016/j.cej.2018.10.037.
[4] Ghiorghita C. A., Lazar M. M., Platon I. V., et al. "Feather-Weight Cryostructured Thiourea-Chitosan Aerogels for Highly Efficient Removal of Heavy Metal Ions and Bacterial Pathogens". International Journal of Biological Macromolecules, Vol. 235, No. 30, 123, 910, 2023. https://doi.org/10.1016/j.ijbiomac.2023.123910.
[5] Meyer N. K. J. and Nodwel J. R. "Biology and Applications of Co-Produced, Synergistic Antimicrobials from Environmental Bacteria". Nature Microbiology, Vol. 6, pp. 1, 118-1, 128, 2021. https://doi.org/10.1038/s41564-021-00952-6.
[6] Chen M. J., Li Y., Liu H. M., et al. "High-Value Valorization of Lignin as Environmental Benign Antimicrobial". Materials Today Bio, Vol. 18, 100, 520, 2023. https://doi.org/10.1016/j.mtbio.2022.100520.

Journal of Physics: Conference Series **2961** (2025) 012033 doi:10.1088/1742-6596/2961/1/012033

# Repair of enamel by calcium fluorophosphate ion clusters

**Tiantian Ma [1], Kegong Ning [2] and Yongjun He [1*]**

[1] College of Chemistry and Chemical Engineering, Xi'an University of Science and Technology, Xi'an, 710054, Shaanxi, China

[2] Yunnan Baiyao Group Health Products Co., LTD, Kunming, 650216, Yunnan, China

[*] Corresponding author.mail addresses: yongjunhe@xust.edu.cn

**Abstract**: Artificially restoring tooth enamel's distinct structure is still difficult because of its intricate hierarchical structure, which spans from nanoscale to microscale. In this paper, triethylamine-stabilized calcium fluorophosphate ion clusters were prepared in an ethanol medium and investigated for the restoration of tooth enamel. Fourier transform infrared spectroscopy (FTIR) was utilized to analyze the structure of ion cluster transition intermediates. Ethanol volatilization removed triethylamine from the calcium fluorophosphate ion clusters, converting them to amorphous calcium phosphate. The calcium fluorophosphate ion clusters were added dropwise to the enamel window surface with a pipette gun and waited for 20 minutes for the coating to form. Then, the samples were placed in artificial saliva and incubated for 48 h. The SEM results showed that the enamel surfaces were denser after being restored by the calcium fluorophosphate ion clusters and that the distinctive fish-scale structure of the natural enamel was accurately replicated. Nanoindentation tests showed that the mechanical properties of the damaged enamel were restored after remineralization. Based on this study, calcium fluorophosphate ion clusters will potentially become a material for restoring enamel for clinical use.

## 1. Introduction

Teeth are an important biomineral that consists of highly mineralized tissues of enamel, dentin, and pulp [1]. In vertebrates, the hardest and most mineralized tissue is the enamel, which covers the outside of the tooth. It consists of more than 95% hydroxyapatite minerals and small amounts of water and organic materials [2]. As a result, tooth enamel has a high degree of hardness and excellent flexural strength and is often used to transmit and withstand the tremendous mechanical forces of chewing [3]. However, tooth enamel is vulnerable to acid damage in a variety of acidic situations, which can result in dental caries [4]. An imbalance between the mineralization and demineralization of tooth minerals is the primary cause of dental caries [5]. In most cases, mineral gain and loss are balanced. Nevertheless, the calcium phosphate in the enamel dissolves and causes demineralization when the pH of the enamel surface falls below 5.5 [6].

$$Ca_{10}(PO_4)_6(OH)_2(s) + 8H^+(aq) \leftrightharpoons 10Ca^{2+}(aq) + 6HPO_4^{2-}(aq) + 2H_2O \quad (1)$$

The equilibrium chemical equation for hydroxyapatite dissolution obtained from (1) states that hydroxyapatite crystals dissolve in terms of both pH and calcium and phosphate ion concentration. Remineralization is the process of creating calcium phosphate crystals on the surface of tooth enamel by adding calcium and phosphate ions [7]. Controlling the hydroxyapatite crystals' nucleation and development is crucial for the remineralization of tooth enamel.

Posner and Betts [8] originally suggested the existence of precursor amorphous phase production of hydroxyapatite based on their observations of prenucleated clusters of $Ca_9(PO_4)_6$ with dimensions about 1 nm, which are also referred to as calcium phosphate ion clusters. Research has demonstrated that calcium phosphate ion clusters may serve as both the precursors and catalysts for the transformation of amorphous calcium phosphate into hydroxyapatite [9]. Rather than going through full ionolysis, amorphous calcium phosphate is merely dissociated into these ionic clusters. Dental enamel might grow epitaxially when hydroxyapatite crystals made of calcium phosphate ionic clusters were discovered by Shao et al. [10]. As a result, calcium phosphate ion clusters present a viable material for tooth enamel remineralization.

Fluoride is commonly used to prevent enamel demineralization and remains the most effective caries preventive agent to this day [11]. This is because fluoride forms fluorapatite, which strengthens and increases the tooth's resistance to acidic conditions by binding to the tooth's hydroxyapatite crystals [12]. Fluoride is currently added to toothpaste for caries prevention. Although fluoride has been questioned for its toxicity, the fluoride ion concentration used in our study was 60-600 $ppm^{-1}$, which is a lower concentration of fluoride compared to fluoride-containing toothpaste. Prior research has described the use of calcium phosphate ion clusters for dental enamel repair, utilizing triethylamine, a tiny organic molecule, as a stabilizer. However, the impact of calcium fluorophosphate ion clusters on tooth enamel repair has not been examined. Therefore, we prepared calcium fluorophosphate ion clusters stabilized by triethylamine to investigate their remineralization effect on damaged enamel.

## 2. Materials and methods

### *2.1. Making of calcium fluorophosphate ion clusters*

This is how calcium fluorophosphate ion clusters were made. First, solution A is prepared by adding 0.1 g of $CaCl_2 \cdot 2H_2O$ and 380 μl of triethylamine to 8 ml of ethanol, followed by 5 minutes of sonication. Subsequently, solution B is prepared by adding 1 ml of 1200 $ppm^{-1}$ NaF solution and 7 μl $H_3PO_4$ to 2 ml of ethanol and vigorously mixing. After adding solution B to solution A and gently stirring, fluorinated calcium phosphate ion clusters (10 mg/ml containing 120 $ppm^{-1}F^{-1}$) started to develop in the mixture. Precipitation was used to gather the calcium fluorophosphate ion clusters that resembled gel. By altering the concentration of NaF solution, fluoride content can be regulated.

### *2.2. In vitro mineralization for enamel repair*

The subsidiary hospital of the University of Kunming Medical School offered noncarious human third molars. Human dental tissue samples may only be used with patient permission and in accordance with a program that has been approved by the hospital ethics committee. After polishing enamel windows with silicon carbonate paper (numbers 1200 and 2000), $H_3PO_4$ (37wt%) was used to etch the glass for 180 seconds. To guarantee the removal of any remaining pollutants, all samples are ultrasonically treated for 20 minutes in deionized water before being allowed to air dry.

In contrast, half of the tooth surfaces served as a control area, covered in acid-resistant nail lacquer, and the remaining half served as an experimental region for calcium fluorophosphate ion cluster restoration. Using a pipette gun, calcium fluorophosphate ion clusters (10 mg/ml) were dropped six times at a volume of 100 μl onto the tooth surface. To create a coating of calcium phosphate ion clusters on the enamel, the sample was air-dried at 25°C for 25 minutes following each drop. At that time, the layer of calcium phosphate ion clusters spontaneously transformed into an amorphous calcium phosphate layer, and the next drop was applied. The tooth modules were submerged in modified simulated oral fluids (10 ml each sample containing 1.5 mM $CaCl_2$, 0.9 mM $K_2HPO_4$, 15 ppm $F^{-1}$, 130 mM KCl, and 1 mM $NaN_3$) at different pH (7.50, 8.00, 8.50, 9.00) for 24 h at 37°C. The above treatments were repeated for the samples in order to fully mineralize the enamel. The treated samples were air-dried and washed with water at the conclusion of the experiment before being tested.

Journal of Physics: Conference Series **2961** (2025) 012033 doi:10.1088/1742-6596/2961/1/012033

*2.3. Characterization*

The calcium fluorophosphate ion clusters induced bulk ACP were prepared into pellets by thoroughly mixing them with KBr powder. Thirty scans were made using FTIR (IRAffinity$^{-1}$, Japan) at 4000 to 400 cm$^{-1}$ with a resolution of 4 cm$^{-1}$. A blank KBr plate was used to establish the background. A polarized light microscope (59XC-PC, China) was used to observe the acid erosion of the enamel window and the progress of remineralization. Scanning electron microscope images were obtained by scanning electron microscope (SU8010, Japan). The teeth of the alloy were sputtered with platinum at an accelerating voltage of 20 kv. EDS was collected under the same instrument. A nanoindentation machine (G200, USA) was used to conduct the nanoindentation testing.

## 3. Results and discussion

Calcium fluorophosphate ion clusters are considered to be precursors of amorphous calcium phosphates, which are thermodynamically unstable substances that are converted to hydroxyapatite in water within a short period of time [10]. As shown in Figure. 1a, the morphology of calcium fluorophosphate ion clusters is consistent with the gel-like form described in the literature and can be stabilized in ethanol solution for four days. Among them, triethylamine is an effective ion cluster stabilizer. The calcium fluorophosphate ion clusters were detected after being left for half an hour, and the infrared spectra exhibited a large peak packet at 570 cm$^{-1}$, indicating that the intermediate was amorphous calcium phosphate (Figure. 1b). After leaving the calcium fluorophosphate ion clusters for four days, their IR spectra showed that they were still amorphous, but there was a divergent peak at $2\theta=30°$, indicating that the ion clusters had begun to transform to hydroxyapatite. The ion clusters were stabilized for about four days at a lower fluorine content of 60 ppm$^{-1}$F$^{-1}$ and only for about 30 minutes at a higher fluorine content of 600 ppm$^{-1}$F$^{-1}$.

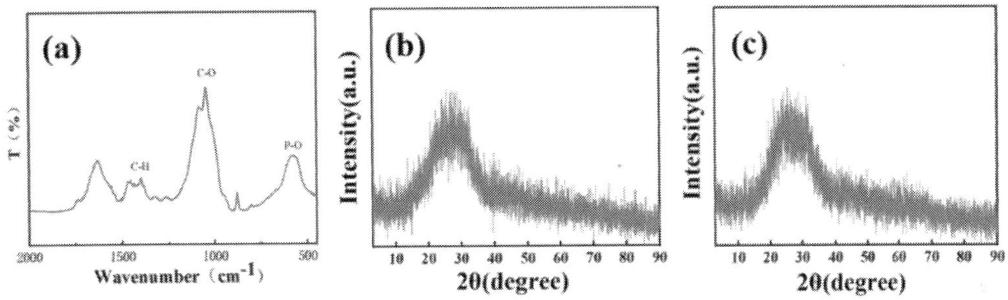


**Figure 1.** (a) FTIR spectra of intermediate structures transformed with fluorinated calcium phosphate ion clusters; (b) XRD spectra of fluorinated calcium phosphate ion clusters (30 min); (c) XRD spectra of fluorinated calcium phosphate ion clusters (4 days).

We prepared tooth enamel windows. Before the remineralization experiments began, the samples needed to be acid-etched. There are more solutions for acid etching tooth enamel. Currently, the 37% phosphoric acid is commonly used for acid etching of enamel. To simulate early caries, the acid etching solution used in the laboratory is 20% phosphoric acid. The enamel window was immersed in 20% phosphoric acid to ensure adequate exposure. The enamel samples were acid etched at different times to ensure that the optimal depth of acid etching was selected. At 30 s of acid etching, the enamel had a more fragmented surface compared to before acid etching (Figure. 2a). After 1 min of acid etching, the enamel structure showed significant damage with enamel columns protruding (Figure. 2b). After 3 minutes of acid etching, the enamel columns of tooth enamel were severely damaged, and the top of the enamel was damaged (Figure. 2c). After 5 minutes of acid etching, the enamel column of the enamel was exposed as a whole, and the interstitium of the enamel column was obvious (Figure. 2d). We selected the sample with the more pronounced enamel columns in order to investigate the restoration effect in greater detail.

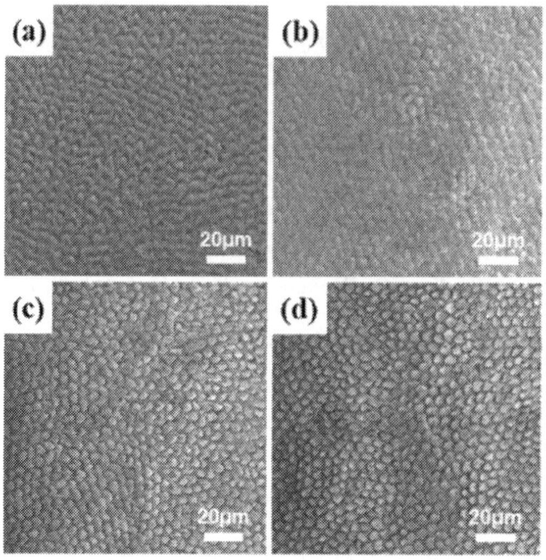


**Figure 2.** SEM image of tooth enamel surface after acid etching. Etching time: (a)30 s, (b)1 min, (c)3 min, (d)5 min.

Human tooth enamel has a complex and well-arranged structure with a typical fish-scale structure. Although there are many enamel remineralization materials, most of them cannot reproduce the special structure of enamel. We tested the restorative effect of calcium fluorophosphate ion clusters as a remineralizing material in vitro and evaluated its potential application in enamel remineralization. In dental enamel remineralization experiments, we usually use polarized light microscopy to observe the progress of experimental remineralization. The surface of unmineralized enamel did not show obvious crystalline particles under a polarized light microscope, and after remineralization, crystalline particles appeared on the surface of the enamel, which appeared white under a polarized light microscope. When the crystalline particles are evenly distributed on the enamel columns of the enamel and form a mineralized layer, it is possible to determine whether the enamel is ready for testing. At two weeks of mineralization with calcium fluorophosphate ionic clusters, a mineralized layer had already appeared on the enamel, and the concentration of fluoride ions played a major role.

Tooth enamel samples were characterized after the removal of nail polish after two weeks of mineralization, distinguishing between before and after mineralization (Figure. 3). After remineralization of the tooth enamel, it is clear that nanoparticles appeared in the mineralized area growing on the enamel prisms. The newly deposited crystals were consistent with the shape of the natural crystals in the characteristic fish scale structure of tooth enamel. The remaining structure of the damaged enamel tissue guided the remineralization, and $Ca^{2+}$, $PO_4^{3-}$, and $F^-$ released by the remineralized material continued to grow on the original enamel prisms and were present in the interstitial spaces between the prisms. This suggests that the calcium fluorophosphate ion clusters can restore the enamel to its original structure and are suitable for enamel remineralization. The addition of fluoride ions accelerates the process of remineralization because of its ability to form crystals quickly. This is a significant advantage over other materials, as it is difficult for other remineralization materials to accurately replicate the structure of natural tooth enamel.

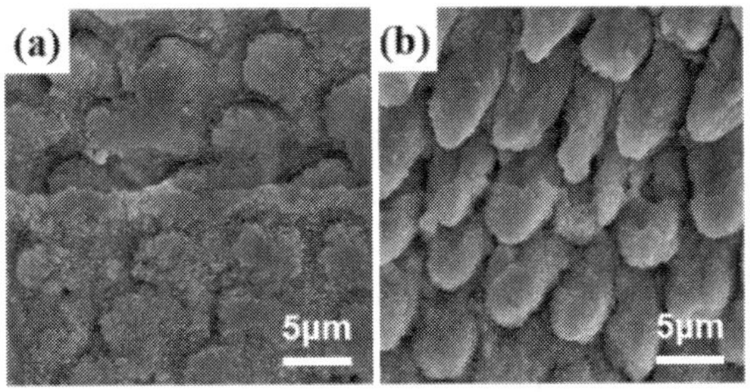


**Figure 3.** High-power SEM image of enamel surface after remineralization.

EDS analysis showed (Figure. 4.a-c) that after being repaired by calcium fluorophosphate ionic clusters, the enamel surface contained mineralized substances of Ca, P, O, and F. The Ca/P values before and after the repairs were basically unchanged at around 2.0, which was close to the Ca/P of hydroxyapatite. The fluorine content of the acid-etched portion of the enamel was 0.65 (Wt%), which increased to 3.14 (Wt%) after remineralization, while the content of other elements remained essentially unchanged. Mechanical properties are essential in the evaluation of enamel restorations and were examined by nanoindentation test for each group of enamel samples (Figure. 4d). Load-displacement curves were used to compute the hardness (H) and elastic modulus (E). Natural enamel has an H value of 4.28±0.08 GPa and an E value of 189.73±15.79 GPa. Enamel's mechanical qualities were greatly reduced after demineralization. The mechanical properties of enamel were improved with remineralization restoration, exhibiting H and E values of 3.95±0.26 and 173.82±13.13 GPa, respectively. After being treated for a brief length of time with ionic clusters containing calcium fluorophosphate, demineralized enamel's mechanical characteristics were markedly improved.

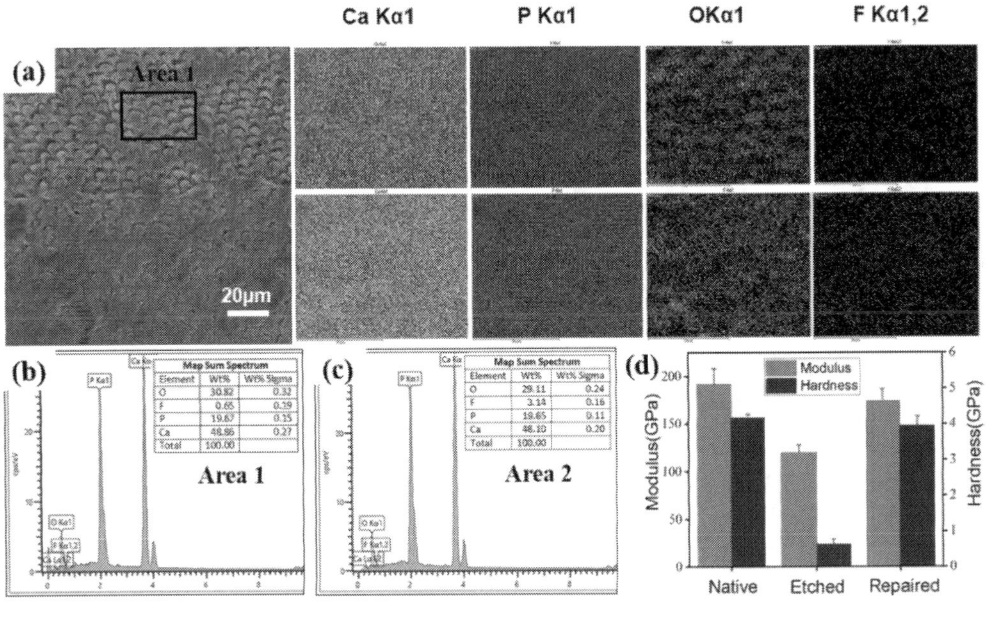


**Figure 4.** (a) EDS spectra of enamel after remineralization; (b) EDS map of area1 region before mineralization; (c) EDS map of area2 region after mineralization; (d) Map of the mechanical properties of enamel: strength and modulus of elasticity of natural enamel, after acid etching and after restoration.

## 4. Conclusions

Calcium fluorophosphate ion clusters are able to repair damaged enamel surfaces, promote crystal growth, and at the same time replicate the unique fish-scale structure of enamel and improve its mechanical properties. During the remineralization process, calcium fluorophosphate ion clusters undergo a transformation into amorphous calcium phosphate, followed by hydroxyapatite. The process of mineralization is accelerated by the presence of trace amounts of fluoride ions. Therefore, calcium fluorophosphate ion clusters are expected to be an effective material for the clinical restoration of dental enamel.

## References

[1] Okamoto Y, Hidaka S. Studies on calcium phosphate precipitation: effects of metal ions used in dental materials. [J]. Journal of Biomedical Materials Research Part A, 2010, 28(12): -.DOI:10.1002/jbm.820281204.

[2] Burwell, A.K., Litkowski, L.J., Greenspan, D.C., 2009.Calcium sodium phosphosilicate (NovaMin): remineralization potential. [J]. Advances in Dental Research, 2009, 21(1): 35.DOI:10.1177/0895937409335621.

[3] Zhao X, O "Brien S, Shaw J, et al. The origin of remarkable resilience of human tooth enamel[J]. Applied Physics Letters, 2013, 103(24): 55.DOI:10.1063/1.4842015.

[4] Pitts, N.B., Zero, D.T., Marsh, P.D., et al. Dental caries[J]. Nature Reviews Disease Primers, 2017, 3: 17030.DOI:10.1038/nrdp.2017.30.

[5] Ensanya A N, Anas A, Adam S, et al. Demineralization–remineralization dynamics in teeth and bone[J]. International Journal of Nanomedicine, 2016, 11: 4743-4763.DOI:10.2147/IJN. S107624.

[6] Jiang S, Jin W, Wang Y N, et al. effect of the aggregation state of amorphous calcium phosphate on hydroxyapatite nucleation kinetics[J]. 2019.DOI:10.1039/c7ra02208e.

[7] Chen, Z., Duan, Y ., Shan, S., et al. Deep and compact dentinal tubule occlusion via biomimetic mineralization and mineral overgrowth[J].Nanoscale, 2022, 14:642–652. DOI: 10.1039/d1nr05479a.

[8] Betts F, Posner A S. An X-ray radial distribution study of amorphous calcium phosphate - ScienceDirect[J]. Materials Research Bulletin, 1974, 9(3): 353-360.DOI:10.1016/0025-5408(74)90087-7.

[9] Wang L, Nancollas G H. Calcium orthophosphates: crystallization and dissolution. [J]. Chemical Reviews, 2010, 40(5): 4628-69.DOI:10.1002/chin.200905233.

[10] Shao C, Jin B, Mu Z, et al. Repair of tooth enamel by a biomimetic mineralization frontier ensuring epitaxial growth[J]. Science Advances, 2019, 5(8): eaaw9569.DOI:10.1126/sciadv. aaw9569.

[11] González-Cabezas, C. & Fernández, C. E. Recent advances in remineralization therapies for caries lesions[J]. Advances in Dental Research, 2018, 29(1):55–59.DOI: 10.1177/ 0022034517740124

[12] Liu Y C, Lee Y T, Huang T C, et al. In Vitro Bioactivity and Antibacterial Activity of Strontium-, Magnesium-, and Zinc-Multidoped Hydroxyapatite Porous Coatings Applied via Atmospheric Plasma Spraying[J]. ACS Applied Biomaterials, 2021.DOI:10.1021/acsabm. 0c01535.

Journal of Physics: Conference Series **2961** (2025) 012034 doi:10.1088/1742-6596/2961/1/012034

# Glucose assisted preparation of $Cu/ZnO/Al_2O_3$ catalyst for the synthesis of neopentyl glycol

**Wanbang Zhu[1], Hui Liu[1], Bing Wei[1] and Fang Wang[1*]**

[1] College of Chemical Engineering, Qingdao University of Science & Technology, Qingdao 266042, PR China

*Corresponding author's e-mail: wfchem@qust.edu.cn (F. Wang)

**Abstract:** The preparation of ternary Cu/ZnO/Al2O3 catalysts with different gradients of glucose mass fraction was carried out using a distributed and concurrent precipitation method. The catalytic promotion effect on the hydrogenation synthesis of neopentyl glycol (NPG) from hydroxypropionaldehyde (HPA) was analyzed and characterized. It was found that adding a certain amount of glucose to the catalyst preparation process can enhance the distribution of surface-reduced copper, increase the amount of active $Cu^+$, and improve the $Cu^+/Cu^0$ ratio. Evaluation results demonstrated that the glucose-modified catalyst has a promoting effect on the conversion of HPA to NPG.

## 1. Introduction

Neopentyl glycol (NPG) is a diol with excellent chemical and thermal stability, which is used in the chemical, pharmaceutical and polyester resin industries.[1, 2] NPG is primarily synthesized via the condensation hydrogenation method, which heavily relies on the crucial role of catalysts. Copper-based catalysts are distinguished for their exceptional catalytic efficiency and cost-effectiveness in catalyzing carbon-oxygen bond hydrogenation, making them highly advantageous in this reaction.[3] They have been utilized in the hydrogenation process of HPA.[4] However, these catalysts do have certain drawbacks, such as poor stability and susceptibility to sintering during individual reactions. Researchers have endeavored to improve the ability of catalysts for aldehyde hydrogenation through various means. Non metallic additives play an important role in catalyst modification. Studies have revealed that $Cu^{2+}$ exhibits a strong propensity for interaction with a wide range of organic compounds, including small molecules, polycarboxylic acids, and sugars, which in turn influences the dispersal characteristics and crystal growth of the catalyst. Cyclodextrin is an organic polymer derived from starch, with glucose as its basic structural unit. Ye[5] et al. prepared low-cost cyclodextrin-modified $Cu/SiO_2$ catalysts via ammonia vapor method and found that $Cu/SiO_2$ catalysts coated with an appropriate amount of cyclodextrin exhibited a more compact nanostructured Cu with favorable formation of Cu-O-Si units. The hydrogenation performance was enhanced by preparing $Cu/SiO_2$ catalysts doped with molecular β-cyclodextrin (β-CD) via sol-gel method.[6] The impact of β-CD loading on the catalyst performance followed a volcano-shaped trend, with the $0.1CD\text{-}CuSiO_2$ catalyst achieving a yield of 95% within 240 hours, indicating that β-CD facilitated the formation of smaller and highly dispersed copper particles. Moreover, β-CD has been employed in the preparation of other multiphase catalysts via impregnation method.[7, 8] The modification of catalysts with sugar compounds has been extensively studied, with glucose playing a crucial role in the preparation of certain catalysts. For instance, glucose can serve as an electron donor in the preparation of catalysts via redox precipitation method and also act as a coating agent to regulate the growth of uniform and small-sized metal oxide nanoparticles during synthesis.[9] Glucose-assisted crystallization of ZSM-5 zeolite results in larger prismatic crystals with increased intracrystalline mesoporosity, decreased acidity and strength, and facilitates the generation of porous catalyst structures.[10]

Glucose, being a natural, non-toxic, and inexpensive substance, holds inherent advantages for industrial applications. In this study, the effects of glucose concentrations on the hydrogenation ability of $Cu/ZnO/Al_2O_3$ catalysts were examined. Additionally, the physical and chemical characteristics of these catalysts were elucidated through a series of advanced characterization methods.

## 2. Experimental

### *2.1. Preparation of the catalysts.*

A series of $Cu/ZnO/Al_2O_3$ catalysts with varying glucose contents were prepared using a distributed and concurrent precipitation method with $NaHCO_3/Na_2CO_3$ buffer solution as the precipitant. In brief, the buffer solution was slowly dripped into $Al(NO_3)_3$ at 348 K until the pH stabilized around neutral. Then, under the same conditions, the copper-zinc mixture solution was precipitated. Following the precipitation process, glucose was introduced in a precise quantity, and the resultant mixture was aged for 1.5 hours. The catalyst precursor was then subjected to filtration, washing, and drying overnight at 378 K, followed by calcination at 773K. The amounts of glucose added was 0, 7.5wt%, 15wt%, and 30wt%, respectively, and the resulting catalysts were represented as P0, P1, P2, and P3, respectively.

### *2.2. Catalyst characterization.*

The XRD spectrum of the prepared catalyst was obtained by Rigaku D/max-2500V PC type (Cu Kα radiation), and the FTIR spectrum of the catalyst precursor was obtained by infrared spectrometer (IS10, Nicolet). The XPS of the reduced sample was analyzed by ESCALAB Xi+.

### *2.3. Catalyst activity testing*

The performance of the catalyst was evaluated through the hydrogenation reaction of HPA. Four grams of the catalyst were loaded into the reaction tube, and place it on a fixed bed reactor. A small amount of $H_2$ was introduced into the system under the protection of $N_2$ to reduce the catalyst at a certain temperature. The concentrations of various substances in the products were detected using GC9860 gas chromatograph, and the catalytic ability was obtained according to the standard curve method. Formula is as follows:

$$Conversion\ (\%) = \frac{mole\ of\ HPA(input) - mole\ of\ HPA(output)}{mole\ of\ HPA(input)} \times 100\% \quad (1)$$

$$Selectivity\ (\%) = \frac{mole\ of\ NPG(product)}{mole\ of\ HPA(input) - mole\ of\ HPA(output)} \times 100\% \quad (2)$$

## 3 Results and discussion

### *3.1 XRD*

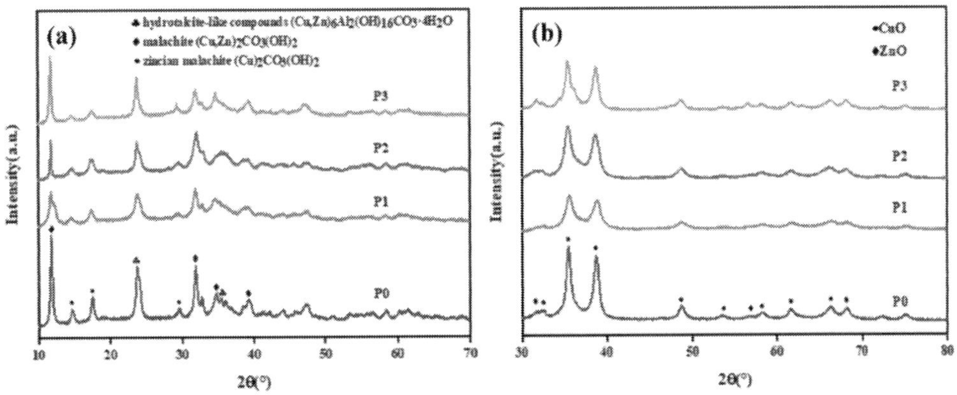


**Figure 1** XRD of catalyst precursors (a) & catalysts (b) with different glucose addition amounts.

Investigated the influence of glucose addition on the formation of various crystalline phases in the catalyst precursors during the preparation process. Figure 1(a) illustrates the XRD patterns of the catalyst precursors. Upon the introduction of glucose, there were no significant changes observed in the crystalline phases of the catalysts. All samples of the precursors exhibited characteristic peaks of $(Cu,Zn)_6Al_2(OH)_{16}CO_3\cdot 4H_2O$, $(Cu,Zn)_2CO_3(OH)_2$ and $Cu_2CO_3(OH)_2$ phases. With increasing glucose content, the peak intensity of the precursor phases initially decreased and then increased. This phenomenon markedly influences the physicochemical properties of the catalysts post-calcination.

Figure 1(b) depicts the peaks at $2\theta$=35.5°, 38.7°, 48.7°, etc., correspond to characteristic peaks of copper oxide, while weaker peaks at $2\theta$=31.8°, 36.3°, and 56.6° are attributed to characteristic peaks of zinc oxide. No diffraction peaks of aluminum oxide were observed, indicating its possible highly dispersed or amorphous state within the catalyst. Based on the Scherrer equation, yielding sizes of 8.7 nm, 4.8 nm, 5.1 nm, and 6.9 nm for the four samples, respectively. The addition of glucose resulted in smaller particle sizes for the catalysts compared to the P0 catalyst, which exhibited higher crystallinity. Therefore, the addition of glucose improved the dispersion of the catalysts, and the samples with added glucose showed better catalytic performance.

### *3.2 FTIR analysis*

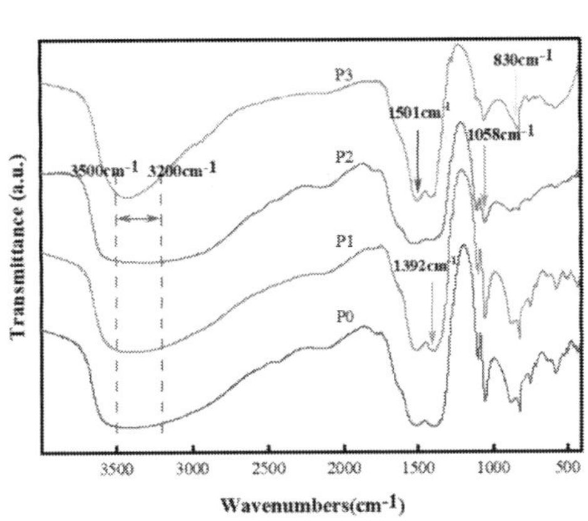


**Figure 2** FTIR of catalyst precursors.

The XRD analysis conducted earlier on the catalyst precursors revealed that the addition of glucose promoted the dispersion of precursor phases. Here, FTIR characterization was employed to the effect of introducing glucose on the catalyst precursors.[11] Figure 2 reveals a broad absorption peak in the 3500-3200 $cm^{-1}$ range, attributable to O-H. This corresponds to the O-H bonds in spinel and zinc spinel phases within the precursor. With an increase in the amount of introduced glucose, the peak width narrows. Additionally, pronounced peaks at 1501 $cm^{-1}$ and 1392 $cm^{-1}$ are associated with carbonates. The peak observed at 830 $cm^{-1}$ further confirms the presence of carbonates.

*3.3 XPS analysis*

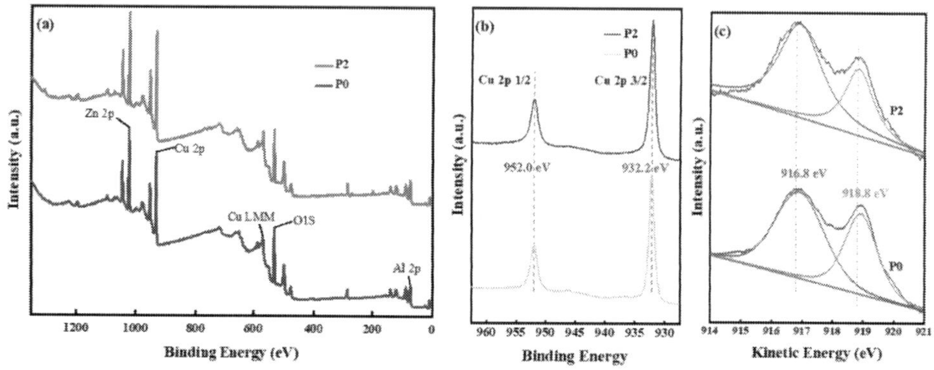


**Figure 3** The XPS full spectrum (a), Cu 2p spectrum (b), and AES spectrum of CuLMM (c) in reduced P0 and P2 samples.

The surface chemical states of the active components of the P0 and P2 catalysts were analyzed through XPS measurements, as shown in Figure 3(a). The XPS survey spectra exhibited characteristic peaks for Cu 2p, Zn 2p, Al 2p, and O1s. By comparing the binding energy positions of these peaks with standard spectra, it was observed that there were no significant shifts in the binding energies of the elements, indicating that the addition of glucose did not affect the electronic structure of these elements.

The Cu 2p spectrum displayed distinct spin-orbit splitting peaks. The satellite feature peak at 943 eV in the Cu 2p spectrum was used to differentiate the oxidation states of copper. According to Figure 3(b), the binding energy at 932.2 eV corresponds to $Cu^+$ and $Cu^0$ formed during the reduction process. There was no shake-up peak of $Cu^{2+}$ at 933.1 eV, and no satellite feature peak was observed at 943 eV, indicating that the samples reduced with 10vol% $H_2/N_2$ did not contain $Cu^{2+}$.[12] Since the binding energy positions of $Cu^+$ and $Cu^0$ in Cu2p 3/2 and Cu2p 1/2 are almost identical and difficult to distinguish, AES analysis was performed on the state of Cu.

Typically, the binding energy positions of $Cu^+$ and $Cu^0$ are around 916.8 eV and 918.8 eV, respectively. As shown in Figure 3(c), the binding energy positions correspond well with the valence states of copper. As shown in Table 1, the ratio of $Cu^+/Cu^0$ in P2 was higher than in P0. Adding appropriate amounts of glucose increased the ratio of $Cu^+/Cu^0$. In the reduced catalysts, a significant amount of $Cu^+$ was present, which is beneficial for the activation of aldehyde groups in the reaction, thereby enhancing the hydrogenation performance of the catalyst.

**Table 1** Cu species measured by XPS.

| Catalyst | BE/ eV | KE/ eV | | X($Cu^+$) [a]/% |
|---|---|---|---|---|
| | $Cu2p_{3/2}$ | $Cu^+$ | $Cu^0$ | $Cu^+/(Cu^++Cu^0)$ |
| P0 | 932.2 | 916.9 | 919.0 | 62.5 |
| P2 | 932.2 | 916.9 | 918.8 | 67.9 |

[a] Obtained by Cu LMM AES spectra

## 4. Hydrogenation performance of catalysts

### *4.1 Catalysts activity and selectivity*

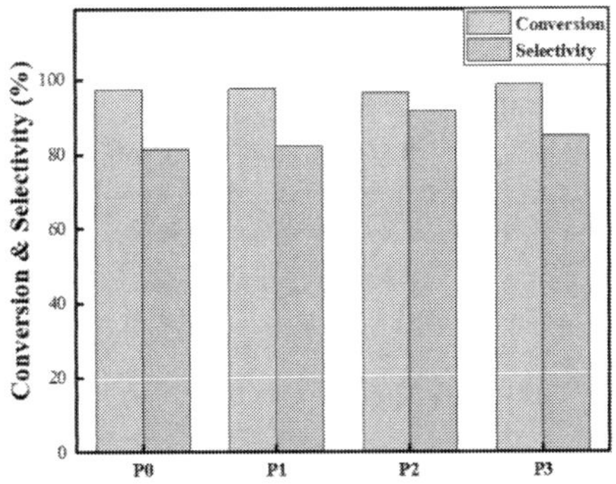


**Figure 4** Activity and selectivity of the P0,P1,P2,P3.

The catalytic hydrogenation performance of P0,P1,P2,P3 catalyst was evaluated under the same reaction conditions. As shown in Figure 4, on the P0 catalyst, the conversion rate of HPA was 97.6%, with a relatively lower NPG selectivity of 81.4%. However, with the introduction of glucose, the NPG selectivity significantly increased. When the amount of glucose added was 15wt%, the NPG selectivity reached 91.9%. Nonetheless, the catalytic performance did not consistently change with increasing glucose content. As the amount of glucose introduced further increased, the selectivity of NPG decreased accordingly. The P2 sample exhibited a 96.5% HPA conversion rate and a 91.9% NPG selectivity. Evaluation results indicate that adding an appropriate amount of glucose is beneficial for improving catalytic performance.

### *4.2 Stability of catalysts*

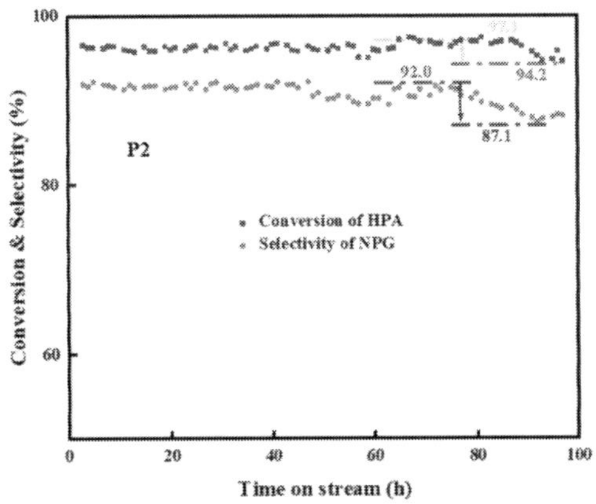


**Figure 5** Catalytic activity of P2 continuous testing.

Stability is an important parameter for industrial catalysts. Long term hydrogenation performance was continuously tested under the identical reaction conditions. The results of the evaluation are illustrated in Figure 5. It is evident that over the course of more than 100 hours of reaction, the catalyst's conversion rate decreased by approximately 2.9%, and its selectivity decreased by 4.9%. This decrease may be attributed to catalyst sintering, resulting in reduced catalyst activity and an increased likelihood of side reactions. Nevertheless, overall, the catalyst exhibited relative stability.

## 5. Conclusion

This study synthesized ternary $Cu/ZnO/Al_2O_3$ catalysts with different gradients of glucose mass fraction using distributed and concurrent precipitation method, and investigated the catalytic effect on the hydrogenation reaction of HPA. As the amount of glucose introduced increases, the NPG selectivity of the catalyst initially increases before subsequently decreasing. The $Cu/ZnO/Al_2O_3$ catalyst with a glucose content of 15wt% exhibits excellent activity and selectivity. The addition of glucose can regulate the formation of precursor crystal phase and the dispersion of CuO in the catalyst, increase the amount of active $Cu^+$, Increasing the $Cu^+/Cu^0$ ratio enhances the catalyst's selectivity for NPG.

## References

[1] Monasterska E, Chrobok A, Pankalla E, Siewniak A (2021) *J. Molecules*, 26(19).
[2] Meng D, Zhao K, Wang A, Zhao W (2020) *J. IOP Conference Series: Materials Science and Engineering*, 735(1).
[3] Murthy P S, Liang W, Jiang Y, Huang J (2021) *J. Energy & Fuels*, 35(10): 8558-84.
[4] Zhu J, Liu H, Zhou X, Wang Z, Bing L, Han D, Wang G, Wang F (2023) *J. Catalysts*, 13(4): 673.
[5] Ye R-P, Lin L, Yang J-X, Sun M-L, Li F, Li B, Yao Y-G (2017) *J. Journal of Catalysis*, 350: 122-32.
[6] Ye R P, Lin L, Liu C Q, Chen C C, Yao Y G (2017) *J. ChemCatChem*, 9(24): 4587-97.
[7] Bai L, Wyrwalski F, Lamonier J-F, Khodakov A Y, Monflier E, Ponchel A (2013) *J. Applied Catalysis B: Environmental*, 138-139: 381-90.
[8] Jean-Marie A, Griboval-Constant A, Khodakov A Y, Monflier E, Diehl F (2011) *J. Chem Commun (Camb)*, 47(38): 10767-9.
[9] Phan N H T, Nguyen C C, Nguyen Dinh M T (2023) *J. RSC Adv*, 13(20): 13354-64.
[10] Nguyen Dinh M T, Nguyen C C, Truong Vu T L, Ho V T, Truong Q D (2020) *J. Applied Catalysis A: General*, 595.
[11] Abbas I, Kim H, Shin C-H, Yoon S, Jung K-D (2019) *J. Applied Catalysis B: Environmental*, 258.
[12] Yang X, Meng Q, Ding G, Wang Y, Chen H, Zhu Y l, Li Y W (2018) *J. Applied Catalysis A: General*, 561: 78-86.

# Effect of carbonization temperature on supported CoMo catalysts with honeycomb structure for thiophene hydrodesulfurization

**Mingyu Song[1], Fuhui Yang[1], Yeqiang Du[1], Guangjian Wang[1], Fang Wang[1], Liancheng Bing[1] and Dezhi Han[1,*]**

[1]State Key Laboratory Base of Eco-chemical Engineering, College of Chemical Engineering, Qingdao University of Science and Technology, Qingdao 266042, China

[*]Corresponding author's e-mail: handzh@qust.edu.cn

**Abstract:** The use of transportation fuels with high sulfur content can lead to the production of a large amount of sulfur oxides, which will have a serious impact on the earth's ecological environment. Hydrodesulfurization (HDS) technology can effectively remove sulfur-containing compounds from transport fuels to ultra-low sulfur levels. In this paper, the supported cobalt-molybdenum HDS catalysts with honeycomb structure were prepared using activated carbon support derived from cypress wood as raw material. We explored how varying carbonization temperatures influence catalysts' structural and chemical properties using low-temperature nitrogen adsorption-desorption, The conversion efficiency of these HDS catalysts in transforming thiophene was examined using X-ray difraction(XRD) and scanning electron microscopy (SEM), within a high-pressure micro-fixed-bed reactor setup. Our observations revealed that all catalysts maintained their honeycomb-like structure, featuring both micropores and mesopores. This specialized pore arrangement helps to alleviate the resistance encountered during the mass transfer between large reactant molecules and the resulting products. The larger specific surface area promotes the dispersion of active ingredients in the catalyst, exposing more active sites. The maximum thiophene conversion over the CoMo-800 catalyst reached 93.28%.

## 1. Introduction

In recent years, governments worldwide have enacted laws to prioritize environmental protection and sustainable development, and clean and green energy has played an increasingly important role in the energy system. However, most industrial processes still rely on fossil fuels, with petroleum remaining the most crucial fuel for world industrial development. Nevertheless, global fossil fuel reserves are limited, and sour crude oil (with sulfur content ranging from 0.5-1%) and high-sulfur crude oil (sulfur content >2%) together account for over 75% of total crude oil production [1]. Sulfides in low-quality petroleum and its products cause serious environmental pollution. Therefore, developing effective technologies to deal with these sulfides is currently a problem that needs to be addressed. Hydrodesulfurization (HDS) has many advantages compared to other desulfurization technologies, notably its substantial feedstock processing capability[2-4]. The core of this technology lies in developing HDS catalysts with high performance, long life, and low cost. Traditional industrial HDS catalysts typically refer to Co(Ni)Mo(W)/$Al_2O_3$ [5]. However, $Al_2O_3$ support strongly interacts with active components, affecting the catalyst pre-sulfurization process. The HDS catalyst support currently can be divided into three main categories: metal oxides, zeolites, and carbon-based supports [6]. The prominent advantages of carbon-based supports are low relative interaction between the support and metals, high degree of pre-sulfurization, excellent hydrothermal stability, etc [7]. Therefore, in this study, activated carbon support derived from cypress wood was used to fabricate the supported CoMo catalysts for thiophene hydrodesulfurization.

## 2. Experimental section

### *2.1. The preparation of catalysts*

Individual cypress wood chips were positioned in porcelain containers and underwent carbonization in a tubefummace, exposed to nitrogen, at different temperatures-600'c, 700'c, 800'c, and 900'c-each for a duration of1.5 hours. After carbonization, the samples were activated by soaking in a KOH solution for 6 hours, then drying at 120°C. They were then heated again at 850°C for 1.5 hours in the tube furnace. After being dried at 120°C for 24 hours, the resultant catalyst precursor underwent further heat treatment in a tubular furnace at 500°C for 8 hours. The catalysts derived from activated carbon carbonized at various temperatures—600°C, 700°C, 800°C, and 900°C—were designated as CoMo-600, CoMo-700, CoMo-800, and CoMo-900, respectively.

To prepare the precursor solution, we dissolved 0.50 g of $(NH_4)_6Mo_7O_{24}{\cdot}4H_2O$ and 0.566 g of $Co(NO_3)_2{\cdot}6H_2O$ in deionized water and stirred the mixture at 35°C for 30 minutes until it was completely dissolved. Using an equal volume impregnation technique, the activated carbon carrier was then sequentially impregnated with the ammonium molybdate solution and the cobalt nitrate solution. The resulting catalyst precursor was dried at 120°C for 24 hours, followed by heat treatment at 500°C in a tubular furnace for 8 hours. The catalysts obtained from activated carbon carbonized at 600°C, 700°C, 800°C, and 900°C were labeled as CoMo-600, CoMo-700, CoMo-800, and CoMo-900, respectively..

### *2.2. Characterization of catalysts*

The crystalline composition of the synthesized catalysts was scrutinized through X-ray diffraction (XRD) employing a Bruker AXS-D8 Advance diffractometer, which utilized Cu-Kα radiation with a wavelength of 0.15406 nm. Furthermore, the porosity and pore architecture of the catalysts were characterized using nitrogen adsorption-desorption techniques on a Micromeritics TriStar II 3020 analyzer. Additionally, the surface topology of the catalysts was inspected via a HITACHI S-4800 scanning electron microscope (SEM).

### *2.3. Evaluation of catalyst activity*

Before evaluating their HDS performance, the oxidized catalysts underwent a pre-sulfidation step. During the assessment, a solution comprising 0.5 wt% thiophene in n-octane (serving as simulated oil) was introduced into the reactor using a plunger pump. The process was carried out with a weight hourly space velocity (WHSV) of 6.3 $h^{-1}$, a hydrogen-to-oil volume ratio of 500:1, and a hydrogen pressure set at 1 MPa. The reaction temperatures ranged from 300°C to 360°C. The liquid products collected during the test were analyzed via gas chromatography (GC) to determine the thiophene conversion rates.

$$Z(\%)=\frac{C_{in}-C_{out}}{C_{in}}\times 100\% \quad (1)$$

In this context, Z represents the conversion rate of thiophene, where $C_{in}$ denotes the initial thiophene concentration in the simulated oil, and $C_{out}$ indicates the thiophene concentration found in the final product.

## 3. Results and discussion

### *3.1. Crystalline structure analysis*

In **Figure 1**, due to the high Co content in the final catalysts, weak diffraction peaks of the $Co_3O_4$ phase can be detected at 36.7°, 43.5°, and 64.3° for all catalysts. Several peaks around 14.24°, 28.36°, 32.16°,

32.96°, 46.96°, and 53.52° correspond to the α-$CoMoO_4$ phase, indicating the presence of octahedral coordinated Mo atoms in the catalysts. Peaks detected at 12.7°, 19.0°, 23.4°, 26.5°, 27.3°, 33.7°, 45.0°, and 52.0° are attributed to the β-$CoMoO_4$ phase with tetrahedral coordinated Mo atoms [8]. The presence of the β-$CoMoO_4$ phase indicates higher HDS activity after sulfurization. Diffraction peaks observed at 12.7°, 23.3°, 25.7°, 27.3°, and 33.7° correspond to $MoO_3$.

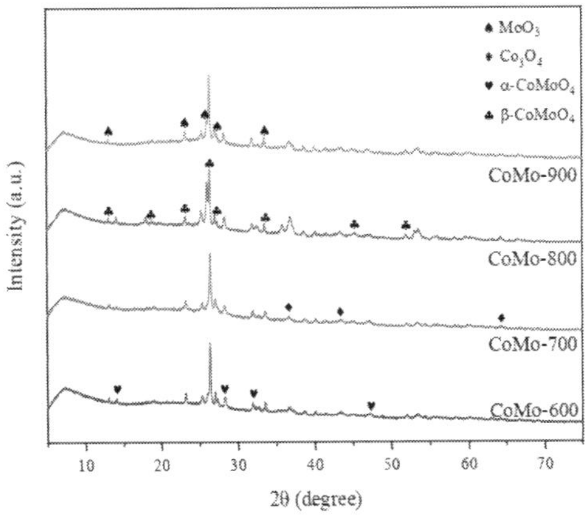


**Figure 1**. The XRD patterns of the prepared catalysts.

*3.2. Morphology analysis*

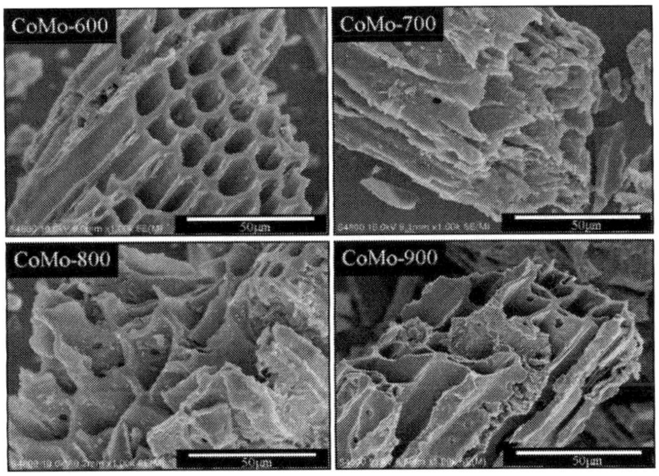


**Figure 2**. The SEM images of prepared catalysts.

**Figure 2** shows the SEM images of the prepared catalysts. It can be observed from the SEM images that the final catalysts still retain the original honeycomb structure of the cypress wood after the processes of carbonization treatment, KOH activation, acid-base neutralization, metal impregnation, and calcination. With the increase in carbonization temperature, the cellulose, hemicellulose, and lignin in the cypress wood were progressively degraded, gradually thinning the honeycomb framework and cell walls, and the formation of some internal pores. This could elevate the catalyst surface area for dispersing the active components. However, the catalyst structure of CoMo-900 was partially destroyed due to the excessively high temperature of 900°C. Moreover, it can be observed that with increasing temperature, some small pores gradually appeared on the honeycomb framework. The presence of these small pores could reduce the transport resistance of the reactant/product. At the carbonization temperature of 800°C, abundant small pores were observed on the catalyst framework, consistent with the $N_2$ adsorption-desorption characterization and performance evaluation result analysis.

### *3.3. $N_2$ adsorption-desorption isotherms*

From Figure 3(a), it is evident that the adsorption-desorption isotherms for all four catalysts display Type IV characteristics with H4 hysteresis loops. A sharp increase in nitrogen adsorption is observed within the relative pressure range of P/P0 = 0 to 0.2, indicating a substantial presence of micropores. The hysteresis loops emerging at P/P0 > 0.4 suggest an uneven distribution of mesopores. At higher pressures (P/P0 > 0.9), the adsorption curves continue to rise without leveling off, implying a few macropores in the catalysts. This phenomenon may be due to the formation of new pores in the pore walls at elevated calcination temperatures. Figure 3(b) presents the pore size distribution, showing that the pore sizes of all catalysts fall between 0.5 and 10 nm, confirming that the structures are primarily microporous with some mesopores. According to Table 1, the specific surface area of the catalysts first increases and then decreases with rising temperature. This trend is linked to the development of new pores within the framework, as observed in the SEM images. Higher temperatures cause the cypress wood to degrade, creating an additional surface area for active component dispersion and thus enhancing catalytic activity. However, excessively high carbonization temperatures, such as those used for CoMo-900, can reduce the BET-specific surface area due to the adverse effects of extreme temperatures on the catalyst structure.

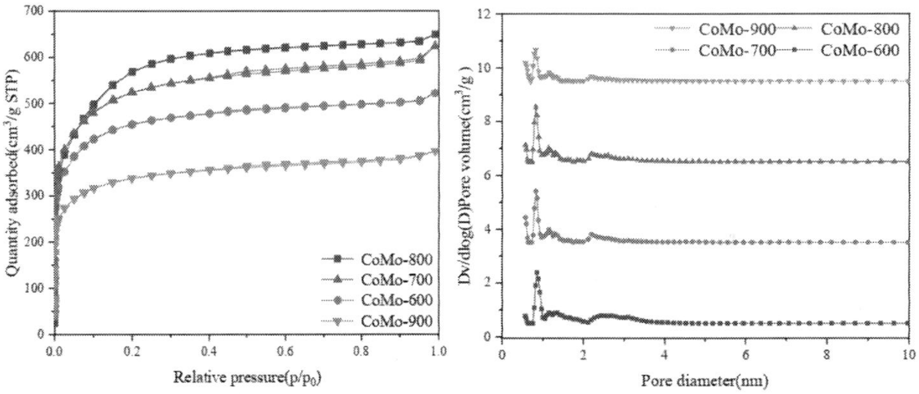


**Figure 3**. (a) The $N_2$ adsorption-desorption isotherm and (b) the pore size distribution curves of catalysts at different carbonization temperatures.

**Table 1**. The pore structure data of catalysts.

| Catalyst | Specific surface area ($m^2/g$) | Pore volume ($cm^3/g$) |
|---|---|---|
| CoMo-600 | 1684.34 | 3.06 |
| CoMo-700 | 1911.00 | 3.41 |
| CoMo-800 | 2032.56 | 3.41 |
| CoMo-900 | 1256.84 | 3.06 |

### *3.4. HDS performance tests*

The thiophene conversion rates over the catalyst were assessed using a high-pressure micro-fixed-bed reactor, as illustrated in **Figure 4**. Before the evaluation, the oxidized catalysts underwent a pre-sulfidation process. Reaction products were collected after 3 hours of testing at temperatures 300°C, 320°C, 340°C, and 360°C, with each temperature tested in triplicate to ensure accuracy. The results revealed that all four catalysts demonstrated a similar pattern of increased HDS activity with rising reaction temperature. The HDS activity followed the order: CoMo-800 > CoMo-700 > CoMo-600 > CoMo-900. Based on the $N_2$ adsorption-desorption analysis, the optimal carbonization temperature of

800°C for cypress wood preserved the honeycomb structure, maintaining a rich network of micropores and mesopores. This structural feature provided CoMo-800 with a high surface area for effective dispersion of active components, contributing to its superior HDS performance with a thiophene conversion rate of 93.28%.

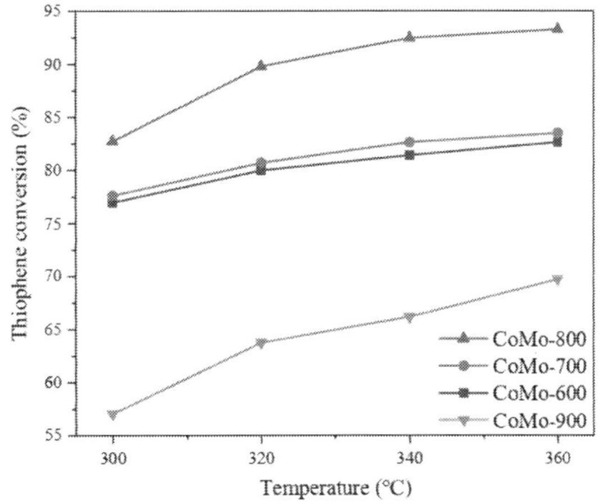


**Figure 4**. The thiophene conversions over the prepared catalysts.

## 4. Conclusion

The activated carbon support derived from cypress wood was employed to prepare CoMo catalysts for thiophene hydrodesulfurization. The prepared catalysts possessed the honeycomb structure with abundant micropores and mesopores. The unique pore structure of catalysts promotes rapid mass transfer of reactants and products in the pore channels. The appropriate carbonization treatment of 800 °C could supply the high catalyst surface area for dispersing the active components. The CoMo-800 catalyst demonstrated an outstanding thiophene conversion of 93.28% at 1 MPa and 360°C. This work provides new ideas for developing HDS catalysts using activated carbon as a carrier in the future.

## References

[1] Steven, Chu, Arun, Majumdar (2012) *J. Nature*, 488(7411): 294-303.
[2] Morales-Valencia E M, Vargas-Montaez O J, Monroy-García P A, Avendao-Barón L G, Quintero-Quintero E A, Elder-Bueno C, Santiago-Guerrero A Y, Baldovino-Medrano V G (2021) *J. Journal of Catalysis*, 404: 204-9.
[3] Zhang X, Zhang B, Chen Y, Li L, Sun H, Liu X, Guo R, Meng B, Liu P, Song W (2022) *J. Fuel*, 307: 121787.
[4] Han D, Li Q, Wang E (2022) *J. Catalysis Today*, (405): 329-36.
[5] Weng X, Cao L, Zhang G, Chen F, Xu C (2020) *J. Industrial & Engineering Chemistry Research*, 59(49): 21261-74.
[6] Ali I, Al-Arfaj A A, Saleh T A (2019) *J. Journal of Molecular Liquids*, 304: 112376.
[7] Hajjar Z, Kazemeini M, Rashidi A, Soltanali S (2018) *J. Fullerenes Nanotubes and Carbon Nanostructures*, 26(9): 557-69.
[8] Shi Y, Liu G, Jin R, Xu H, Wang Q, Gao S (2019) *J. Carbon Energy*, 1(2): 253-75.

# $LaAl_{11.76-x-y}(Ni_xTi_y)O_{19}$ ($x,y$=0.2-0.8) microwave dielectric ceramics exhibit $\tau_f$ nearly zero

**Mian Jiang**[1], **Hua He**[1,*] **and Chuyun Huang**[1]

[1]School of Science, Hubei University of Technology, Wuhan 430068, China

* Corresponding author's e-mail: 309362619@qq.com

**Abstract:**The development of mobile communication technology urgently needs microwave dielectric ceramic materials with high-temperature stability. Herein, $LaAl_{11.76-x-y}(Ni_xTi_y)O_{19}$ (LANT, $x,y$=0.2-0.8) ceramic samples were produced using the solid-state reaction technique. The experimental results show that after sintering at 1300 °C for 2 h, the excellent microwave dielectric properties are gained at $x$=0.4 and $y$=0.6, $\varepsilon_r$=18.1, Q×f=23677 GHz, and $\tau_f$=+0.15 ppm/°C. The $\varepsilon_r$ and Q×f values first rise and then diminish as the amount increases $x$ due to the change in the structural stability of the material and the symmetric tensile pattern of the Al/NiTi-O bond. It is speculated that the addition of $[Ni_xTi_y]^{3+}$ increases the inclination angle of the Oxygen octahedron, and thus promotes the orientation of $\tau_f$ to zero. This work verifies the great potential of LANT ($x,y$=0.2-0.8) ceramics in improving the temperature stability of microwave devices.

## 1. Introduction

Recent times have attested to a rapid progression in the evolution of fifth-generation (5G) wireless networking technology. The widely used RF devices in communication equipment have great demand increases for high-performance microwave dielectric ceramic materials[1]. As three important vital performance characteristics of microwave dielectric ceramics, the value of the $\varepsilon_r$ affects the integration of the device, the quality factor reflects the power consumption of the device, and the $\tau_f$ reflects the stability of the equipment in different working environments[2]. The parameter $\tau_f$ is the key factor to measure the performance of microwave dielectric filters, which directly reflects the sensitivity of material resonant frequency with temperature. Therefore, reducing the $\tau_f$ and improving the stability of ceramic filters under temperature change holds a key position in improving the performance and reliability of the communication system.

$LaAl_{11.76}O_{19}$ exhibits an alright Q×f value and a great permittivity in the microwave frequency scope. However, the $\tau_f$ is still a significant deviation from the optimal value of zero[3]. Therefore, its applicability in some 5G applications is hindered. The researchers modified the microwave dielectric ceramic materials by different methods and achieved some results. It is proved that the partial replacement of the B-position ions in various materials offers an effective strategy for improving the attributes of microwave dielectric materials. Based on these findings, attempts were made to optimize the main properties of the material to a greater extent by partially replacing $Al^{3+}$ with $[Ni_xTi_y]^{3+}$[4].

## 2. Experimental procedure

LANT samples were fabricated utilizing the orthodox solid-state reaction technique. The constituents were measured out based on the stoichiometric ratio, blended with deionized water, and subsequently subjected to ball milling for a duration of 90 min. Afterward, the mixture was dried in a drying oven is kept at a thermostatic of 60 °C for a period of 24 h, before undergoing a screening process. A cylindrical specimen with a diameter of 15 mm was then molded under a compression force equal to approximately 10 MPa. The final sample was obtained after sintering for 2 h in an air atmosphere of 1250-1350 °C.

The phase composition was analyzed using the X-ray powder diffractometer of the LANT ceramic samples. The density was assessed using the Archimedean principle. The surface microstructure was inspected via the scanning electron microscope (SEM). Raman spectroscopic analysis was performed at room temperature, with a wavelength of 532 nm being utilized. The dielectric characteristics were obtained using an Agilent PNA-N5234A vector network analyzer, manufactured in the United States, inside a TE011 shielded cavity. Equation (1) provided outlines the method for calculating $\tau_f$ within the temperature interval of 25 to 85 °C:

$$\tau_f = \frac{f(T_1) - f(T_2)}{(T_1 - T_2) \times f(T_2)} \times 10^6 (ppm / ℃) \quad (1)$$

The sample's resonant frequency at 25 °C is designated as $f(T_1)$, and at 85 °C, it is designated as $f(T_2)$.

### 3. Results and discussion

Figure 1 displays the XRD peaks of LANT ceramics sintered at 1300 °C for 2 h with varying Ni and Ti doping ratios. As can be seen from the figure, multiple phases were detected. The reason for this phenomenon may be that when the doping amount of titanium and nickel exceeds the saturation value, the excess Ti or Ni reacts to produce other substances[5]. As Ni doping increased, so did the diffraction peaks of other phases, potentially caused by high-temperature reactions with other elements.

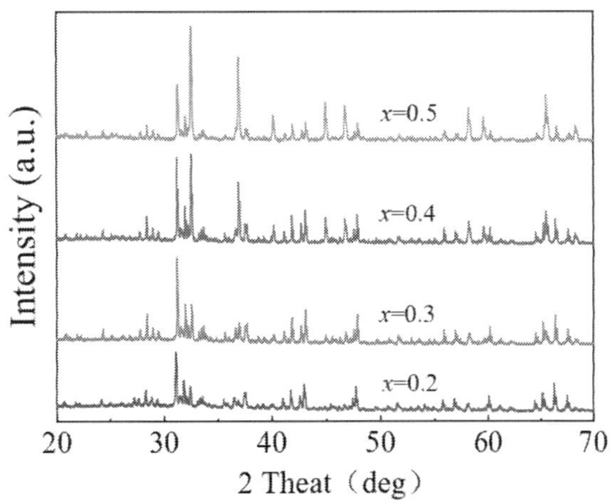


**Figure 1.** XRD pattern of LANT ceramics fritted at 1300 °C.

Figure 2 illustrates the microstructure of LANT ceramics that were sintered at 1300 °C for 2 h with different doping ratios of Ni and Ti. The grains display a somewhat random, rod-shaped arrangement, characterized by irregular growth and the presence of pores. As the doping level of Ni increases, the ceramic's grain size progressively diminishes from 1.25 μm in Figure 2 (a) to 0.78 μm in Figure 2 (d). As the doping level of Ni rises, the grain size decreases, decreasing from 1.25 μm in (a) to 0.78 μm in (d). The photographs in Figures 2 (c) and (d) show a scattered grain distribution and diminished grain sizes, potentially due to irregular grain growth. This indicates that adjustments in doping ratios have an influence on the microwave dielectric attributes of LANT ceramics.

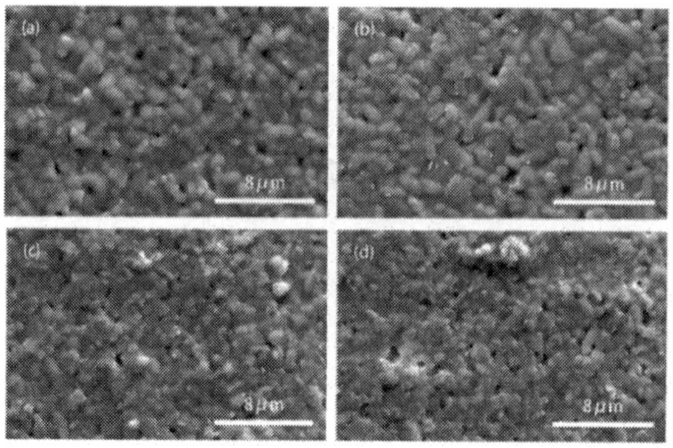


**Figure 2.** SEM of LANT ceramics fritted at 1300 °C for 2 h. (a) $x$=0.2; (b) $x$=0.3; (c) $x$=0.4; (d) $x$=0.5.

Figure 3 exhibits the Raman spectral analysis of LANT ceramics after they have been sintered at 1300 °C under different doping ratios of Ni and Ti with 2 h wave number range of 100-700 $cm^{-1}$. In $ABO_3$ perovskites with R-3c structure, the frequency of $A_{1g}$ mode is mainly related to the diamond distortion and is less related to the change of A-O and B-O distance[6]. LANT has something in common with this structure. In particular, $A_{1g}$ mode can be utilized as a marker to evaluate the degree of B-site ordering within complex perovskite materials, which does not appear in Figure 3, indicating that B-site is partially replaced by Ni and Ti, so that B-site cations are rearranged from disorder to order, indicating that the introduction of Ni and Ti increases the order of B-site ions. It is worth noting that when $x$ is 0.2, two new Raman vibration peaks emerge, approximately 440 $cm^{-1}$ and 600 $cm^{-1}$, which may indicate a symmetrical stretching move of $BO_6$.

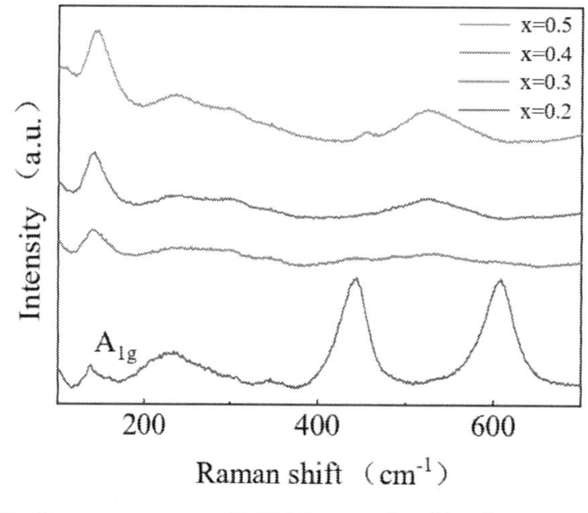


**Figure 3.** The Raman spectra of LANT ceramics fritted at 1300 °C for 2 h.

Figure 4 is the density change curve of LANT sintered for 2 h at 1300 °C under different doping ratios of Ni and Ti. The diagram indicates a corresponding augmentation in the density of LANT with an increase in the Ni doping content, implying progressive compaction of the sample. Considering the negligible changes in density, it is conjectured that density does not play a pivotal role in determining the performance.

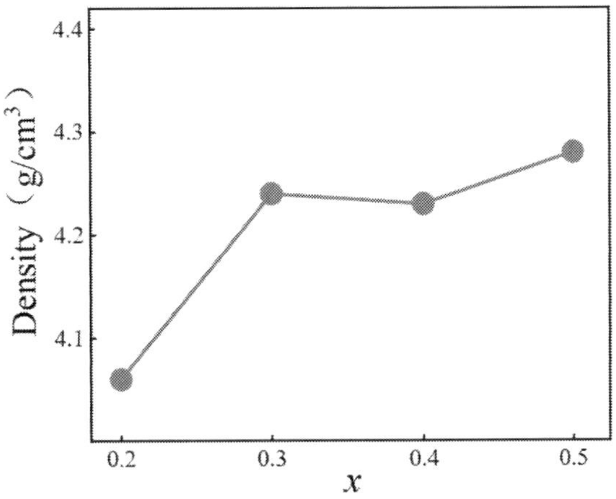


**Figure 4.** The density variation of LANT ceramics with $x$, post-fritting at 1300 °C for 2 h.

Figure 5 depicts the pattern of change in the $\varepsilon_r$ of LANT across various sintering temperatures. Clearly, an elevation in sintering temperature results in an augmentation of the relative dielectric constant when the doping amount remains unchanged. At the best sintering temperature, an increment in $x$ from 0.2 to 0.4 is observed, and the $\varepsilon_r$ value of LANT ceramics experiences a slight increase from 18.8 to 19. However, when $x$ exceeds 0.4, the $\varepsilon_r$ value of LANT ceramics gradually decreases to 15.4 with an increase in $x$. Due to the introduction of $[Ni_xTi_y]^{3+}$, the density of the ceramic increases as $x$ increases, causing the $\varepsilon_r$ of LANT ceramics to initially rise and then decline. It is considered that the introduction of $[Ni_xTi_y]^{3+}$ is one of the reasons for the change of $\varepsilon_r$ value in ceramics.

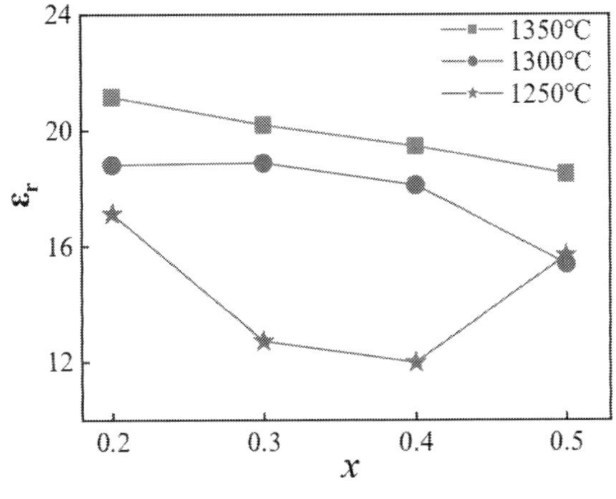


**Figure 5.** The $\varepsilon_r$ values of LANT vary with x across different temperature ranges (1250-1350 °C).

Figure 6 presents the variation curve of Q×f numeric of LANT ceramics sintered for 2 h at 1300 °C with different Ni and Ti doping ratios. As illustrated in the figure, an increase in sintering temperature results in an elevation of the Q×f value when in the same doping ratio, but when $x$ is 0.3, and $x$ is 0.4, the Q×f numeric at 1300 °C is larger than the Q×f value at 1350 °C, combined with XRD phase analysis, may be due to the reason for the reduction of secondary phase content. At the same temperature, with the increase of Ni doping amount Q×f value showed a non-linear change trend, but when $x$ was 0.5, Q×f numerics appeared at 1300 °C and 1350 °C were the opposite trend. Combined with XRD analysis, this phenomenon may be due to the increase of Ni doping amount, causing the occurrence of the secondary phase. It is among the factors that impact the Q×f value.

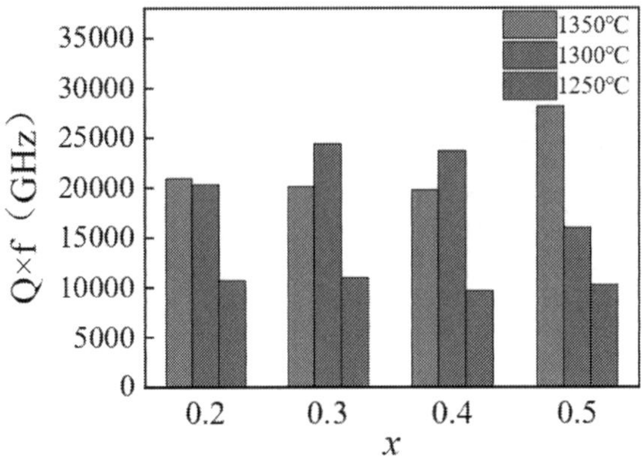


**Figure 6.** The Q×f values of LANT exhibit changes based on $x$ at various temperature levels (1250-1350 °C).

Figure 7 shows the resonance frequency temperature coefficient ($\tau_f$) of LANT sintered at 1300 °C for 2 h with different Ni and Ti doping ratios. It is evident from the figure that, at the same temperature, the $\tau_f$ initially decreases and subsequently increases with the addition of Ni doping amount. Under the same doping amount, the $\tau_f$ increases with the increase of Ni doping amount. The $\tau_f$ is closer to zero, which reflects the good thermal stability of the material. It can be seen from the figure that when $x$ is 0.4, the $\tau_f$ of the sample is closest to the zero axis. As the proportion of Ni in the Ni and Ti ion content gradually increases, it may be because of the increase of the inclination of the oxygen octahedron, and the crystalline structure's symmetry decreases, so the structure of LANT ceramics gradually tends to be symmetrical and stable, and the $\tau_f$ gradually tends to zero.

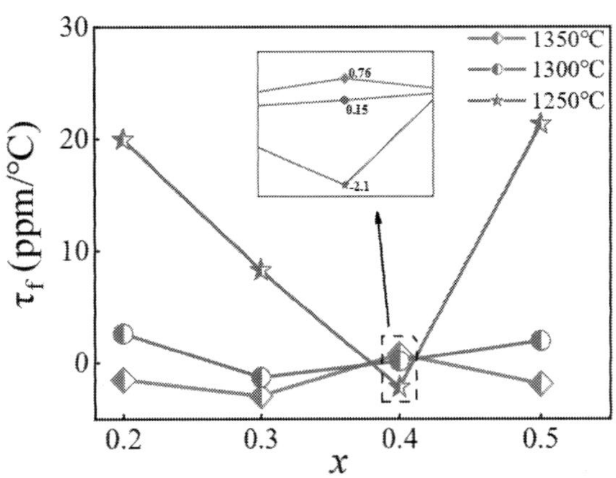


**Figure 7.** The $\tau_f$ values of LANT ceramics vary with $x$ at varying temperature levels (1250-1350 °C).

## 4. Conclusion

The microwave dielectric properties of LANT were improved by co-doping Ni and Ti with $LaAl_{11.76}O_{19}$. The experimental results show that with the change of Ni and Ti doping ratio, the $\varepsilon_r$ first rises and then falls, while the Q×f value displays a non-linear pattern of change, and the $\tau_f$ approximates zero. The practical application field of LANT is increased. When $x$ is 0.4, LANT ceramics are sintered at 1300 °C for 2 h to obtain the optimal microwave dielectric properties. It can be seen that the incorporation of both Ni and Ti as co-dopants exhibits a notable influence on enhancing the microwave dielectric properties of LANT ceramics.

## Acknowledgments

This work endeavor received financial and logistical support from multiple sources, including the Opening Project of the State Key Laboratory of Advanced Float Glass Technology (Grant Number:

2022KF05), the Outstanding Talent Program aimed at fostering green industry leadership at HBUT under the grant JCRC2021003.

**References**

[1] G. Wang, D. Zhang, J. Li, G. Gan, Y. Rao, X. Huang, Y. Yang, L. Shi, Y. Liao, C. Liu, L. Jin, H. Zhang (2020) Crystal structure, bond energy, Raman spectra, and microwave dielectric properties of Ti-doped $Li_3Mg_2NbO_6$ ceramics. J. Am. Ceram. Soc. 103(8), 4321-43322.
[2] C. Du, D. Zhou, SZ. Hao, H.T. Chen, J. Zhang, S.K. Sun, T. Zhou, M.Z. Dang, S.F. Wang, H.W. Liu, L. Li (2021) High-uality-factor AlON transparent ceramics for 5 GHz Wi-Fi aesthetically decorative antennas. ACS Appl. Mater. Interfaces 13(39), 46866-46874.
[3] S. Kustov., Iu. Liubimova, E.K.H. Salje (2018) $LaAlO_3$: A substrate material with unusual ferroelastic properties. Appl. Phys. Lett. 112, 042902.
[4] Q. Zhang, L. Xu, X. Tang, F. Huang, X. Wu, Y. Li, Y. Jing, Z.K. Han, H. Su (2021) Electronic structure, Raman spectra, and microwave dielectric properties of Co-substituted $ZnWO_4$ ceramics. J. Alloys Compd. 874, 159928.
[5] S.B. Narang, S. Bahel (2010) Low loss dielectric ceramics for microwave applications: a review. J. Ceram. Process. Res. 11(3), 316-321.
[6] S. Zhang, H. Sahin, E. Torun, F. Peeters, D. Martien, T. DaPron, N. Dilley, N. Newman (2017) Fundamental mechanisms responsible for the temperature coefficient of resonant frequency in microwave dielectric ceramics. J. Am. Ceram. Soc. 100(4), 1508-1516.

Journal of Physics: Conference Series **2961** (2025) 012037 doi:10.1088/1742-6596/2961/1/012037

# In situ fabrication of solid-state carbon dots for random lasers

**Zefeng Wu, Wenfei Zhang*, Shaofeng Zhang, Xin Yao, Ye Lin and Zhenhui Chen**

Key Laboratory of Optoelectronic Devices and Systems of Ministry of Education and Guangdong Province, College of Physics and Optoelectronic Engineering, Shenzhen University, Shenzhen 518060, China

*Corresponding author E-mail addresses: zhangwf@szu.edu.cn

**Abstract:** Random microlasers with convenient laser cavities preparation are vital for various photonic applications. In addition, the method of in situ fabricating carbon dots utilizes the structures of precursors hindering aggregation-caused quenching and provides high CD yield with easy-to-design random laser cavities. Inspired by the above, a facile solvothermal route was reported for the in situ controllable synthesis of blue-emitting solid-state carbon dots (CDs with TEOS) with the addition of tetraethoxysilane (TEOS). The fluorescence spectrum of CDs with TOES ranges from 390-650 nm. The carbon dots (CDs without TEOS) without TEOS addition are similar to those with TEOS possessing the same fluorescence spectrum range. What's more, the emission intensity of CDs with TEOS can be enhanced 37 times than that without TEOS, and the PL lifetimes of CDs with TEOS and CDs without TEOS are 1.23 ns and 7.26 ns, respectively. As a result, CDs with TEOS were employed to gain materials realizing blue emissive random lasers with a threshold of about 17 mJ/cm$^2$ due to their intense emission.

## 1. Introduction

Solid-state microlasers have become important in communications, military weapon devices, biomedical therapy, and material processing. In the last few years, solid-state microlasers using colloidal quantum dots, organic semiconductors, and perovskites as gain materials have demonstrated distinct advantages, including unique optical properties, flexibility, and ease of integration [1, 2]. These attributes contribute to their extensive range of applications. The swift advancement of solid-state microlasers has led to increased demand for gain media, thereby highlighting cost, toxicity, and environmental sustainability concerns. Therefore, carbon dots (CDs), as a novel class of nanomaterials, demonstrate significant potential, rendering them highly suitable candidates for applications as laser gain materials. Yu's research group achieved blue laser emission from organic solvents [3]. The research group of Lu synthesized a solid-state laser by combining the R-CDs solution and microcavity [4]. These studies have substantially advanced the progress of CDs-based lasers. However, all the mentioned CDs lasers were in solution or synthesized by mixing CDs solution and matrix, possibly resulting in aggregation-caused quenching (ACQ). Hahn et al. obtained ACQ-resistant CDs by in situ fabrication due to the surface's residual polymer chains disrupting the CDs aggregation [5]. Zhang et al. in situ synthesized thermally stable CDs by conveniently designing random laser cavities [6]. As mentioned above, in situ fabrication possesses the advantages of hindering CDs from aggregation, high yield, and conveniently designing random laser cavities [5, 6].

In this study, we report a one-step solvothermal route to in situ fabricate blue-emitting solid-state T-CDs and C-CDs, using citric acid (CA), anilinomethyl triethoxysilane (AMTOS), tetraethoxysilane

(TEOS) as raw materials with advantages of resist-aggregation, high yield and conveniently designing random laser cavities. The photoluminescence (PL) characteristics of CDs are influenced by the proportion of TEOS. The emission intensity of CDs with TEOS is 37 times higher than those without TEOS. At the same time, the morphology of CDs with TEOS is transformed to amorphous nanoparticles, compared to crystalline nanoparticles of CDs without TEOS. Finally, CDs with TEOS are employed as gain media to produce random laser emission from 390 to 600 nm, preserving a 17 mJ/cm$^2$ threshold at different angles.

## 2. Experimental section

### *2.1. Synthesis of C-CDs and Synthesis of T-CDs.*

Generally, 0.83 mmol of CA and 0.93 mmol of AMTOS were mixed in a solution of 5 mL of water and 5 mL of ethanol, stirred for 5 minutes, and then placed into a 50 mL Teflon-lined stainless-steel autoclave. The mixture was heated to 150 °C and maintained for 3 hours. Then, the products were gathered by rinsing them three times with a mixed solution of ethanol and water in a 2:1 volume ratio, and then they were dried to produce C-CDs. For T-CDs, 0.83 mmol CA, 0.93 mmol AMTOS and 12.5 mmol TEOS were mixed in a mixture of water and ethanol. Other processes are the same as a synthesis of C-CDs.

## 3. Result and discussion

As shown in Figure 1, the yellowish powder of T-CDs was produced from CA, TEOS and AMTOS via a solvothermal reaction in a mixture solution of ethanol and $H_2O$ at 150 °C. Solid-state C-CDs were synthesized from CA and AMTOS by the solvothermal reaction at 150 °C. TEM and HR-TEM were employed to study the CDs to gain the nanostructure of these unique carbon dots. The morphology of the C-CDs and T-CDs was characterized by TEM. In Figure 2a, the synthesized C-CDs are crystalline quasi-spherical particles with average size of 1.92 nm. The HRTEM pictures in the upper-right inset of Figure 1a demonstrate that these particles exhibit distinct lattice fringes. The observed lattice spacing measures 0.21 nm, corresponding to graphite's (100) plane spacing [6]. Figure 2b presents TEM images and histograms depicting T-CDs' particle size distribution.

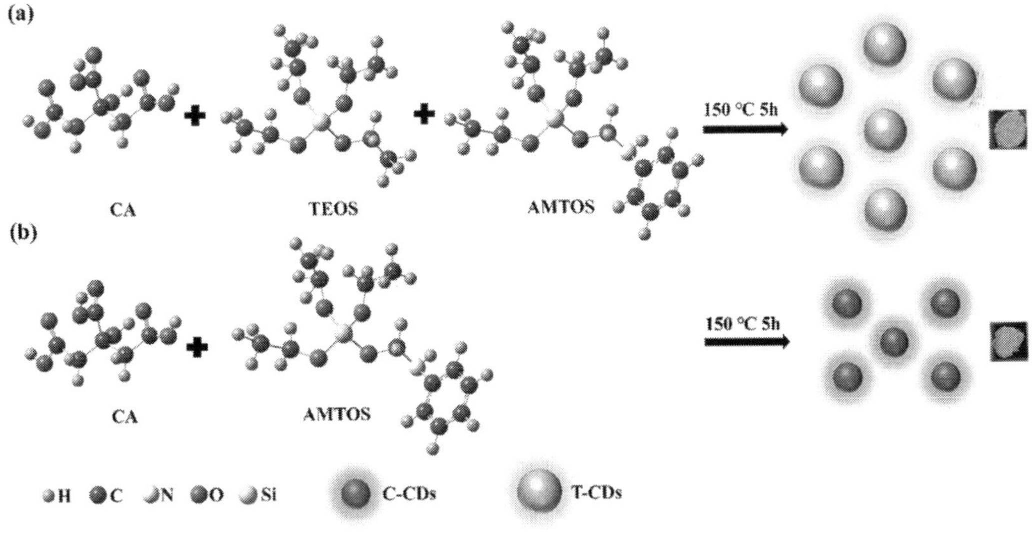


**Figure 1.** (a) Synthesis route for T-CDs. (b) Synthesis route for C-CDs.

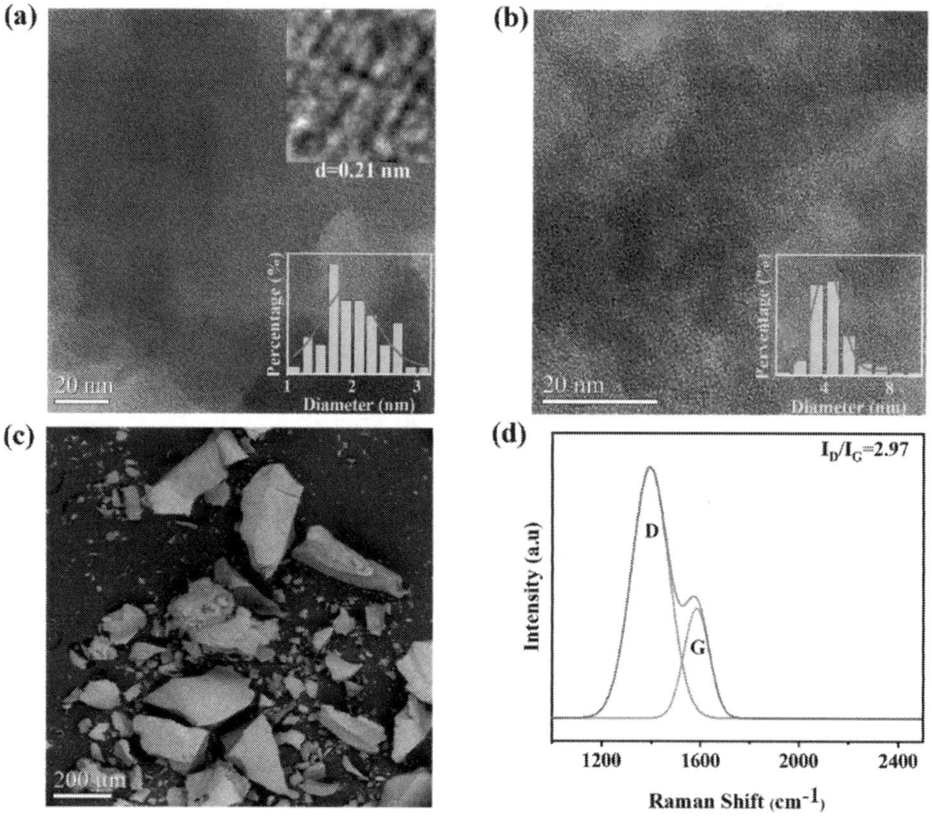


**Figure 2.** (a) TEM of C-CDs, the insets are the HRTEM image and distribution histogram of C-CDs. (b) TEM of T-CDs, the inset presents a distribution histogram of T-CDs. (c) SEM of T-CDs. (d) Raman of T-CDs.

It is observed that amorphous CDs with an average size range from 1.5 nm to 9.5 nm are obtained in T-CDs. Most of the CDs were amorphous. This reveals that the structure of CDs was changed by the silica matrix. The SEM image shows the morphology of silica particles with CDs, which are rough microscale particles (Figure 2c). In Figure 2d, Raman spectroscopy characterization shows the disordered $sp^3$ carbon (D-band) and the $sp^2$ graphitized carbon (G-band) for T-CDs. The ratio between the intensities of D and G bands ($I_D/I_G$) of T-CDs is 2.97, indicative of its disordered state corresponding to amorphous CDs in the TEM picture.

The FL spectra of T-CDs with different reaction temperatures in Figure 3a demonstrate that all samples possess similar FL behaviors with peaks around 475 nm. Furthermore, the intensity of blue emission exhibited an increase of 1.6 times as the reaction temperature was elevated from 110 to 150 °C, caused by high reaction temperatures promoting the formation of luminescent centers. The blue emission decreased by 65% from 150 to 190 °C, possibly caused by excessive carbonization. In Figure 3c, the optimal excitation peak wavelengths of T-CDs are redshifted from 349 to 385 nm as the reaction temperature increases from 110 to 190 °C. FL and excitation spectra of T-CDs with varying concentrations of TEOS are presented in Figures 3b and 3d. It is evident that the T-CDs show the best emission when the TEOS content is adjusted to 86.4 wt%, which is 37 times higher than that of without TEOS. Luminescence Decay Curves of C-CDs and T-CDs show their measured lifetimes to be 1.23 and 7.26 ns, respectively (Figures 3e-f). The decay time of excited electrons of C-CDs was shorter than that of T-CDs due to TEOS maintaining the stability of energy levels to reduce the nonradiative transition in T-CDs compared to C-CDs. Furthermore, the colors of C-CDs and T-CDs are faint yellow in daylight, as observed in the Luminescence Decay Curves in the middle-left insets of Figures 3e-f, respectively. The middle-right insets show photographs of weak blue emissions of C-CDs and intense blue emissions of T-CDs under 365 nm excitation, respectively.

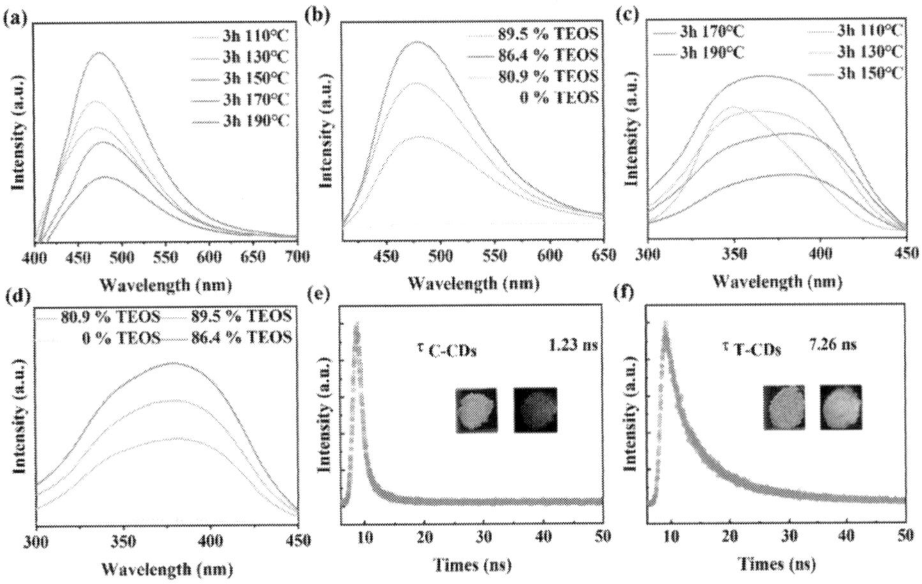


**Figure 3.** FL (a) and excitation spectra (c) of different reaction temperatures. FL (b) and excitation spectra (d) of different content of TEOS. Luminescence Decay Curves of two samples (e) (f). Left inset: images of two samples under daylight. Right inset: images of two samples under ultraviolet light (365 nm).

Considering the emission intensity of C-CDs and T-CDs, T-CDs powder is used as the optical gain medium for lasing. As illustrated in the inset located on the right side of Figure 4b, the T-CDs were stimulated by a 365 nm laser to acquire a blue-emitting random laser detected at different angles. In this system, T-CDs have the capacity to enhance optical scattering effectively, functioning as a type of scattering medium. Additionally, they can serve as optical resonators to prolong the feedback path. Conversely, T-CDs also act as optical gain media to promote stimulated emission (yellow section of the inset in Figure 4b). As shown in Figure 4a, the emergence of distinct sharp peaks in the spectra was observed when the excitation power exceeded the threshold, as indicated by the inflexion points in the excitation-emission curves (excitation power versus emission power) presented in Figure 4b. More lasing modes are observed when the pump power is further increased, covering the entire gain bandwidth. It can be noted that the sharp peaks are distributed randomly throughout the spectra. It was caused by random scattering of light by the nonuniform distribution of silica. Figure 4 presents the laser spectra corresponding to various detection angles ranging from 30° to 75°. Thresholds of different detection angles fluctuate in the range of 16 to 21 mJ/cm$^2$. The laser can be detected from various directions as a result of the random pathways generated within the laser cavity.

## 4. Conclusion

In summary, solid-state blue light-emitting T-CDs have been in situ synthesized via a one-step solvothermal reaction from CA, AMTOS, and TEOS, by adding TEOS, the morphology of CDs with TEOS morphs into amorphous nanoparticles from crystalline quasi-spherical nanoparticles of CDs without TEOS, with the average size increasing from 1.9 nm to 4.1 nm. Meanwhile, the PL lifetime of CDs with TEOS is longer than those without TEOS, increasing from 7.26ns to 1.23 ns. The emission intensity of CDs with TEOS can be increased 37 times by properly adding TEOS compared to those without TEOS. Remarkably, solid-state CDs-based random lasers are achieved under 365 nm laser excitation. With the pump power threshold of about 21.54 mJ/cm$^2$, 16.42 mJ/cm$^2$, 16.89 mJ/cm$^2$, and

17.72 mJ/cm$^2$ in different detection angles (30°, 45°, 60°, and 75°, respectively), laser emissions were collected. The irregular distribution of lasing modes and similar thresholds demonstrate that the as-prepared solid-state CDs are able to derive random laser as a laser gain medium. This work provides a straightforward, environmentally friendly, and effective approach for the in situ fabrication of solid-state CDs with potentials in solid-state micro/nano random lasers.

**Acknowledgement**
This research received funding from Science and Technology Projects of Shenzhen (20231121100725001), Shenzhen Key Laboratory of Photonics and Biophotonics (ZDSYS20210623092006020), and the National Natural Science Foundation of China (61875127).

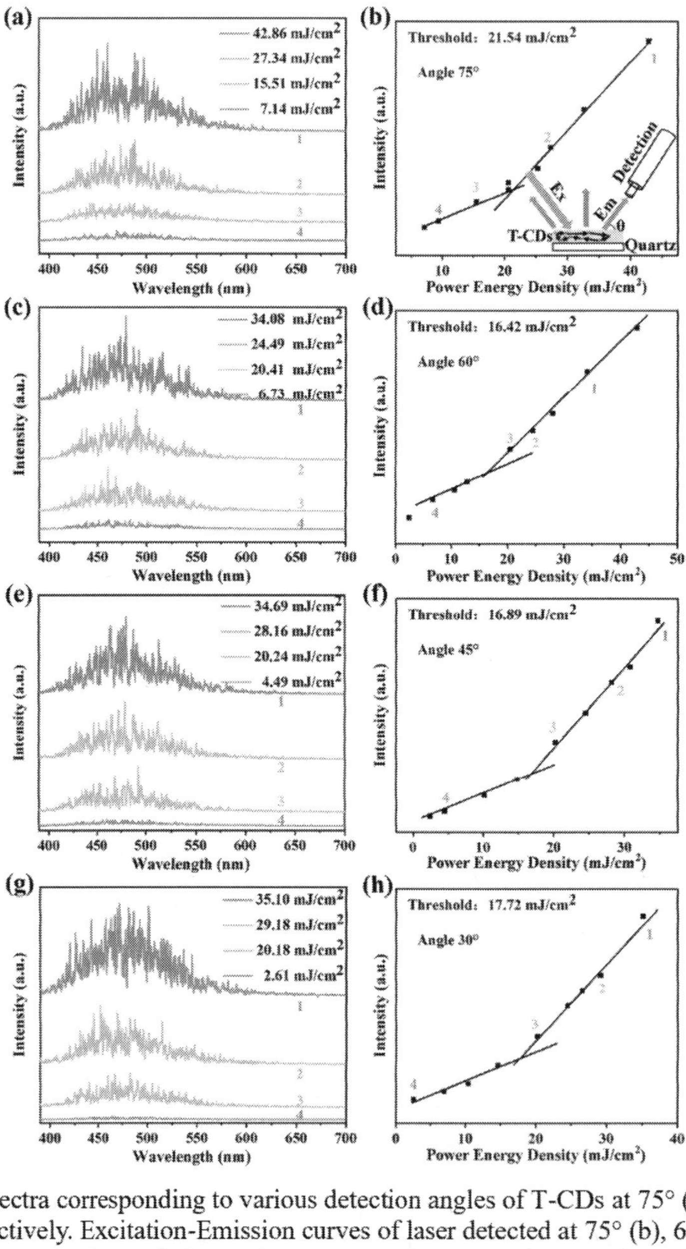


**Figure 4.** Laser spectra corresponding to various detection angles of T-CDs at 75° (a), 60° (c), 45° (e), and 30° (g), respectively. Excitation-Emission curves of laser detected at 75° (b), 60° (d), 45° (f), and 30° (h), respectively. The inset of Figure 4b presents a schematic representation of the excitation and detection processes in a random laser system, with θ meaning the detection angle.

**References**

[1] A.J. Kuehne, M.C. Gather, Organic Lasers: Recent Developments on Materials, Device Geometries, and Fabrication Techniques, Chem Rev, 116 (2016) 12823-12864.
[2] Y. Jiang, Y.Y. Liu, X. Liu, H. Lin, K. Gao, W.Y. Lai, W. Huang, Organic solid-state lasers: a materials view and future development, Chem Soc Rev, (2020).
[3] W.F. Zhang, H. Zhu, S.F. Yu, H.Y. Yang, Observation of lasing emission from carbon nanodots in organic solvents, Adv Mater, 24 (2012) 2263-2267.
[4] Y. Zhang, H. Song, L. Wang, J. Yu, B. Wang, Y. Hu, S.Q. Zang, B. Yang, S. Lu, Solid-State Red Laser with a Single Longitudinal Mode from Carbon Dots, Angew Chem Int Ed Engl, 60 (2021) 25514-25521.
[5] M. Hahn, J.M. Kim, H. Hong, C. Lee, D. Kim, M.Y. Han, H.S. Kim, Y. Piao, Sterically Stabilized Carbon Dots as Solid-State Phosphors for White-Light-Emitting Diodes, ACS Applied Nano Materials, 5 (2022) 11896-11905.
[6] S. Zhang, J. Wang, Z. Chen, J. Chen, Z. Wu, L. Li, P. Yan, W. Zhang, Multi-core@shell double confinement of carbon dots for giant enhancement of thermal Stability: Photoluminescence and lasing up to 1000 K, Chemical Engineering Journal, 493 (2024).

# Carbon dots with recoverable photo-bleaching for reusable UV detectors

**Zhenhui Chen, Shaofeng Zhang, Zefeng Wu, Zhenbo Lin, Xin Yao, Jiong Liu, Zhihao Huang and Wenfei Zhang***

Key Laboratory of Optoelectronic Devices and Systems of Ministry of Education and Guangdong Province, College of Physics and Optoelectronic Engineering, Shenzhen University, Shenzhen 518060, China

*Corresponding author E-mail addresses: zhangwf@szu.edu.cn

**Abstract:** Carbon dots (CDs) have emerged as a promising category of nanomaterials for sensing applications, primarily due to their cost-effectiveness, thermal stability, low toxicity, and unique luminescence properties. CDs sensors have been widely used in ion detection, but rarely in ultraviolet (UV) light detection. In this paper, we synthesized CDs via one one-step pyrolysis process. The exceptional thermal stability of CDs is exhibited by their ability to maintain 100 % initial photoluminescence (PL) intensity over a temperature range from 300 K to 700 K. The bleached CDs can recover almost 100 % of PL intensity through simple thermal treatment at 800 K. The photoluminescence of CDs retains approximately 80% of their initial intensity following five cycles of photo-bleaching and thermal recovery. Moreover, the fluorescence quenching degree maintains a good linear relationship with the UV photo-bleaching time. These indicate that CDs can be effectively used in reusable UV detectors.

## 1. Introduction

Carbon dots (CDs) are carbon nanofluorescent materials that are usually less than 10 nanometers in size, exhibiting low toxicity, low cost, and unique optical properties [1]. The aforementioned intriguing properties render them potential nanomaterials for diverse applications across various domains, including light-emitting diodes (LEDs), information anti-counterfeiting, biolabeling, drug delivery, biosensing, photothermal and fluorescent sensing [2-6].

As mentioned above, CDs are widely used in fluorescence sensing. Researchers have developed two distinct varieties of fluorescence sensors: solution and solid phase sensors. On the one hand, the CD-based solid phase sensors are used for fingerprint and test paper. Peng et al. fabricated a highly photoluminescent CD for the purpose of multicolor patterning and printing ink applications [6, 7]. On the other hand, CDs-based solution sensors are often used in detecting the PH, ions, other organics, and detections of ultraviolet (UV) rays. Sun and Lei synthesized high-quality GQDs (Graphite-carbon quantum dots) from monolithic 3D graphene applied for $Fe^{3+}$ sensing [8]. ANANTHANARAYANAN et al. synthesized iron-doped CDs (Fe-doped CDs) to serve as NIR-IIfluorescence probes for PH detection [9, 10]. Ci et al. created carbon quantum dot (CQD)-based fluorescence resonance energy transfer sensors for the purpose of detecting organophosphate pesticides [10, 11]. Park et al. reported muti-color emitting CDs that change color under UV excitation [12]. However, the muti-color emitting CDs are easily photobleached, which limits their application in UV detectors.

Here, we design cost-effective reusable UV detectors using CDs with recoverable photo-bleaching as sensors. The CDs originated from the one-step pyrolysis process with the raw materials tetraethyl orthosilicate (TEOS), SBA-15, and (3-amino-propyl) trimethoxysilane (APTMS). The CDs demonstrate extraordinary thermal stability, maintaining almost 100% of their initial photoluminescence (PL) intensity from 300 K to 700 K. The CDs are stable with little change in size and spectral wavelength after photo-bleaching or thermal treatment. The photobleached CDs possessing 40% of their initial PL intensity are able to recover 100% of the original intensity by thermal treatment at 800 K. The PL intensity of CDs can remain at about 80% of the original intensity after five cycles of photo-bleaching-thermal recovery. Interestingly, the fluorescence quenching degree maintains a good linear relationship with UV photo-bleaching time. These unique optical properties can be applied in reusable UV detectors.

## 2. Experiment

### *2.1. Synthesis of the three recoverable CDs and CDs*

The recoverable CDs were synthesized via one-step pyrolysis. Tetraethyl orthosilicate (TEOS) and APTMS were combined in a mass ratio of 2:3 through stirring to create a homogeneous mixture. SBA-15 was then mixed into the mixture and subsequently heated to 973 K for a duration of 15 minutes. Upon reaching room temperature, the initial carbon dots (I-CDs) were synthesized. The bleached CDs (B-CDs) are synthesized through UV photo-bleaching of I-CDs at an intensity of 250 kW/cm$^2$. The recovery CDs (R-CDs) originated from B-CDs, subjecting them to a thermal treatment at 800 K for a duration of 10 minutes.

### *2.2. Characterizations*

High-resolution transmission electron microscopy (HR-TEM) utilizing the Figu FEI Tecnai G2 F30 was employed to examine the morphological characteristics of the three CDs. UV-Vis spectra were collected using an FS5 spectrometer from Edinburgh Instruments, UK. A neodymium-doped yttrium aluminum garnet (Nd: YAG) pulsed laser (355 nm, 6 ns, 10 Hz, Continuum Surelite, California, USA), in conjunction utilizing an optical parametric oscillator (Continuum Horizon, California, USA), was utilized to examine the photo-bleaching effects on CDs. PL spectra were acquired at various temperatures using a HORIBA iHR320 spectrophotometer (Minami-ku, Kyoto, Japan) by an optical fiber.

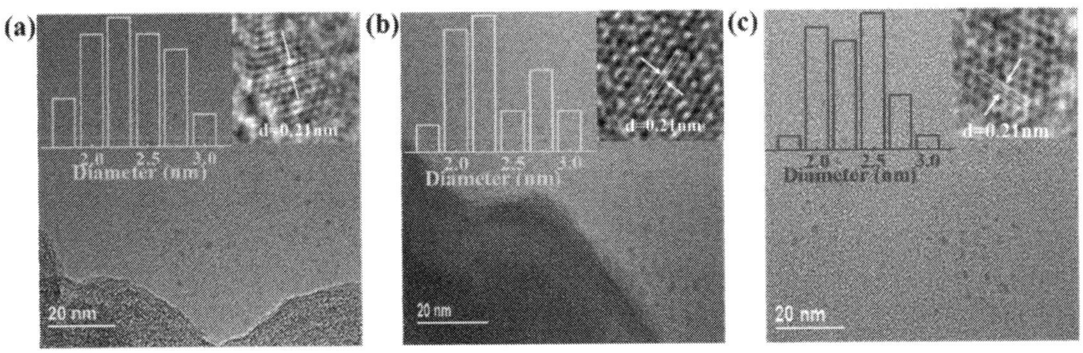


**Figure 1.** TEM image of (a) I-CDs. (b) B-CDs. (c) R-CDs, the upper left insets illustrate the size distribution histograms of the CDs, while the upper right insets feature HRTEM images of the CDs.

## 3. Results and discussion

Figure 1 presents TEM images that demonstrate the uniform spherical morphology of the three CDs, which exhibit an average size ranging from 2.3 to 2.4 nm, as indicated in the inset located in the upper left corner of the TEM images. No noticeable difference in size distribution is observed among these three types of CDs. HRTEM images (shown in the upper right corner of TEM images) reveal that all CDs retain a well-crystallized graphitic carbon-like structure. The CDs also possess comparable lattice fringes with a noticeable lattice spacing of 0.21 nm, which is associated with the (100) facet of the graphite plane.

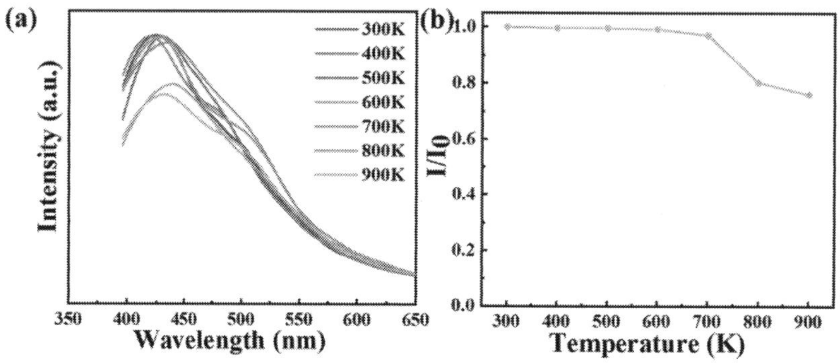


**Figure 2.** (a) PL spectra (b) PL intensity of recoverable CDs obtained at various temperatures.

PL characteristics of the CDs at various heating temperatures are illustrated in Figure 2. It is exhibited that the PL of CDs maintain approximately 97% of initial intensity as the temperature increases from 300 K to 700 K. Then the PL intensity starts to decay slowly when the temperature exceeds 700 K. Notably, the CDs exhibit the capacity to retain approximately 80% of their initial photoluminescence intensity when heated to 800 K. These are manifested the fantabulous thermal stability of CDs, which is the premise of thermal recovery of bleached CDs.

Figure 3 depicts a broad excitation peak, exhibiting a maximum of around 365 nm within the wavelength range of 320 to 390 nm. PL excitation peak of CDs redshifts from 365 to 371 nm undergoing photo-bleaching, while blue shifts from 371 to 360 nm after heating at 800K. In addition, the emission peaks of three CDs are measured to be 425, 431, and 422 nm, respectively (Figure 3a). The impact of photo-bleaching or heating on PL wavelength remains minimal.

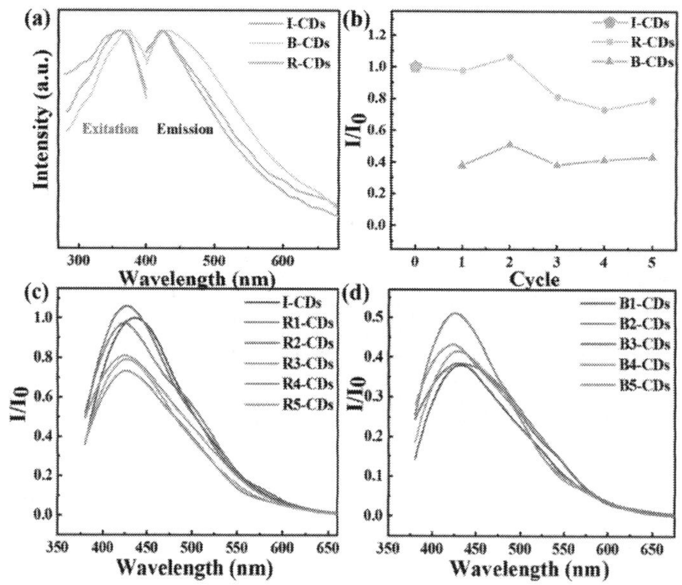


**Figure 3.** (a) Excitation spectra and PL spectra (collected at 300k) of three CDs. (b) the PL intensity of recoverable CDs in five consecutive cycles of photo-bleaching process and thermal treatment. (c) PL spectra of I-CDs and R-CDs, (d) PL spectra of B-CDs subjected to five successive cycles of photobleaching and thermal treatment.

Figure 3b plots the change of PL intensity with five consecutive cycles between photo-bleaching and thermal treatment. The CDs undergo a photobleaching process, resulting in a reduction to approximately 40% of their initial PL intensity. The subsequent thermal treatment at a temperature of 800 K, however, enables recovery of the PL intensity. PL intensity can remain at about 80 % of the original intensity after five cycles of photo-bleaching - thermal recovery. These results manifest that reversible photo-bleaching maintains potential applications.

In Figure 4, the optical sensing capabilities of the synthesized CDs are assessed by examining the variations in PL intensity of CD under 365 nm UV irradiation. The results indicate that the degree of fluorescence quenching (Corresponding to the proportion of the decline in fluorescence intensity, ($I_0$ - I)/$I_0$) of CDs rises with the increase of UV photo-bleaching time. Furthermore, a strong linear correlation is observed between the degree of fluorescence quenching and the duration of UV photo-bleaching. The linear equations were performed in Figure 4, with the linear range of 5-25 mins, where y is the fluorescence quenching degree. These linear equations presented suggest that the overall intensity of UV light can be calculated by analyzing the variation in PL intensity of CDs. This relationship demonstrates that the intensity decreases in a linear fashion as the duration of UV photobleaching increases. Moreover, the fluorescence from CDs is reversible by the simple thermal treatment. In total, the linear relationship and the property of recoverable photobleaching are proven to be the feasibility of these CDs for use in reusable UV detectors.

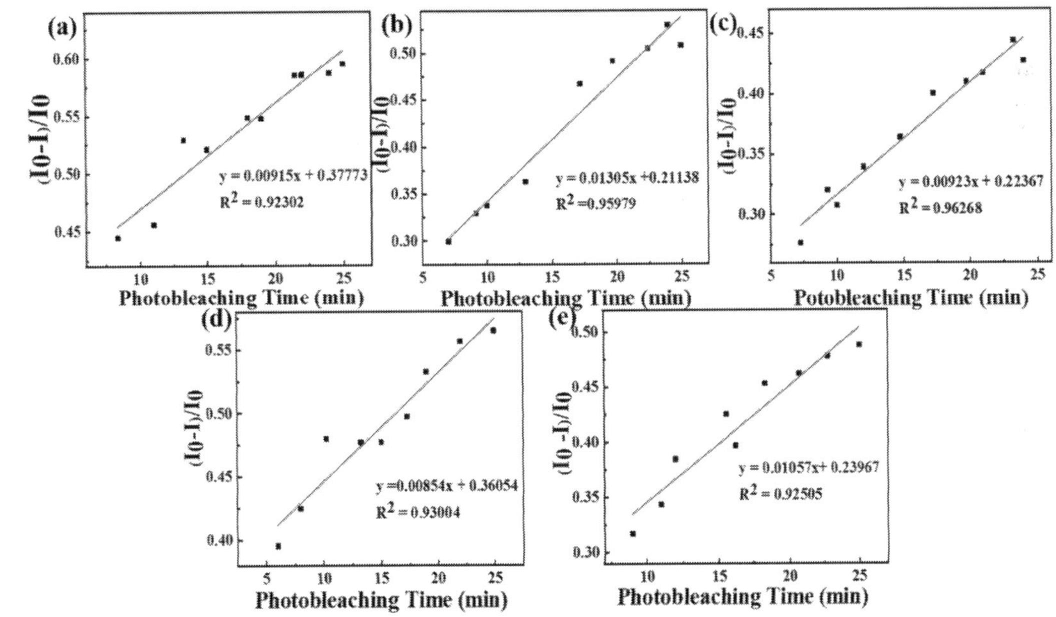


**Figure 4.** Standard working curve of fluorescence quenching degree versus UV photobleaching time in five photobleaching processes.

## 4. Conclusion

In this study, the novel CDs synthesis strategy is designed. The CDs exhibit remarkable thermal stability to retain 100 % initial PL intensity at 700 K and nearly 80 % at 800 K. Moreover, the CDs can recover about 100% of the initial intensity via thermal treatment at 800 K when CDs are photobleached to about 40% of their initial PL intensity. After undergoing five consecutive cycles of photobleaching-thermal treatment, the R-CD can retain 80% of the PL intensity observed in the original CDs. More significantly, it can keep a good linear relationship between fluorescence quenching degree and photo-bleaching time during five photo-bleaching process, and the fluorescently bleached CDs can be reused after simple thermal treatment. In summary, the distinctive optical characteristics of CDs can be utilized in the development of reusable ultraviolet (UV) detectors. This research represents a crucial step towards the advancement of CDs-based reusable UV detection systems, which holds considerable potential for future research endeavors focused on high-performance optoelectronic devices.

**Acknowledgment**
This research was funded by the Science and Technology Projects of Shenzhen (20231121100725001), Shenzhen Key Laboratory of Photonics and Biophotonics (ZDSYS20210623092006020), National Natural Science Foundation of China (61875127).

**References**

[1]LI S, LI L, TU H, et al. The development of carbon dots: From the perspective of materials chemistry [J]. Materials Today, 2021, 51: 188-207.
[2]HE C, XU P, ZHANG X, et al. The synthetic strategies, photoluminescence mechanisms and promising applications of carbon dots: Current state and future perspective [J]. Carbon, 2022, 186: 91-127.
[3]LI J, GONG X. The Emerging Development of Multicolor Carbon Dots [J]. Small, 2022, 18(51).
[4]LI X, FU Y, ZHAO S, et al. Metal ions-doped carbon dots: Synthesis, properties, and applications [J]. Chemical Engineering Journal, 2022, 430.
[5]LI B, ZHAO S, HUANG L, et al. Recent advances and prospects of carbon dots in phototherapy [J]. Chemical Engineering Journal, 2021, 408.
[6]PENG Z, HAN X, LI S, et al. Carbon dots: Biomacromolecule interaction, bioimaging and nanomedicine [J]. Coordination Chemistry Reviews, 2017, 343: 256-77.
[7]ZHU S, MENG Q, WANG L, et al. Highly Photoluminescent Carbon Dots for Multicolor Patterning, Sensors, and Bioimaging [J]. Angewandte Chemie International Edition, 2013, 52(14): 3953-7.
[8]SUN X, LEI Y. Fluorescent carbon dots and their sensing applications [J]. TrAC Trends in Analytical Chemistry, 2017, 89: 163-80.
[9]ANANTHANARAYANAN A, WANG X, ROUTH P, et al. Facile Synthesis of Graphene Quantum Dots from 3D Graphene and their Application for $Fe^{3+}$ Sensing [J]. Advanced Functional Materials, 2014, 24(20): 3021-6.
[10]CI Q, WANG Y, WU B, et al. Fe-Doped Carbon Dots as NIR-II Fluorescence Probe for In Vivo Gastric Imaging and pH Detection [J]. Advanced Science, 2023, 10(7).
[11]WU X, SONG Y, YAN X, et al. Carbon quantum dots as fluorescence resonance energy transfer sensors for organophosphate pesticides determination [J]. Biosensors and Bioelectronics, 2017, 94: 292-7.
[12]PARK J Y, RAMA RAJU G S, HONG W T, et al. Photobleach effect of multi-color emitting carbon dots for UV-light sensing [J]. Chemical Engineering Journal, 2023, 464.

# Thermal stability of Pd-Au core-shell nanostructures in an in-situ liquid TEM

**Dehuan Meng[1, 2]*, Shuai Liu[1, 2], Weiping Wang[1, 2], Zixin Wang[1, 2], Xiaoyan Hu[1, 2]* and Xue Zhang[3]**

[1]Information Science Academy of China Electronics Technology Group Corporation, Beijing, P. R. China
[2]National Key Laboratory of Integrated Circuits and Microsystems, Beijing 100042, P. R. China
[3]North China Research Institute of Electro-optics (NCRIEO)

*Corresponding author's e-mail: dehuanmeng@126.com, huxiaoyan@cetc.com.cn

**Abstract:** The investigation of the thermal stability and evolution of Pd-Au core-shell nanostructures across various temperatures proceeded with in situ heating transmission electron microscopy (TEM) along with spectral analysis techniques. In situ heating transmission electron microscopy (TEM) was utilized to examine the morphology and elemental distribution of core-shell nanostructures composed of $Pd_{0.5}$-$Au_{0.5}$ and $Pd_{0.9}$-$Au_{0.1}$, which were synthesized via the seed growth technique. Compared with the $Pd_{0.5}$-$Au_{0.5}$ core-shell nanostructures, the $Pd_{0.9}$-$Au_{0.1}$ core-shell nanostructures have better thermal stability.

## 1. Introduction

Recently, there has been considerable interest in noble bimetallic nanostructures owing to their remarkable properties, leading to a wide range of applications in catalysis [1], optics [2], information storage [3], sensing [4], biomedicine [5], and imaging [6]. The composition, morphology, and size of noble bimetallic nanostructures can be adjusted to modulate their properties [7-9]. Therefore, exploring the surface composition and structure of nanoparticles can essentially contribute to the understanding of nanoparticles and expand their industrial applications. Currently, continuous reduction and simultaneous reduction are the main methods for synthesizing bimetallic nanostructures. The seed growth reaction is increasingly favoured by researchers due to the convenience of controlling the parameters (shape, size, etc.) of the nanostructures [10].

The annealing process can be utilized to adjust the catalytic characteristics and recycling capabilities of nanoparticles, while many catalytic reactions necessitate elevated temperatures to commence [11-12]. For bimetallic nanostructures, high temperatures may lead to phase separation and alloying [13-14]. In addition, the surface composition and morphology are also believed to affect the catalytic performance and selectivity of the catalysts. Therefore, investigating the evolution of bimetallic nanostructures at different temperatures will contribute to a comprehensive understanding of their thermal stability and physical properties. Theoretical calculations and experimental methods have been used to analyze the structural thermodynamics and kinetics of bimetallic nanostructures at different temperatures [15-16]. Ding et al. explored the annealing process of Au-Pd droplets with sizes of 10-20 nm on a Ni grid, where Shockley sectioning and compact alloying were observed [16]. It is challenging to observe the nanostructural changes in real time during the annealing process due to uncertain variations. More and more scientists have started to apply in-situ heating TEM techniques to realize real-time observation at different temperatures.

The Au-Pd bimetallic systems are the most attractive systems with unique catalytic properties compared to pure clusters and can be used in a variety of reactions such as $H_2O_2$ synthesis [17], CO oxidation [18], and cinnamaldehyde hydrogenation [19]. Not only Pd and Au but also bimetals and reactants have a strong synergistic cooperative effect. Many researchers have been investigating the significance of the catalytic effects. Currently, research on the preparation methods and thermal stability of Pd-Au core-shell nanostructures remains limited. These bimetallic core-shell nanostructures display distinct stability characteristics in comparison to pure clusters. Hence, investigating the stability of Pd-Au core-shell nanostructures is crucial, and in situ analytical techniques can effectively reveal the changes occurring in catalysts during chemical reactions. To enhance the comprehension of the alteration process and thermal stability of Pd-Au core-shell nanostructures at varying temperatures, we conducted detailed in situ heated TEM studies on $Pd_{0.5}$-$Au_{0.5}$ core-shell nanostructures and $Pd_{0.9}$-$Au_{0.1}$ core-shell nanostructures.

## 2. Materials and methods

### *2.1. Materials*

L-ascorbic acid (AA, 99%) and hexadecyl trimethyl ammonium bromide (CTAB, 99%) were obtained from Aladdin. Chloroauric acid ($HAuCl_4$, 99.995%) was purchased from Sigma-Aldrich. Potassium palladium (II) chloride ($K_2PdCl_4$, Pd 32.0%) was purchased from Maklin. Potash iodide (KI) was obtained from Tianjin GuangFu technology development company.

### *2.2. Method*

Pd-Au core-shell nanostructures were synthesized using a method previously described in the literature [20]. To synthesize palladium (Pd) particles, 5 mL of a 0.01M $H_2PdCl_4$ solution was added to 25 mL of a 0.1M CTAB aqueous solution. Following this, 5 mL of 0.04M potassium iodide (KI), 5 mL of 0.02M ascorbic acid (AA), and 10 mL of deionized water were mixed into the solution. The resulting mixture was then stirred at 90°C for one hour. For the formation of Pd-Au core-shell nanostructures, 0.5 mL of the Pd particles were dispersed in 5 mL of a 0.02M CTAB aqueous solution. Subsequently, 24 μL of a 10 mM $HAuCl_4$ solution and 125 μL of a 2.5 mM AA solution were introduced. The mixture was maintained in a water bath at 60°C for 30 minutes.

Characterizations using both TEM and STEM techniques were conducted using a FEI Tecnai F20 microscope. The situ-heated TEM experiments were performed using a DENS solutions Wildfire S3 heating holder, which allows specimen temperatures of up to 1300°C on the silicon nitride chip.

## 3. Results and discussion

We chose to study Pd-Au core-shell nanostructures (which have the same Pd core) with composition ratios of Pd: Au =1: 1 ($Pd_{0.5}$-$Au_{0.5}$ core-shell nanostructures) and Pd: Au=9: 1 ($Pd_{0.9}$-$Au_{0.1}$ core-shell nanostructures). In the in-situ heating experiments, we combined EDAX and TEM to help distinguish the Pd-Au core-shell nanostructures and observe the alloying process at different temperatures. The morphology of the two nanoparticles is slightly different due to the difference in Au element content and the selective growth of the planes. However, the cores of both specimens are identical Pd nanocubes. In the $Pd_{0.5}$-$Au_{0.5}$ core-shell nanostructures, the Au content in the faces is higher than that of the corners. On the contrary, the $Pd_{0.9}$-$Au_{0.1}$ core-shell nanostructures have a higher Au content in the corners than in the faces.

We captured images of individual nanoparticles at various stages of heating, with each stage maintained for different time intervals as the temperature rose. Figures 1 and 2 show the evolution of the $Pd_{0.5}$-$Au_{0.5}$ core-shell nanostructure and the $Pd_{0.9}$-$Au_{0.1}$ core-shell nanostructure at different temperatures. Once the temperature reaches 400°C, the Au on the surface of the Pd cube tends to equilibrate, and the Au on the surface diffuses to the corners. At the same time, the polygonal corners on the $Pd_{0.5}$-$Au_{0.5}$ core-shell nanostructures change from circular to spherical. The mutual diffusion of the two elements is very obvious, and the distinction between the core and the shell becomes increasingly indistinct at 500°C. The activation energy for the diffusion of Au into Pd is measured at 0.9 eV, while the activation energy for Pd diffusing into Au is 0.6 eV. This indicates that the diffusion rate of Pd into Au exceeds that of Au into Pd [20]. This phenomenon is not obvious due to the limitation of the EDAX resolution. The pure Pd cubic became spherical and melted at 500°C has been reported in the literature [21]. In our study, the Pd core lost its morphology at a temperature of 500°C. The Au element plated with {100} facets had no obvious effect on stabilizing the core-shell structure. The expressed facets disappeared at 700°C, and a complete solid solution sphere was observed. For the $Pd_{0.5}$-$Au_{0.5}$ core-shell nanostructures, the Au shells have no significant effect on the stabilization of the shape of the Pd core.

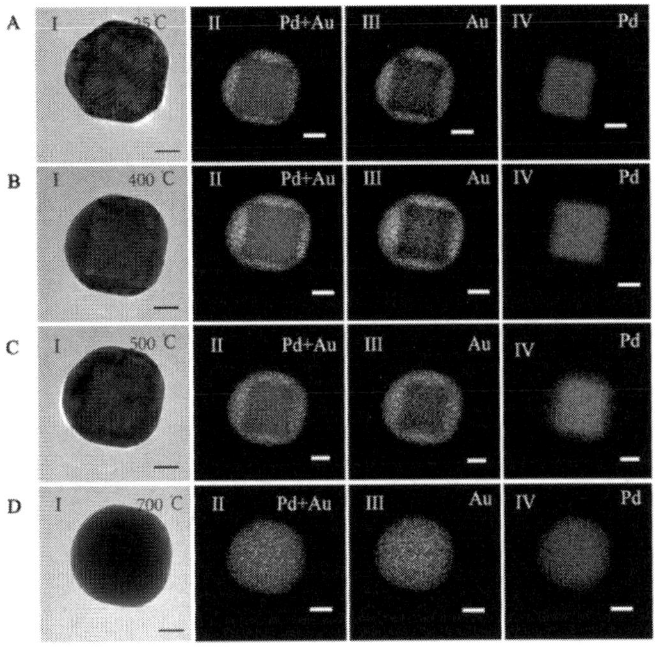


**Figure 1.** The evolution of $Pd_{0.5}$-$Au_{0.5}$ core-shell nanostructure at different temperatures. Scale bar 20 nm.

To comprehend the correlation between thermal stability and the distribution of elements, we performed similar experiments on the $Pd_{0.9}$-$Au_{0.1}$ core-shell nanostructures, as shown in Figure 2. Due to the different element composition, the shape of $Pd_{0.9}$-$Au_{0.1}$ core-shell nanostructures is close to cubic, with a large amount of Au elements deposited on the {110} plane. Similar to the $Pd_{0.5}$-$Au_{0.5}$ core-shell nanostructures, the Au shell tends to equilibrate, but the Au elements start to transfer from the {110} plane to the {100} plane. The diffusion rate on the plane is related to the concentration of Au elements. The minor morphological changes in the nanostructures may be due to the disappearance of surface-coated CTAB under heating conditions [22]. Dispersion of inhomogeneous Au shells can also lead to this phenomenon. When heated to 500°C, the shape of $Pd_{0.9}$-$Au_{0.1}$ core-shell nanostructures begins to show obvious deformation, while the shape and facets of the Pd cubes do not change significantly even when heated at 500°C. The distribution of Au elements on the surface tends to be uniform. When the temperature reaches up to 700°C, the shape of the cube is observed. Our analysis indicates that gold is distributed fairly uniformly across the surface of the palladium cube. The mapping spectral distribution of Au also confirms this conclusion. When the temperature was increased to 800°C, distinct truncated cubic particles were found. With the extension of the heating time, the elements become more active,

Journal of Physics: Conference Series **2961** (2025) 012039 doi:10.1088/1742-6596/2961/1/012039

and the mutual diffusion of the core and shell becomes more intense. When the temperature reaches 900°C, the shape is nearly spherical, and the core-shell structure is completely alloyed. In comparison to the $Pd_{0.5}$-$Au_{0.5}$ core-shell nanostructures, the $Pd_{0.9}$-$Au_{0.1}$ core-shell nanostructures exhibit greater stability in their core-shell architecture. Furthermore, both the morphology of the Pd core cubes and the cubic shape of the $Pd_{0.9}$-$Au_{0.1}$ core-shell nanoparticles are enhanced.

A
B
C
D
E
F

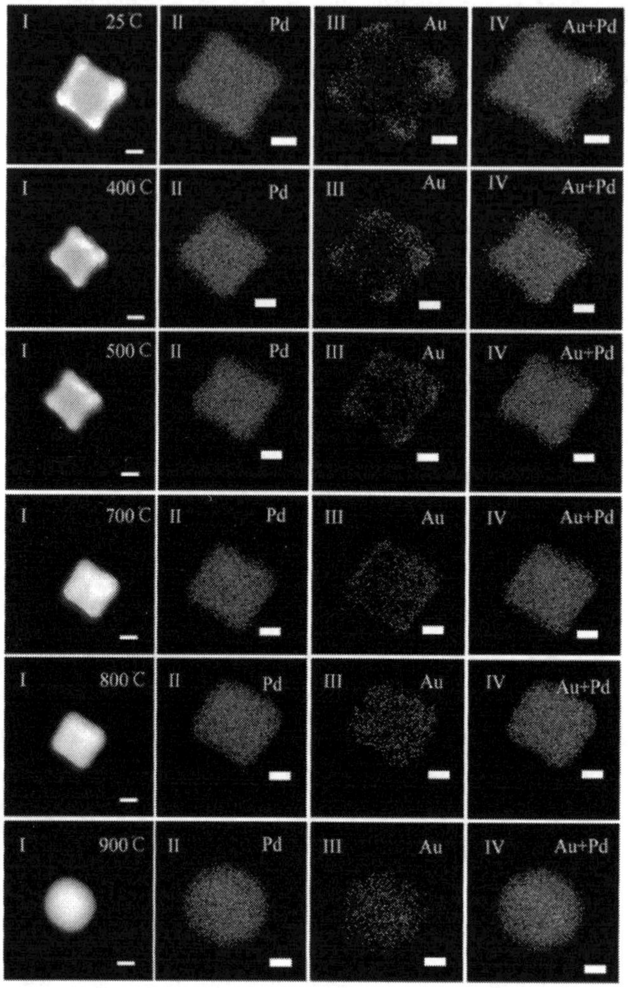


**Figure 2.** The evolution of $Pd_{0.9}$-$Au_{0.1}$ core-shell nanostructure at different temperatures. Scale bar 20 nm.

Different diffusion paths of Au on the surface of different proportions of nanoparticles lead to different thermal stability of different proportions of nanoparticles. In the core-shell structure (with Pd cube as the nucleus), the diffusion on the {100}-face is fatal. A significant quantity of gold is accumulated on the surface of the {100} facet, leading to a strong interdiffusion of Pd and Au elements on this surface. Conversely, the movement of Au from the {110} facet to the {100} facet enhances the thermal stability of both the shape and the core-shell configuration. At the same time, the relatively low interdiffusion of Au and Pd on the {100} plane is also a factor that ensures its thermal stability. The distribution of Au tends to be homogeneous when heated to 700°C, and the diffusion on the {100} plane dominates the alloying process.

To elucidate the importance of the {100} plane, we observed the evolution of agglomerated Pd-Au core-shells in an in situ heated TEM. As shown in Figure 3, the shape and structure of $Pd_{0.9}$-$Au_{0.1}$ core-shell nanostructures disappeared when the temperature was increased to 800°C, which is lower than the temperature of the individual particles. It can be observed from the Mapping image that the distance between the Pd core becomes smaller. Therefore, the nanoparticles increase with the temperature, and then they contact with the surface. The contact with {100} plane strengthens the diffusion effect, and the $Pd_{0.9}$-$Au_{0.1}$ core-shell could not remain unchanged at 800°C. Similar to the individual $Pd_{0.9}$-$Au_{0.1}$ core-shell nanoparticles, the structure of agglomerated nanoparticles disappeared at 800°C. This observation suggests that Au's content is deposited on the Pd cubes.

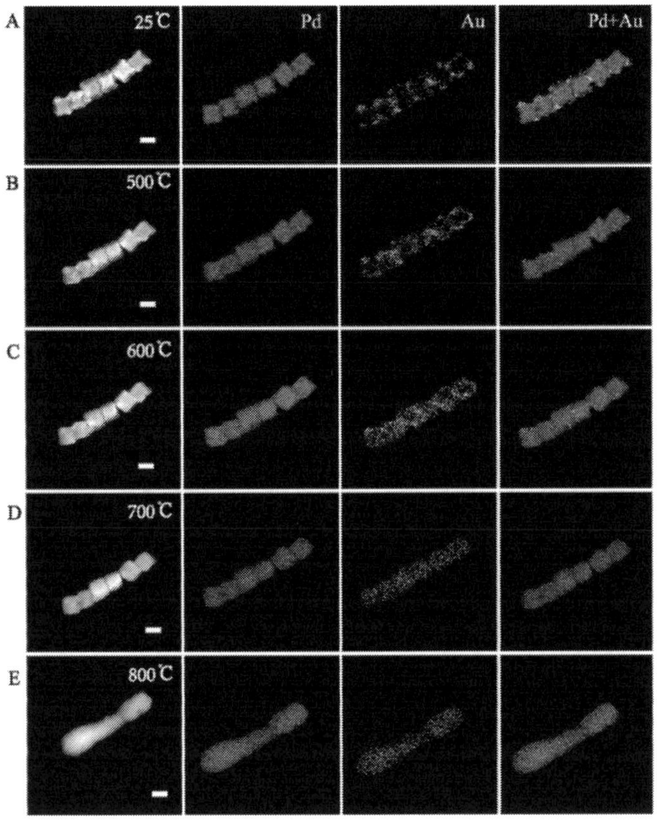


**Figure 3.** The evolution of agglomerated $Pd_{0.9}$-$Au_{0.1}$ core-shell nanostructure at different temperatures. Scale bar 50 nm.

## 4. Conclusion

In summary, the thermal stability of bimetallic nanostructures composed of gold and palladium was examined using in situ heated transmission electron microscopy (TEM) coupled with energy dispersive X-ray analysis (EDAX). The Au shell enhances the stability of cubic Pd, and it is particularly important to control the amount of Au and the deposition location. Similar to the melting point of pure Pd, the $Pd_{0.5}$-$Au_{0.5}$ core-shell nanostructures show a change in morphology at the temperature of 500°C. The $Pd_{0.9}$-$Au_{0.1}$ core-shell nanostructures have a more stable core-shell structure. The cubic $Pd_{0.9}$-$Au_{0.1}$ core-shell nanostructures were able to withstand temperatures up to 800°C. This study elucidates the development and thermal activation processes of Pd-Au core-shell nanoparticles, offering insights into the regulation of these nanoparticles and facilitating enhancements in their catalytic performance.

## References

[1]Agarwal N, et al., 2017 Science, 358: 223-227.
[2]Silva K N, et al., 2024 ACS Nano, 18: 24391-244-3.

[3]Zijlstra P, Chon J W M and Gu M 2009 Nature, 459: 410-413.
[4]Li G J, et al., 2021 Nanomaterials, 11: 1926.
[5]Ahmed H B and Emam H E 2021 Journal of Molecular Liquids, 321: 114660.
[6]Rajeev R, et al., 2021 Microchemical Journal, 163: 105910.
[7]Ramos R M C R and Regulacio M D 2021 ACS Omega, 6: 7212-7228.
[8]Loza K, Heggen M and Epple M 2020 Adv. Funct. Mater., 30:1909260.
[9]Maize M, et al., 2021 Journal of Colloid and Interface Science, 585:480-489.
[10]Rej S, et al. 2016 Angew. Chem. Int. Ed., 55:7222-7226.
[11]Gao C B, Lyu F L and Yin Y D 2020 Chemical Reviews, 121: 834-881.
[12]Frey H, et al., 2022 Science, 376: 982-987.
[13]Bondesgaard M, et al., 2019 Adv. Funct. Mater., 29:1905933.
[14]Huang L L, et al., 2020 J. Am. Chem. Soc., 142:4570-4575.
[15]Liu H B, Pal U, Perez R, and Ascencio J A 2006 J. Phys. Chem. B., 110:5191-5195.
[16]Ding Y, Fan F R, Tian Z Q, and Wang Z L 2010 J. AM. Chem. Soc., 132:12480-12486.
[17]Ricciardulli T, et al., 2021 J. Am. Chem. Soc., 143: 5445-5464.
[18]Khder A S, et al., 2023 ACS Applied Nano Materials, 6:4243-4252.
[19]Pinto J, et al., 2023 Catal. Sci. Technol., 13:4082-4091.
[20]Debonte W J and Poate J M 1975 Thin Solid Films, 25:441-448.
[21]Lu N, Wang J G, Xie S F, Xia Y N and Kim M J 2013 Chem. Commun., 49: 11806-11808.
[22]Choi B S, Iqbal M, Lee T, Kim Y H and Tae G 2008 J. Nanosci. Nanotechnol., 8:4670-4674.

# Development of new corrosion and scale and salt inhibitors and research on the effect of field application

**Xiaolong Li[1*], Zigang Lei[2], Cong Li[3], Yuan Wang[1], Xufeng Shang[2] and Jie Zhou[2]**

[1]China National Key Laboratory of Oil and Gas Drilling and Transportation Equipment, PetroChina Engineering Materials Research Institute Co., Ltd., Xi'an 710077, Shaanxi, China;
[2]Xi'an Dexincheng Science and Technology Co., Ltd., Xi'an 710065, Shaanxi, China;
[3]PetroChina Tarim Oilfield Branch, Korla 841000, Xinjiang, China

[*]Corresponding author's e-mail: 695602620@qq.com

**Abstract**: Several puncture leaks and perforation failures occurred in the water injection pipeline outside a joint station in a western oilfield, seriously affecting the pipeline's integrity. By analyzing the clogging in the pipeline, it was found that the corrosion caused by salt and scale inside the pipeline was the main reason for the perforation failure of the water injection pipeline outside the station. Given the serious corrosion and scaling of the water injection pipeline outside the station, the multi-functional agent of corrosion inhibitor, scale inhibitor, and salt suppressant was developed. It was determined that the agent had the best effect when the ratio of the single-agent additive corrosion inhibitor, scale inhibitor, and salt suppressant was 2:3:5. The sum of the three is 87% of the formula. At 40°C, when the addition amount was 150 mg/L, the corrosion rate of P110 steel was only 0.0307 mm/a, the scale inhibition rate for $CaCO_3$ can reach 83.5%, and the salt inhibition rate can get 4.5%. Field corrosion monitoring test piece evaluation test results show that the uniform corrosion rate of the coupons monitored by the three monitoring points is greatly reduced after the corrosion and scale and salt inhibitor are added, and the higher the injection concentration is, the lower the uniform corrosion rate is. When the concentration is high, there are almost no salt crystals on the surface of the monitoring coupon, and a small amount of salt crystals are precipitated after the concentration is reduced.

## 1. Introduction

As some domestic oilfields enter the middle and late stage of exploitation, the development of residual oil in old blocks of high-water content and low permeability reservoirs and the development of thick oil and other difficult-to-use reservoirs have become the focus of increasing reserves and production [1]. Therefore, secondary or tertiary development has become important to increase production and ensure benefits. Scientific and effective secondary oil recovery measures can increase the recovery rate of old oilfields to more than 10% [2]. The measures to increase recovery rate during secondary or even tertiary development, such as acidification and gas injection drive recovery, will lead to further increase in the mineralization of formation water, and increase the risk of corrosion, scaling, and salting of the recovered fluids at the same time [3-5].

The sewage output of a joint station of an oil field in the west is 3, 000 $m^3/d$, the sewage is transported to the transfer station, water is supplied to the water injection valve group after being upgraded by the external water pump, and then water is provided to a single well after being upgraded. The sewage water injection pipelines outside the station are all repaired oil pipes with specifications of Φ89 mm×6.45 mm and made of P110 carbon steel pipes. In the past two years, the problem of pipeline leakage, puncture, and failure accidents caused by salt and scale inside the water injection pipeline has been severe. There are problems such as single function (only corrosion/corrosion and scale inhibition effect or only salt inhibition effect) and poor field application of the existing agents, and there are very few research reports about the composite agents with triple effects of corrosion inhibition, scale inhibition, and salt inhibition. In this paper, based on a large number of literature reviews, we chose to use triazine-containing N-heterocyclic molecules as the corrosion inhibitor and compounded with epoxy succinic acid copolymer salt inhibitor and polyphosphoric acid scale inhibitor, to form a new type of agent with the efficacy of corrosion inhibition, scale inhibition, salt inhibition [6-12]. The agent was applied in the western oilfield from September 2023 to October 2023, which can significantly reduce the corrosion rate of on-site corrosion monitoring specimens, improve the salt formation on the surface of the coupons, and achieve good on-site application results.

## 2. Materials and test methods

### *2.1. Experimental apparatus and drugs*

Experimental instruments: Correst high temperature and high-pressure dynamic reactor from the United States, Shanghai Jingke MP502B analytical balance, Heidolph MR3001 magnetic stirrer, FEI INSPECT F50 field emission scanning electron microscope from the United States, and Shimadzu XRD-6100 X-ray diffractometer from Japan.

Drugs: NaCl, $CaCl_2$, KCl, $MgCl\cdot6H_2O$, $NaHCO_3$, $Na_2SO_4$, $BaCl_2$, $SrCl_2$, and NaOH, all analytically pure, are provided by Sinopharm Chemical Reagent Co., Ltd. Epoxy succinic acid copolymer and polyphosphoric acid substances are commercially available, and triazine derivatives are laboratory synthesized products. $H_2S$, $CO_2$, and $N_2$ are Tianjin's special gases. Experimental water is deionized water.

### *2.2. Specimen processing*

The specimen material is P110 steel, with the specification of 50 mm×10 mm×3 mm, grinding with sandpaper step by step to 800# before the test. Then, it was dehydrated with anhydrous ethanol, degreased with acetone, blown dry with cold air, placed in a desiccator for 24 hours, and then weighed (accurate to ±0.1 mg) and measured (accurate to ±0.02 mm).

### *2.3. Analysis of the composition of the clogging*

The blockages in the pipelines and water injection pipelines outside the station of an oil field in the west were retrieved for drying and grinding. The XRD-6100 X-ray diffractometer produced by Japan's Shimadzu Company was used for physical analysis. The analysis range (2θ) was 10~80°, the test voltage was 40 kV, and the target material was Cu.

### *2.4. Laboratory corrosion simulation tests*

Referring to SY/T 5273-2014 *Technical Specifications and Evaluating Methods of Corrosion-inhibitors for Oilfield-produced Water*, the high temperature and high-pressure dynamic reactor was used to evaluate the corrosion inhibition performance of the agent. The corrosion medium was an on-site simulation solution. The specific composition is shown in Table 1, and the test conditions are shown in Table 2. The additional amount of corrosion, scale, and salt inhibitor is 150 mg/L. After the test, we use a film removal solution (the preparation method is 100 ml of analytically pure hydrochloric acid and 5 to 10 g of hexamethylenetetramine, and distilled water is added to 1, 000 ml) to clean the corrosion products on the surface of the specimen by ethanol. Then it is blown dry by cold air and placed in a desiccator for 24 hours. We weigh and calculate the uniform corrosion rate according to Equation (1).

$$R = \frac{87600\times(m_1-m_2)}{A\times t\times\rho} \tag{1}$$

where $R$ is the corrosion rate (mm/a); $m_1/m_2$ is the mass of the specimen before and after corrosion (g) respectively; $A$ is the surface area of the specimen ($cm^2$); $t$ is test time (h); $\rho$ is material density ($g/m^3$).

**Table 1.** Composition of corrosion simulation solution (mg/).

| Ingredient | $Cl^-$ | $SO_4^{2-}$ | $HCO_3^-$ | $Ca^{2+}$ | $Mg^{2+}$ | $Ba^{2+}$ | $Sr^{2+}$ | $K^+$ | $Na^+$ |
|---|---|---|---|---|---|---|---|---|---|
| Quantity contained | 114, 204 | 621 | 177 | 7, 411 | 494 | 23.3 | 306 | 1, 342 | 59, 476 |

**Table 2.** Test conditions.

| Temperature (°C) | Total pressure (MPa) | $P_{CO2}$ (MPa) | $P_{H2S}$ (MPa) | Flow rate (m/s) | Material | Time (h) |
|---|---|---|---|---|---|---|
| 40 | 1 | 0.015 | 0.0028 | 0.08 | P110 | 168 |

*2.5. Laboratory evaluation of scale inhibition properties*

According to the standard Q/SY 126-2014, the scale inhibition rate of $CaCO_3$-type scale samples was tested by titration method.

*2.6. Laboratory evaluation of salt suppression performance*

Salt suppression effect evaluation method: (1) We add 100 mL deionized water and 40 g NaCl to the beaker, heat it to boiling while stirring with a thermostatic water bath, take the upper layer of the clear liquid to get saturated NaCl solution, and reserve it at a constant temperature of 40°C.

(2) We take two clean watch glasses marked 1# and 2#, dry them in a constant temperature blast drying oven at 105°C for 20 minutes, weigh them when cooled to room temperature, record the mass of the watch glasses as $m_1$, $m_2$, and then put them into a 40°C constant temperature water bath for spare use after weighing.

(3) We add 5 mL of upper saturated NaCl solution to each of watch glasses 1# and 2#, add 1 mL of salt suppressor solution to watch glass l#, and let them stand at 40°C for 12 hours. We put the two watch glasses into a drying oven and dry them at 100°C for 4 hours. After drying is completed, we should wait until the surface blood drops to room temperature. We use a brush to gently sweep away the loose materials in the watch glasses, and then clean the two watch glasses. The weighing scale is $m_3$ and $m_4$.

(4) We calculate the salt suppression rate [13] according to Equation (2).

$$P = \frac{(m_4-m_2)-(m_3-m_1)}{m_4-m_2} \times 100\% \quad (2)$$

where P is salt suppression rate, %; $m_1$ is the mass of watch glass 1 #, g; $m_2$ is the mass of watch glass 2#, g; $m_3$ is the mass of watch glass 1# and its solid attachments, g; $m_4$ is the mass of watch glass 2# and its solid attachments, g.

*2.7. Field trials*

The pipeline from a joint oil field station in the west through the oil transfer station to the single water injection well was selected as the field test section. The total length of the pipeline is about 35 km, and a total of 3 corrosion monitoring points (the outlet of the water feeding pump in the station, the oil transfer station, and the single water injection well) were selected to be loaded into 3 corrosion monitoring specimens, and the material of the specimen was kept the same as that of the on-site pipeline. The blank test was not filled with corrosion inhibitor, the first round of test filling concentration was 300 mg/L in the station+100 mg/L in the oil transfer station and the second round of test filling concentration was 150 mg/L in the station + 50 mg/L in the oil transfer station, the corrosion monitoring specimens were loaded after 3 days of corrosion inhibitor dosing, and each round of test cycle was 7 days.

Journal of Physics: Conference Series **2961** (2025) 012040
doi:10.1088/1742-6596/2961/1/012040

After the field test is completed, we take out the corrosion monitoring specimens, gently dry the moisture on the surface of the specimens with filter paper, wrap the specimens with filter paper and put them into a vacuum bag, and use a hand pump to evacuate the specimens to ensure that the specimens are in a vacuum state. We select one specimen from each monitoring point for surface analysis. After the analysis is completed, we remove the film, weigh it, and calculate the corrosion rate as soon as possible according to the method specified in the SY/T 5273-2014 standard.

## 3. Results and discussion

### *3.1. Physical analysis of blockages at the site*

XRD tests were performed on the blockage taken from the interior of the water injection pipeline outside the western oilfield station, and the results were shown in Figure 1.

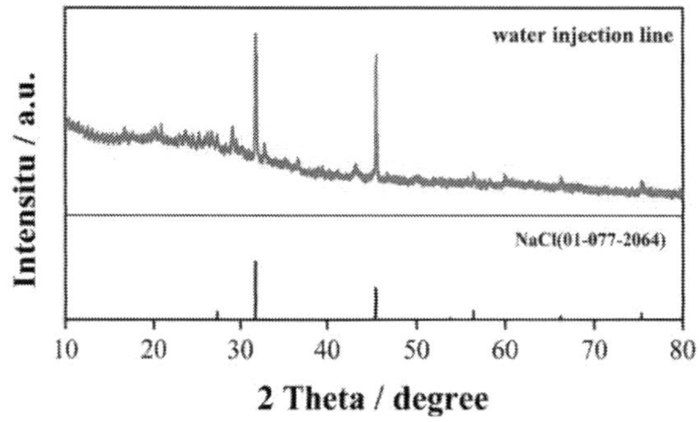


**Figure 1.** XRD pattern of blockage in water injection pipeline.

The XRD analysis results of the blockage show that the composition of the blockage in the water injection pipeline outside the station was mainly NaCl and that salt deposition on the inner surface of the pipeline was the main cause of the blockage. To further determine other components in the blockage, the blockage was washed with deionized water several times, then dried and ground, and then subjected to XRD testing.

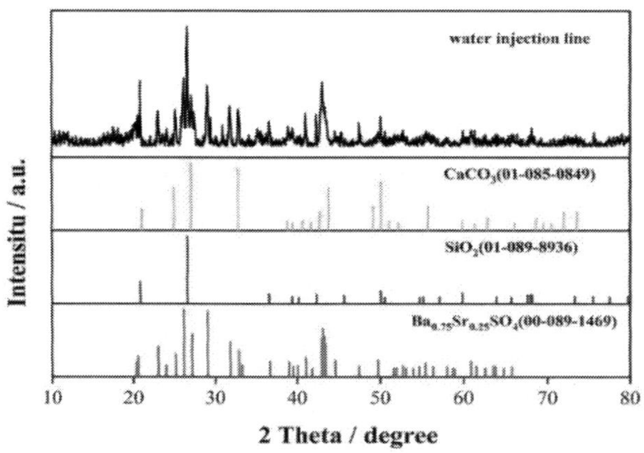


**Figure 2.** XRD pattern of water injection pipeline blockage after cleaning.

It can be seen from Figure 2 that after the salt in the blockage is removed, the remaining substances are mainly $CaCO_3$, $SiO_2$, and barium strontium scales, of which $SiO_2$ is caused by sand content in the produced water.

*3.2. Laboratory evaluation tests*

*3.2.1 Corrosion and scale and salt inhibitor compounding.* A triazine derivative is used as the main agent of the corrosion inhibitor, a polyphosphoric acid scale inhibitor and an epoxy succinic acid copolymer salt inhibitor are used as auxiliaries, and liquid alkali is used as a solvent to formulate the compound. When the ratio of corrosion inhibitor, scale inhibitor, and salt inhibitor in the formula was 2:3:5, and the sum of the three accounted for 87% of the formula, the compounded corrosion and scale and salt inhibitor had the best effect, and the optimized formula was 17.4% triazine derivatives+26.1% polyphosphoric acid scale inhibitor+43.4% epoxy succinic acid copolymer salt inhibitor+13% liquid alkali.

*3.2.2 Corrosion inhibition properties.* As can be seen from Table 3, when the concentration of the corrosion and scale and salt inhibitor is 150 mg/L, the average corrosion rate of P110 specimens is only 0.0307 mm/a, which has excellent corrosion inhibition performance. This is mainly because the triazine derivative molecules have a cyclic structure containing multiple N atoms. The lone pair of electrons contained in the N atoms can enter the hybrid orbit of the metal atom and form a dense layer of adsorption on the metal surface through the formation of coordination bonds, preventing the corrosive medium and the contact of the metal surface to inhibit the corrosion of the further occurrence of corrosion, to play a role in corrosion inhibition.

**Table 3.** Corrosion rate of P110 specimens in simulated environment.

| Specimens | Concentration of the agent mg/L | Corrosion rate mm/a | Average corrosion rate mm/a |
|---|---|---|---|
| 1# | | 0.0305 | |
| 2# | 150 | 0.0321 | 0.0307 |
| 3# | | 0.0296 | |

*3.2.3 Scale inhibition properties.* The scale inhibition performance of the corrosion and scale and salt inhibitor on $CaCO_3$ was tested at 40°C, and the test results are shown in Table 4.

**Table 4.** Test result of scale inhibition performance of corrosion and scale and salt inhibitor on $CaCO_3$.

| Serial number | Temperature °C | Concentration of the agent mg/L | Scale inhibition rate % |
|---|---|---|---|
| 1 | 40 | 150 | 83.5 |

As can be seen from Table 4, the scale inhibition rate of the composite agent on $CaCO_3$ reached 83.5% at 40°C, and the scale inhibition effect was obvious. On the one hand, the agent in the epoxy succinic acid copolymer and polyphosphoric acid can form water-soluble chelates with scale-forming ions under alkaline conditions. On the other hand, epoxy succinic acid copolymer and polyphosphoric acid undergo hydrolysis reaction in a water solution, dissociating a large number of negatively charged ions. These ions can be adsorbed on the surface of the crystal or enter the interior of the crystal, hindering the normal and orderly growth of the crystal, thus playing a scale inhibition role.

*3.2.4 Salt inhibition properties.* The results of analyzing the blockage of the pipeline at the site showed that the content of salt scale in the blockage was very high. To slow down the salt scaling cycle and reduce the risk of pipeline failure, a salt inhibitor was added to the formula and tested for salt inhibition properties, and the results are shown in Table 5.

**Table 5.** Test results of salt inhibition properties of the agent.

| Serial number | Temperature °C | The concentration of the agent mg/L | Salt inhibition rate % |
|---|---|---|---|
| 1 | 40 | 150 | 4.5 |

As can be seen from Table 5, when the temperature is 40°C, the salt inhibition rate can reach 4.5% when the dosage of the compounded corrosion and scale inhibitor is 150 mg/L, mainly because the anion dissociated from the groups epoxy succinic acid copolymer salt inhibitor can be adsorbed on $Na^+$, which can inhibit the inter-adsorption of $Na^+$ and $Cl^-$, and reduce the formation of salt crystals. At the same time, the anionic group can also be adsorbed on the formed NaCl crystals destroying the nucleus structure, making the face-centred cubic cell structure of regular and orderly growth turn into a loose dendritic structure, which can be more easily dispersed into the solution.

### *3.3 Field test results*

*3.3.1 Corrosion rate of on-site monitoring specimen.* Laboratory evaluation test results show that the developed corrosion and scale and salt inhibition agent has better corrosion, scale, and salt inhibition properties when the addition amount is 150 mg/L. To evaluate its on-site applicability, a western oilfield station through the station outside the transfer station outside the transmission of the water injection pipeline to a single well carries out two rounds of field tests. Due to the high cost of on-site corrosion inhibitor filling points, corrosion monitoring point modification, and testing, more on-site influencing factors were considered, and the working conditions were more complex. There was a certain deviation between the laboratory evaluation results and the field test results, and we chose the first round of test-filling high-concentration agents. If the filling concentration is higher, the agent field applicability is better, and the fill concentration optimization test and the corrosion rate of on-site monitoring specimens are shown in Figure 3.

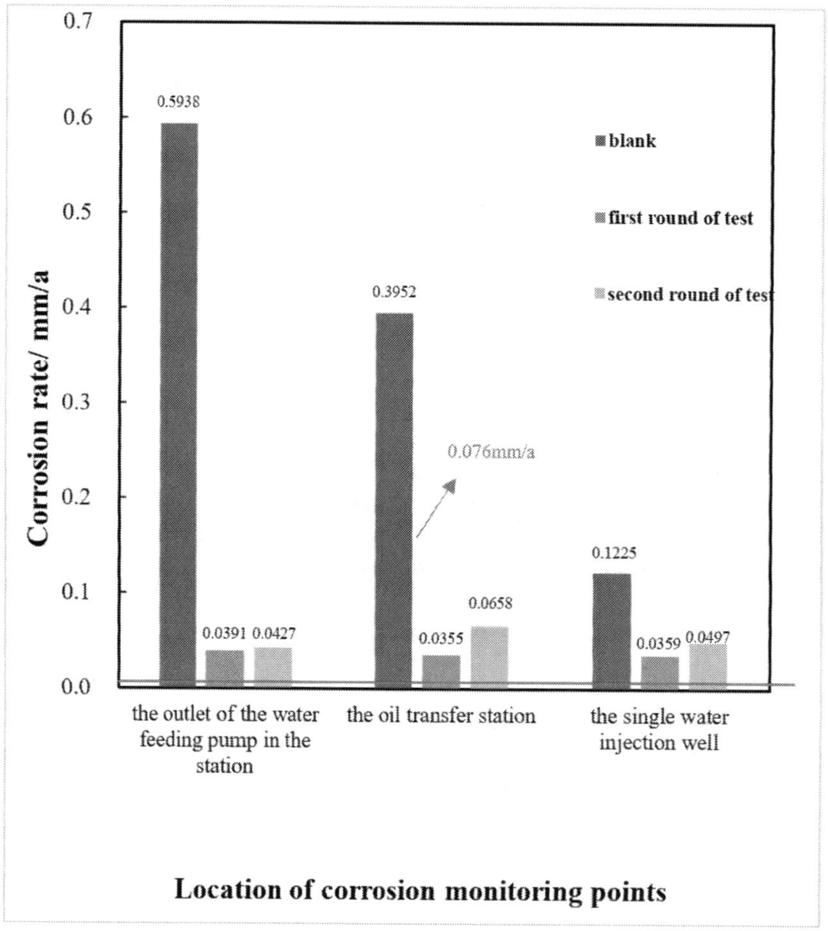


**Figure 3.** In-situ monitoring of the corrosion rate of hanging sheets.

As can be seen from Figure 3, after adding the corrosion and scale and salt inhibitor, the corrosion rate of the specimens at different monitoring points has dropped significantly, the average corrosion rate of the specimens at all monitoring points is <0.076 mm/a, and the concentration of the agent injection is the higher. The higher the spiking concentration is, the lower the corrosion rate of the on-site monitoring specimens is. It indicates that the developed agent has better on-site applicability.

*3.3.2 Micro-morphology of on-site monitoring specimen.* By observing the on-site corrosion monitoring coupons with a scanning electron microscope, the morphology of the deposits on the surface of the coupons can be directly observed. Figure 4 shows the macroscopic morphology of the corrosion monitoring coupon before the film was removed at the oil transfer station after 7 days of testing. To further determine the surface deposits on the coupon, EDS elemental analysis was further performed on the coupon. The results are shown in Table 6.

**Figure 4.** SEM morphology of the corrosion monitoring coupon surface in the oil transfer station (a-blank, b-first round of test, c-second round of test, and d-EDS position map).

**Table 6.** EDS data of corrosion test pieces (Wt %).

| Serial number | C | O | Na | Cl | Ca | Fe |
|---|---|---|---|---|---|---|
| 1# | / | 5.62 | 40.36 | 47.60 | / | 6.42 |
| 2# | 8.19 | 1.50 | 36.32 | 48.89 | 0.35 | 4.75 |

It can be seen from Figure 4 that after the blank test, a large number of regular cubes precipitated on the surface of the coupon and were deposited on the surface of the coupon. Energy spectrum analysis showed that the cubes were NaCl. When the filling concentration of on-site corrosion, scale, and salt inhibitor is 150 mg/L in the station + 50 mg/L in the oil transfer station, the number of NaCl crystals deposited on the surface of the coupon is greatly reduced. When the filling concentration is increased to 300 mg/L in the station + oil transfer station, the concentration reaches 100 mg/L, and there is almost no crystal deposition on the surface of the coupon. It shows that this corrosion, scale, and salt inhibitor not only has good laboratory evaluation results but also has an excellent salt suppression effect on oilfield on-site water injection pipelines. The higher the injection concentration is, the better the salt suppression effect is.

*3.3.3 XRD analysis of on-site monitoring coupons.* To further determine the composition of the deposits on the surface of the on-site corrosion monitoring coupons, samples were taken for XRD analysis, and the results are shown in Figure 5.

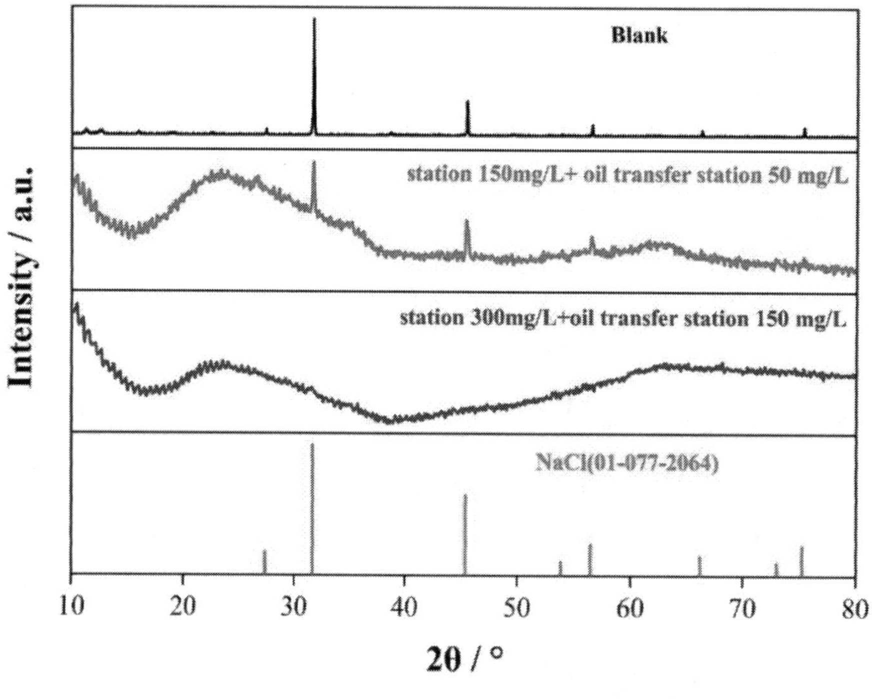


**Figure 5.** XRD pattern of corrosion monitoring coupon in the oil transfer station.

The results show that after the blank test, the deposits on the surface of the coupon are mainly NaCl, and the absorption peak of NaCl is very strong; when the on-site corrosion, scale, and salt inhibitor filling concentration is 150 mg/L in the station+50 mg/L in the oil transfer station, the coupon NaCl can still be detected on the surface. However, the absorption peak is relatively weak; when the filling concentration is increased to 300 mg/L in the station + 100 mg/L in the oil transfer station, the absorption peak of NaCl is not detected. On the one hand, the epoxy succinic acid copolymer and polyphosphoric acid substances in the corrosion, scale, and salt inhibitor can be adsorbed on Na * in the solution, inhibiting the mutual adsorption of Na * and Cl. Reducing the formation of salt crystals increases the solubility of NaCl; on the other hand, epoxy succinic acid copolymers and polyphosphoric acid substances can be adsorbed on the surface of NaCl crystals and doped in the crystal lattice, interfering with the regular and orderly growth of the crystals. It is the dispersion of the crystal nuclei or the occurrence of lattice distortion, which turns the regular, dense, and hard cubic structure into defective, soft, small-grain crystals [14], which are then washed away by the water in the pipeline.

## 4. Conclusion

(1) The main blockage inside the on-site water injection pipeline is NaCl, and salt deposition is the main cause of the blockage. After removing the salt in the blockage through washing, the remaining materials are mainly $CaCO_3$, $SiO_2$, and barium strontium scale.

(2) In response to the multiple failure accidents of the water injection pipeline outside an oilfield station in the west, a compound agent with corrosion and scale and salt inhibition properties was developed. The formula of the agent is 17.6% triazine derivatives + 26.1% polyphosphoric acid scale inhibitor +43.5% epoxy succinic acid copolymer salt inhibitor + 13% liquid alkali. When the addition amount is 150 mg/L, the corrosion rate of P110 steel is only 0.0307 mm/a, the scale inhibition rate for $CaCO_3$ can reach 83.5%, and the salt inhibition rate can reach 4.5%.

(3) The results of two rounds of field tests show that after adding corrosion and scale and salt inhibitors, the corrosion rate of the monitoring coupons is greatly reduced, and when the concentration of the agent is higher, the corrosion rate of the on-site monitoring coupons is smaller, and there are a large number of regular cubic NaCl on the surface of the coupon. When the filling concentration is 300 mg/L in the station + 100 mg/L in the transfer station, there is almost no NaCl precipitation on the surface of the coupon for on-site corrosion monitoring. The filling concentration is reduced to 150 mg/ in the station + 100 mg/L in the transfer station. When the oil station is 50 mg/L, only a small amount of NaCl precipitates on the surface of the coupon. It shows that the compound corrosion inhibitor, scale inhibitor, and salt inhibitor also have excellent salt inhibitory effects on oilfield on-site water injection pipelines.

**Acknowledgments**

The paper is funded by the National Key Research and Development Program of China (2021YFB3700800).

**References**

[1] AliPanah R., Aboutaleb S. P. M., and Mehdi F. Efficiency of ionic liquid/polymer flooding combined with smart water injection on oil recovery through secondary and tertiary patterns using Iranian carbonate rock [J]. Petroleum Science and Technology, 2023, 41 (10): 1, 081-1, 098.

[2] Runtao Yu. The Research of Geological Reserves Distribution on the Date of High Water in the SaErTu Oilfield of Daqing [D]. Northeast Petroleum University, 2006.

[3] Mubarak Ghadeer, Elkhodbia Mohamed, Gadala Ibrahim, AlFantazi Akram, and Barsoum Imad. Failure analysis, corrosion rate prediction, and integrity assessment of J55 downhole tubing in ultra-deep gas and condensate well [J]. Engineering Failure Analysis, 2023, 151: 107, 381.

[4] Akinsanya Akinyemi Olugbenga, Qin Jianjun, Guan Yue, Sørensen John Dalsgaard, and Faber Michael Havbro. Risk-informed integrity management of sub-surface well production tubings subject to combined scale and corrosion degradations [J]. International Journal of Pressure Vessels and Piping, 2021, 192: 104, 424.

[5] Junzheng Yang, Yunan Zhang, Honglan Zou, Dapeng Li, Qinghua Wang, Lei Zhang, Xiuyun Wang, and Minxu Lu. Study on Corrosion Risk of Injection Well in Oil Field [J]. Materials Protection, 2019, 52 (03): 51-54.

[6] Mohammad Salman, Kashif R. Ansari, Jiyaul Haque, Vandana Srivastava, Mumtaz A., Quraishi, and Mohammad A. J. Mazumder. Ultrasound-assisted synthesis of substituted triazines and their corrosion inhibition behavior on N80 steel/acid interface [J]. Journal of Heterocyclic Chemistry, 2020, 57 (5): 2, 157-2, 172.

[7] Wei Linjie, Lin Yuanhua, Li Chunyue, Wang Shaoyi, Gong Chengmeng, Jiang Yangang, and Li Zhonghui. Performance and mechanism study of PESA-IA as a green oilfield scale inhibitor: experimental and molecular dynamics simulation [J]. Journal of Polymer Research, 2022, 29 (12): 525.

[8] Zhou X., Sun Y., and Wang Y. Inhibition and dispersion of poxysuccinate as a scale inhibitor [J]. Journal of Environmental Sciences, 2011, 23: S159- S161.

[9] Jie Xiao, Xueqin Deng, and Lin Zhu. Study on synergistic effect of muticompound inhibitor system [J]. Applied Chemical Engineering, 2007 (05): 520-522+525.

[10] Chao Ma, Lun Li, Jie Xiao, and Xin Zhang. Compound Inhibiting Agent and Effect of Inhibiting Salinization in High Salt Formation Water [J]. Science Technology and Engineering, 2016, 16 (02): 143-147.

[11] Yong Liu. Research on a salt-inhibiting and anti-scaling agent for oil fields [D]. Southwest Petroleum University, 2015.

[12] Yanmin Du, Hongjie Chen, Wenqiang Zhao, and Xiangchun Yan. Development of CHD Series Corrosion & Salt Inhibitors for Oil Well and Evaluation [J]. Petrochemical Corrosion and Protection, 2014, 31 (04): 57.

[13] Yan Chen, Chao Ma, Wendong Ren, Xingyu Liu, Xin Liu, and Wen Gu. Development and performance of a sustained-release solid salt inhibitor [J]. CHINA SCIENCE PAPER, 2023, 18 (01): 50-56.

[14] Haipeng Jin. A Study on Salt Deposition Mechanism and Development Technology in High Salinity Oilfield [D]. China University of Geosciences (Beijing), 2016.

# Thermal runaway hazards of Lithium-Ion batteries in enclosed spaces

**Xiaoqiang Liang[1*], Na Liu[1], Kun Zhang[1], Jun Tian[1], Wenhao Sun[1], Cuijun Tian[1], Daozhong Hu[1], Lei Tong[1] and Hongbo Gao[1]**

[1] China North Vehicle Research Institute, Beijing, 100072, China

*Corresponding author's e-mail: liangxiaoqiang15@mails.ucas.ac.cn

**Abstract:** Lithium-ion batteries, celebrated for their high voltage, durability, and cleanliness, are the cornerstone of today's new energy field. Their thermal stability in confined, explosion-proof settings is crucial for high-risk industries like military and mining, directly impacting worker safety and equipment reliability. Insight into thermal runaway triggers and progression is vital for crafting preventive strategies and enhancing battery system safety. This research, involving thermal runaway tests on three lithium-ion batteries in an explosion-proof enclosure, shows that while the enclosure doesn't halt runaway, it can lower temperatures, especially peak temperatures, easing the runaway's impact. Moreover, the batteries' differing stability levels lead to varied outcomes in these tests, underscoring the importance of battery design in critical applications.

## 1. Introduction

Lithium-ion batteries are now widely used across numerous energy storage scenarios, owing to their benefits like greater energy density, advanced safety mechanisms, expanded service duration, and the capability for rapid charging at high rates. The application fields involve various aspects of life, such as biology, medicine, military, entertainment, construction, and transportation [1-2]. In recent years, pure electric vehicles in China have developed rapidly with strong policy support. In the quest for maximizing the lithium-ion batteries' energy capacity for electric vehicles to achieve greater travel distances and easier operation, concerns about their safety risks have grown more pronounced [3]. Fire and explosion incidents of power batteries are occurring frequently both domestically and internationally. Investigations into the thermal safety of power batteries have emerged as a prominent, pivotal, and challenging topic within scientific inquiry. Exploring high-safety lithium-ion power battery systems is critically important for enhancing the overall safety of electric vehicles and safeguarding the lives of passengers [4-5].

Despite the country's implementation of rigorous mandatory testing standards [6], there is a notable lack of comprehensive understanding among many researchers regarding the mechanisms behind battery thermal runaway. Three primary causes lead to thermal runaway: mechanical abuse, electricity abuse, and heat abuse. The teams led by Choi [7], Lai [8], and Shim [9] have conducted research on the mechanical characteristics of batteries under low-speed stretching conditions and have also examined the impact of temperature variations on their mechanical responses.

Zhang and colleagues [10] studied the plastic deformation of the casing and the fracture characteristics of lithium-ion cylindrical battery cells. Established plastic and breakage models can be validated through experimental results under various load conditions. The research groups headed by

Leising [11] and Saito [12] have investigated that heat generation is predominantly directly proportional to the current, suggesting that Electroheating is a primary source of heat during the overcharging process, revealing the mechanism of side reactions caused by overcharging. The underlying reason for all three issues is that once the battery's internal temperature surpasses a critical threshold, it results in a hastened temperature rise and the accumulation of significant heat [13]. The battery will release uncontrollable heat, leading to a temperature threshold for combustible gases and even explosions. Faced with the safety requirements of lithium-ion batteries for specific vehicle platforms and energy storage grids, addressing the issue of thermal runaway of batteries, strengthening the reliability of battery systems, and establishing testing and evaluation technologies for thermal runaway of power and energy storage batteries have important foundational support for the development of new generation power energy technologies [14-19].

A multitude of investigations have been conducted regarding thermal runaway incidents in unconfined areas. Still, scant attention has been given to the thermal runaway characteristics of lithium-ion batteries deployed in specialized machinery within confined spaces. This research concentrates on examining how four-centimeter-thick steel plates in confined spaces alter the thermal runaway dynamics of lithium-ion batteries. It also investigates the effects of thicker steel plate barriers on the heat-induced failures of three sorts of lithium-ion batteries for sale that share identical capacities and dimensions. We explore the potential for thick steel plates to suppress thermal runaway in batteries. Moreover, we determine if the rise in pressure or temperature within confined spaces contributes to the exacerbation of thermal runaway, whether the temperature changes in all three types of lithium-ion batteries are consistent. By research on thermal runaway phenomena, the design of battery structures and thermal management systems can be optimized, improving battery performance and reliability, reducing the risk of thermal runaway, promoting the healthy development of these fields, and improving the safety level and competitiveness of the entire industry.

## 2. Experimental section

Thermal runaway test is mainly divided into sample preparation, experimental equipment preparation, battery excitation, monitoring and recording, and data analysis. Details are as follows:

(1) Sample: first, we need to select three kinds of lithium iron phosphate batteries (3.2 V/250 Ah) with the same theoretical capacity and mature industrialization to ensure the same size, shape, and consistent initial state. These batteries are selected to meet the actual needs of the pre-development project. In addition, a battery with this size and capacity is a common battery monomer in the mine. The purpose is to explore the impact of this flameproof shell on the three batteries and compare their stability differences. Before the experiment, the surface of the battery was treated and cleaned, the pet blue film with poor thermal conductivity was torn off, and the basic data, including battery quality and voltage, were recorded.

(2) Experimental equipment: It includes lithium battery thermal runaway test device, temperature sensors, gas analyzers, etc. We ensure that all equipment adheres to safety standards and is equipped with appropriate protective measures.

(3) Battery excitation: We place the battery inside an explosion-proof enclosure and employ a 1000 A current for overcharging to trigger the battery's thermal runaway reaction. Temperature sensors are placed at the cathode, the center, and the anode, recording the temperature data as T1, T2, and T3, respectively. We install pressure sensors within the sealed stainless steel enclosure and continuously monitor the temperature fluctuations. The entire process takes place within a thermal runaway safety test chamber.

(4) Monitoring and recording: We employ temperature sensors and gas analyzers to record changes in battery temperature, gas emissions, and other pertinent data in real time throughout the experiment. These data serve as critical parameters for assessing batteries' thermal runaway characteristics and safety.

(5) Data analysis: After the experiment, the collected data were organized and analyzed to understand the characteristics of battery thermal runaway, the triggering conditions, and possible preventive measures.

Caution: The thermal runaway test poses a risk. It is imperative to strictly adhere to safety protocols to ensure the safety of personnel and equipment stability. We prepare safety plans and ensure that fire extinguishers and first-aid kits are readily available. Moreover, the specific procedures for lithium battery thermal runaway testing may differ based on the equipment's conditions and the type of battery.

## 3. Result and analysis

The theoretical capacity of the three types of lithium-ion batteries is 250 mAh/g, with the same size and initial discharge state. We perform surface treatment, remove PET coating with poor thermal conductivity, and record basic data, including battery quality and voltage. Figure 1 shows the charge and discharge curves of three different lithium iron phosphate batteries (LFP-1, LFP-2, LFP-3) at 25 °C. According to the experimental data, the discharge capacities for LFP-1, LFP-2, and LFP-3 are 244.3 mAh/g, 243.2 mAh/g, and 243.6 mAh/g, respectively. Correspondingly, their charge capacities are 243.6 mAh/g, 242.4 mAh/g, and 242.9 mAh/g, respectively. Three charge-discharge curves are basically consistent with the actual charge-discharge capacity. It can be seen that the LFP battery has high stability and reliability. Although it comes from different manufacturers and uses different processes and materials, it ensures similar performance. The consistency of charge and discharge lays an essential foundation for thermal runaway design and performance optimization.

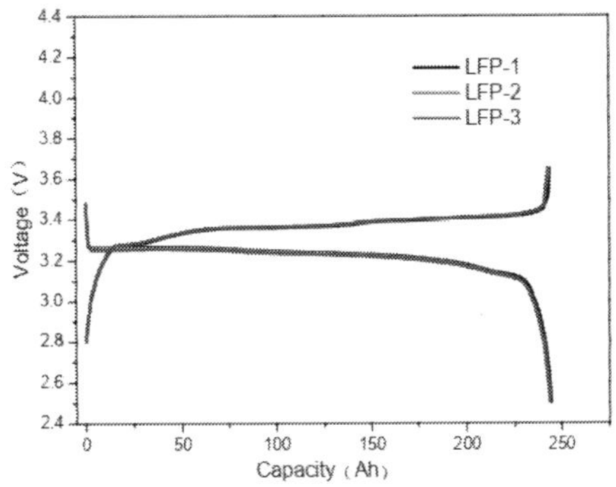


**Figure 1.** At room temperature, the charge and discharge curves of LFP-1, LFP-2, and LFP-3 batteries.

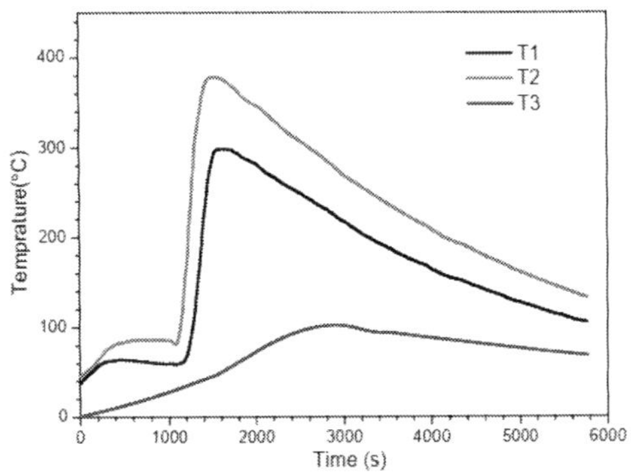


**Figure 2.** The temperature variation curves of the positive pole (T1), central position (T2), and negative pole (T3) of LFP-1 in the thermal runaway test.

Journal of Physics: Conference Series **2961** (2025) 012041 doi:10.1088/1742-6596/2961/1/012041

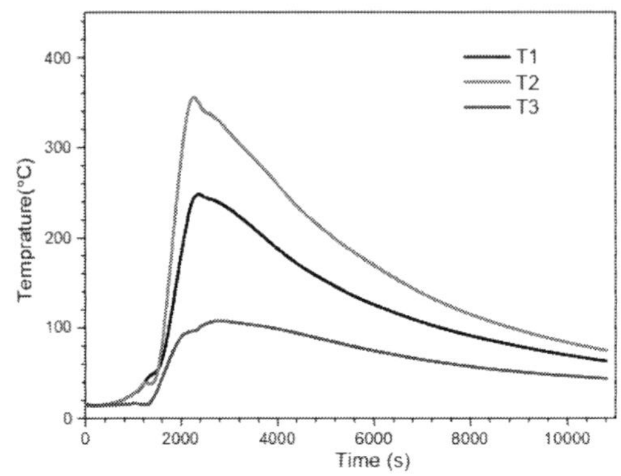


**Figure 3.** The temperature variation curves of the positive pole (T1), central position (T2), and negative pole (T3) of LFP-2 in the thermal runaway test.

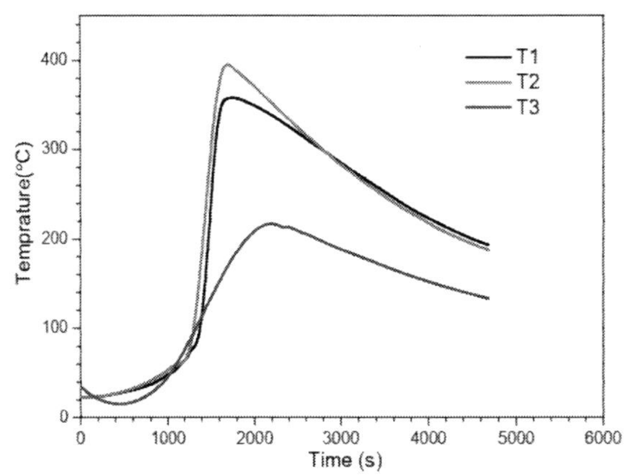


**Figure 4.** The temperature variation curves of the positive pole (T1), central position (T2), and negative pole (T3) of LFP-3 in the thermal runaway test.

Figures 2, 3, and 4 show the temperature change curves of three batteries during the thermal runaway test. In that order, T1, T2, and T3 denote the temperature alterations for positive, intermediate, and negative electrodes. They have the following common points: (1) The progression of thermal runaway adheres to a fundamental pattern, encompassing phases such as the initiation phase, decomposition of the SEI (Solid Electrolyte Interphase) film and electrolyte reactions, breakdown of the positive electrode material, intense electrochemical reactions between negative electrode and electrolyte, and the thermal runaway explosion, followed by a decrease in temperature; (2) The temperature change at the center is the most obvious, indicating that the thermal runaway occurs first at the center of the battery, which is due to the high lithium ion concentration, high shuttle efficiency of positive and negative electrodes, sufficient electrolyte, and the most obvious temperature rise; (3) The temperature at the positive pole is the second, and the temperature change at the negative pole is the smallest. (4) With the increase of time, enough heat accumulates inside the battery, and the temperature increases sharply in a short time, reaching the highest point in an instant. (5) The temperature at T3 is significantly lower than that at T2 and T1.

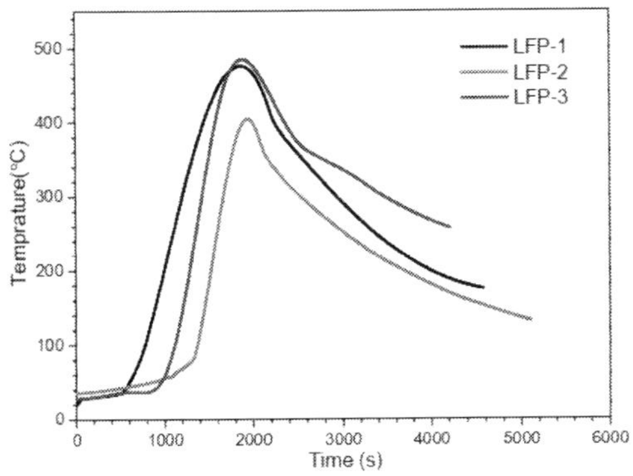


**Figure 5.** Temperature change curve LFP-1, LFP-2, and LFP-3 batteries center in the flameproof enclosure when thermal runaway occurs.

As shown in Figure 5, there are also differences during the thermal runaway incidents in the trio of batteries: (1) The time of thermal runaway of the positive, negative, and central positions of the three batteries are different. The occurrence time is LFP-2>LFP-3>LFP-1, and the time is 2268s, 1701s, and 1529s, respectively. (2) The highest temperature trigger points of T2 are different. According to the occurrence temperature, the order is LFP-3>LFP-1>LFP-2, which are 395.08 °C, 378.41 °C, and 355.57 °C, respectively. The rule of T1 and T3 is the same as above. In addition, the peak temperatures of T1 and T3 of the LFP-2 battery are significantly lower than those of T1 and T3 of LFP-1 and LFP-3, indicating that the LFP-2 battery is the most stable.

Figure 6 shows the experimental curve at LFP-1, LFP-2, and LFP-3's center without flameproof enclosure. The figure indicates that (1) At the center, the occurrence time of thermal runaway was ranked as LFP-2>LFP-3>LFP-1, and the time was 1946s, 1887s, and 1867s, respectively. (2) The highest temperature trigger point was LFP-3>LFP-1>LFP-2, which were 485.01 °C, 476.17 °C, and 404.15 °C, respectively. It showed that the moment of incidence was consistent with the law of temperature change. After adding the flameproof shell, the specific values are different. The occurrence time of LFP-2 is prolonged, and the time increment is 322s. The occurrence time of LFP-3 and LFP-1 is shortened, and the shortening amounts are 186s and 338, respectively. The peak temperature during thermal runaway showed a substantial decline, and the temperature drops of LFP-3, LFP-1, and LFP-2 were 89.93 °C, 97.76, and 48.48 °C, respectively.

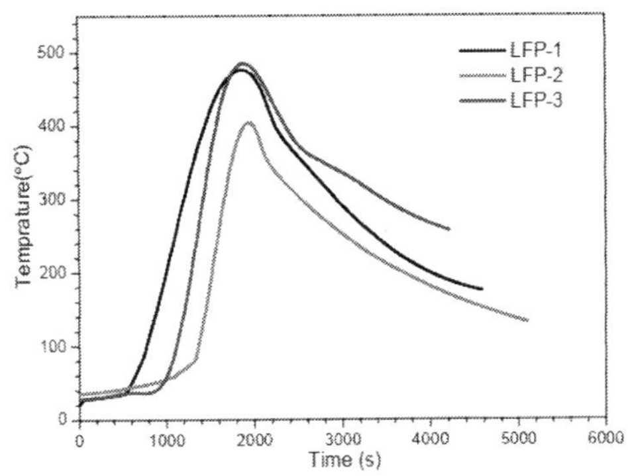


**Figure 6.** The temperature change curve at the center of LFP-1, LFP-2, and LFP-3 batteries without flameproof enclosure when thermal runaway occurs.

## 4. Conclusion

In a word, the lithium battery was placed in the flameproof enclosure for this test. It was found that the flameproof enclosure could not prevent the occurrence of thermal runaway but could reduce the occurrence temperature and its severity. The occurrence time of the battery in the flameproof enclosure is advanced, which may be due to the narrow space in the enclosure, the increase in air pressure and temperature, and heat accumulation. LFP-2 has the latest thermal runaway and the most stable performance, followed by LFP-3 and LFP-1. For LFP-2, the occurrence time of thermal runaway increased slightly. Still, the maximum temperature also decreased significantly, indicating that the casing effectively alleviates the intensity of thermal runaway events, and the lengthened duration could be attributed to its inherent stability.

**Acknowledgments**

This work was partly supported by the National Natural Science Foundation, China, under Grant 22209156 and 52075028, respectively.

## References

[1] Cheng XB, Zhang R, Zhao CZ, Zhang Q. Chem Rev. 2017 Aug 9;117(15):10403-10473.
[2] Shichao Zhang S. C., Shen ZY, Lu YY, Acta Physico-Chimica Sinica, 2021(01):61-78.
[3] Abada, S., Marlair, G., Lecocq, A., Petit, M., Sauvant-Moynot, V., & Huet, F. (2016). Journal of Power Sources, 306, 178-192.
[4] Chen Y, Kang Y, Zhao Y, et al., Journal of Energy Chemistry, 2021, 59(8):83-99.
[5] Wang, Hsin, Lara-Curzio, et al., [2024-05-13].
[6] H.-J. Noh, S. Youn, C. S. Yoon, Y. -K. Sun, J. Power Sources 233 (2013) 121-130.
[7] H. Y. Choi, I. Lee, J. S. Lee, Y. M. Kim, H. Kim, 23rd International Technical Conference on the Enhanced Safety of Vehicles (ESV). 2013: 13-0115.
[8] W. -J. Lai, M. Y. Ali, J. Pan, J. Power Sources 248 (2014) 789-808.
[9] K. H. Shim, S. K. Lee, B. S. Kang, S. M. Hwang, J. Mater. Process. Tech. 155 (2004) 1935-1942.
[10] X. Zhang, T. Wierzbicki, J. Power Sources, 280 (2015) 47-56.
[11] R. A. Leising, M. J. Palazzo, E. S. Takeuchi, K. J. Takeuchi, J. Electrochem. Soc. 148(8) (2001) A838-A844.
[12] Y. Saito, K. Takano, A. Negishi, J. Power Sources, 97 (2001) 693-696.
[13] Lin L S, Wang H, Li J L, et al., Journal of Energy Storage, 2023.
[14] Meng H , Yang Q , Zio E ,et al.An integrated methodology for dynamic risk prediction of thermal runaway in lithium-ion batteries[J].Process Safety and Environmental Protection, 2023, 171:385-395.
[15] Zheng-Hai L , Guo-Qiang Z .Progress on early warning technology for thermal runaway of lithium-ion battery[J].Advanced Technology of Electrical Engineering and Energy, 2019.
[16] Wang TH, Zhai HJ, Qin P, et al. Study on thermal runaway and propagation behavior of LiFePO4 battery in module box space [j] Fire science, 2022,31 (1): 25-34 DOI:10.3969/j.issn.1004-5309.2022.01.04.
[17] Lin L S , Wang H , Li J L ,et al.Mechanically induced thermal runaway severity analysis for Li-ion batteries[J].Journal of Energy Storage, 2023.
[18] Sun Y , Jin Y , Jiang Z ,et al.A review of mitigation strategies for li-ion battery thermal runaway[J].Engineering failure analysis, 2023.
[19] Yang M , Rong M , Pan J ,et al.Thermal runaway behavior analysis during overheating for commercial LiFePO4 batteries under various state of charges[J].Applied thermal engineering: Design, processes, equipment, economics, 2023(Pt.B):230.

Journal of Physics: Conference Series **2961** (2025) 012042 doi:10.1088/1742-6596/2961/1/012042

# High-performance coatings for corrosion protection of power equipment and their performance analysis

Sihao Huang[1], Lilong Pang[2,3,4,5*], Zhiguang Wang[2,3,5]

[1]Guangzhou City University of Technology, Guangzhou, China
[2]Institute of Modern Physics, Chinese Academy of Sciences, Lanzhou, China
[3]School of Nuclear Science and Technology, University of Chinese Academy of Sciences, Beijing, China
[4]Huizhou Research Center of Ion Sciences, Huizhou, China
[5]Advanced Energy Science And Technology Guangdong Laboratory, Huizhou, China

*Corresponding author's e-mail: panglilong@impcas.ac.cn

**Abstract**—Power equipment of new energy and novel power systems, exposed to harsh corrosive environments such as high temperatures, saline, high flow rates, and oxidation, pose a significant risk to their safe, efficient, and normal operation. In this study, an $Al_2O_3$/Fe-Al coating was prepared on the surface of stainless steel, and its high-performance anti-corrosion effect was verified by extreme corrosion tests. Additionally, the study analyzed the micro-structural changes and elemental migration of the coating in the corrosion process. The coating's excellent resistance to extremely high temperatures and oxidative environments indicates its potential for application to the outdoor power equipment of various new energy and novel power systems.

## 1. Introduction

The development of new energy and the construction of novel power systems are conducive to reducing global carbon emissions and protecting the global environment. Therefore, the development of new energy, including solar, wind, and nuclear power generation, is becoming a global trend. The construction and research of key applications in novel power systems, such as prefabricated cabin substations and molten salt energy storage systems, are also advancing simultaneously. However, in the process of technological development, engineering application, and commercial promotion, the material problem has been identified as the key bottleneck restricting the development of new energy power generation and novel power systems. The structural materials of the equipment for these applications in harsh environments, such as high temperature, high flow rate, saline, high pressure, etc., are challenged by corrosion attacks, including thermal penetration, oxidation, scouring, stress, and so on[1-3].

The corrosion of critical components of power equipment can result in significant alterations to their characteristics. This can lead to the degradation of various aspects of the material's properties, including mechanical, electromagnetic, and thermal properties. Additionally, corrosion can break the integrity of the equipment components, leading to a notable decline in the equipment's performance and an increased risk of safety issues. Therefore, research on high-performance corrosion-resistant technology needs to be conducted to guarantee the safe, efficient and normal operation of the new energy equipment and the novel power system throughout its entire life cycle.

At present, among the materials corrosion-resistant methods, coating technology can be achieved on the basis of corrosion resistance to ensure that the excellent inherent properties of the substrate materials do not require significant changes in the operating environment caused by the loss of system performance. It is known that coating technology is one of the most effective and commercially available methods of corrosion resistance [4-6].

The corrosion resistance of a range of materials, including ceramics, Fe-Al-based alloys, refractory metals, and high-entropy alloys, has been extensively investigated in recent years[7-10]. Among these materials, Fe-Al-based alloys show good compatibility with structural steel substrates[11]. $Al_2O_3$ ceramic materials currently demonstrate promising results in the high-temperature condition tests[12]. The $Al_2O_3$/Fe-Al coating is an effective utilization of the beneficial properties of $Al_2O_3$ and Fe-Al alloys[13-15]. However, further investigation is required to fully understand the corrosion resistance behavior and mechanism of the $Al_2O_3$/Fe-Al coating. This should include an in-depth analysis of the coating's composition, structure, element diffusion, and other factors that may influence its corrosion resistance and failure phenomenon.

This study applies $Al_2O_3$/Fe-Al coatings to anti-corrosion processes in new energy and new power systems. The samples of $Al_2O_3$/Fe-Al coating were tested in heavy liquid metals, with the aim of investigating the resistance to extreme temperatures and oxidative environments. Also, the rules of element diffusion of the coating under varying harsh conditions were primarily revealed.

## 2. Experimental procedures

### *2.1 Preparation of $Al_2O_3$/Fe-Al gradient coating*

In this study, $Al_2O_3$/Fe-Al gradient coating is prepared on the surface of SIMP steel (steel designed by IMP-CAS and IMR-CAS). SIMP steel is a ferritic/ martensitic steel with enhanced corrosion resistance. The dimensions of the steel sample are $35 \times 15 \times 1.5$ mm$^3$.

The coating preparation process described in this paper employs a combination of ionic liquid Al-electroplating, heat treatment, and selective oxidation. The ionic liquid plating process utilizes an $AlCl_3$-EMIC ionic liquid with a molar ratio of 2:1. The activation current density of the steel surface to be plated is 20 mA/cm$^2$, the current density of aluminum deposition is 10 mA/cm$^2$, and the deposition time is 10 minutes. Subsequently, the Al-plated samples are subjected to a heat treatment at 700 °C for 2 hours and followed by selective oxidation at 720 °C for 80 hours, where the oxygen partial pressure is less than 100 Pa.

Our previous report[16] confirmed that this process can achieve uniform coating preparation on the surface of special-shaped parts, such as the inner surface of a tube. In addition, the prepared coating has pure composition, good bonding performance, and controllable thickness and structure.

### *2.2 Corrosion test*

The corrosion test in this study was conducted in a lead-bismuth eutectic (LBE) corrosion tank. The coating samples and the bald samples were fixed to a steel bracket and subsequently immersed in molten saturated oxygen LBE simultaneously. With the K-type thermocouples immersed into the tank and the heating wires situated outside the tank, the temperature of the LBE remained stable throughout the course of the corrosion test, with fluctuations in temperature remaining below ±5 °C.

**Table 1.** Corrosion parameters of the numbered samples

| Samples' numbers and their test parameters | 400 °C | 500 °C | 600 °C |
|---|---|---|---|
| 300 h | 4-3 | 5-3 | - |
| 600 h | 4-6 | 5-6 | 6-6 |
| 1000 h | 4-10 | 5-10 | 6-10 |
| 2000 h | 4-20 | 5-20 | 6-20 |
| 3000 h | - | - | 6-30 |

The temperature range of the corrosion tests conducted on the samples in this study is 400-600 °C, with a total testing duration of 3000 hours. The test parameters of the samples, together with the corresponding numbers, are listed in Table 1.

### *2.3 Characterization methods*

This paper uses a scanning electron microscope (SEM, Thermo Fisher Scientific Apreo S, ZEISS Crossbeam 350, and JEOL JSM-7800F) to observe the microstructure of the cross-section of the samples after the corrosion test. The changes in element distribution of the samples undergoing the corrosion test were characterized by energy dispersive X-ray spectroscopy (EDX, EDAX Octane Super-A, and Aztec Live Lite Ultim Max 40) assembled on SEM. The microstructure and phase composition of the sample surface were also characterized using grazing incidence X-ray diffraction (GIXRD, PANalytic B.V. X'Pert [3] Powder, Cu $K_{\alpha 1}$, 1°).

## 3. Result and discussion

### *3.1 Remarkable corrosion resistance of the coating*

The typical SEM cross-sectional image and EDX mapping images of the bald SIMP steel samples after LBE corrosion are shown in Figure. 1. The results show that there is an obvious enrichment stratification of Fe and O elements on the surface of the sample, which indicates that a significant oxidative corrosion layer has been generated. The corrosion of stainless steel in LBE is a common occurrence, which is harmful to the material's thermal conductivity and mechanical properties.

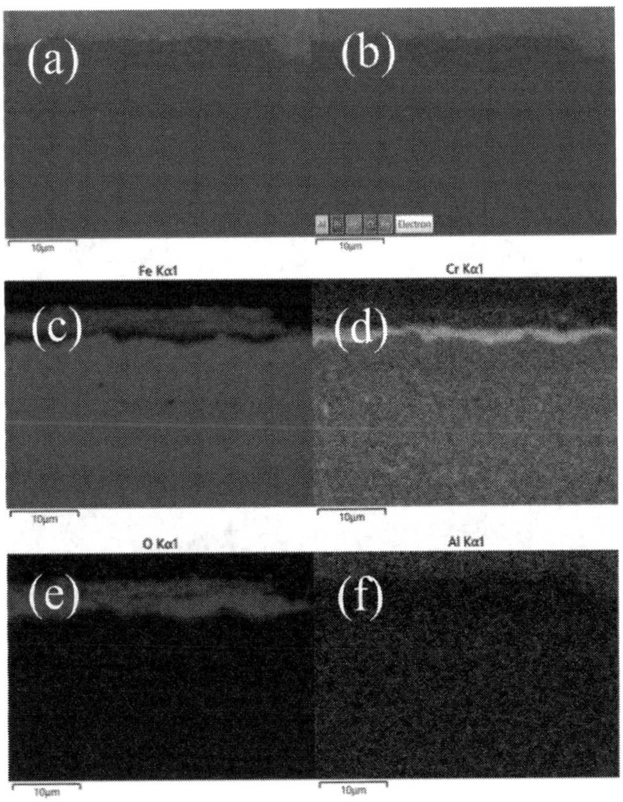

Journal of Physics: Conference Series **2961** (2025) 012042 doi:10.1088/1742-6596/2961/1/012042

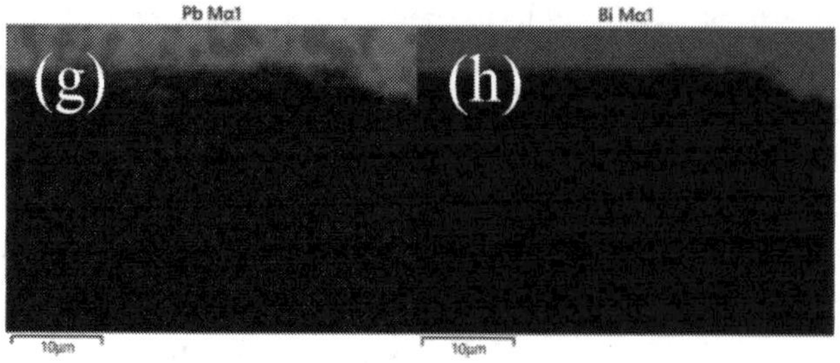


**Figure 1.** (a) Cross section and its (b) EDX mapping images of SIMP steel corroded in saturated-oxygen LBE at 400 °C for 1000 h; (c)-(h) Individual EDX mapping images of elements Fe, Cr, O, Al, Pb, Bi elements, respectively.

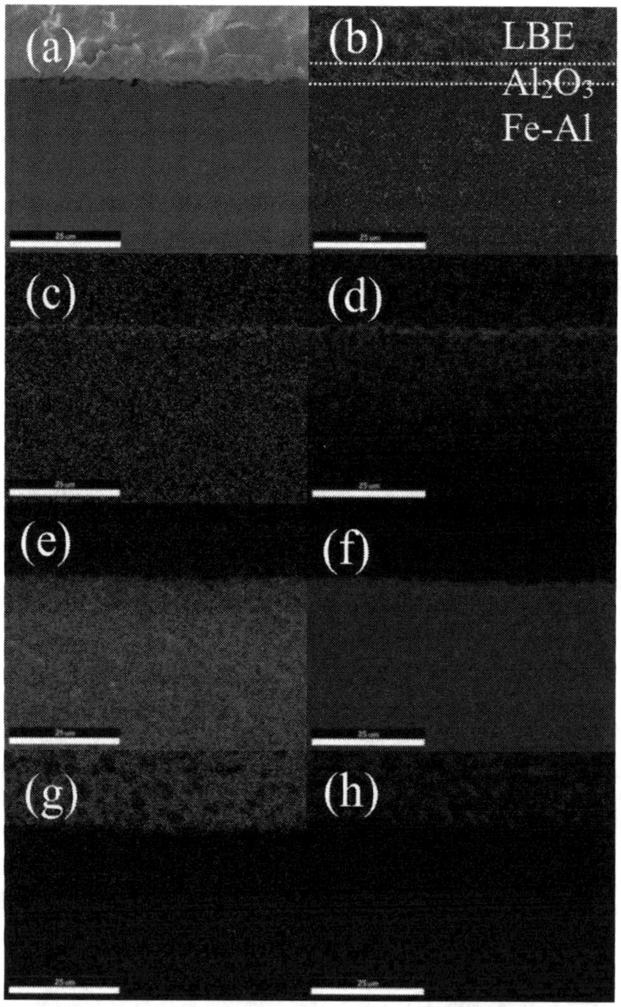


**Figure 2.** (a) Cross-section of sample 5-10 and its (b) EDX mapping image; (c)-(h) Individual EDX mapping images of the O, Al, Cr, Fe, Pb, Bi elements, respectively.

However, following the application of the coating described in this study, the sample exhibited distinguished corrosion resistance, as evidenced by the absence of observable corrosion on the surface.

The $Al_2O_3$/Fe-Al gradient coating applied to the SIMP steel substrate preserves the structural integrity of the surface during the LBE test.

Figure. 2 shows the typical SEM cross-sectional image and EDX mapping images of the coated samples. It can be observed that there is a dense, thin layer of $Al_2O_3$ on the surface of the sample, and LBE is blocked above it. No obvious enrichment of O, Pb, and Bi is observed in the material below the thin layer. These indicate that the coating hinders the penetration and oxidation of LBE into the substrate, which initially proves the corrosion resistance of the coating.

Figure. 3 shows the cross-sectional SEM morphology image of the sample that underwent corrosion testing at 500 °C for up to 2000 hours. It can be observed from Figure. 3(a) that long-term high-temperature LBE corrosion causes an obvious oxidation corrosion layer to form on the surface of the bald SIMP steel. In contrast, under the same corrosion parameters, the surface of sample 5-20 still maintains uniformity and integrity, as shown in Figure. 3(b). These further prove that the coating significantly improves the corrosion resistance of the material, allowing the substrate steel to remain intact in high-temperature LBE corrosion conditions for a long time.

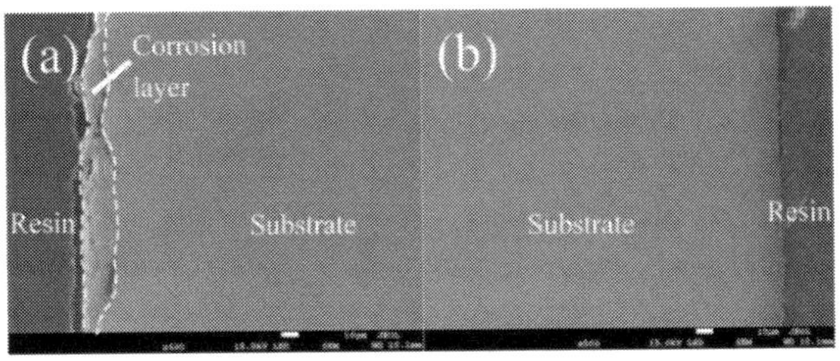


**Figure 3.** Typical SEM cross-sectional morphology of (a) bald and (b) $Al_2O_3$/Fe-Al coated samples corroded corroded in 500 °C saturated-oxygen LBE for 2000 h.

*3.2 Effect of corrosion duration*

The SEM image with EDX line scanning analysis allows for the observation of changes in element distribution within the Fe-Al gradient region after corrosion tests. Figure. 4 shows the SEM images of the cross-sections of samples 4-3 and 4-10, accompanied by their EDX line scanning results. It is evident that the element signal peaks of Al and O are present in the $Al_2O_3$ region near the surface of the sample. Subsequently, in the Fe-Al gradient region, the signal of the O element has almost disappeared, while the signal of the Al element continues to decline at a gentle slope. In contrast, the signals of the Fe and Cr elements are seen to rise rapidly.

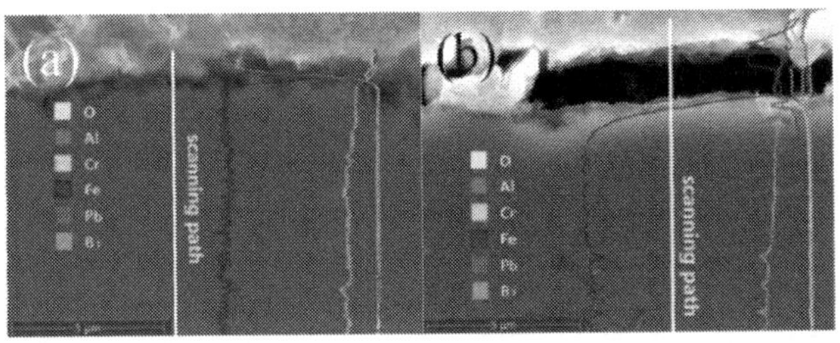


**Figure 4.** Cross-sectional SEM and EDX line scan analysis of the sample (a) 5- and (b) 5-10.

The essential points illustrated in Figure. 4 are that the length of the area where the element signal of Al gradually diminishes from a sustained decline to a level approaching zero is only approximately 8 μm in sample 5-3, but upon reaching 1000 hours of corrosion duration, the element content of Al exhibits a persistent downward trend across the entire scanning path. The area of change in the element distribution is markedly extended, indicating an increase in the thickness of the gradient region. It can, therefore, be inferred that the long-term thermal effect of the corrosion process causes the Al element to continue to diffuse, resulting in a gradual increase in the thickness of the Fe-Al gradient region.

Figure. 5 illustrates the GIXRD pattern of samples 5-3, 5-6, and 5-10. In comparison to sample 5-3, an increase in corrosion time results in the detection of a weaker FeAl peak in sample 5-6 at approximately 44-45°. In the case of sample 5-10, it can be observed that the strongest Fe-Al peak is essentially confined to the position of $Fe_3Al$. These above phenomena indicate that the FeAl phase of the $Al_2O_3$/Fe-Al gradient coating diminishes with the progression of corrosion duration, whereas the $Fe_3Al$ phase is progressively generated. It can be therefore proven that, as a consequence of the thermal effect of the corrosion process, the Al element in the Fe-Al gradient structure continues to diffuse into the substrate, which results in a reduction of the FeAl phase and an increase of the $Fe_3Al$ phase.

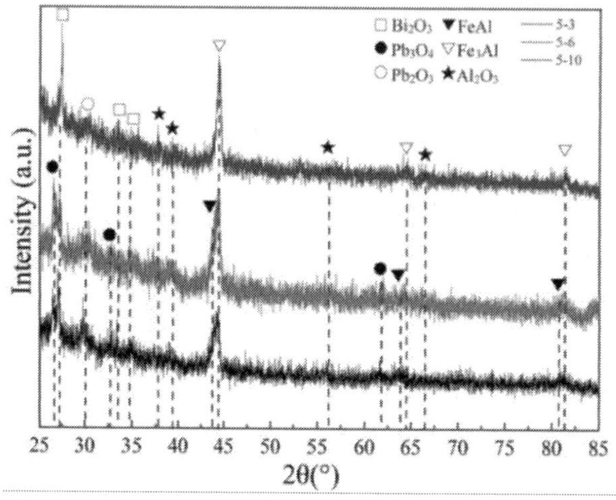


**Figure 5.** GIXRD patterns of $Al_2O_3$/Fe-Al gradient coatings corroded in saturated-oxygen LBE at 500 °C for 300-1000 h.

*3.3 Effect of corrosion temperature*

Figure. 6 shows the cross-sectional SEM morphology and EDX line scan images of samples 4-6 and 5-6. The thin white lines in the images represent the path of the EDX line scan. Table 2 presents the results of the EDX point analysis conducted at the position designated white crosses along the specified path.

From Figure. 6, it can be seen that the distribution of element content along the depth of samples 4-6 and 5-6 is relatively similar. This indicates that the diffusion rules of the elements in the samples at different temperatures exhibit certain similarities. Furthermore, the Al content at positions II, III, and IV (Fe-Al region) of sample 5-6 is consistently less than that of sample 4-6. It indicates that an increase in corrosion temperature results in a faster diffusion rate of Al elements in the coating to the substrate, causing a corresponding decrease in Al content at the Fe-Al region position, which is consistent with the principles of common sense.

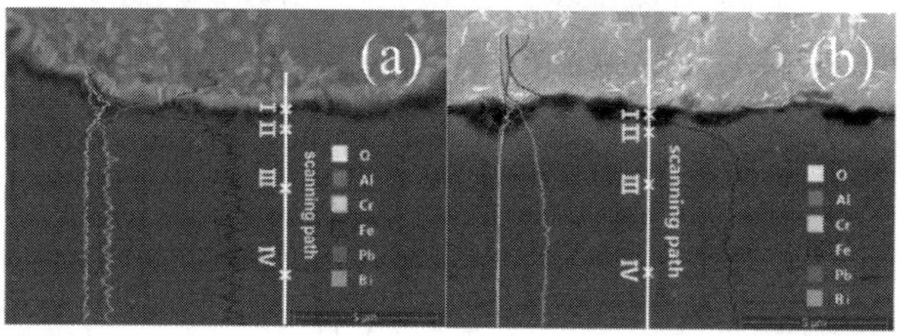


**Figure 6.** SEM morphology and EDX line scan images of the cross sections of sample (a) 4-6 and (b) 5-6.

**Table 2.** Results of EDX point analysis along depth for samples 4-6 and 5-6.

| Element | 4-6 | | | | 5-6 | | | |
|---|---|---|---|---|---|---|---|---|
| | Piont I | Piont II | Piont III | Piont IV | Piont I | Piont II | Piont III | Piont IV |
| O% | 33.1 | 1.8 | 1.4 | 1.8 | 36.1 | 4.6 | 3.1 | 3.1 |
| Al% | 24.0 | 8.4 | 7.8 | 5.7 | 29.3 | 8.0 | 4.3 | 4.8 |
| Cr% | 4.2 | 8.2 | 8.9 | 10.2 | 3.9 | 9.5 | 12.4 | 14.6 |
| Fe% | 34.2 | 81.6 | 81.9 | 82.3 | 25.1 | 77.6 | 80.0 | 77.2 |
| Pb% | 2.5 | 0.1 | 0.0 | 0.0 | 2.7 | 0.1 | 0.1 | 0.2 |
| Bi% | 2.0 | 0.0 | 0.0 | 0.0 | 2.9 | 0.2 | 0.1 | 0.1 |

However, Table 2 also reveals that the Al content at positions II, III, and IV (Fe-Al region) of sample 5-6 initially decreases and then increases slightly. This suggests that the diffusion of elements within the coating may also be influenced by an additional mechanism.

In our previous study, it was demonstrated that in the LBE corrosion test at a higher temperature of 600 °C, the Al element diffused out of the coating[16]. High-temperature LBE corrosion effects result in the consumption of the $Al_2O_3$ layer of the $Al_2O_3$/Fe-Al gradient coating, but the self-repair effect of the coating makes up for this loss. During this process, the Al element continues to diffuse outward. It can thus be inferred that the diffusion behavior of the Al element at 500 °C is likely to be affected by the combined effect of thermal diffusion and self-repair.

Nevertheless, it may ultimately result in the lack of Al element within the Fe-Al region when the Al element in the Fe-Al gradient structure persists in diffusing over an extended period. It is hard to see the obvious Al signal on the Fe-Al region of sample 6-10 from the EDX line scan, as shown in Figure. 7. It seems that the Al element is almost being consumed by diffusion both outward and inward.

Figure. 8 shows the cross-sectional SEM and the corresponding EDX mapping images of sample 6-30. It can be observed that the distribution of elements in the Fe-Al region is no longer homogeneous. Local Al-poor areas appear near the surface of the coating, accompanied by pores marked by red arrows. These observations show that following prolonged corrosion of the coating, the Fe-Al gradient region exhibits localized structural deterioration. The formation of the pores can be attributed to the following mechanism. In the Fe-Al region of the Fe-Cr-Al ternary alloy, the outward diffusion rate of the Al and Cr elements is faster than that of the Fe element. Consequently, in accordance with the reverse process of the Kirkendall effect, the shallow region of the Fe-Al region becomes somewhere enriched in Al and Cr, while the Fe element is missing. This results in the formation of local reverse Kirkendall pores.

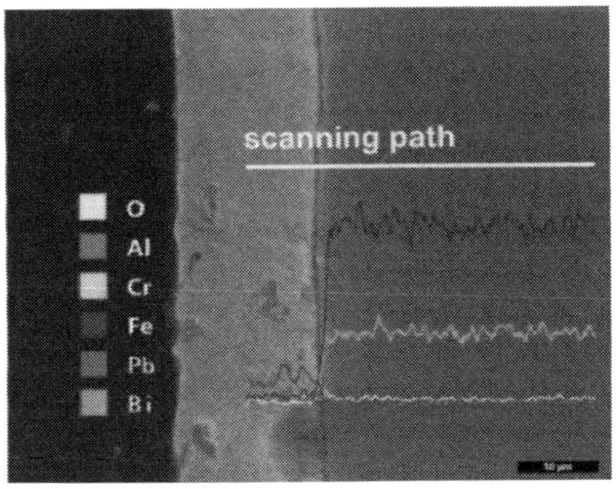


**Figure 7.** SEM cross-sectional morphology and EDX line scan analysis of sample 6-10.

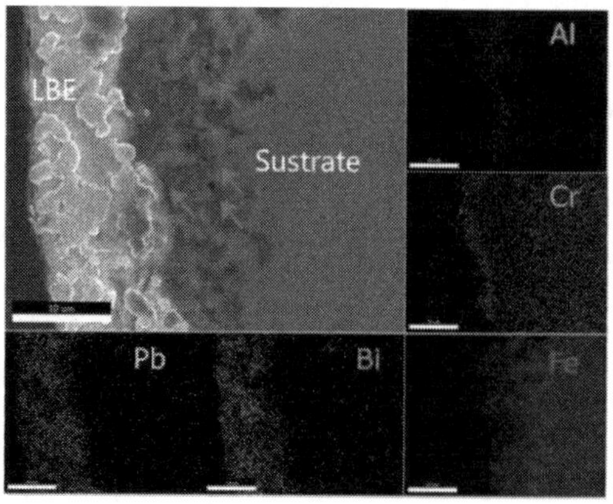


**Figure 8.** Localized SEM cross-sectional morphology and EDX mapping of sample 6-30.

Overall, no penetration of LBE was observed in the coated samples. The above results conclude that the coating retains its efficacy as a barrier against corrosion at high temperatures and an oxidizing environment for a duration of up to 3000 hours.

## 4. Conclusion

The $Al_2O_3$/Fe-Al gradient coating shows excellent corrosion resistance to extreme temperatures and oxidative environments. It has great development potential in the field of outdoor power equipment in a variety of new energy and novel power systems.

The study also found the rules of the structural evolution and element diffusion of the coating. With the increase of corrosion duration, the Fe-Al gradient region generally becomes thicker due to the continuous element inter-diffusion. At higher corrosion temperatures, especially when it reaches 600 °C, the self-healing effect of the Fe-Al gradient region causes the Al element to diffuse outward, ensuring the integrity of the $Al_2O_3$ layer.

In the future, the coating needs to be tested under more different corrosion conditions, and its preparation process also needs to be optimized according to different application requirements in new energy and novel power systems.

## Acknowledgment

This work was supported by the National Natural Science Foundation of China (Grant Nos. 12075292, U22B2066, U1832206), the Natural Science Foundation of Guangdong Province, China (Grant No. HNY20301GJT), West Light Foundation of the Chinese Academy of Science (Grant No. [2022]4), Engineering Research Centre for Digital Grid Technology for Coordinating New Energy (Grant Nos. 2021GCZX003).

## References

[1] J. Zhang, N. Li, Review of the studies on fundamental issues in LBE corrosion, J. Nucl. Mater., 373 (2008) 351-377.
[2] X. Liu, Z.J. Jia, Corrosion Damage and Protective Measures of Offshore Photovoltaic, Environmental Technology, 42 (2024) 38-41.
[3] H.B. Guo, C. Ma, Z.Q. Wen, Key technologies and prospect of prefabricated cabin substation, Electrical Engineering, 24 (2023) 1-10.
[4] S. Hu, H. Finklea, X. Liu, A review on molten sulfate salts induced hot corrosion, J. Mater. Sci. Technol., 90 (2021) 243-254.
[5] Handbook on Lead-bismuth Eutectic Alloy and Lead Properties, Materials Compatibility, Thermal-hydraulics and Technologies: 2015 Edition, OECD Nuclear Science, 2015.

[6] J.C. Zhou, H. Li, J.P. Fan, W.F. Liang, J.L. Li, Performance Evaluation of Anti - Corrosive Materials for Prefabricated Cabin in Substations, Materials Protection, 56 (2023) 219-224.

[7] A. Jianu, R. Fetzer, A. Weisenburger, S. Doyle, M. Bruns, A. Heinzel, P. Hosemann, G. Mueller, Stability domain of alumina thermally grown on Fe–Cr–Al-based model alloys and modified surface layers exposed to oxygen-containing molten Pb, J. Nucl. Mater., 470 (2016) 68-75.

[8] C. Fazio, I. Ricapito, G. Scaddozzo, G. Benamati, Corrosion behaviour of steels and refractory metals and tensile features of steels exposed to flowing PbBi in the LECOR loop, J. Nucl. Mater., 318 (2003) 325-332.

[9] W. Zhang, R. Tang, Z. Yang, C. Liu, H. Chang, J. Yang, J. Liao, Y. Yang, N. Liu, Preparation, structure, and properties of an AlCrMoNbZr high-entropy alloy coating for accident-tolerant fuel cladding, Surf. Coat. Technol., 347 (2018) 13-19.

[10] F. García Ferré, M. Ormellese, F. Di Fonzo, M.G. Beghi, Advanced Al2O3 coatings for high temperature operation of steels in heavy liquid metals: a preliminary study, Corros. Sci., 77 (2013) 375-378.

[11] A. Weisenburger, A. Heinzel, G. Müller, H. Muscher, A. Rousanov, T91 cladding tubes with and without modified FeCrAlY coatings exposed in LBE at different flow, stress and temperature conditions, J. Nucl. Mater., 376 (2008) 274-281.

[12] F.G. Ferré, A. Mairov, D. Iadicicco, M. Vanazzi, S. Bassini, M. Utili, M. Tarantino, M. Bragaglia, F.R. Lamastra, F. Nanni, L. Ceseracciu, Y. Serruys, P. Trocellier, L. Beck, K. Sridharan, M.G. Beghi, F. Di Fonzo, Corrosion and radiation resistant nanoceramic coatings for lead fast reactors, Corros. Sci., 124 (2017) 80-92.

[13] S. Majumdar, B. Paul, P. Chakraborty, J. Kishor, V. Kain, G.K. Dey, Formation of $Al_2O_3$ /FeAl coatings on a 9Cr-1Mo steel, and corrosion evaluation in flowing Pb-17Li loop, J. Nucl. Mater., 486 (2017) 60-65.

[14] H. Chenhao, Z. Shunlin, H.W.J.M. Jun, A.P.M. Materials Transactions, M. Science, Three Composite Coatings Al+$Al_2O_3$, Fe–Al and Fe–Al+$Al_2O_3$ as a Barrier Against LBE Alloys on F/M Steel by Multi-arc Ion Plating, 54A (2023) 4701-4715.

[15] Y. Zhong, W. Zhang, Q. Chen, J. Yang, C. Zhu, Q. Li, J. Yang, N. Liu, J. Yang, Effect of LBE corrosion on microstructure of amorphous $Al_2O_3$ coating by magnetron sputtering, Surf. Coat. Technol., 443 (2022).

[16] S. Huang, L. Pang, P. Tai, Z. Wang, T. Shen, J. Chai, Z. Ma, H. Chang, Study on $Al_2O_3$/Fe-Al gradient coating assisted prepared by ionic liquid for Lead-Bismuth eutectic corrosion resistance, Nucl. Mater. Energy, 38 (2024).

Journal of Physics: Conference Series **2961** (2025) 012043
doi:10.1088/1742-6596/2961/1/012043

# Facile synthesis of 1D perovskite microwire arrays with ultrafast photoresponse

**Yisen Li[1], Haomiao Xu[1], Guishuang Jiang[1], Xinghao Zhao[1] and Lifan Shen[1*]**

[1]Key Laboratory of Optoelectronics Technology, College of Information Science and Technology, Beijing University of Technology, Beijing 100124, China

[*]Corresponding author's e-mail: shenlifan@bjut.edu.cn

**Abstract:** Perovskite materials are ideal candidates for the activity layers in electrooptic components due to their effective light absorption and high carrier mobility. However, perovskites produced through traditional solution methods often suffer from poor crystallinity and high defect density and are susceptible to damage under high temperature and light exposure, which limits device performance. In this work, high-quality 1D single-crystal $CsPbBr_3$ microwire (MW) arrays are successfully synthesized on muscovite substrates using chemical vapor deposition (CVD) by optimizing growth parameters such as growth time, temperature, and substrate position. The photodetector based on the as-grown MWs exhibits outstanding photoelectric performance, achieving a light-to-dark current ratio of up to $5.53 \times 10^2$ at a working voltage of 5 V and a photoresponsivity of up to $9.4 \times 10^{-4}$ A $W^{-1}$ at a small illumination power intensity of 1 mW $cm^{-2}$. Additionally, this photodetector demonstrates an ultrarapid photoresponse time with a fall time of 140 μs and a rise time of 92 μs under a 365 nm laser. This research presents a facile and controllable strategy for synthesizing high-quality 1D perovskites with ultrafast photoresponse, which access the mean for developing high-performing 1D perovskite photodetectors.

## 1. Introduction

Photodetectors transform optical signals into electronic signals for use in communication, image sensing, and environmental monitoring [1-3]. Currently, photodetectors in the ultraviolet to near-infrared spectral range primarily rely on crystalline inorganic elements or compound semiconductors, such as III-V and silicon semiconductors [4-5]. Recently, perovskite materials have gained prominence due to their excellent light absorption, adjustable bandgap, excellent current carrier mobility, and wide gamut, making them a good choice for the new round of photodetectors [6]. Metal halide perovskites have the general formula $ABX_3$ and can be classified into all-inorganic perovskites and organic-inorganic hybrid perovskites [7]. We can substitute the organic positive ion in $ABX_3$ with more stable inorganic cations Cs+, leading to higher stability and making them suitable for active layers in optoelectronic devices [8].

The unique crystal structures of 3D and 2D metal halide perovskites provide remarkable properties. While 3D devices offer excellent performance and low costs, their large surface area leads to high defect densities that hinder carrier transport and result in damage under high temperatures [9-10]. In contrast, 1D perovskites have superior optoelectronic characteristics due to their anisotropy and efficient transport of photo-generated carriers, improving charge extraction in photodetectors [11-13]. Most 1D perovskites are synthesized via solution methods, which can result in poor crystallinity [14].

Journal of Physics: Conference Series **2961** (2025) 012043 doi:10.1088/1742-6596/2961/1/012043

Thus, developing a low-cost method for synthesizing large, high-quality 1D single-crystal perovskites is essential for advancing their physical properties and enabling high-performing electrooptic components.

This paper synthesizes superior 1D single-crystal $CsPbBr_3$ with specific growth orientation on muscovite using chemical vapor deposition (CVD). We promote rapid crystallization by slowly heating the precursor powder in a dual-zone tube furnace. The photodetector is made using the $CsPbBr_3$ as the sensing channel and Au as the electrode. The photodetector achieved a light-to-dark electric current rate of $5.53 \times 10^2$ at 5 V and fast response times (rise time:92μs; fall time:140μs) under a 365 nm laser. This research provides insights for developing electrooptic components based on perovskite.

## 2. Experimental part

### *2.1. Synthesis of 1D Single-Crystal $CsPbBr_3$*

1D single-crystal $CsPbBr_3$ was synthesized using a solid-source CVD method in a dual-zone tube furnace. First,170 mg of CsBr and 100 mg of $PbBr_2$ were grinded and pre-annealed for 30 minutes at 370°C(Figure 1a). Second, this precursor was heated to 480°C and maintained 60 minutes in the upstream area, while the muscovite substrate was warmed to 370°C and maintained 60 minutes in the downstream area(Figure 1b). Argon gas (100sccm) was used as the carrier and maintained at 0.5 Torr. After synthesis, both zones cooled under argon.

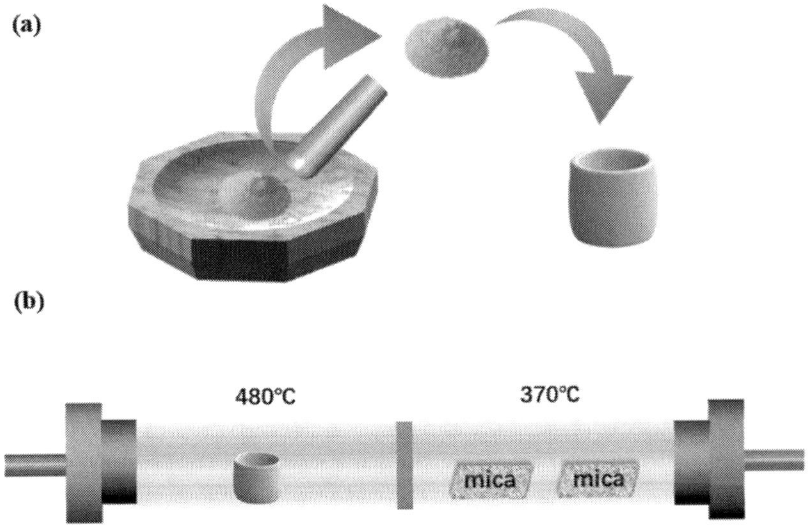


**Figure 1.** a) Powder mixing and grinding in an agate mortar, followed by the transfer into a boron nitride crucible; b) CVD growth in a dual temperature zone furnace.

### *2.2. Fabrication of photodetectors*

A copper mesh with a channel width of 30 μm is used as a mask to define the shape of the electrodes. Subsequently, a 50 nm thick Au electrode is deposited on the surface of the 1D single-crystal $CsPbBr_3$ using a magnetron sputtering technique. This resulted in the fabrication of the photodetector based on the 1D single-crystal $CsPbBr_3$.

### *2.3. Characterization and measurement of materials and devices*

The surface morphology of the single crystals is analyzed using optical microscopy and SEM (Hitachi SU8020). Stoichiometric compositions of 1D single-crystal $CsPbBr_3$ are determined by EDS. Optical properties are assessed with PL spectroscopy. I-V characteristics and I-t curves are measured using a Keithley B1500, and response time is evaluated with a pulsed 365 nm laser.

## 3. Results and discussion

### *3.1. Surface configuration and characterizations of 1D $CsPbBr_3$ MW arrays.*

1D single-crystal $CsPbBr_3$ microwires (MWs) are synthesized using the CVD method in a dual-zone tubular furnace, with growth times ranging from 30 minutes to 2 hours. After 30 minutes, small MWs, about 1 µm wide and several hundred micrometers long, show up on the muscovite supporting base in Figure 2a. As the progression of growth times, the MWs interconnect to form a network. Figure 2b shows the MWs exhibit parallel or hexagonal symmetry, influenced by van der Waals forces and aligned with the (001) muscovite surface. After 2 hours, the line width increases (Figure 2c), indicating further growth potential. Figure 2d reveals that MWs grown for 1 hour have smooth surfaces and a triangular prism structure, well arranged on the mica substrate.

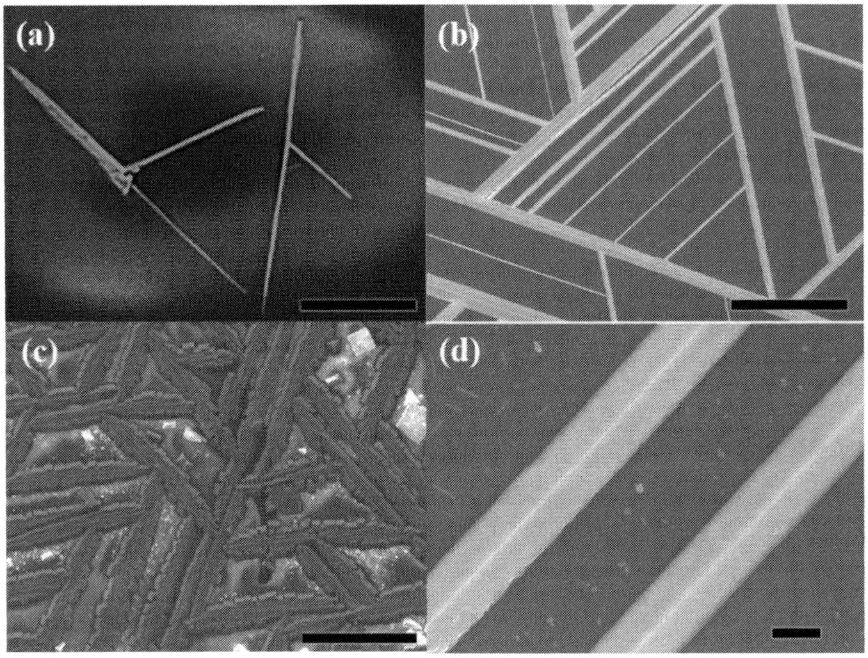


**Figure 2.** SEM images of $CsPbBr_3$ MW at different growth times from 30 minutes to 2 hours. a) Growing for 30 minutes; b) 1 hour; c) 2 hours (Scale bar: 20µm); d) $CsPbBr_3$ MW grown for 1 hour with triangular structure (Scale bar: 500nm).

To analyze the elemental distribution, we use the energy-dispersive X-ray microanalysis (EDS) technique to execute upon the MWs, as portrayed in Figures 3a-3e, which indicates that the distribution of Cs, Pb, and Br is consistent throughout the MWs frame. The corresponding EDS line spectral in Figure 3f shows the proportion of Cs, Pb, and Br is approximately 1:1:3, which is consistent with the chemical quantities of $CsPbBr_3$. Additionally, steady-state photoluminescence measurements are performed on the grown samples. The PL spectrum in Figure 3g shows that the photoluminescence peak is located at 523 nm, with a half-value width of approximately 20 nm, further confirming the good crystalline feature of $CsPbBr_3$ MWs.

 

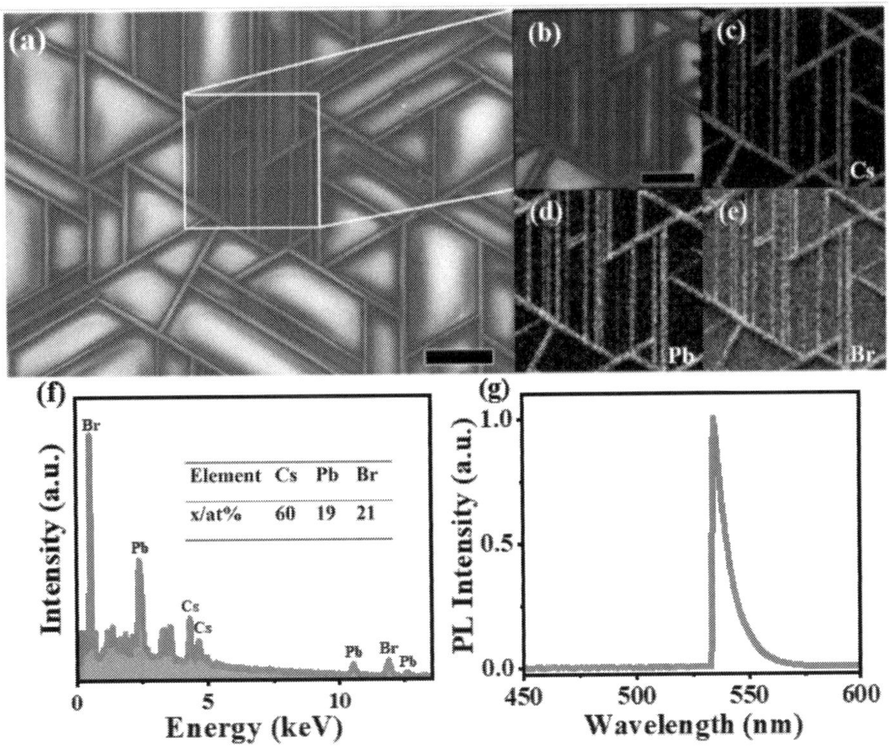


**Figure 3.** a) SEM image of $CsPbBr_3$ MWs (Scale bar: 20μm); b)-e) EDS component allocation chart shows even distribution of Cs, Pb, and Br components across space; f) EDS energy spectrum of $CsPbBr_3$ MWs; g) The PL spectrum of $CsPbBr_3$ MWs.

### *3.2. Performance characterization of $CsPbBr_3$ MW array photodetectors*

To investigate the photoelectric properties of 1D single-crystal $CsPbBr_3$, a 50 nm thick layer of Au was deposited via magnetron sputtering to create electrodes, fabricating a metal-semiconductor-metal photodetector based on the $CsPbBr_3$ micro-wire (MW) array(Figure 4a). We tested the I-V characteristic response of the $CsPbBr_3$ MW parallel array photodetector below dark and illuminated conditions (365 nm, 1 mW $cm^{-2}$). The contact between the $CsPbBr_3$ MWs and the Au electrodes showed ohmic behavior. Due to the low defect state density in the $CsPbBr_3$ MW array, the dark current remained low at 77 pA at a working voltage of 5 V. Under laser at the same voltage, the photocurrent reached 47 nA, resulting in a photo-dark current ratio of $5.53 \times 10^2$, highlighting the excellent photoelectric characteristics(Figure 4b). Below a low illumination power intensity of 1 mW $cm^{-2}$, the device achieved a photoresponsivity of up to $9.4\times10^{-4}$ $AW^{-1}$. We also tested the time-resolved photoresponse by periodically turning the laser on and off. The resulting current variations demonstrated stable photoresponse characteristics(Figure 4c). In a complete on/off cycle, the photogenerated current rose from 10% to 90% of the maximum signal in 92 μs and fell from 90% to 10% in 140 μs, indicating this photodetector's rise and fall times(Figure 4d). These results confirm the device's capability to capture rapidly changing light signals.

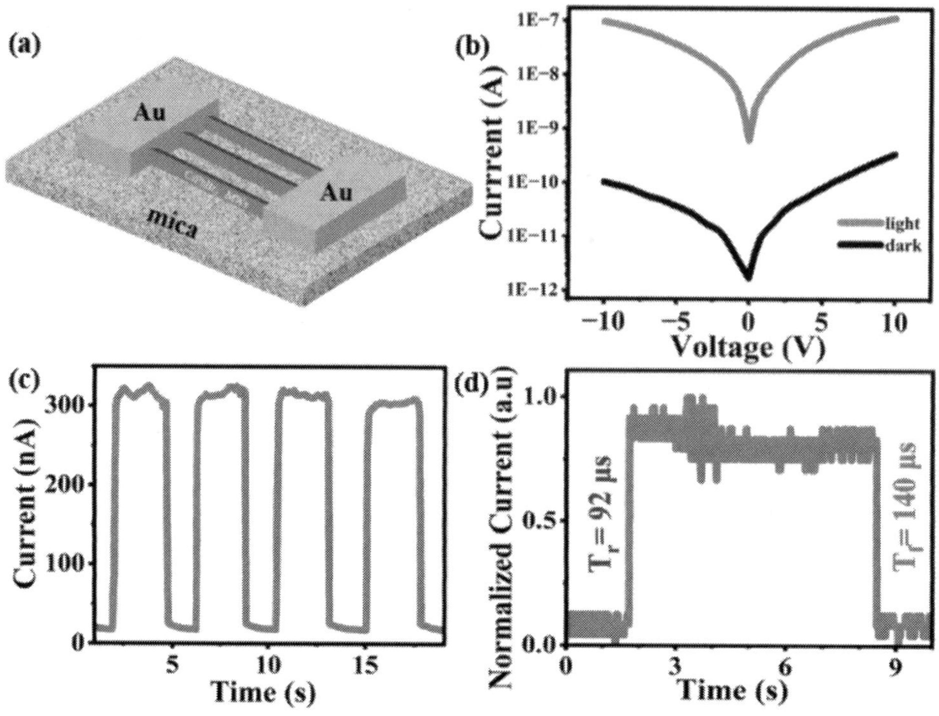


**Figure 4.** a) The sketch map of the photodetector based on $CsPbBr_3$ MW array; b) I-V response of the photodetector; c) I-t response of the photodetector below 365 nm illumination with 1 mW $cm^{-2}$ power density and 5 V voltage; d) The fall and rise times of this photodetector.

## 4. Conclusion

In summary, we successfully grow oriented and size-controllable 1D single-crystal $CsPbBr_3$ MWs on the surface of mica using chemical vapor deposition. By utilizing 1D single-crystal $CsPbBr_3$ as the sensitive channel and Au as the electrodes, a metal-semiconductor-metal photodetector based on 1D single-crystal $CsPbBr_3$ MW arrays is constructed. The photodetector based on the as-grown MWs exhibits outstanding photoelectric performance, achieving a light-to-dark current ratio of up to $5.53 \times 10^2$ at a working voltage of 5 V and a photoresponsivity of up to $9.4 \times 10^{-4}$ A $W^{-1}$ at a small illumination power intensity of 1 mW $cm^{-2}$. Additionally, this photodetector demonstrates an ultrafast photoresponse time with a fall time of 140 μs and a rise time of 92 μs under 365 nm irradiation. This work presents a facile and controllable method for synthesizing high-quality 1D perovskites with ultrarapid photoresponse, paving the way for the advancement of top-performing 1D perovskite photodetectors. By demonstrating effective synthesis techniques, this research contributes to advancing photonic technologies and enhancing the overall performance of photodetectors in practical applications.

## Acknowledgments

The authors convey their appreciation for the financial contribution from the Beijing Municipal Natural Science Foundation (Grant No. 4204090).

## References

[1]Dey A, Ye J, De A, Debroye E.(2021)State of the art and prospects for halide perovskite nanocrystals .J. ACS. nano, 15:10775-10981.
[2]Wang H, Li S, Liu X, Shi Z.(2021)Low-dimensional metal halide perovskite photodetectors .J. Advanced Materials, 33:2003309.

[3]García de Arquer F, Armin A, Meredith P, Sargent E.(2019)Solution-processed semiconductors for next-generation photodetectors .J. Nature Reviews Materials, 2:1-17.

[4]Chen W, Du J, Zhang H, Wang H.(2024)Surface treatment of GaN nanowires for enhanced photoelectrochemical water-splitting .J. Chinese Chemical Letters, 35:109168.

[5]Marchenko E, Belich N, Iosimovska A . (2024)Theoretical assessment of chloride ion influence on grain growth of hybrid perovskites .J. Mendeleev Communications, 34:321-324.

[6]Wang M, Chen K, Xie L.(2024) Polarized emission of $Cs_3Cu_2I_5$ nanowires embedded in nanopores of an anodic aluminum oxide template .J. Optics Letters, 49:1349-1352.

[7]Wang Y, Liu P, Ru F.(2019)Epitaxial growth of large-scale orthorhombic $CsPbBr_3$ perovskite thin films with anisotropic photoresponse property .J. Advanced Functional Materials, 29:1904913.

[8]Wang M, Tian W, Cao F, Wang M.(2020)Flexible and self-powered lateral photodetector based on inorganic perovskite $CsPbI_3$–$CsPbBr_3$ heterojunction nanowire array .J. Advanced Functional Materials, 30:1909771.

[9]Mihalyi-Koch W, Folpini G, Roy C.(2023)Tuning Structure and Excitonic Properties of 2D Ruddlesden Popper Germanium Tin and Lead Iodide Perovskites via Interplay between Cations .J. Journal of the American Chemical Society, 145:28111-28123.

[10]Najarian A, Dinic F, Chen H, Sabatini R.(2023)Homomeric chains of intermolecular bonds scaffold octahedral germanium perovskites .J. Nature, 620:328-335.

[11]DuBose J, Kamat P.(2022)Efficacy of perovskite photocatalysis: challenges to overcome .J. ACS Energy Letters, 7:1994-2011.

[12]Hu Z, Liu Z, Bian Y, Li S.(2018) Enhanced two-photon-pumped emission from in situ synthesized nonblinking $CsPbBr_3/SiO_2$ nanocrystals with excellent stability .J. Advanced Optical Materials, 6:1700997.

[13]Miah M, Khandaker M, Islam M, Nur-E-Alam M.(2024) Perovskite materials in X-ray detection and imaging: recent progress challenges and future prospects .J. RSC advances, 14:6656-6698.

[14]Hanqi B, Jiang M, Lin C, Liu M.(2022) Flexible $CsPbBr_3$ microwire photodetector with a performance enhanced by covering it with an Ag nanolayer .J. CrystEngComm, 24:7620-7631.

# Study on characteristics of ceramic-based 3D printed catalyst supports used in methanol steam reforming for hydrogen production

**Hongwei Zou[1], Zhuangdian Liang[2], Bingchuan Han[1], Gang Wang[2] and Yongdong Chen[1]***

[1]Hefei General Machinery Research Institute Co., Ltd. Hefei, China
[2]Shanghai Advanced Research Institute Shanghai, China

* Corresponding author: cyd@hgmri.com

**Abstract:**3D printing technology has the advantages of constructing ceramic-based catalyst supports with a complex three-dimensional(3D) flow field, outstanding thermostability and resistance to corrosion, which is suitable for methanol steam reforming (MSR). In order to master the chemical characteristics of ceramic-based 3D printed catalyst supports used in MSR, the mathematical models of reaction rate based on dynamic reaction partial pressure were established, and the accuracy of the model was verified by the experimental methods. The comparison between reaction rate models from the literature and the models derived from experiments was performed in detail. With the introduction of contact time and specific area of active component particles, a lower deviation of the mathematical models can be reached with a minimum of 10%. The model generated by 3D printing technology can be utilized for the high-precision design of ceramic-based 3D printed catalyst supports used for MSR.

## 1. Introduction

Hydrogen energy is one of the most promising renewable energies, which is beneficial to solve the environmental problems caused by fosil fuels[1]. Due to the high cost and difficulties in storing and transporting hydrogen, on-site hydrogen production and utilization based on fossil fuels play an important role in preparations for hydrogen energy's extension and application. At present, MSR is regarded as one of the most potential routes to acquire hydrogen due to the high carbon-to-hydrogen ratio (1:4) and excellent transportability of methanol.

Catalyst is the key factor affecting the performance of MSR. The ceramic-based catalysts have delightful prospects in enhancing chemical processes since their inherent characteristics of excellent mechanical strength, temperature and corrosion resistance are inherited from ceramic materials[2]. However, there are major challenges to constructing the ceramic-based catalyst supported by traditional processing methods, in that the texture of the ceramic materials is hard and brittle[3]. Up to now, 3D printing technology has become a popular method to overcome problems such as difficult manufacturing and poor manufacturing accuracy caused by traditional methods of processing ceramic-based supports[4]. 3D printing technology has the advantage of tailoring catalyst supports layer by layer, during which it skillfully manages to fabricate supports with complex 3D flow fields more directly and conveniently without additional structure processing instead. So far, experimental methods are mostly focused on studying the structural characteristics of catalyst supports for MSR. For example, Zheng et al. [5] used metal laser powder bed fusion 3D printing technology to create integral porous catalysts with body-centered cubic and face-centered cubic structures, enhancing the internal heat and mass transfer processes. Liang et al. [6] used direct light processing 3D printing technology(DLP 3D printing) to create fractal structure catalysts. They found that the radial crush strength of these 3D-printed catalysts was 152.4 N/mm, which is four times higher than traditional cylindrical ones. Li et al. [7] utilized metal laser powder bed 3D printing technology along with dealloying methods to fabricate novel three-dimensional gradient nanoporous catalysts and achieved a methanol conversion rate of up to 98.3%.

Compared to research on the structure of ceramic-based catalyst supports, establishing a mathematical model of reaction rate is beneficial to a clearer understanding of the main characteristics of the MSR. At present, the reaction rate model of the MSR with ceramic-based catalyst supports has not been fully clarified. Therefore, the ceramic-based catalyst supports were fabricated through digital light processing (DLP) 3D printing technology. Effects of temperature and gas hourly space velocity (GHSV) on methanol conversion, CO selectivity and hydrogen rate were evaluated experimentally. Analysis of the deviations among reaction rate models from literature and experimental results was conducted. Reaction rate models with the introduction of contact time and specific surface area of active components were established. This study has important guiding significance for the fabrication of ceramic-based 3D printed catalyst supports.

## 2. Experiment descriptions

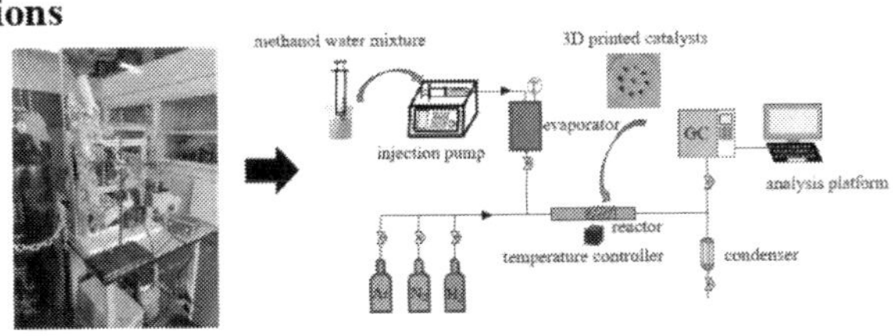


**Figure 1.** Testing system for methanol steam reforming.

The catalyst used in the experiment was $Cu/ZnO/Al_2O_3$, and the ceramic material was $Al_2O_3$. Figure 1 depicts the experimental setup for methanol steam reforming to produce hydrogen. It consists of an injection pump, evaporator, single-tube reactor, temperature control device, and analysis platform. Both the evaporator and the temperature control device are equipped with a spring resistance wire and a power controller. The reactor is a single tube with an inner diameter of 13.5 mm, containing three 3D-printed catalysts with an active component mass of 0.16 g. The experiment parameters are shown in Table 1. The steam-to-carbon ratio was 1:1.

**Table 1.** Experimental parameters of methanol steam reforming.

| Group | Temperature (°C) | GHSV (mL/$g_{cat}$/h) |
|---|---|---|
| 1 | 250,270,290,310,330 | 10000 |
| 2 | 250 | 5000,10000,15000,20000 |
| 3 | 270 | 5000,10000,15000,20000 |
| 4 | 290 | 5000,10000,15000,20000 |

The porosity of the catalysts was obtained by the Archimedes method, and the mean porosity of the catalyst was calculated as 43.1%. The density of the catalyst was calculated as 3578 kg/m$^3$. The specific surface area of the active component was calculated as 54794 m$^2$/kg.

Dual-rate models, including methanol steam reforming and reverse water-gas shift reaction(rWGS), were selected since they can simplify the description of the overall reaction process and product distribution[8] [9] [10] [11]. The reaction equation was shown as follows:

Journal of Physics: Conference Series **2961** (2025) 012044 doi:10.1088/1742-6596/2961/1/012044

MSR:

$$CH_3OH + H_2O \rightarrow CO_2 + 3H_2 \quad (1)$$

rWGS:

$$CO_2 + H_2 \leftrightarrow CO + H_2O \quad (2)$$

Methanol conversion, $H_2$ production rate and selectivity of CO were used as the evaluation of the reaction performance, which were calculated by the following formulas.

Conversion of methanol:

$$X_{CH_3OH}(\%) = \frac{F_{CH_3OH,in} - F_{CH_3OH,out}}{F_{CH_3OH,in}} \times 100\% \quad (3)$$

$H_2$ production rate:

$$Y_{H_2} = \frac{F_{H_2}}{F_{CH_3OH,in}} \quad (4)$$

Selectivity of CO:

$$S_{CO}(\%) = \frac{F_{CO,out}}{F_{CO,out} + F_{CO_2,out}} \times 100\% \quad (5)$$

where $F$ is the molar flow rate of each component.

## 3. Results and discussion

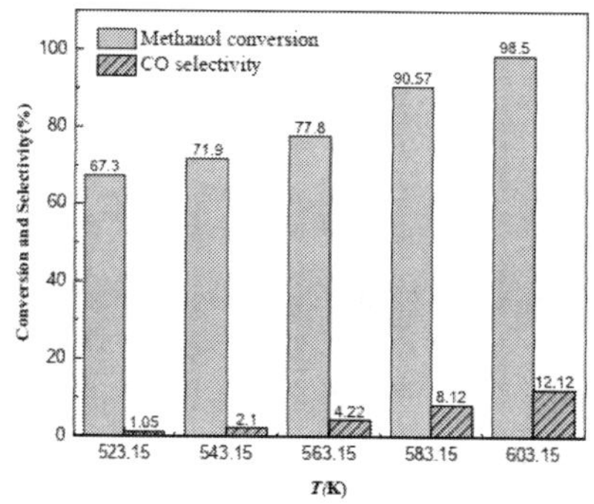


**Figure 2.** Diagram of methanol conversion rate and CO selectivity at different temperatures (GHSV=10000ml/$g_{cat}$/h).

Figure 2 shows the effects of temperature on the methanol conversion and CO selectivity at different temperatures. Since the MSR and rWGS are endothermic, both methanol conversion and CO selectivity increased with the increase in temperature. The methanol conversion rose from 67.3% to 98.5% when the temperature increased from 523.15K to 603.15K as well as the CO selectivity increased from 1.05% to 12.12% with the temperature rise.

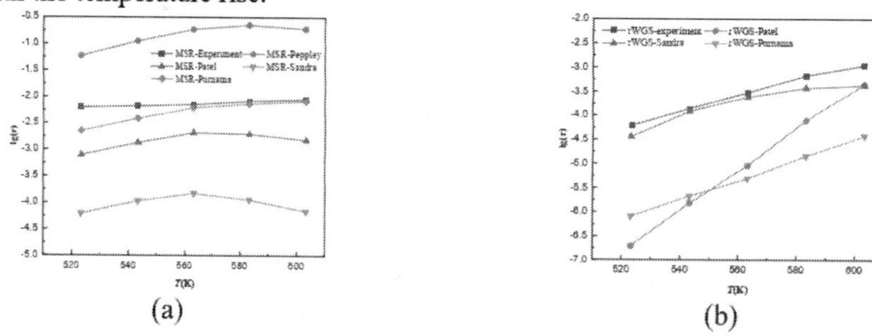


**Figure 3.** Logarithm of reaction rate from experiment and literature (GHSV=10000ml/$g_{cat}$/h, S/M=1.0, T=523.15K~603.15K).

The experimental reaction rate was depicted in Figure 3, as well as the reaction rate calculated from literature (e.g., Peppley [12, 13], Patel [14], Purnama [11], and Sandra[15]). The model from Peppley is a three-rate model, which includes methanol steam reforming, methanol decomposition (MD) and water-gas shift reaction (WGS), while the other models are dual-rate models for MSR and rWGS.

Since the model from Peppley is a three-rate model and regarded the WGS reaction as one of the side reactions, only the prediction of the MSR rate was compared to the experimental results. It was shown that the Peppley model had the largest deviations in MSR reaction rates' predictions; that probably was because the three-rate model is more complex with numerous input parameters, which inevitably introduces more errors in fitting data, making it difficult to solve and be challenging to apply in practical engineering design [8]. Among the dual-rate models, Purnama's MSR rate values are closest to the experimental values, though the values were lower than the experimental results when temperatures were below 563.15 K. The same situation happened to Sandra's model as well. The model from Sandra's was observed to have the closest predicted values to the experimental results, but it showed a decrease in reaction rate at 583.15 K. That is probably because ceramic-based 3D-printed catalyst supports have better temperature resistance, allowing it to be used when the temperature exceeds 573.15. In contrast, the model from Sandra's was only obtained within the temperature range of 503 K $\leq$ T $\leq$ 573 K and cannot be fully used in MSR with ceramic-based 3D printed catalyst supports. As for the lower values predicted from Purnama's model, it was inferred that the positive effects of the 3D flow field in ceramic-based catalysts should be considered in reaction models as well. In summary, the models from the literature models provide guidance for predicting the methanol-reforming reaction rate with 3D-printed catalysts but still showed limitations in predicting reaction rates at temperatures below 563.15 K and above 573.15 K. Therefore, it is necessary to refit the reaction rate models with ceramic-based 3D-printed catalysts.

To further improve the accuracy of the fitting model, methanol-reforming experiments for hydrogen production were conducted in a wider range, which includes a wider range of temperatures (i.e., from 523.15 K to 563.15 K) and wider GHSV (i.e., from 5000 to 20000 mL/gcat/h). Figure 4 illustrates the methanol conversion rate and CO selectivity as functions of GHSV when the temperature ranges from 523.15 K to 563.15 K. Both the methanol conversion rate and CO selectivity decrease with the increase in GHSV. As shown in Figure 5, higher space velocities contribute to increased hydrogen production.

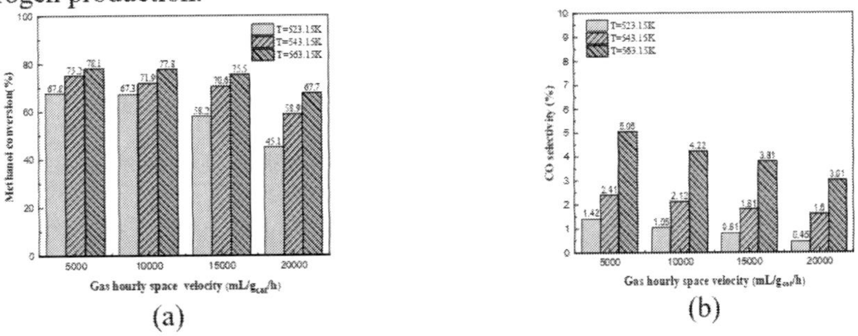


**Figure 4.** Methanol conversion rates and CO selectivity at different temperatures and inlet gas hourly space velocity.

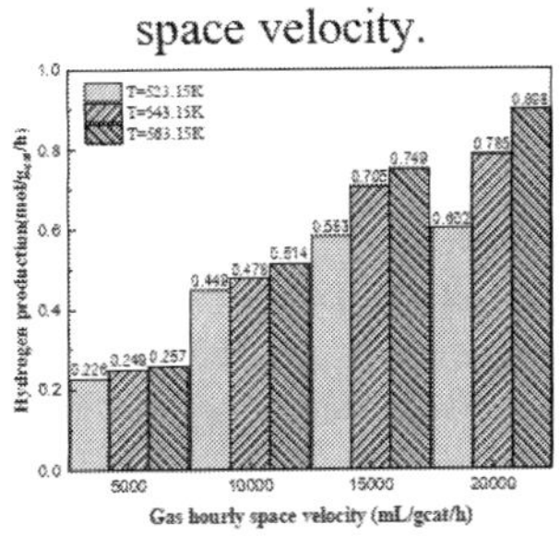


**Figure 5.** Hydrogen production at different temperatures and inlet gas space velocity.

After extending the experimental data points, the reaction rate model with ceramic-based 3D-printed catalysts was refitted based on the Purnama and Sandra models. Both equilibrium constants from MSR and rWGS were reorganized into temperature-dependent correlations:

$$K_R = e^{-\frac{(-5.024\times10^4+170.98T+2.644\times10^{-2}T^2)}{RT}} \tag{6}$$

$$K_W = e^{-\frac{(-4.1735\times10^4+46.66T-7.55\times10^{-3}T^2)}{RT}} \tag{7}$$

The influence of the active component in ceramic-based 3D printed catalyst and the contact time(W/F), $g_{cat}$·s/mL (i.e., the ratio of the mass of the effective catalyst component to the flow rate of the feed gas), was taken into account. Based on the experimental results, the pre-exponential factor was refitted through the Levenberg-Marquardt algorithm. The results were detailed in Equations (8) and (9).

$$k_{pre(msr)} = \exp(3.245e^{-6}T^3 \text{-} 5.316e^{-3}T^2 + 2.884T \text{-} 3.058(\frac{W}{F})^2 + 2.322\frac{W}{F} \text{-} 516.668) \tag{8}$$

$$k_{pre(rwgs)} = \exp(-5.969e^{-6}T^3 + 0.0104T^2 - 6.074 + 2.996(\frac{W}{F})^2 - 3.922\frac{W}{F} + 1171.58) \tag{9}$$

The other three equilibrium constant parameters are shown in Table 2.

**Table 2.** Reaction kinetics model parameters.

| Parameters | Values |
|---|---|
| $K_{CH_3O^{(1)}}$ | $e^{-29/R+2\times10^4/(RT)}$ |
| $K_{HCOO^{(1)}}$ | $e^{199/R-1\times10^5/(RT)}$ |
| $K_{OH^{(1)}}$ | $e^{-85/R+2\times10^4/(RT)}$ |
| $k_{MSR}$ | $k_{pre(msr)}\exp(\frac{7.4\times10^4}{RT})$ |
| $k_{rWGS}$ | $k_{pre(rWGS)}\exp(\frac{7.0\times10^4}{RT})$ |

The final refitted model was shown in Equations (10) and (11), with reaction rate units in kmol/m$^3$/s:

$$r_{MSR} = k_{MSR}p_{CH_3OH}^{0.6}p_{H_2O}^{0.4}S_c\rho_b\times10^{-3} \tag{10}$$

$$r_{rWGS} = \frac{k_{rWGS}K_{HCOO^{(1)}}p_{H_2}^{1/2}p_{CO_2}(1-\frac{p_{CO}p_{H_2O}}{K_{rWGS}p_{H_2}p_{CO_2}})}{(1+K_{CH_3O^{(1)}}\frac{p_{CH_3OH}}{p_{H_2}^{1/2}}+K_{OH}^{(1)}\frac{p_{H_2O}}{p_{H_2}^{1/2}}+K_{HCOO^{(1)}}p_{CO_2}p_{H_2}^{1/2})^2}S_c\rho_b\times10^{-3} \tag{11}$$

where $S_c$ is the specific area of the active component, m$^2$/kg, and $\rho_b$ is the density of the catalysts, kg/m$^3$.

Based on experimental data at different GHSV and temperatures, the results were compared with the predictions from the refitted models, as shown in Figure 6. It is shown that the maximum deviation between the predicted and experimental reaction rate values is within 10%, which means the refitted correlations are applicable for ceramic-based 3D-printed catalyst structures under GHSV= 5000~20000 mL/$g_{cat}$/h, T=523.15~603.15 K, S/M=1, and W/F=0.18~0.72 $g_{cat}$·s/mL.

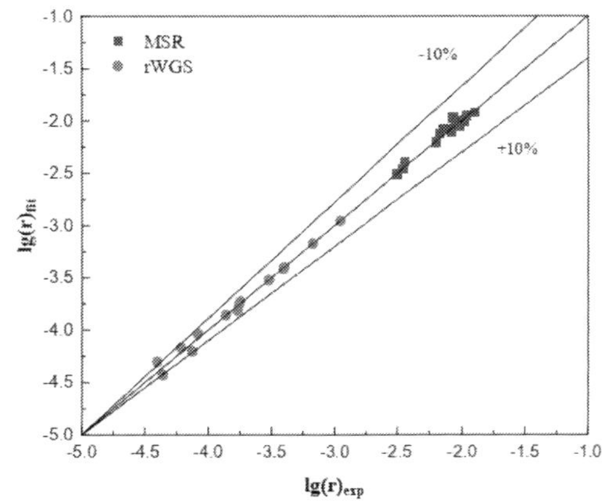


**Figure 6.** Comparison between experimental and predicted values of reaction rate (inlet gas space velocity 5000-20000mL/gcat/h, temperature 523.15-603.15K, S/M=1, W/F=0.18-0.72$g_{cat}$·s/mL).

## 4. Conclusion

This study focused on the preparation of 3D-printed catalysts and their application in methanol reforming for hydrogen production under conditions of T=523.15~603.15 K, GHSV=5000 to 20000 mL/$g_{cat}$/h, and S/M=1. The experimental results were compared with literature models, and modifications were conducted based on Purnama's MSR rate model and Sandra's rWGS rate model. The main conclusions are as follows:

(1) At a space velocity of 10000 mL/$g_{cat}$/h, increasing the temperature from 523.15 K to 603.15 K improved the methanol conversion rate from 67.3% to 98.5%, while the selectivity for carbon monoxide rose from 1.05% to 12.12%, which complies with the law of endothermic reaction.

(2) Within the temperature range from 523.15 K to 563.15 K, the increase of the GHSV from 5000 to 20000 mL/$g_{cat}$/h will decrease the methanol conversion rate as well as the CO selectivity. However, hydrogen production will increase with the increase in the GHSV and the temperature.

(3) The comparison between the experimental results and the value predicted by literature models was conducted. It was found that Purnama's MSR rate model and Sandra's rWGS rate model closely matched the experimental data. However, both models predicted lower values at temperatures below 563.15 K and above 573.15 K. The deviations were probably brought from ceramic materials and 3D printing technology. Therefore, a revised reaction rate model was obtained. The refitted model was validated through a comparison between experiment results and predictions from refitted models. The maximum deviation within 10% was achieved. It is hoped that the model can provide insights for the design of 3D-printed catalysts in the future.

**Acknowledgments**

This work was supported by Key Research and Development Project of Anhui Province (No.2023z04020020) and Youth Science and Technology Fund of Hefei General Machinery Research Institute Co., Ltd. (No.2023010773).

**References**

[1] V. A. Danilov, M. Wichert, G. Kolb, 2D model of the transfer processes for CO2 methanation in a microchannel reactor, Chem. Eng. J. 450 (2022).
[2] O. Ovalle-Encinia, J.Y.S. Lin, Water-gas shift reaction in ceramic-carbonate dual-phase membrane reactor at high temperatures and pressures, Chem. Eng. J. 448 (2022).
[3] Y. Zhang, S. He, X. Jiang, M. Xiong, Y. Ye, X. Yang, Three-dimensional multi-phase simulation of different flow fields with cooling channel in proton exchange membrane fuel cell, International Journal of Hydrogen Energy 47 (2022) 37929-37944.

[4] D. Fu, X. Li, X. Zhu, D. Yuan, W. Chen, T. Fu, X. Chu, T. Luo, Y. Gao, W. Zhou, Development of ceramic-based 3D staggered microchannel catalyst support used in MSR microreactor for hydrogen production, International Journal of Hydrogen Energy 49 (2024) 680-695.
[5] T. Zheng, W. Zhou, D. Geng, Y. Li, Y. Liu, C. Zhang, Methanol steam reforming microreactor with novel 3D-printed porous stainless steel support as catalyst support, International Journal of Hydrogen Energy 45 (2020) 14006-14016.
[6] Z. Liang, G. Wang, G. Zeng, J. Zhang, Z.J.J.o.E.C. Tang, Highly controlled structured catalysts for on-board methanol reforming, Journal of Energy Chemistry 68 (2022) 19-26.
[7] C. Li, X. Yao, R. Zhang, H. Zheng, S. Yuan, X. Yu, B. Li, M. Zhu, S.-T. Tu, Structured nanoporous Cu/ZnO catalysts for on-board methanol steam reforming prepared by laser powder bed fusion and dealloying, Chem. Eng. J. 487 (2024).
[8] X.X. Xiaoru ZHUANG, Xin XIA, Lun LI, Wenfu XU, Review of reaction kinetics of methanol steam reforming forhydrogen production, Chemical Industry and Engineering Progress 39 (2020) 152-165.
[9] M.P. LI Yanhao, SU Xu, HU Kongcheng, HAO Shuren, CHENG Yuchun, Study on Process of Methanol Steam-Reforming to Hydrogen over CuO / ZnO / Al2O3Catalyst by in Situ Infrared Spectroscopy, Chinese Journal of Catalysis 24 (2003) 93-96.
[10] J. Agrell, H. Birgersson, M. Boutonnet, I. Melián-Cabrera, R.M. Navarro, J.L.G. Fierro, Production of hydrogen from methanol over Cu/ZnO catalysts promoted by ZrO2 and Al2O3, J. Catal. 219 (2003) 389-403.
[11] H. Purnama, T. Ressler, R.E. Jentoft, H. Soerijanto, R. Schlögl, R. Schomäcker, CO formation/selectivity for steam reforming of methanol with a commercial CuO/ZnO/Al2O3 catalyst, Applied Catalysis a-General 259 (2004) 83-94.
[12] B.A. Peppley, J.C. Amphlett, L.M. Kearns, R.F. Mann, Methanol-steam reforming on Cu/ZnO/Al2O3. Part 1: The reaction network, Applied Catalysis a-General 179 (1999) 21-29.
[13] B.A. Peppley, J.C. Amphlett, L.M. Kearns, R.F. Mann, Methanol-steam reforming on Cu/ZnO/Al2O3 catalysts. Part 2: A comprehensive kinetic model, Applied Catalysis a-General 179 (1999) 31-49.
[14] S. Patel, K.K. Pant, Experimental study and mechanistic kinetic modeling for selective production of hydrogen via catalytic steam reforming of methanol, Chem. Eng. Sci. 62 (2007) 5425-5435.
[15] S. Sa, J.M. Sousa, A. Mendes, Steam reforming of methanol over a CuO/ZnO/Al2O3 catalyst, part I: Kinetic modelling, Chem. Eng. Sci. 66 (2011) 4913-4921.

Journal of Physics: Conference Series **2961** (2025) 012045 doi:10.1088/1742-6596/2961/1/012045

# Study on the influence of silane coupling Agent-Modified BN on the thermodynamic properties of polytetrafluoroethylene

**Fangcheng Lv and Wenqi Zhang***

School of Electrical and Electronic Engineering, North China Electric Power University, Beijing, China

*Corresponding author: 751546868@qq.com

**Abstract:** To study the effect of different silane coupling agents modified boron nitride (BN) fillers on the thermodynamic properties of PTFE, this work selects KH550, KH560, and KH580 silane coupling agents to modify BN, and constructs a PTFE composite model in *Materials Studio*. The thermal conductivity, mechanical properties, and cohesive energy density of the composite models are analyzed using molecular dynamics methods. The results indicate that coupling agent modification can further improve the thermodynamic stability of the composite. Among them, KH550 has the optimal modification effect, with a thermal conductivity of 0.192 W/(m·K), an increase of 66.9% compared to the pristine. The underlying mechanism for the improvement of thermodynamic properties is elucidated from the perspectives of molecular thermal motion and intermolecular interaction forces through the calculation of cohesive energy density as well as the fractional free volume. This study provides theoretical guidance for the screening of coupling agent modification and the design of high-performance polymer composite based on molecular simulation.

## 1. Introduction

Polytetrafluoroethylene (PTFE), as an artificially synthesized polymer compound, not only possesses excellent electrical insulation properties but also exhibits remarkable heat resistance, chemical corrosion resistance, and an extremely low coefficient of friction [1-2]. It is widely applied in various fields such as electricity, chemicals, machinery, aerospace, and more. In recent years, PTFE has been used in the fabrication of insulating protective layers for wires and cables, as a substrate for microwave circuits, and in various types of electrical devices. Nevertheless, the inherently low thermal conductivity of its polymer structure limits the further application of PTFE in electrical and electronic systems. With the continuous increase in voltage levels and the trend towards miniaturization in electrical devices, the thermal stability of insulating materials has gradually become a crucial challenge for their safe and stable operation. Specifically, the low thermal conductivity hinders the dissipation of heat accumulation during the long-term operation of electrical equipment, which can easily lead to equipment aging and even thermal breakdown failures [3].

Over the past decades, the incorporation of inorganic fillers has proven to be an effective method for enhancing the thermomechanical properties of polymers. Among these fillers, boron nitride (BN) stands out as an ideal candidate for doping due to its exceptional combination of high thermal conductivity and insulating properties [4]. However, polymer modification frequently encounters interface compatibility challenges. Rigid inorganic fillers often struggle to achieve good interface compatibility with flexible matrices, and the resulting defects can easily elevate interface thermal resistance and evolve into potential weak points for breakdown. To overcome this challenge, surface modification employing coupling agents is regarded as an effective strategy for enhancing dispersion and interfacial compatibility. Serving as a "molecular bridge" between polymers and fillers, coupling agents are capable of not only reacting with the surface functional groups of inorganic fillers to establish bonds but also exhibiting strong intermolecular interactions with organic functional groups within polymers, thereby reinforcing the bonding strength of the composite. Yang *et al.* investigated the modification effects of different coupling agents on PMIA/AlN composite [5]. The results showed that appropriate coupling agents facilitated the uniform dispersion of inorganic fillers and established heat transfer pathways that promoted heat dissipation. Liu *et al.* employed KH550 for surface grafting of BN improving the dispersion of fillers and suppressing interface phonon scattering [6]. The composite epoxy resin exhibited an increase of 193.7% in thermal conductivity. However, considering the wide variety of coupling agents, different kinds of functional groups often present distinct physicochemical properties and have varying effects on the modification of composite. Hence, rational selection is crucial for optimizing the performance of composite materials. Compared to traditional experiments, molecular simulation has been extensively performed in material modification simulation and performance analysis. In this work, various PTFE composite models doped with boron nitride fillers, being modified with different silane coupling agents, are constructed using *Materials Studio* (MS). Based on the molecular dynamics method, thermal conductivity, glass transition temperature, and Young's modulus of these composite models are calculated. Additionally, by calculating the cohesive energy density, free volume fraction, and so on, the underlying enhanced performance is revealed from the perspective of molecular thermal motion and microstructure. This provides theoretical guidance for the preparation of high-performance composite insulation materials in the future.

## 2. Experimental methods

The molecular model of PTFE is built according to its chemical structure in MS. Considering that the subsequent construction of the composite model and the dynamics calculation involve both organic and inorganic components, the COMPASS III force field is selected. Andersen and Berendsen methods are used for temperature and pressure control, respectively. For the control of van der Waals and electrostatic interactions, Atom Based and Ewald methods are employed, and the atomic starting velocity in the model is randomly specified by the Maxwell distribution. The system energy under this force field is:

$$E_{tol} = \left(E_b + E_\theta + E_f + E_\omega + E_c\right) + \left(E_h + E_e + E_v\right) \quad (1)$$

where $E_b$ represents the energy of bond length stretching vibration; $E_\theta$ represents bending vibration energy; $E_f$ is the dihedral angle torsional energy; $E_\omega$ represents the out-of-plane angular displacement vibration energy; $E_c$ is the cross valence bond interaction energy; $E_h$ represents the hydrogen bond interaction energy; $E_e$ represents the electrostatic interaction energy between molecules; $E_v$ represents the van der Waals interaction energy.

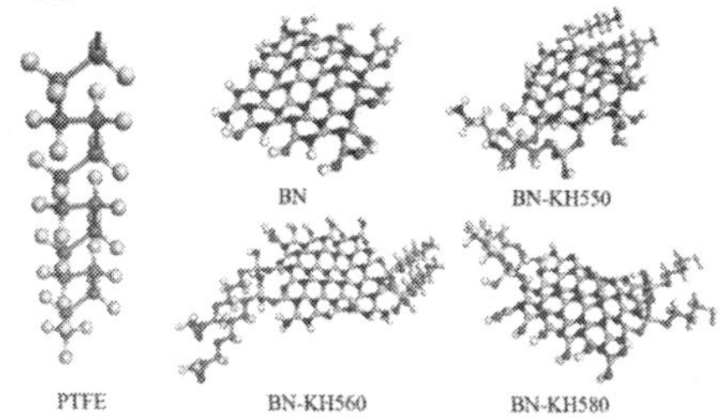


**Figure 1.** Molecular models of PTFE and modified BN.

In Figure 1, PTFE monomer is constructed according to its molecular structure, and then subjected to amorphous calculation. The polymerization degree is set as 10 to obtain the PTFE molecular model. The BN crystal structure is imported from the MS database, sliced along the (110) orientation, and

Journal of Physics: Conference Series **2961** (2025) 012045
doi:10.1088/1742-6596/2961/1/012045

saturated the edge unsaturated N and B atoms with hydrogen and hydroxyl atoms, respectively. Then, three kinds of silane coupling agents are grafted onto the BN edge.

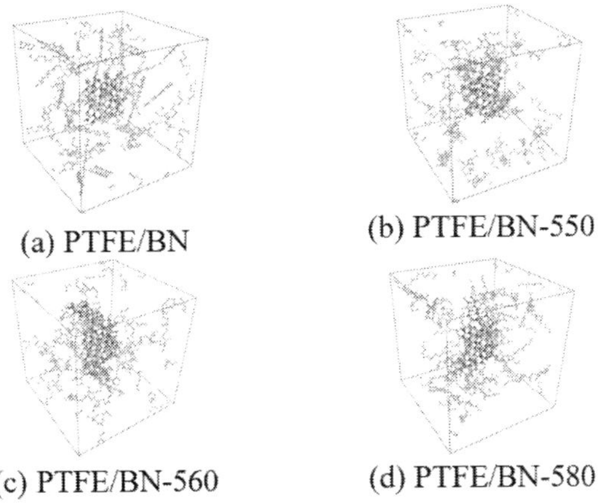

(a) PTFE/BN (b) PTFE/BN-550

(c) PTFE/BN-560 (d) PTFE/BN-580

**Figure 2.** PTFE/BN composite models.

By combining PTFE models with various BN and modified BN models, different composite models are obtained in Figure 2, denoted as PTFE/BN, PTFE/BN-550, PTFE/BN-560, and PTFE/BN-580 models. To rationalize the conformation, geometric optimization is performed on the molecular model. The number of optimization steps is set to 50,000, and the distribution of force field, charge, and other parameters is the same as above. Subsequently, dynamics is carried out for 300 ps in both NPT (constant temperature and constant pressure) and NVT (constant temperature and constant volume) ensembles, with a step of 1 fs. The changes in temperature and energy of the models *via* time are depicted in Figure 3.

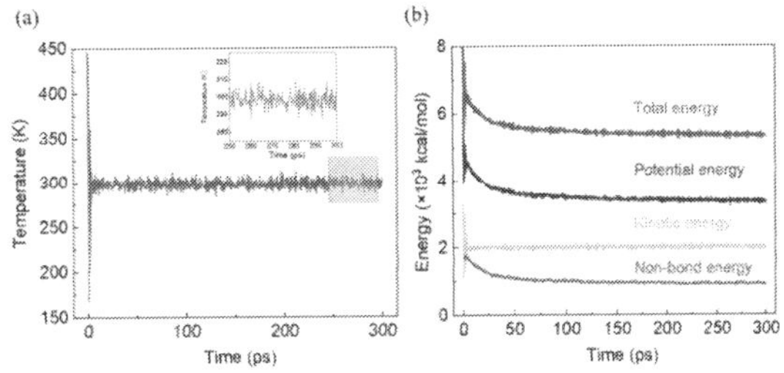


**Figure 3.** Relaxation changes of (a) temperature and (b) energy.

As the relaxation time increases, the temperature and energy of the model gradually stabilize, and the temperature fluctuation within the last 50 ps is less than 15 K, confirming that the model has developed to a stable state and can be further analyzed for performance calculations.

## 3. Result

In this work, the NEMD method is applied to accurately simulate the thermal conductivity of various models. By using the local hot bath method to construct the temperature gradient, the molecular model is uniformly divided into 100 layers along a specific direction (the *x*-direction is selected). Among them, the leftmost layer of the system is defined as the heat source layer, while the rightmost layer is defined as the heat sink layer. To maintain the two layers at a preset temperature, *the Langevin* temperature control method is employed, with the heat source layer set at a preset temperature of 350 K and the heat sink layer set at 250 K, as shown in Figure 4.

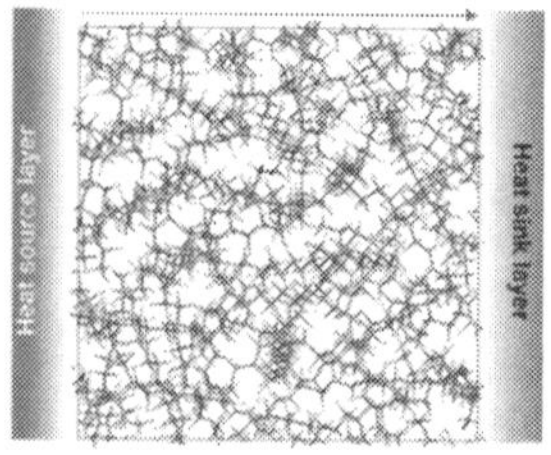


**Figure 4.** NEMD heat transfer diagram.

During the heat transfer process, cold particles in the heat source layer will continuously exchange energy with hot particles in the heat sink layer, eventually forming a stable temperature gradient. The thermal conductivity equation based on Fourier heat transfer is as follows:

$$\kappa = -\frac{J}{A \cdot \nabla T} \quad (2)$$

$$J = \frac{\partial E}{\partial t} \quad (3)$$

$$\nabla T = \frac{\partial T}{\partial x} \quad (4)$$

where $J$ represents heat flux; $A$ is the heat transfer area; $\nabla T$ is the temperature gradient; $E$ represents the average absolute value of the energy of the heat source and heat sink; $t$ denotes the simulation time; T is the temperature; $x$ is the direction of heat transfer.

**Table 1.** Calculation of thermal conductivity.

| Molecular Models | $\lambda$ (W/(m·K)) |
|---|---|
| PTFE | 0.115 |
| PTFE/BN | 0.156 |
| PTFE/BN-550 | 0.192 |
| PTFE/BN-560 | 0.186 |
| PTFE/BN-580 | 0.178 |

As shown in Table 1, the pure PTFE model possesses a relatively low thermal conductivity with only 0.115 W/(m·K), and the incorporation of BN effectively increases the thermal conductivity of the composite models. This can be attributed to two reasons: firstly, the intrinsic high thermal conductivity of BN, and secondly, the introduction of BN occupies the free volume of the molecular system, reducing the average free path of phonon heat transfer. Additionally, the thermal conductivity of composite models containing BN modified with various coupling agents is further improved. Among them, the PTFE/BN-550 composite model exhibits the maximum thermal conductivity of 0.192 W/(m·K), which is 66.9% higher than the pure PTFE model. This is mainly because the introduction of a coupling agent further reduces the interfacial thermal resistance between the fillers and polymer chains, accelerating the heat transport process. To access the mechanical properties of the composite models, the static constant strain method is performed to calculate the stress-strain curves of the materials. The Mechanical Properties task is selected in the Forcite module of MS, and the Constant strain force application method is set. Before calculation, the Optimize structure is first selected to ensure reasonable conformation. The relationship between solid strain $j$ and stress $i$ satisfies Hooke's law:

$$\sigma_i = C_{ij} \cdot \varepsilon_j \quad (5)$$

where $C_{ij}$ represents the stiffness coefficient matrix. Commonly, all mechanical properties of materials can be solved by $C_{ij}$. Considering that the elastic strain of most materials is symmetric, $C_{ij}=C_{ji}$ is satisfied. Hence, the simplified matrix is expressed as Equation (6)

$$C=\begin{bmatrix} \lambda+\mu & \lambda & \lambda & 0 & 0 & 0 \\ \lambda & \lambda+\mu & \lambda & 0 & 0 & 0 \\ \lambda & \lambda & \lambda+\mu & 0 & 0 & 0 \\ 0 & 0 & 0 & \mu & 0 & 0 \\ 0 & 0 & 0 & 0 & \mu & 0 \\ 0 & 0 & 0 & 0 & 0 & \mu \end{bmatrix} \tag{6}$$

$$\lambda=\frac{1}{3}(C_{11}+C_{22}+C_{33})-\frac{2}{3}(C_{44}+C_{55}+C_{66}) \tag{7}$$

$$\mu=\frac{1}{3}(C_{44}+C_{55}+C_{66}) \tag{8}$$

$$E=\mu\cdot\frac{3\lambda+2\mu}{\lambda+\mu} \tag{9}$$

The $\lambda$ and $\mu$ are the Lame constant, which can be calculated by Equations (7) and (8). The Young's modulus ($E$) can be solved through Equations (9).

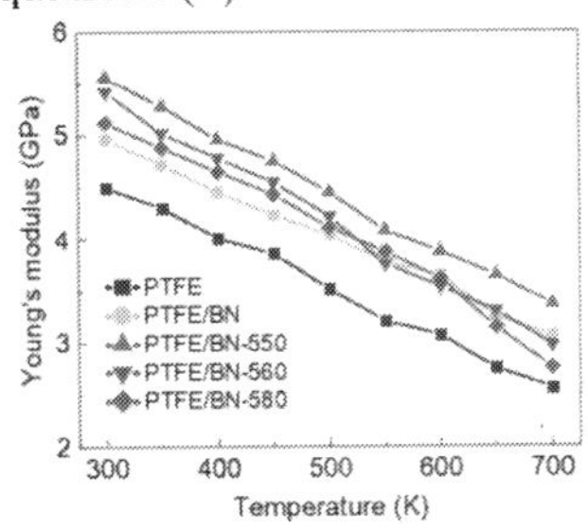


**Figure 5.** Young's modulus of various composite models at different temperatures.

In Figure 5, the mechanical properties of each model show a decreasing trend with increasing temperature, which is because temperature rise promotes the activation movement of molecular chains. Among them, the KH550-modified composite system exhibits optimal mechanical properties. The total internal volume of a polymer composite consists of occupied volume and free volume, where occupied volume refers to the spatial volume occupied by the molecular chains, and free volume refers to the spatial volume in which the molecular chain can move freely. The equation for calculating the fraction of free volume (*FFV*) is defined as follows:

$$FFV=\frac{V_f}{V_0+V_f}\times 100\% \tag{10}$$

where $V_f$ represents the free volume within the model, and $V_0$ represents the occupied volume. According to Equation (10), the *FFV* of PTFE and composite models are obtained.

(a) PTFE

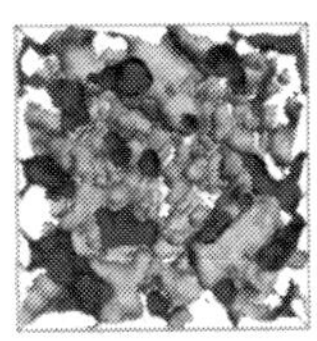

(b) PTFE/BN

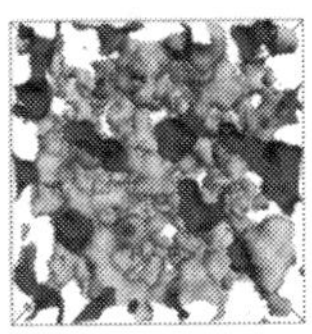

(c) PTFE/BN-550

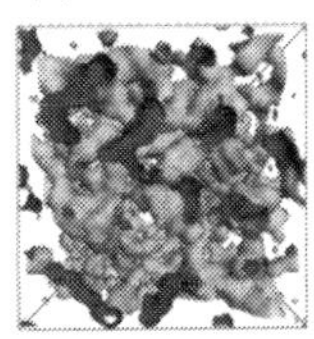

(d) PTFE/BN-560

(e) PTFE/BN-580

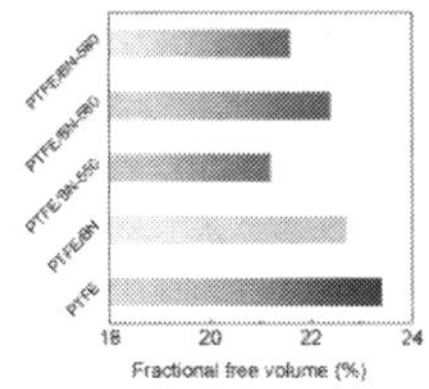


**Figure 6.** Fraction of free volume of various models.

As shown in Figure 6, compared to the pure PTFE, the *FFV* of composite models containing BN filler is significantly reduced, owing to the occupation of composite system space by BN. Specifically, the rigidity of the BN molecular structure hinders the rotation of flexible PTFE chains, hence constraining the free movement of the segments. Moreover, the PTFE/BN-550 composite model presents the maximum value of *FFV* with 21.2%, suggesting that such a coupling agent contributes to strengthening the interactions between the fillers and the polymers. Cohesive energy density (CED) refers to the energy required to eliminate intermolecular forces per unit volume. It is an important indicator of the magnitude of intermolecular forces, and its calculation equation is as follows:

$$E_{\mathrm{coh}} = E_{\mathrm{all}} - E_{\mathrm{molecule}} \tag{11}$$

$$\delta_{CED} = \frac{E_{\mathrm{coh}}}{V} \tag{12}$$

where $E_{\mathrm{coh}}$ is the cohesive energy of molecular segments within the system; $E_{\mathrm{all}}$ is the total energy of the system; $E_{\mathrm{molecule}}$ represents the sum of all forces between different molecules; $V$ is the volume of the system.

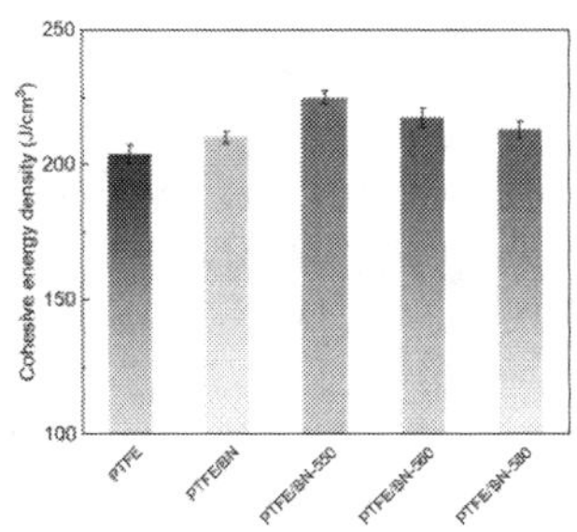


**Figure 7.** Cohesive energy density of various models.

As shown in Figure 7, the addition of BN will increase the CED of the composite system, suggesting stronger thermal stability of the composite. Different coupling agents have different effects on the stability improvement of the composite system, among which BN modified with KH550 exhibits optimal stability. Notably, this variation is also consistent with the change in *FFV*.

## 4. Conclusions

This work explores the influence of different silane coupling agents modified by BN on the thermodynamic properties of PTFE using molecular simulation. The results indicate that the introduction of BN is beneficial for increasing the thermal conductivity and mechanical properties of the PTFE composite. The thermodynamic stability of the composite system modified with KH550 is further enhanced, mainly due to the stronger interaction between KH550 and PTFE molecular chains, which reduces the free volume of the composite system and increases the cohesive energy density. This study provides theoretical guidance for the design of high-performance polymer composites.

**References**

[1] E. Dhanumalayan and G. M. Joshi, "Performance properties and applications of polytetrafluoroethylene (PTFE)—a review", Advanced Composites and Hybrid Materials, vol. 1, no. 2, pp. 247-268, 2018/06/01 2018.
[2] S. Wang et al., "A self-adjusting PTFE/TiO2 hydrophobic double-layer coating for corrosion resistance and electrical insulation", Chemical Engineering Journal, vol. 402, p. 126116, 2020/12/15/ 2020.
[3] M. Feng, Y. Feng, C. Zhang, T. Zhang, Q. Chen, and Q. Chi, "Ultrahigh energy storage performance of all-organic dielectrics at high-temperature by tuning the density and location of traps", Materials Horizons, 10.1039/D2MH00912A vol. 9, no. 12, pp. 3002-3012, 2022.
[4] K. Kim, H. Ju, and J. Kim, "Vertical particle alignment of boron nitride and silicon carbide binary filler system for thermal conductivity enhancement", Composites Science and Technology, vol. 123, pp. 99-105, 2016/02/08/ 2016.
[5] R. Yang et al., "High thermal conductivity insulating AlN/poly(m-phenylenedicarbonyl-m-phenylenediamine) paper realized by enhanced compatibility: a selection of appropriate coupling agent", Journal of Physics D: Applied Physics, vol. 56, no. 44, p. 445501, 2023/08/02 2023.
[6] Z. Liu, J. Li, and X. Liu, "Novel Functionalized BN Nanosheets/Epoxy Composites with Advanced Thermal Conductivity and Mechanical Properties", ACS Applied Materials & Interfaces, vol. 12, no. 5, pp. 6503-6515, 2020/02/05 2020.

# In-situ modification of nickel nanoparticles based on laser-induced graphene and their electromagnetic Properties regulation

**Hu Ge**[1] **and Daming Gao**[1,*]

[1]Hefei University, Hefei, Anhui, 230601, China

[*]Corresponding author's e-mail: dmgao@hfuu.edu.cn

**Abstract:** The investigation of flexible and high-performance materials for shielding against electromagnetic interference (EMI) has consistently remained a prominent subject in the realm of electromagnetic protection. In this study, a porous graphene/nickel nanoparticle (Ni NPs) composite (LIG/Ni) with in-situ modification was fabricated via secondary laser-induced graphene (LIG). The concentration of the $NiSO_4 \cdot 6H_2O$ solution could be adjusted to modify the size and amount of the Ni NPs, which would adapt the electromagnetic characteristics of the composite material. The results demonstrate that when the concentration of $NiSO_4 \cdot 6H_2O$ is 0.4 mol/L, the composite material exhibits a conductivity 3.6 times that of LIG, a saturated magnetization strength 23.2 times that of LIG, and delivers an impressive 56.33 dB shielding effectiveness (SE) in the X-band. Furthermore, even after 1000 bending cycles, it retained 82.23% of its SE. The proposed secondary laser scanning method in an inert atmosphere holds promising application prospects for efficient and rapid large-scale production of flexible electromagnetic shielding materials. It is foreseeable that by altering the precursor solution and the laser scanning environment, this technique could be further expanded to achieve rapid in situ doping of various metals or metal oxides within graphene.

## 1. Introduction

Amid the exponential growth of modern communication technologies and apparatus, the unavoidable spread of electromagnetic wave contamination poses a serious risk [1][2], resulting in severe interference between electronic components, leakage of electromagnetic information, and potential health hazards for individuals, including various forms of cancer, Alzheimer's disease, and profound psychological issues [3]. To address the increasing need to mitigate electromagnetic interference, it is crucial to develop dependable devices for electromagnetic shielding. These devices, which are known as electromagnetic shielding materials, can effectively absorb and reflect electromagnetic waves to achieve electromagnetic protection. The electromagnetic spectrum's mid-to-high frequency region (8.2-12.4 GHz) is home to the X-band, which is widely used in communication systems and is where a lot of radar and navigational equipment operate [4]. These devices exhibit high sensitivity to external electromagnetic interference. Establishing electromagnetic interference shielding for the X-band can effectively safeguard electronic equipment and communication systems in the military and aerospace fields. This endeavor seeks to thwart eavesdropping and disruptions to communication signals, thus ensuring the secure and steady operation of electronic devices. Consequently, research endeavors have focused on crafting efficient EMI shielding materials tailored for the X-band, which carry substantial significance.

Journal of Physics: Conference Series **2961** (2025) 012046 doi:10.1088/1742-6596/2961/1/012046

Currently, efficient EMI materials for the X-band primarily rely on metallic materials. However, owing to the rapid evolution of new generations of flexible electronic products and communication devices, traditional EMI shielding materials represented by metals are gradually being phased out because of limitations, such as high density, difficulty in processing, and lack of deformability [5]. Carbon materials, conductive polymers, and metal nanomaterials are emerging as a new generation of EMI shielding materials, offering lightweight properties and excellent mechanical performance. Notably, graphene exhibits outstanding physical and chemical properties, with a high theoretical electrical conductivity ($10^6$ S/m) and large specific surface area (2630 $m^2/g$), making it a formidable candidate for electromagnetic shielding applications. For instance, a series of reduced graphene oxide aerogels synthesized via methods such as freeze-drying and chemical reduction have demonstrated superior electromagnetic SE in the X-band, achieving over 20 dB shielding efficiency, translating to 99% SE [6]. This underscores the potential of graphene in electromagnetic shielding. Unfortunately, current methods for graphene synthesis often involve high-temperature processing or multi-step chemical routes, leading to the aggregation and stacking of graphene particles upon application [7], thereby limiting its widespread use. Laser-induced graphene (LIG) has recently become a viable substitute [8]. This one-step process offers rapid production of graphene with continuous three-dimensional porous structures, presenting advantages over traditional methods of three-dimensional graphene synthesis [9]. Furthermore, the increasing demand for advanced electronic devices and highly secure military aviation sectors necessitates an enhanced electromagnetic shielding performance. Graphene alone may not be sufficient to meet the shielding requirements of high-precision equipment during operation. Therefore, there is a pressing need to explore composite approaches to improve the electromagnetic shielding efficacy of graphene. However, methods that effectively enhance electromagnetic shielding performance through uniform doping remain to be developed.

According to the principles of the electromagnetic field theory, an electromagnetic field is typically composed of mutually perpendicular magnetic and electric fields. The electric and magnetic fields of high-frequency electromagnetic waves, which define radiation intensity, are inherently interdependent. As such, shielding the electric field may result in effective shielding of the magnetic field and vice versa [10]. Therefore, electromagnetic coupling EMI shielding, magnetic shielding, and electric shielding are the three primary EMI shielding techniques. Conventional shielding materials against electromagnetic interference are mostly composed of conductive metals. The efficacy of shielding materials against electromagnetic interference is improved when incident electromagnetic waves interact with the electrons or holes on their surfaces. This process is known as conductive loss and results in the reflection of electromagnetic waves. Furthermore, magnetic materials can generate eddy currents and magnetic resonance under a magnetic field, leading to magnetic loss and increased absorption of electromagnetic wave energy [11]. Therefore, the key synergistic strategy to improve EMI SE is to blend high-permeability and high-conductivity fillers to impose both magnetic loss and conductive loss collectively.

Through the laser-induced technique in this work, a flexible electromagnetic shielding film composed of three-dimensional graphene and a Ni composite was synthesized. The intricate porous structure of LIG facilitates multiple reflections and scattering, thereby increasing the ability of the material to dissipate the energy of electromagnetic waves and lengthen the path along which they can propagate. Furthermore, Ni nanoparticles can significantly enhance the electrical conductivity and magnetic properties of composite materials. Additionally, the remarkable difference in conductivity between LIG and Ni leads to a wealth of heterogeneous interfaces, resulting in substantial polarization losses. Based on these synergistic internal mechanisms, the LIG/Ni composite material exhibits excellent electromagnetic shielding performance. This study focused on investigating the microstructure, electrical properties, magnetic saturation, and electromagnetic shielding performance of graphene-based composite materials. By optimizing these parameters, a high-performance flexible LIG/Ni electromagnetic interference shielding composite film was rapidly prepared.

## 2. Experimental

### *2.1. Experimental materials and methods*

A commercial polyimide (PI) film (200 μm thickness) as the precursor for LIG and a $CO_2$ laser (Morn Laser, 10.6 μm wavelength) were directly employed at room temperature to produce graphene on the PI film. The initial laser scan was performed with a power of 4.8 W, scanning speed of 150 mm/s, line spacing of 0.125 mm, and bidirectional horizontal scanning mode with a laser spot size of approximately 60 μm. Subsequently, 500 μl of 0.2, 0.4, and 0.6 mol/L $NiSO_4 \cdot 6H_2O$ solutions were dropped onto a 10 mm × 10 mm area of the LIG surface. After thorough impregnation of LIG with the $NiSO_4 \cdot 6H_2O$ solution, the samples were dried in an oven at 60°C for 5 min. This was followed by a secondary laser processing step in an argon atmosphere with the following parameters: laser power of 3.6 W, scanning speed of 150 mm/s, line spacing of 0.125 mm, and bidirectional vertical scanning mode. The resulting samples were named LIG/Ni-X, where X is the concentration of the $NiSO_4 \cdot 6H_2O$ solution. For instance, a sample prepared by adding 100 μl of 0.2 mol/L $NiSO_4 \cdot 6H_2O$ solution according to the procedure described above is denoted as LIG/Ni-0.2. After sample preparation, the samples were rinsed with deionized water to prevent residue from surface compounds.

### *2.2. Characterization*

A Gemini SEM 500 scanning electron microscope (SEM), which enabled the simultaneous observation of the elemental distribution features of the samples by energy-dispersive X-ray spectroscopy (EDS) mapping during scanning, was used to investigate the morphologies of the LIG and Ni NPs. A JEM-2100 F transmission electron microscopy (TEM) was used to examine the nanostructure of the LIG material and Ni NPs, and high-resolution transmission electron microscopy (HRTEM) was used to further characterize the crystal structure of Ni. The crystallinity of LIG and LIG/Ni was assessed by X-ray diffraction (XRD) using Cu Kα radiation. Raman spectroscopy (532 nm laser) was employed to analyze the crystal quality of the graphene composite materials. X-ray photoelectron spectroscopy (XPS; Thermo ESCALAB 250Xi) was used for elemental analysis of the samples. A Suzhou Jingge Electronics Co., Ltd. four-point probe instrument (ST2258C) was used to measure the conductivity and surface resistance of the materials. A vector network analyzer (VNA) with waveguide measurements was used to test the scattering parameters of the samples (S11 and S21) at room temperature after the apparatus had been calibrated. Subsequently, the electromagnetic scattering effectiveness (SE) was calculated using the provided formula.

$$\mathrm{SE}_A = -10\lg\left(\mathrm{T}/(1-\mathrm{R})\right) \tag{1}$$

$$\mathrm{SE}_R = -10\lg\left(1-\mathrm{R}\right) \tag{2}$$

$$\mathrm{SE}_T = \mathrm{SE}_A + \mathrm{SE}_R \tag{3}$$

Here, $R = |S_{11}|^2$, T= $|S_{21}|^2$, and A=1-R-T. R represents the coefficient of reflection, T denotes the coefficient of transmission, and A stands for the coefficient of absorption. The calculation of the SE is based on the "computational theory". Performance comparisons were conducted using commercially purchased shielding films.

## 3. Results and discussions

The preparation process of LIG/Ni is illustrated in Figure 1 (a), where a laser is applied to rapidly transform a carbon-containing polymer (PI) into graphene. After a certain volume of $NiSO_4 \cdot 6H_2O$ solution was applied to the graphene surface and allowed to dry, another $CO_2$ laser scan was performed to achieve in situ growth and even dispersion of the Ni NPs in the graphene. As shown in Figure 1 (b), graphene has exceptional electromagnetic characteristics owing to the homogenous doping of Ni NPs in the LIG phase.

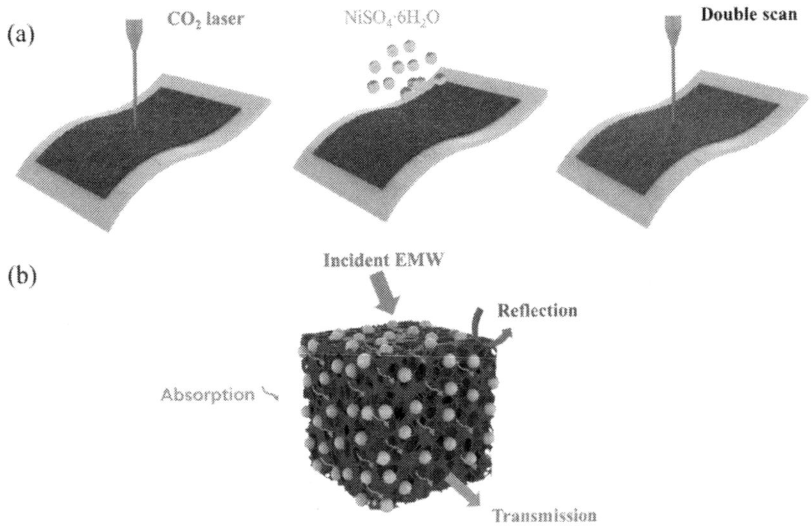


**Figure 1.** (a) Manufacturing of LIG/Ni composite using double scanning. (b) Electromagnetic interference shielding mechanism of the LIG/Ni composite materials.

### *3.1. Morphology and structure*

When the laser interacts with the polymer, it undergoes a localized, transient process of high temperature and pressure followed by rapid cooling without annealing. As a result, the PI precursor polymer can transition to $SP^2$ carbon atoms without the need for a catalyst [12]. Figure 2(a) shows an SEM image of graphene obtained from the laser scan, revealing its surface with three-dimensional interconnected structures and large pores. The magnified image in Figure 2(b) reveals a multitude of mesoporous and microporous structures inside the LIG, mostly due to the quick gas release that occurs during the LIG creation process [9]. A cross-sectional SEM image of the LIG obtained from the first scan is shown in Figure 2(c). This indicates that the LIG has a thickness of approximately 27 μm and a complex porous structure on its side. The $NiSO_4 \cdot 6H_2O$ solution can permeate inside the graphene material because of the three-dimensional porous structure of LIG, allowing for an equitable distribution of $Ni^{2+}$ throughout the LIG. Figures 2(d-f) show the SEM images of LIG/Ni-0.2, LIG/Ni-0.4, and LIG/Ni-0.6, revealing the uniform loading of spherical Ni NPs within the graphene pores, indicating the successful penetration of the $NiSO_4 \cdot 6H_2O$ solution into the LIG pores. Additionally, the particles were uniformly and individually dispersed on the graphene nanosheets without aggregation, even at the nanoscale level. This suggests that the second scanning laser method can achieve a high degree of dispersion at the nanoscale, which is beneficial for improving the performance of composite materials as electromagnetic shields. Furthermore, the amount of Ni NPs increased progressively as the concentration of $NiSO_4 \cdot 6H_2O$ solution increased. It is noteworthy that with an increase in the $NiSO_4 \cdot 6H_2O$ solution concentration, there is a significant variation in the size of the Ni NPs. The size of the Ni NPs in LIG/Ni-0.6 is significantly larger than that in LIG/Ni-0.2 and LIG/Ni-0.4, possibly because of the excessively high concentration of $Ni^{2+}$ ions within the graphene pores, leading to the growth and aggregation of the Ni NPs. Figure 2(g) shows the elemental distribution characteristics of the carbon wall, demonstrating the uniform distribution of carbon and nickel within the LIG/Ni composite, thus confirming the successful deposition of Ni NPs onto the LIG framework. TEM images of LIG/Ni are presented in Figure 2(h). The TEM image reveals the uniform distribution of the LIG/Ni NPs adsorbed onto the LIG flakes.

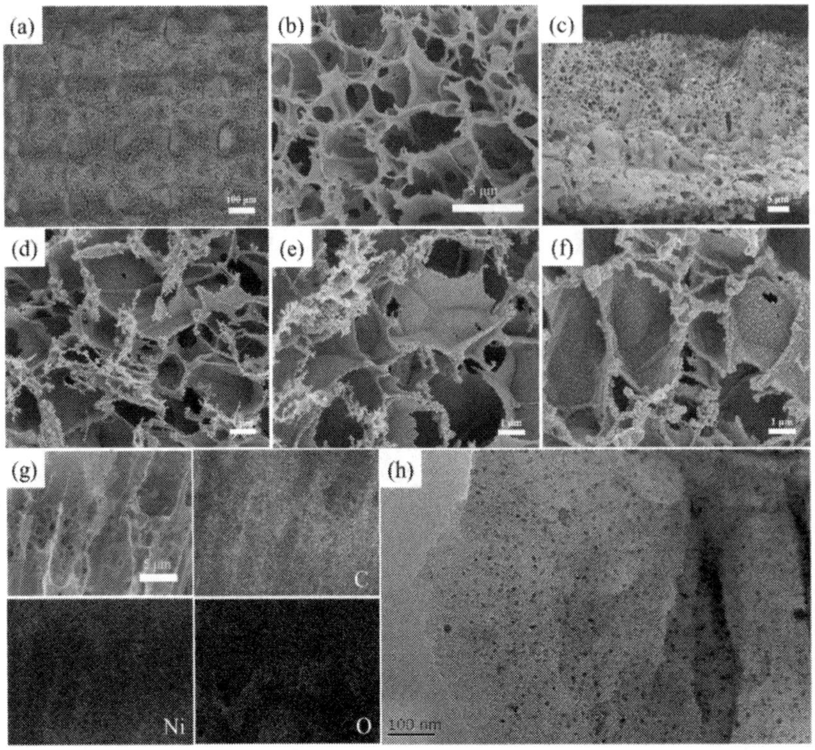


**Figure 2.** (a) LIG SEM images; (b) LIG mesoporous and microporous structures. (g) LIG/Ni EDS mapping images (c) SEM images of LIG cross-section (d–f) SEM images of LIG/Ni-0.2, LIG/Ni-0.4, and LIG/Ni-0.6, respectively. (h) LIG/Ni TEM image.

The Raman spectrum shown in Figure 3(a) accurately shows three characteristic peaks of graphene: the G peak (1581 $cm^{-1}$) representing the sp2 hybridized carbon structure, the D peak (1347 $cm^{-1}$) indicating carbon lattice defects, and the 2D peak (2696 $cm^{-1}$) originating from the second-order two-phonon resonance Raman peak [13]. However, Ni doping leads to a decrease in the peak intensity of graphene, which may be due to the interference of the metal particles with the interaction between the Raman light and graphene. The XRD pattern in Figure 3(b) further reveals the crystal structure of LIG/Ni. The strong, broad peak at $2\theta = 22.40°$ represents the (002) crystal plane of graphene (JCPDS 75-1621) [14], indicating a relatively high degree of carbonization. The X-ray diffraction pattern of the LIG/Ni composite at $2\theta$ values of 45°, 52°, and 78° is aligned with the (110), (200), and (220) planes of Ni (JCPDS No. 04-0850) [15]

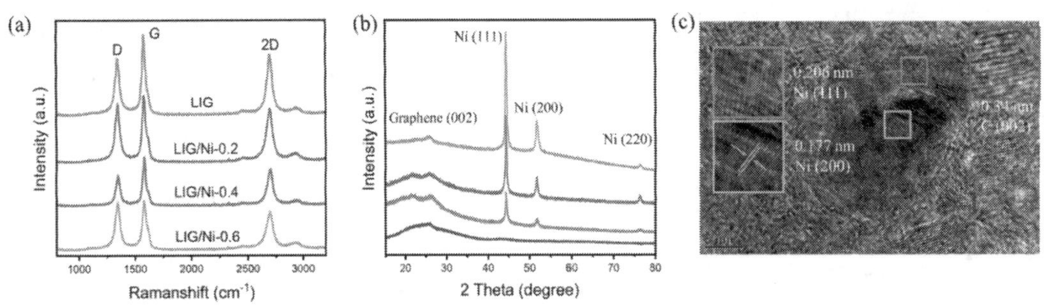


**Figure 3.** (a) Raman spectra of LIG and LIG/Ni (b) XRD patterns of LIG and LIG/Ni (c) HRTEM of LIG/Ni.

Additionally, the lattice spacings of the nanoparticles at 0.178 and 0.205 nm, which correspond to the (200) and (111) crystal planes of Ni, respectively, are visible in the HRTEM image in Figure 2(h). This was consistent with the crystal structure analysis of the XRD data.

Subsequently, LIG/Ni was subjected to XPS analysis. The overall spectrum is shown in Figure 4(a), where the peaks correspond to carbon (C), oxygen (O), and nickel (Ni). The Ni 2p spectrum is shown in Figure 4(b). The presence of metallic Ni is indicated by the peak at 852.5 eV. Ni $2P_{3/2}$ and Ni $2P_{1/2}$ are represented by the peaks at 855.59 eV and 873.39 eV, respectively. Furthermore, the satellite peaks of Ni $2p_{3/2}$ and Ni $2p_{1/2}$ are connected to the peaks at 861.21 eV and 879.29 eV [16]. The O1s spectrum in Figure 4(c) shows that graphene is the source of the binding energies of C-O and C=O, which are 533.0 eV and 531.47 eV, respectively. However, the existence of nickel oxide is indicated by the binding energy of Ni-O, which is 530.4 eV with a lower peak area. While the absence of a NiO peak in the XRD pattern suggests the presence of an amorphous oxide layer on the surface of the Ni NPs, the presence of Ni indicates oxidation of Ni on the surface. It can be deduced that the degree of oxidation is relatively low based on the smaller area of the Ni-O bonding energy peak in the low O1s region in Figure 4(c) and the shallower probing depth of XPS compared to XRD.

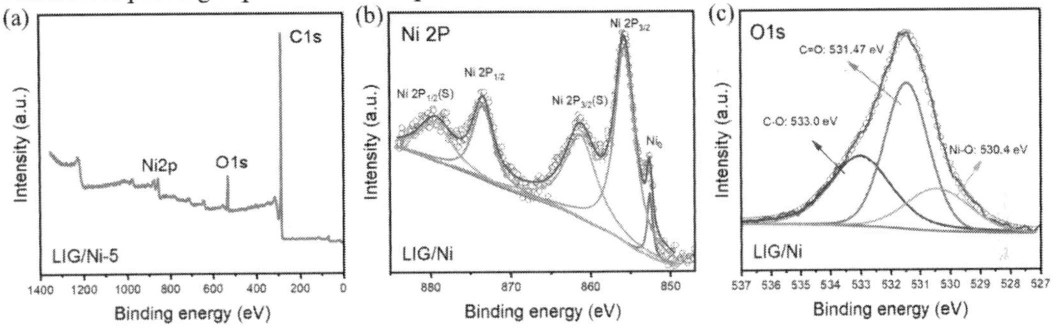


**Figure 4.** (a) XPS spectra of LIG and LIG/Ni (b) Ni 2p spectra of LIG/Ni (c) O1s spectra of LIG/Ni.

*3.2. Electromagnetic properties and electromagnetic shielding performance of LIG/Ni.*

The $SE_T$ value is the sum of $SE_A$ and $SE_R$, with $SE_A$ and $SE_R$ primarily contingent on the electrical conductivity and magnetic permeability of the materials, as shown in Equations (4) and (5):

$$SE_R(dB) = 39.5 + 10\log\sqrt{\frac{\sigma}{2\pi f\mu}} \tag{4}$$

$$SE_A(dB) = 8.7d\sqrt{\pi f\mu\sigma} \tag{5}$$

where σ (S/m), f (MHz), and μ (H/m) denote the electrical conductivity, electromagnetic wave frequency, and magnetic permeability of the material, respectively. Thus, doping LIG with Ni NPs to increase its electrical conductivity is a useful way to increase the effectiveness of the material as an electromagnetic shield [17]. This work suggests the use of a secondary laser to quickly and uniformly introduce Ni NPs into LIG. The abundance of free electrons provided by a large number of Ni NPs to graphene causes a two-fold increase in the electrical conductivity of the LIG/Ni composite material. As shown in Figure 5(a), the electrical conductivity rises with increasing Ni nanoparticle doping concentration. The electrical conductivity of the LIG was measured to be 875.57 S/m. When doped with a relatively low concentration of Ni ions (0.2 mol/L), the limited quantity of Ni NPs results in only a slight increase in the conductivity of the LIG, reaching 1749.82 S/m. However, as the Ni ion concentration increases beyond 0.4 mol/L, the electrical conductivity of the material rises to 3151.44 S/m (LIG/Ni-0.4) and further to 5909.38 S/m (LIG/Ni-0.6). Compared with LIG, the electrical conductivity of LIG/Ni-0.6 was enhanced by approximately 6.75-fold. Improving the electrical conductivity of composite materials is essential to enhance their electromagnetic shielding properties.

The saturation magnetization of the material also increased with increasing Ni-NP content in the LIG, as shown in Figure 5(b). As the number of magnetic particles increases, the magnetization intensity of the material increases, thereby enhancing the magnetic saturation strength. The increase in the saturation magnetization corresponds to an increase in the interfacial magnetic permeability. When the concentration of Ni ions is low (0.2 mol/L), LIG/Ni contains fewer nanoparticles, and the saturation magnetization only increases from 0.2 emu/g in pristine graphene to 2.27 emu/g. The magnetic induction intensity of LIG/Ni-0.4 and LIG/Ni-0.6, respectively, increases dramatically to 4.64 emu/g and 6.38 emu/g with a further increase in the concentration of Ni ions. This implies that the LIG/Ni composite materials can effectively absorb electromagnetic waves through eddy current loss, natural resonance, and exchange resonance magnetic losses.

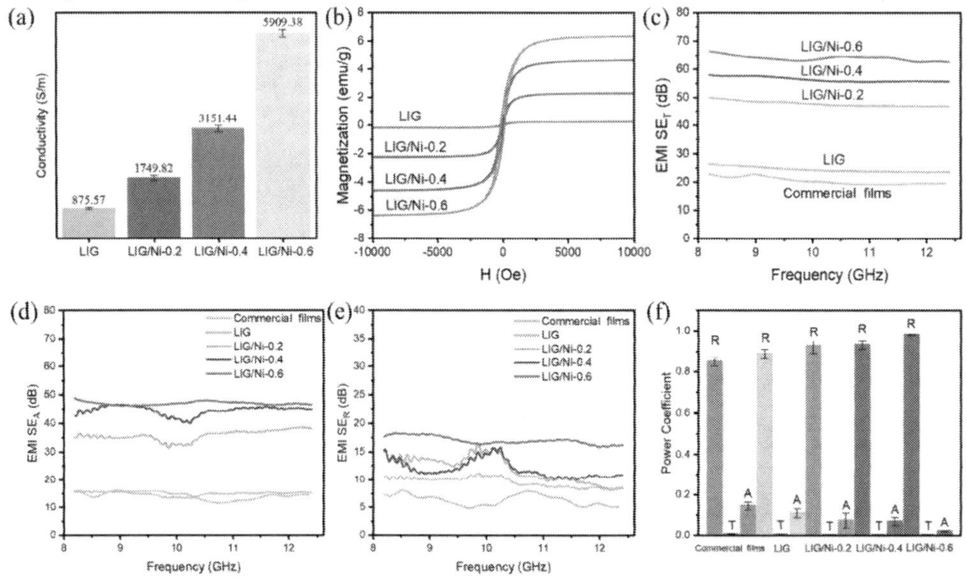


**Figure 5.** (a) Electrical conductivity of LIG and LIG/Ni (b) Magnetic hysteresis loops for the LIG and LIG/Ni (c) EMI SE$_T$ of LIG, LIG/Ni and commercial film (d) EMI SE$_A$ of LIG, LIG/Ni and commercial film(e) EMI SE$_R$ of LIG, LIG/Ni and commercial film (f)A, R, T of LIG, LIG/Ni and commercial film.

The electromagnetic shielding performance of the materials was assessed in the X-band (8.2-12.4 GHz), and the electromagnetic SE of the materials was obtained based on Equations (1) – (3). Figure 5(c) illustrates the total SE (SE$_T$) of LIG, LIG/Ni-0.2, LIG/Ni-0.4, and LIG/Ni-0.6. The SE ranges from 20-62 dB, and it gradually increases with the loading of Ni NPs on the LIG. After loading the Ni particles, the SE of the LIG/Ni composite materials was consistently greater than 20 dB across the entire X-band, indicating compliance with commercial SE standards requiring a minimum of 20 dB, achieving over 99% shielding effectiveness. Specifically, the SE$_T$ of commercial shielding composite films is approximately 20.36 dB, whereas the SE$_T$ of laser-induced graphene is 24.34 dB, a difference of only 3.98 dB. However, the electromagnetic interference shielding ability of LIG/Ni was greatly enhanced by the in situ doping of graphene with Ni. The total SE is 47.71 dB, 56.33 dB, and 63.86 dB at Ni ion concentrations of 0.2, 0.4, and 0.6 mol/L, respectively. This was mainly ascribed to the loading of Ni nanoparticles (Ni NPs) in three-dimensional porous graphene, which increased the magnetic permeability and electrical conductivity of the composite material. Conductive losses result from more electrons and holes engaging in the electromagnetic wave interactions. Furthermore, under the influence of an electromagnetic field, the magnetic characteristics of the Ni NPs promoted magnetic losses in the composite material. Furthermore, doping with Ni NPs increased the heterogeneous interfaces in the composite material, enhancing the polarization losses. Based on these synergistic effects, the LIG/Ni composite material exhibited superior electromagnetic interference shielding performance.

Figure 5(d) shows that the $SE_A$ of the composite material increased dramatically as the number of Ni-NPs in the material increased. Although LIG has an average $SE_A$ of 14.78 dB, LIG/Ni-0.2, LIG/Ni-0.4, and LIG/Ni-0.6 have average $SE_A$ of 35.72 dB, 44.42 dB, and 46.85 dB, respectively. This indicates an enhancement in the composite material's ability to absorb electromagnetic waves through magnetic and polarization losses. In contrast, as shown in Figure 5(e), the increase in $SE_R$ was more gradual. The average $SE_R$ of LIG/Ni-0.6 is the highest at 17 dB, which is only 7.45 dB higher than the average $SE_R$ of LIG. Moreover, the average $SE_A$ of the LIG/Ni composite material was consistently higher than the average $SE_R$. This indicates that the electromagnetic waves entering the material primarily dissipate through absorption, which is also attributed to the significant magnetic and polarization losses induced by the presence of Ni NPs. Further analysis of the material's reflection coefficient (A), absorption coefficient (R), and transmission coefficient (T), as shown in Figure 5(f), indicated that the reflection coefficient of the composite material increased significantly with an increase in the number of Ni NPs. This suggests that when electromagnetic waves act on the material surface, their dissipation of the electromagnetic waves is primarily through reflection. The continuous decrease in the absorption coefficient indicates that the electromagnetic waves dissipated within the material are relatively low, which is mainly due to the significant increase in the surface conductivity of the material caused by the doping with Ni NPs.

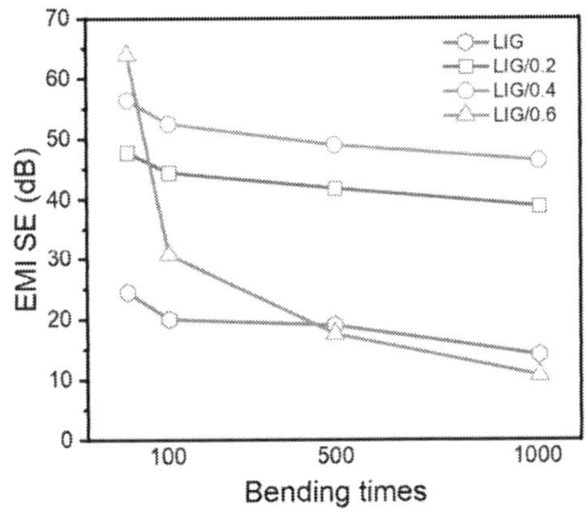


**Figure 6.** EMI SE of LIG and LIG/Ni after bending.

Finally, the LIG and LIG/Ni composite materials were subjected to bending tests (0-90°), where they were bent 100, 500, and 1000 times. The electromagnetic SE measurements were performed after the bending tests. Following 100 bending tests, the SEM SE of the LIG was 13.95 dB, as shown in Figure 6, with a retention rate of 57.31% over the initial value. On the other hand, after 1000 bends, LIG/Ni-0.2 and LIG/Ni-0.4 obtained 38.77 dB and 46.32 dB, respectively, with retention rates of 81.26% and 82.23%. However, the considerable doping of Ni NPs within graphene, which results in deterioration of the mechanical characteristics of the composite material and severe damage during bending, is the reason why the EMI SE of LIG/Ni-0.6 drastically decreased after bending. In contrast, LIG/Ni-0.4 demonstrated superior flexibility and strong electromagnetic interference shielding, which qualified it for use in various flexible electronic components and devices.

## 4. Conclusions

(1) Flexible electromagnetic shielding films of nickel nanoparticles/porous graphene were rapidly fabricated directly on solid materials using a two-step laser method, and by altering the precursor solution and creating a laser scanning environment, the technique can be further expanded to achieve rapid in situ doping of various metals or metal oxides within graphene.

(2) By adjusting the concentration of the $NiSO_4 \cdot 6H_2O$ precursor solution, the electrical conductivity and magnetic permeability of LIG/Ni increased with the increase of Ni loading after laser reduction, with LIG/Ni-0.8 exhibiting the highest electrical conductivity of 5909.38 S/m and the highest magnetic saturation strength of 6.38 emu/g.

(3) The fabricated LIG/Ni-0.4 composite film possessed small-sized Ni NPs with the best dispersion and exhibited excellent electromagnetic shielding and mechanical flexibility. It achieved an SE of 56.33 dB in the X-band, and the SE was maintained at over 80% after 1000 cycles of flexible bending.

## References

[1] Ghaffarkhah A, Hashemi S A, Rostami S, Amini M, Ahmadijokani F, Pournaghshband I A, Mhatre S E., Rojas O J., Kamkar M, Wuttke S, Soroush M, Arjmand M. (2023) Ultra-Flyweight Cryogels of MXene/Graphene Oxide for Electromagnetic Interference Shielding. J. Adv Funct Mater, 33(50): 2304748. https://doi.org/10.1002/adfm.202304748

[2] Lee H, Ryu S H, Kwon S J, Choi J R, Lee S, Park B. (2023) Absorption-Dominant mmWave EMI Shielding Films with Ultralow Reflection using Ferromagnetic Resonance Frequency Tunable M-Type Ferrites. J. Nano-Micro Lett, 15(1): 76. https://doi.org/10.1007/s40820-023-01058-w

[3] Nan Z, Wei W, Lin Z H, Chang J J, Hao Y. (2023) Flexible Nanocomposite Conductors for Electromagnetic Interference Shielding. J. Nano-Micro Letters, 15(1): 172. https://doi.org/10.1007/s40820-023-01122-5

[4] Han M K, Yin X W, Wu H, Hou Z X, Song C Q, Li X L, Zhang L T, Cheng L F. (2016) $Ti_3C_2$ MXenes with Modified Surface for High-Performance Electromagnetic Absorption and Shielding in the X-Band. J. ACS Appl. Mater. Interfaces, 8(32): 21011-21019. https://doi.org/10.1021/acsami.6b06455

[5] Liu H G, Zhu Z L, Tian N, Li Y, You C Y, Islam S M. (2023) Synergistically-enhanced flexible electromagnetic interference shielding nanocomposites. J. Appl. Surf. Sci, 628: 157379. https://doi.org/10.1016/j.apsusc.2023.157379

[6] Bi S G, Zhang L Y, Mu C Z, Liu M, Hu X. (2017) Electromagnetic interference shielding properties and mechanisms of chemically reduced graphene aerogels. J. Appl. Surf. Sci, 412: 529-536. https://doi.org/10.1016/j.apsusc.2017.03.293

[7] Xu Y D, Fei Q H, Page M, Zhao G G, Ling Y, Chen D, Yan Z. (2021) Laser-induced graphene for bioelectronics and soft actuators. J. Nano Res, 14(9): 3033-3050. https://doi.org/10.1007/s12274-021-3441-9

[8] Lin J., Peng Z., Liu Y., Ruiz-Zepeda F., Ye R., Samuel E. L., Yacaman M. J., Yakobson B. I., Tour J. M. (2014) Laser-induced porous graphene films from commercial polymers. J. Nat Commun, 5: 5714. https://doi.org/10.1038/ncomms6714

[9] Ye R Q, James D K., Tour J M. (2019) Laser-Induced Graphene: From Discovery to Translation. J. Adv Mater, 31(1): 1803621. https://doi.org/10.1002/adma.201803621

[10] Xia Y X, Gao W W, Gao C. (2022) A Review on Graphene-Based Electromagnetic Functional Materials: Electromagnetic Wave Shielding and Absorption. J. Adv Funct Mater, 32(42): 2204591. https://doi.org/10.1002/adfm.202204591

[11] Ryu S H, Han Y K, Kwon S J, Kim T, Jung B M, Lee S, Park B. (2022) Absorption-dominant, low reflection EMI shielding materials with integrated metal mesh/TPU/CIP composite. J. Chem. Eng. J, 428: 131167. https://doi.org/10.1016/j.cej.2021.131167

[12] Ye R Q, James D K., Tour J M. (2018) Laser-Induced Graphene. J. Accounts of Chemical Research,51(7): 1609-1620. https://doi.org/10.1021/acs.accounts.8b00084

[13] Silva D L., Campos J L E., Fernandes T F. D., Rocha J N., Machado L R. P., Soares E M., Miquita D R., Miranda H, Rabelo C, Vilela N O P., Jorio A, Cançado L G. (2020) Raman spectroscopy analysis of number of layers in mass-produced graphene flakes. J. Carbon, 161: 181-189. https://doi.org/10.1016/j.carbon.2020.01.050

[14] Yang Q., Yang J., Tang L. L., Zhang H., Wei D. P., Shi H. F., Wei X. Z., Zhang Y. P., Su B. (2023) Superhydrophobic graphene nanowalls for electromagnetic interference shielding and infrared

photodetection via a two-step transfer method. J. Chem. Eng. J, 454:140159. https://doi.org/10.1016/j.cej.2022.140159

[15] Ma Z N, Li J W, Zhang J J, He A, Dong Y Q, Tan G G, Ning M Q, Man Q K, Liu X C. (2021) Ultrathin, flexible, and high-strength Ni/Cu/metallic glass/Cu/Ni composite with alternate magneto-electric structures for electromagnetic shielding. J. J Mater Sci Technol, 81: 43-50. https://doi.org/10.1016/j.jmst.2020.12.012

[16] Yin J, Zhang J X, Zhang S D, Liu C, Yu X L, Chen L Q, Song Y P, Han S, Xi M, Zhang C L, Li N, Wang Z Y. (2021) Flexible 3D porous graphene film decorated with nickel nanoparticles for absorption-dominated electromagnetic interference shielding. J. Chem. Eng.J, 421(1): 129763. https://doi.org/10.1016/j.cej.2021.129763

[17] Feng J, Zong Y, Sun Y, Zhang Y, Yang X, Long G k, Wang Y, Li X H, Zheng X L. (2018) Optimization of porous FeNi3/N-GN composites with superior microwave absorption performance. J. Chem. Eng. J, 345: 441-451.https://doi.org/10.1016/j.cej.2018.04.006

# Synthesis and optical properties of novel $Tm^{3+}$ doped $Gd_3BWO_9$ blue phosphors

**Ruixin Luan***, **Xiaoxia Tian and Zheng Miao**

Department of Basic Science, Air Force Engineering University, Xi'an, Shaanxi 710051, China

*Corresponding author's e-mail: 554929341@qq.com

**Abstract:** The synthesis of blue luminous phosphors doped with $Tm^{3+}$ and $Gd_3BWO_9$ was achieved using a high-temperature solid-state process. Their degree of purity and optical characteristics were extensively examined. Results indicated an optimal $Tm^{3+}$ doping concentration of 0.05 mol. Electro-multipolar interactions were deemed the mechanism behind concentration quenching in $Gd_3BWO_9$ materials. Additionally, $Gd_3BWO_9$:0.02$Tm^{3+}$ exhibits blue fluorescence with color coordinates (0.1727, 0.0832), suggesting its suitability as a blue component in w-LED fabrication.

## 1. Introduction

In the field of contemporary luminescent materials, a comparative analysis highlights the exceptional luminescent efficacy, thermal robustness, and straightforward synthetic procedures exhibited by traditional rare-earth-doped inorganic phosphors in contrast to emerging alternatives such as carbon dots and perovskite quantum dots. These inherent characteristics make them exceedingly desirable for applications requiring exceptional light emission performance and stability across various environmental conditions. White light-emitting diodes (WLEDs) are receiving considerable recognition for their energy efficiency and other advantages [1, 2]. Presently, the prevailing method for creating white LEDs (w-LEDs) entails the fusion of a blue InGaN LED with a yellow inorganic phosphor, specifically $Y_3Al_5O_{12}:Ce^{3+}$ or YAG. However, this approach has significant disadvantages, including a diminished color rendering index caused by the absence of a red hue [3]. A new tricolor fabrication method for w-LEDs has been developed to address these issues. Nevertheless, the red and green phosphors absorb the blue light, decreasing blue emission and declining luminous efficiency [4, 5]. Researchers are currently facing a significant and pressing task of discovering blue-emitting phosphors for lighting purposes.

Tungstate compounds are excellent substrate materials due to their outstanding luminescence properties and good thermal and chemical stability. Additionally, charge-transfer transitions between the $(WO_4)^{2-}$ and $(WO_6)^{6-}$ complexes and the metal cause tungstate-based phosphors to exhibit a wide spectrum of excitation and emission spectra [6]. To find novel illuminating materials, tungstate phosphors doped with rare earth ions have been the subject of much investigation. Examples include $K_{0.5}La_{0.5}SrMgWO_6$: $Mn^{4+}$ [7], $CaGd_2(WO_4)_4$: $Eu^{3+}$ [8], and $Y_2WO_6$: $Ln^{3+}$ (where Ln = Dy, Eu, and Sm) [9]. The $Gd_3BWO_9$: $Yb^{3+}$, $Er^{3+}$ phosphors, under the influence of a 980 nm pump laser, produce a luminous green emission peaking at 522 nm and 543 nm [10]. In this study, the $Gd_3BWO_9:Tm^{3+}$ phosphors were synthesized using a solid-state reaction method. X-ray diffraction (XRD) analysis confirmed the products' phase purity and optical characteristics, and CIE color coordinates were extensively examined.

## 2. Experimental

### *2.1. Synthesis*

To obtain the $Tm^{3+}$-doped $Gd_3BWO_9$ phosphors, a high-temperature solid-state reaction method was used. The synthesis employed high-purity $Gd_2O_3$, $WO_3$, $Tm_2O_3$, and $H_3BO_3$ as the basic ingredients. The raw constituents were precisely measured based on a preset proportion and thereafter extensively pulverized in a quartz mortar for 20 minutes at ambient temperature. After gathering the mixture, it was transferred into a crucible and subjected to a temperature of 600°C for a duration of 3 hours using a muffle furnace. Subsequently, the phosphors that had been created were thoroughly ground at room temperature for a further 20 minutes. The mixture was collected, put into a crucible, and cooked in a muffle furnace for 12 hours at 1000°C. Then, the prepared phosphors were ground fully at normal temperature for another 20 minutes. The experimental reaction was expressed through the following equation:

$$1.5(1\text{-}x)Gd_2O_3+WO_3+H_3BO_3+1.5xTm_2O_3 \xrightarrow[(2)\,1000\,℃\times 12\,h]{(1)\,600\,℃\times 3\,h} Gd_{3(1-x)}Tm_{3x}W(BO_3)O_6$$

### *2.2. Characterization method*

Phase purity is of utmost importance in phosphor materials, ensuring the desired crystal structure and composition. The phase purity of $Gd_3BWO_9$:$Tm^{3+}$ was assessed using X-ray diffraction (XRD) with a radiation wavelength of $\lambda$ = 0.154184 nm. The D2 PHASER instrument from Bruker and Germany was employed for this evaluation. The photoluminescence excitation and emission spectra were obtained using the FLS 980 spectrometer, an advanced instrument manufactured by Edinburgh, UK.

## 3. Results and discussion

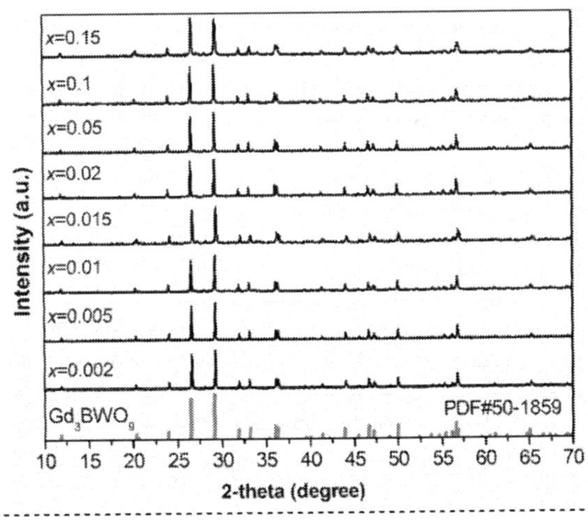


**Figure 1.** The XRD patterns of $Gd_3BWO_9$:$xTm^{3+}$($x$=0.002-0.15) phosphors.

As can be seen in Figure 1, the X-ray powder diffraction pattern of $Gd_3BWO_9$:$Tm^{3+}$ corresponds one to one with the basic peak position of the standard card PDF#50-1859 of matrix $Gd_3BWO_9$. There is basically no shift in the position of the peaks, which indicates that the dopant ions have completely entered the matrix's structure and have not altered the crystal cells. The $Tm^{3+}$ ions in the samples have replaced $Gd^{3+}$ ions in $Gd_3BWO_9$ matrix lattice position in the sample and basically did not change the matrix crystal structure. This is because $Tm^{3+}$ (r = 0.88 Å) and $Gd^{3+}$ (r = 0.938 Å) have similar ionic radii, so the resulting product is pure phase. This suggests that the phase purity of the phosphor is exceptionally excellent.

Equation (1) is used to compute the radius percentage difference ($D_r$) between $Tm^{3+}$ and the possible substituent ions in $Gd_3BWO_9$:

$$D_r = \frac{R_m(CN)-R_d(CN)}{R_m(CN)} \times 100\% \tag{1}$$

A contamination radius is determined as $R_m$ represented by replacement ions, a doped radius, represented by $Tm^{3+}$, and a coordination number, represented by CN. It's evident that the $D_r$ between $B^{3+}$ (CN = 6) to $Tm^{3+}$ (CN = 6) and $Gd^{3+}$ (CN = 6) to $Tm^{3+}$ (CN = 6) is 67.4% and 5.38%, respectively.

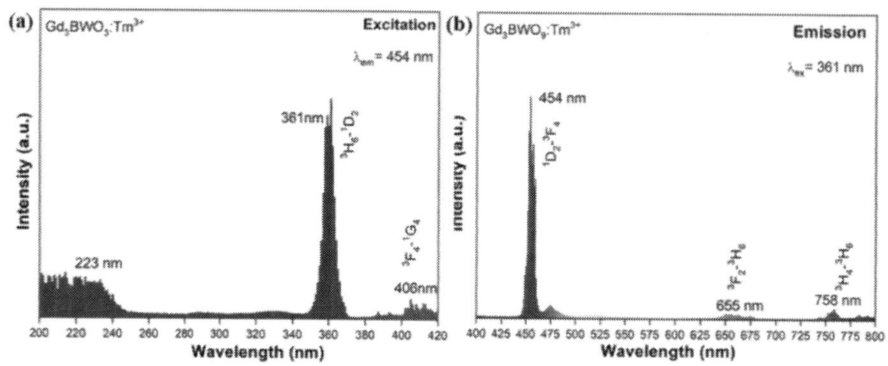


**Figure 2(a)**. The excitation spectrum of sample $Gd_3BWO_9:Tm^{3+}$; (b) The emission spectrum of sample $Gd_3BWO_9:Tm^{3+}$.

From Figure 2(a), it can be seen that 223 nm wavelength is the $(WO_6)^{6-}$ substrate absorption band and 361 nm wavelength violet excitation is a better choice. The $Gd_3BWO_9:Tm^{3+}$ phosphor is suitable for the excitation of violet (361 nm) chips. The 361 nm wavelength is assigned to the $^3H_6 \rightarrow {}^1D_2$ transition of $Tm^{3+}$ ions, while the 406 nm wavelength corresponds to the $^3F_4 \rightarrow {}^1G_4$ transition. Figure 2(b) shows the emission spectrum of $Gd_3BWO_9:Tm^{3+}$. From it, we can see that the peaks of $Gd_3BWO_9:Tm^{3+}$ phosphor under 361 nm UV excitation are located at 454 nm, 655 nm, and 758 nm. From Figure 2(b), it can be seen that the 454 nm emission peak is from $^1D_2 \rightarrow {}^3F_4$, the 655 nm emission peak is from $^3F_2 \rightarrow {}^3H_6$, and the 758 nm emission peak is from $^3H_4 \rightarrow {}^3H_6$. The strongest emission band, 454 nm, is located in the blue region, so the phosphor $Gd_3BWO_9:Tm^{3+}$ is suitable for use as a blue LED phosphor element.

From Figure 3(a), the luminescence intensity of the $Gd_3BWO_9:Tm^{3+}$ phosphor first increases and then decreases as the concentration of $Tm^{3+}$ ions varies. This occurs because, with an increase in the concentration of the dopant ion $Tm^{3+}$, the number of luminescent centers grows, which boosts the emission intensity. However, as the concentration continues to rise, the distance between $Tm^{3+}$ ions decreases, leading to increased energy transfer among the luminescent centers. This heightened energy transfer enhances the likelihood of non-radiative energy transfer, ultimately causing concentration quenching. From Figure 3, it is clear that the optimum $Gd_3BWO_9:Tm^{3+}$ phosphor doping concentration should be around 0.05 mol. Figure 3(b) shows that the upward-pointing arrows indicate excitation from the ground state to various excited states. In contrast, the downward-pointing arrows represent the emission transition from the excited $^1D_2$ to $^3F_4$ levels, resulting in blue emission.

Concentration quenching in rare earth-doped luminous material research is generally ascribed to multipolar contacts and an exchange mechanism, both of which are defined by a critical distance $R_c$. Multipolar interactions become significant when the distance $R_c$ is less than 5 Å; otherwise, these interactions diminish in effect. $R_c$ is derived by Equation (2) [11]:

$$R_c \approx 2\left(\frac{3V}{4\pi\ x_c N}\right)^{1/3} \quad (2)$$

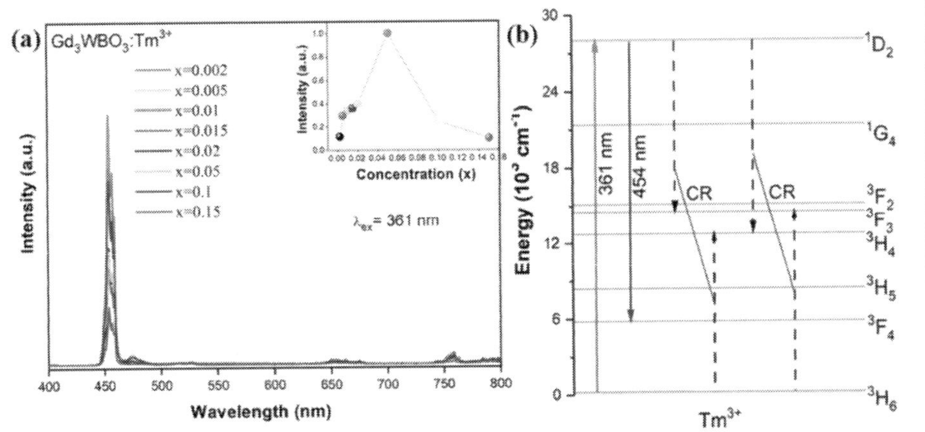


**Figure 3**. (a) PL spectra of $Gd_3BWO_9$:$x$$Tm^{3+}$ ($0.002 \leq x \leq 0.15$) phosphors. Inset: the variation tendency with the doping concentration of $Tm^{3+}$; (b) Energy level transition diagram of $Tm^{3+}$.

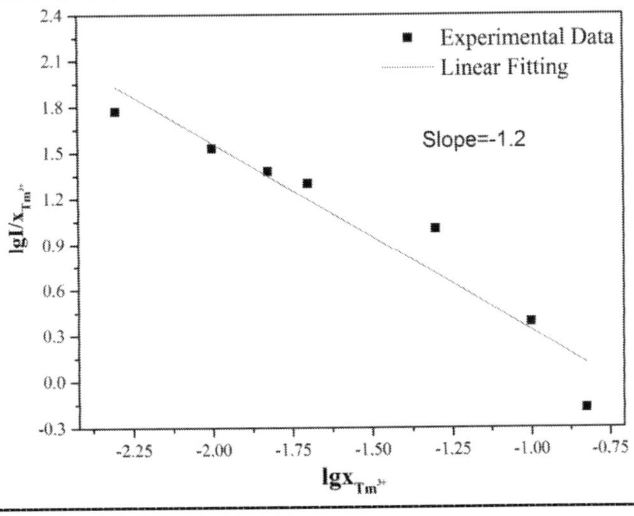


**Figure 4**. Relationship between lgI/$x$ and lg$x$ in $Gd_3BWO_9$:$x$$Tm^{3+}$ ($0.002 \leq x \leq 0.15$) phosphors.

In this context, $N$ is the number of cations that can be used in their place ($N$ = 2), $V$ defines the volume per cell ($V$ = 343.5 Å$^3$), $C$ shows the concentration of $Tm^{3+}$ ions ($x$ = 0.02), and xc refers to the ideal doping concentration. Consequently, the critical distance $R_c$ value is determined to be 25.44 Å. Specifically, since $R_c$ Å, this indicates the predominance of multipolar interactions. $Q$ represents the constant of multipolar interaction, which can take values of 3, 6, 8, or 10. The following equation demonstrates the phenomenon of multilevel interaction, which refers to the intricate interplay and exchange of energy between multiple energy levels within a system. This equation provides a mathematical representation of how these energy levels interact, allowing us to understand and analyze the complex dynamics and transitions occurring within the system. By studying this equation, we can gain insights into the intricate processes that contribute to the overall behavior and properties of the system under investigation [12]:

$$\frac{I}{x} = K\left[1 + \beta\left(x\right)^{Q/3}\right]^{-1} \tag{3}$$

The $x$ indicates the degree of doping in $Tm^{3+}$, the $I$ indicates emission intensity, whereas $K$ and $\beta$ are constants. Figure 4 depicts the correlation between the logarithm of $x$ and the logarithm of $I/x$, given an excitation of 361 nm. The slope is -1.2, corresponding to a coordination number ($Q$ = 3.6) close to 3. This result indicates that the predominant concentration quenching mechanism is the nearest-neighbour interaction, consistent with reports of $Ca_2YTaO_6$:$Tm^{3+}$ and $ZnB_2O_4$:$Tm^{3+}$ [13, 14].

CIE chromaticity coordinates are crucial descriptors for assessing the luminous capabilities of materials. Figure 5 shows the color coordinates of $Gd_3BWO_9$:$Tm^{3+}$ ($0.002 \leq x \leq 0.15$), all of which are consistently positioned within the blue area. Equation (4) can be used to evaluate the color purity of the materials. Equation (4) can be used to assess how pure the color of the materials is.

$$Color\ purity = \sqrt{\frac{(x-x_i)^2+(y-y_i)^2}{(x_d-x_i)^2+(y_d-y_i)^2}} \times 100\% \tag{4}$$

In the formula, ($x_i$, $y_i$) is (0.333, 0.333), and ($x_s$, $y_s$) are the CIE coordinates for different doping concentrations. The intersection point between the straight line connecting the coordinate points ($x_s$, $y_s$) and ($x_i$, $y_i$) and the boundary of the chromaticity diagram is designated as ($x_d$, $y_d$). As illustrated in Figure 5, the products' color coordinates are located in the blue area. The color coordinates of $Gd_3BWO_9$:$x$$Tm^{3+}$ ($0.002 \leq x \leq 0.15$) are (0.1967, 0.1431), (0.1646, 0.0625), (0.1593, 0.0464), (0.1607, 0.0535), (0.1727, 0.0832), (0.1775, 0.0925), (0.1577, 0.0436), and (0.1812, 0.1065), respectively. In the $Gd_3BWO_9$:0.10$Tm^{3+}$ complex, the color purity is estimated to be approximately 94.4%, exceeding the 92% of commercial blue phosphor $BaMgAl_{10}O_{17}$:$Eu^{2+}$ phosphor. When $Gd_3BWO_9$:0.10$Tm^{3+}$ is stimulated by ultraviolet (UV) radiation, it may effectively produce blue light with high color purity.

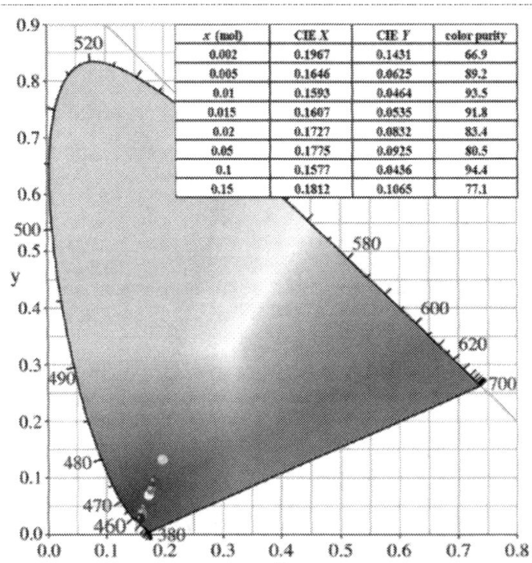


| x (mol) | CIE X | CIE Y | color purity |
|---|---|---|---|
| 0.002 | 0.1967 | 0.1431 | 66.9 |
| 0.005 | 0.1646 | 0.0625 | 89.2 |
| 0.01 | 0.1593 | 0.0464 | 93.5 |
| 0.015 | 0.1607 | 0.0535 | 91.8 |
| 0.02 | 0.1727 | 0.0832 | 83.4 |
| 0.05 | 0.1775 | 0.0925 | 80.5 |
| 0.1 | 0.1577 | 0.0436 | 94.4 |
| 0.15 | 0.1812 | 0.1065 | 77.1 |

**Figure 5**. CIE color coordinates of the $Gd_3BWO_9$:$x$$Tm^{3+}$ ($0.002 \leq x \leq 0.15$) phosphors.

## 4. Conclusions

$Gd_3BWO_9$:$x$$Tm^{3+}$ phosphors were produced for the first time using a typical high-temperature technique, with $x$ ranging from 0.002 to 0.15. Upon stimulation with light at a wavelength of 361 nm, these phosphors displayed prominent emission peaks at 454 nm. A doping concentration of 0.05 mol $Tm^{3+}$ is ideal in $Gd_3BWO_9$. The phenomenon of concentration quenching is proposed as the mechanism for the contacts between ions that are closest to each other. This interaction occurs when the ions are within a threshold of around 25.44 Å. The CIE color coordinates closely approximate the conventional blue position. These results indicate that $Gd_3BWO_9$:$Tm^{3+}$ products have excellent blue-emission capability and color purity of 94.4%, highlighting their significant potential for w-LED applications.

**Acknowledgments**

This work was supported by the Natural Science Foundation of Shaanxi Province (No. 2023-JC-YB-105)

## References

[1]A. V. Egorysheva, P. A. Ryabochkina, E. F. Popova, S. A. Khrushchalina, I. A. Yurlov, S. V. Golodukhina, Opt. Mater. **152**, 115425 (2024) [CrossRef]

[2]J. Liu, L. Lu, X. Liu, X. Mi, J. Lumin. **275**, 120793 (2024) [CrossRef]

[3]Y. Xiang, H. Tang, C. Liao, H. Li, J. Li, H. Zhang, C. Liu, J. Zhu, J. Alloy. Compd. **1005**, 176027 (2024) [CrossRef]

[4]H. Dong, Z. Lei, S. Su, W. Yang, X. Zhang, W. Teng, G. Zu, B. Teng, D. Zhong, Inorg. Chem. Front. **11**, 2784 (2024) [CrossRef]

[5]A. Drozdowski, D. Poelman, M. Runowski, H. Hemmerich, F. Rivera-López, T. Grzyb, J. Mater. Chem. C **12**, 13040 (2024) [CrossRef]

[6]F. Luan, Z. Jiang, Q. Zeng, H. Du, N. Zhang, C. Yin, D. Guo, Opt. Mater. **150**, 115127 (2024) [CrossRef]

[7]J. k. Liu, Z.-h. Long, S. Xiao, X. Yang, Mater. Today Commun. **31**, 103214 (2022) [CrossRef]

[8]G. Li, Y. Wei, Z. Li, Y Yang, G. Xu, Opt. Mater. **66**, 253 (2017) [CrossRef]

[9]M. N. Huang, Y. Y. Ma, F. Xiao, Q. Y. Zhang, Spectrochim. Acta, Part A **120**, 55 (2014) [CrossRef]

[10]W. Wei, Y. Dai, D. Li, X Yao, J. Xu, H. Li, C. Duan, Q. Zhao, J. Solid State Chem. **319**, 123789 (2023) [CrossRef]

[11]G. Blasse, B. Grabmaier, Radiative return to the ground state: Emission, Luminescent Materials, Springer, 1994, pp. 33. [CrossRef]

[12]Y. Wu, X. Chen, Z. Gong, B. Tian, E. Xue, K. Zheng, J. Liang, W. Wu, J. Mater. Chem. C **12**, 11497 (2024) [CrossRef]

[13]H. Lin, S. Fang, T. Lang, J. Yu, H. Cheng, J. Ou, Z. Ye, R. Xu, X. Shui, H. Qu, L. Wang, J. Mater. Chem. C (2024) [CrossRef]

[14]N. Zhang, H. Zhou, T. Wang, X. Ma, N. Lin, X. Fu, Y. Yin, J. Zhang, L. Ye, Z. Jia, Ceram. Int. **50**, 36137 (2024) [CrossRef]

# Analysis of thermal and mechanical properties of transparent nylon modified PA66/GF system

**Enyong Tian [1], Lin Zhang [1, *], Junjie Wang [1] and Qiang Mao [1]**

[1]Guizhou power grid Co., LTD Guiyang power supply bureau, Guiyang, China

[*]Corresponding author's e-mail: 1965468227@qq.com

**Abstract:** PA66 was selected as the reaction matrix for preparing the modified composite system. Nylon 66 was modified using GF and maleic anhydride grafted polypropylene copolymer. Based on compatibilization and blending technology, a PA66/GF modified composite system was prepared by combining toughening agents and compatibilizers. The specific effects of different transparent nylon contents on the thermal and mechanical properties of the composite system were tested and analyzed using sample testing methods. The experimental results show that adding 30% transparent nylon to the modified composite blend system can significantly improve the system's overall performance. Specifically, the mass fraction of the composite material can be increased by 30% to 40%, with the best melting point performance. The crystallization temperature increased by 81.4℃ compared to the sample without transparent nylon and the highest tensile strength. The notch impact strength of the material reaches 8.4kJ · m-2. Compared with the EPDM modification method, this article's melting behavior and tensile strength are better, indicating that the research method can further optimize the comprehensive performance of composite materials.

## 1. Introduction

Nylon PA66 features a large number of amide groups attached to its molecular chains, forming a relatively stable copolymer structure through cross-repeated combinations. It possesses several advantages, including a high melting point, high friction resistance, and high compressive strength. Additionally, it exhibits comprehensive properties such as heat resistance and corrosion resistance [1]. In engineering applications, the demand for Nylon PA66 has consistently been at the forefront of engineering plastics. By incorporating GF (Glass Fiber) into PA66 materials through blending techniques, modified PA66/GF systems can be obtained, leading to further enhancements in composite materials' mechanical, physical, and mechanical properties. To improve various performance aspects, expand its range of applications, and facilitate its broader usage in industries like defense and machinery, this study conducts an in-depth investigation into the influence of varying levels of transparent nylon addition on the comprehensive performance of modified composite systems.

In this research, PA66 and PA6T serve as the primary raw materials. Along with selecting other experimental materials and equipment, the experimental preparations involve using compatibilizers and toughening agents. Utilizing techniques such as melt blending, extrusion granulation, and injection molding, PA66/GF systems are prepared using the compounding method. The study tests and analyzes these systems' thermal and mechanical properties. Specifically, it investigates the impact of transparent nylon content on the composite material's thermogravimetric properties, melt behavior, crystallization behavior, tensile strength, and notch impact strength. The aim is to provide valuable insights for the development and comprehensive performance analysis of similar materials.

## 2. Experimental preparation

The blending modification of PA66/GF systems is a primary method for enhancing modified engineering plastics' physical and chemical properties. Relevant experimental results indicate that by blending PA with other polymers, it is possible to improve dimensional stability to some extent, thus expanding the applicability of PA66/GF systems [2].

To elucidate the specific relationship between the added amount of different transparent nylons and the performance of modified composite systems, PA66, PA6T, and GF were used as the main raw materials in the experiment. Comprehensive performance-enhanced PA66/GF composite materials were prepared using a compatibilization blending method under other additive conditions [3]. The influence of varying the content of transparent nylon on the performance of PA66/GF composite materials was investigated.

### *2.1. Experimental materials and equipment*

The thermoplastic material used in the PA66/GF composite system needs to have characteristics of significantly different strengths and thermodynamic incompatibility, weak interaction forces, and ease of dispersion. Therefore, this experiment used nylon resin and glass fiber as the main reactants [4], with transparent nylon as the filling modification material. Compatibilizers and toughening agents were selected as needed for the experiment to prepare a blend system. The experimental materials and equipment used are shown in Table 1.

**Table 1.** Experimental materials and equipment.

| Material Names | Model Numbers | Instrument and Equipment Names | Model Numbers |
|---|---|---|---|
| PA66 | EPR27 | Vacuum Drying Oven | DZF-6050 |
| PA6T | M5000 | High-Speed Mixer | SHR-10A |
| GF | T538 | Twin-Screw Extruder | SHJ |
| Anhydrous Ethanol | Analytical grade | Injection Molding Machine | HTF86 |
| Polypropylene | F401 | Cantilever Beam Impact Testing Machine | XJ-300A |
| Maleic Anhydride Grafted Polypropylene | CMG9801-GS | Granulator | Fei Quanta 200 |
| Compatibilizer | N190X-71283 | Scanning Calorimeter | DSC214 |
| Calcium Hydroxide | Analytical Grade | Universal Testing Machine | RL-Z1B1 |
| Ibuprofen | Analytical Grade | Micro-impact Testing Machine | XUJ-5.5A |
| Hexane diamine | Chemically Pure | Simultaneous Thermal Analyzer | STA449C |
| Silane Coupling Agent | KH-550 | Heat Press Machine | R-3201 |

### *2.2. Experimental method*

The preparation process of the PA66/GF blend modification system includes three steps: raw material mixing, extrusion granulation, and injection molding. PA66 is a high-strength engineering plastic, while GF is a high-strength and rigidity-reinforcing material. By blending GF with PA66, the material's strength and rigidity can be significantly enhanced while maintaining the excellent performance of PA66. Additionally, PA66 itself exhibits good heat resistance, and adding GF further enhances the material's heat resistance. The choice of using the PA66/GF blending modification system is to utilize the advantages of both PA66 and GF fully. Through adjustments in the formulation and processing techniques, a high-strength, heat-resistant, corrosion-resistant, and relatively lightweight engineering plastic material is achieved. The process flowchart is shown in Figure 1.

In preparing PA66/GF composite materials, the melting temperature of the component polymers should be significantly lower than the heating temperature of the twin-screw extruder to ensure that the polymers can fully melt and mix uniformly. Polypropylene and branched polypropylene are used as compatibilizers between the components, and composite granules are fed into the system at a temperature between the melting points of each component to obtain an isotropic PA66/GF system [5].

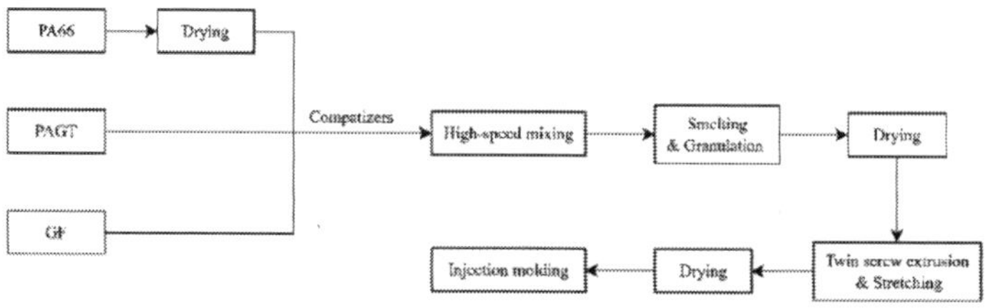


**Figure 1.** Process flowchart for PA66/GF system preparation.

## 3. Composite material preparation

### *3.1. Melt blending*

Mixing materials is an essential process in the preparation of composite materials. The purpose of mixing is to ensure that additives are evenly distributed within the matrix resin, thus ensuring the preparation of a more stable mixture [6]. The formulation for composite materials is shown in Table 2.

**Table 2.** PA66/GF blend material formulation (%).

| Sample | Transparent nylon | PA66 | PA6T | GF | Compatibilizer | Compatibilizer |
|---|---|---|---|---|---|---|
| A1 | 0 | 40 | 8 | 30 | 0 | 5 |
| A2 | 10 | 52 | 8 | 30 | 0 | 5 |
| A3 | 20 | 47 | 8 | 25 | 10 | 0 |
| A4 | 30 | 42 | 10 | 25 | 10 | 0 |
| A5 | 40 | 50 | 10 | 20 | 20 | 10 |
| A6 | 50 | 36 | 10 | 20 | 20 | 10 |

By a 1:2:1 ratio, we take nylon 66 (PA66), glass fiber (GF), and a toughening agent, mixing them in a hot press to create a polymer blend. Afterward, we remove the blend from the hot press and let it sit at room temperature for 18 to 24 hours, allowing it to fully cool and partially solidify to minimize the moisture content in the nylon. This step is essential to avoid any potential influence on the performance test results due to moisture in the nylon. We weigh according to the formula in Table 2 and place it into a high-speed mixer for 6 minutes; Next, we add the mixed particles into a co-rotating twin-screw extruder for blending and extrusion (preheating the twin screws and maintaining them at a constant temperature for 3 hours) [7]. The screws rotate at a speed of 200 revolutions per minute (rpm), and the material is discharged after achieving uniform mixing.

### *3.2. Extrusion granulation*

In the next step, the mixed granules obtained in the previous step are reintroduced into the co-rotating twin-screw extruder for extrusion. Then, they are stretched into fine fibers using a drawing machine.

Journal of Physics: Conference Series **2961** (2025) 012048 doi:10.1088/1742-6596/2961/1/012048

During this process, the dispersed phase PA66 is in a solid or semi-molten state, and after stretching, microfibers are formed [8]. Simultaneously, we adjust the speed of the drawing machine to obtain fine threads with a stretch ratio of $(\lambda = 4)$. The calculation formula for the stretch ratio is as follows:

$$\lambda = \frac{S_1}{S_2} \tag{1}$$

In the above equation, $S_1$ and $S_2$ represent the cross-sectional areas of the die opening and the cylindrical thread, respectively.

We cut the long fibers that meet the desired stretch ratio using a pelletizer and place them in a vacuum dryer for 4 to 6 hours to remove moisture. The dried solid material is transferred to a temperature-controlled instrument, and a catalytic reaction is initiated at a high temperature of 110°C. The mixed material is introduced in the first feed inlet of the twin-screw extruder. Modified glass fibers (GF) are added to the second feed inlet. We control the extruder's temperature in the 120-130°C [9], adjust the screw speed, and modify the GF content as needed. After the material has melted and plasticized, it is extruded, and then it is cooled and granulated.

### *3.3. Injection molding*

We add maleic anhydride-grafted polypropylene, citric acid, hexamethylenediamine, and deionized water into a high-temperature, high-pressure reactor in the proportions specified in Table 2. We inject nitrogen with a concentration of over 80% into the reactor for 15 minutes to completely remove air from the reactor. Then, we close the reactor and start stirring at a rotation speed of 80 rpm, raising the temperature to 180-200°C within 2 hours while maintaining a pressure of 1.5-2.5 MPa [10]. After maintaining pressure, we open the exhaust valve and gradually reduce the pressure to zero within approximately 0.5 hours. We continue heating until the temperature reaches 130-150°C, create a vacuum in the reactor to a pressure of -2.0 Mpa, and dry the extracted copolymer in a convection oven at 100°C for 12 hours to remove surface moisture. This prepares the copolymer for subsequent testing and processing applications.

After replacing the injection molding machine with the mold for testing samples, installing heating rods, and adjusting the mold, once everything is ready, we preheat the injection molding machine for 4 hours, mix the extracted and dried GF/PA66 (mass ratio 3:1) with 3.5 grams of calcium hydroxide (Ca $(OH)_2$) in a high-speed mixer for 15 minutes, and add the premix into the injection molding machine to prepare standard samples [11]. After drying, we inject the extruded pellets into standard dumbbell-shaped tensile bars (40 mm X 15 mm X 5 mm) and rectangular bars (80 mm X 10 mm X 8 mm). The barrel temperatures of the injection molding machine are set at 200°C, 220°C, and 252°C, while the mold temperature ranges from 180°C to 210°C. The injection molding pressure is 65 MPa. We cut one-third of the rectangular specimens with a 35° angled notch with a depth of 2.5 mm, which will be used for subsequent notch impact strength testing [12-14].

Before conducting various performance tests, we allow the samples to sit at room temperature for at least 24 hours to ensure they have fully cooled and crystallized.

## 4. PA66/GF composite system thermal performance testing

The above-prepared PA66/GF composite material samples were used as experimental specimens to test the composite material's thermal performance by varying the transparent nylon content in the system. Among them, the thermal performance includes material thermal weight loss, melting behavior, and crystallization behavior.

### *4.1. Thermal performance testing*

*4.1.1 Thermogravimetric analysis.* We weigh 2-5 mg of the sample and scan PA66 and its flame-retardant composite materials at a heating rate of 10°C/min from 10°C to 500°C. This scanning is

Journal of Physics: Conference Series **2961** (2025) 012048
doi:10.1088/1742-6596/2961/1/012048

conducted under a nitrogen atmosphere with a nitrogen flow rate of 1.2 mL/min and an airflow rate of 2.5 mL/min. In an air atmosphere, both oxygen and nitrogen have an average flow rate of 2.2 mL/min.

For the analysis of thermal weight changes in the modified blend system, samples A1 and A2 were selected. The mass retention rate (TG) and mass loss rate (DTG) of the system were recorded using a thermogravimetric analyzer to analyze the thermal stability of the system [12]. The results are shown in Figure 2.

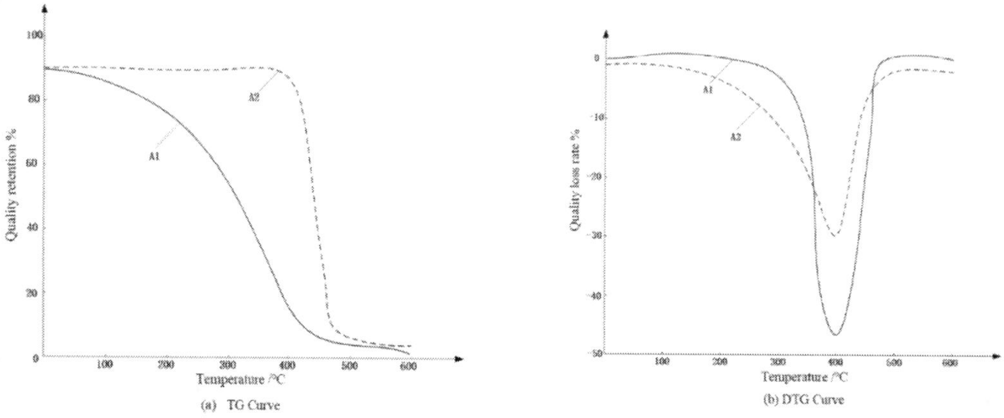


**Figure 2**. Thermal stability curve of the composite material.

In the figure, A1 represents a sample with 0% transparent nylon content, while A2 represents a sample with 10% transparent nylon content. The figure shows that the composite material undergoes two distinct processes during thermal degradation: the thermal degradation process of PA66/GF material derivatives and the initial thermal degradation process of nylon 66 [13]. In Figure 2(b), for A1, the first peak occurs at around 330°C, and the second peak is around 400°C. By analyzing the TG curve results under nitrogen conditions, it can be observed that appropriately increasing the content of transparent nylon can advance the composite material's initial softening temperature while reducing its maximum decomposition temperature. When the content of transparent nylon is increased to 10%, the composite material can retain an additional 30% to 40% mass fraction compared to the composite material without transparent nylon. This indicates that during the degradation process, transparent nylon can absorb a significant amount of energy at the structural cross-section, thereby enhancing the flame-retardant performance of the material. It suggests that introducing transparent nylon into the PA66/GF system can significantly improve its flame-retardant properties.

*4.1.2 Melt behavior analysis.* According to GB/T 1634.1-2004 requirements, a melt analysis of composite materials should be conducted. Test samples should be selected from A1 to A5 samples. Approximately 0.5 to 2.0 mg of PA66/GF sample should be weighed and placed in a crucible. The crucible is then placed inside the instrument furnace. Under a nitrogen environment, the temperature of the composite material should be increased from 0°C to 220°C at a rate of 20°C/min and then decreased back to 20°C while recording the temperature-melting behavior curve of the material.

Transparent nylon can enhance the compatibility of the PA66/GF blend and alter its crystalline morphology, thereby influencing its melt behavior. During the experimental process, a synchronous thermal analyzer was used to analyze the impact of varying amounts of transparent nylon on the melt crystallization of the blend, as shown in Figure 3.

By analyzing the above figure, it can be observed that with an increase in the content of transparent nylon, the melting peak of the PA66/GF system gradually shifts towards lower temperatures. However, the overall shape of the melting curve remains relatively unchanged. The melting peak temperature for sample A1 is 475°C, while for samples A2, A3, A4, and A5, the melting peak temperatures are 412°C, 365°C, 302°C, and 282°C, respectively. This represents a decrease of 63°C to 193°C compared to

Sample A1. It indicates that the addition of transparent nylon does not alter the crystalline structure of the PA66/GF matrix, suggesting that transparent nylon affects its chemical structure energetics. As the mass fraction of transparent nylon gradually increases, the formation of hydrogen bonds between amide groups in transparent nylon is hindered due to polymer chains [15]. This leads to the instability of copolymer molecules formed during the compatibility reaction, increasing the degree of disorder between molecules and consequently raising the system's melting point. Therefore, by analyzing the above data, it can be concluded that the higher the content of transparent nylon, the higher the melting point of the PA66/GF system.

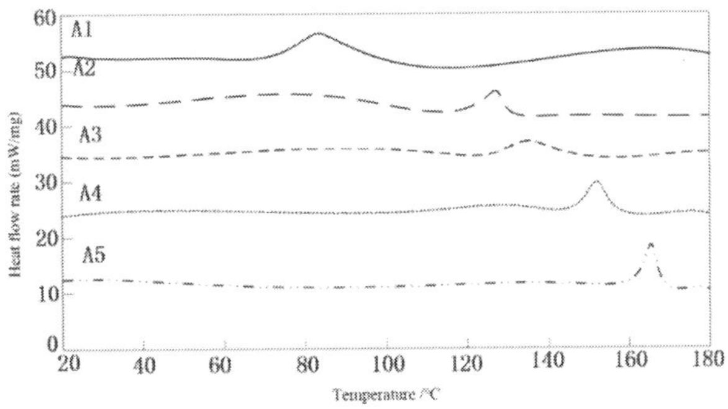


**Figure 3**. Heating melt curve of the composite material.

*4.1.3 Crystallization behavior analysis.* We utilize the DSC 214 Differential Scanning Calorimetry (DSC) method to analyze the crystallization behavior of the sample. Using an electronic balance, we weigh 6-8 mg of the sample from the impact specimen and place it in a crucible, compacting it. The crucible is heated and transferred to the DSC instrument. Initially, we raise the temperature to 120°C and hold it at this temperature for 5 minutes to eliminate any thermal history of the sample. At a rate of 15°C/min, we gradually lower the temperature to 20°C to obtain the crystallization behavior curve of the sample, allowing for the analysis of the sample's crystallization process. Then, we calculate the crystallinity of PA66/GF using the following formula:

$$X = \frac{H_1}{H_2} \times 100\% \qquad (2)$$

In the above equation, $H_1$ represents the material's melting enthalpy in J/g, and $H_2$ represents the instantaneous melting enthalpy when the material crystallizes. In this paper, the value of $H_2$ is taken as 190 J/g.

Based on the SDC readings, an analysis of the influence of different transparent nylon contents on the crystallization behavior of the PA66/GF system was conducted, resulting in the cooling crystallization curve, as shown in Figure 4. The corresponding crystallization parameters are presented in Table 3.

Combining Figure 4 with Table 3, it can be observed that the crystallization temperature of the pure PA66/GF system, without the addition of transparent nylon material, is 87.3°C. After adding transparent nylon with mass fractions of 10%, 20%, 30%, and 40% successively into the composite material, the crystallization temperature of the system gradually increases. It increases by a maximum of 81.4°C compared to sample A1. The main reason for this is that transparent nylon can act as a homogenous nucleating agent in the PA66/GF system. Simultaneously, it can alter the material's structural characteristics, such as anisotropy and the orientation of toughening agents. Moreover, as the content of one of the monomers increases, the material's internal structure gradually tends toward the amorphous phase. This results in the broadening of the peak shape of the composite material and an increase in the crystallization temperature.

**Table 3**. Crystallization parameters of the system.

| Numbering | Melting Temperature /°C | Melting Enthalpy (J/g) | Crystallization Temperature /°C | Crystallinity /% |
|---|---|---|---|---|
| A1 | 475 | 65.9 | 222.6 | 87.3 |
| A2 | 412 | 56.3 | 204.1 | 122.5 |
| A3 | 365 | 44.2 | 168.5 | 138.6 |
| A4 | 302 | 33.0 | 145.2 | 157.6 |
| A5 | 282 | 31.5 | 133.6 | 168.7 |

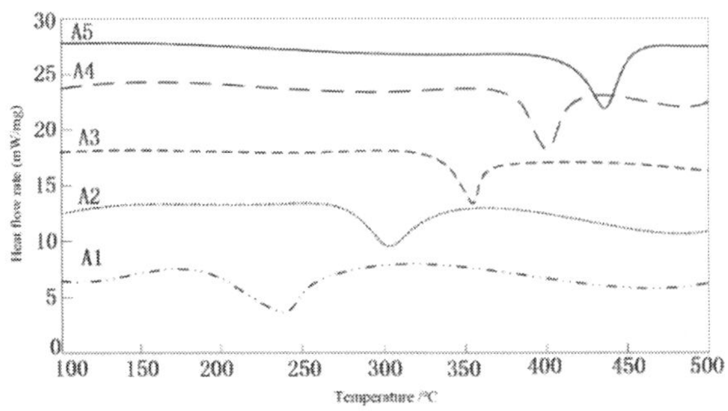


**Figure 4**. Cooling crystallization curve of the composite material.

## *4.2. Mechanical properties testing*

*4.2.1 Tensile strength analysis.* According to the relevant standards for tensile strength testing of the PA66/GF composite system, the tensile strength of the samples is tested using an electronic universal testing machine. The calculation formula is as follows:

$$\varepsilon = \frac{F_m}{S_0} \tag{3}$$

In the above formula, $F_m$ represents the ultimate load-bearing capacity of the material, and $S_0$ represents the cross-sectional area of the material at the point of rupture.

Following the test requirements, the tensile strength of the instrument is set to 40 mm/min. Ten samples are chosen for each formula, using the formula mentioned earlier. The tensile strength for each sample is calculated. The average of the results from the ten measurements is then taken. A curve depicting the variation of tensile strength with nylon content is plotted, as shown in Figure 5.

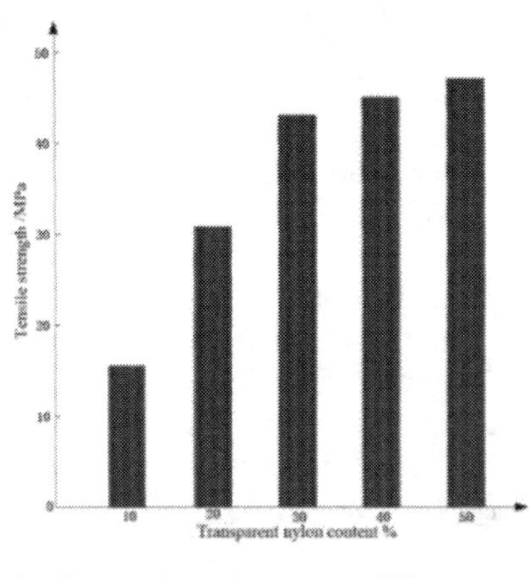


**Figure 5**. Tensile strength of PA66/GF system at different transparent nylon contents.

 IOP Publishing
doi:10.1088/1742-6596/2961/1/012048

Based on the above figure, it can be observed that the greater the amount of nylon added, the higher the tensile strength of the PA66/GF system. This is primarily because the intermolecular entanglement interactions in transparent nylon can increase the forces between the shear plastic groups in the composite material. The material's ultimate tensile strength is enhanced with the functional groups' effects. When the mass fraction of transparent nylon is 10% to 30%, the material's tensile strength increases rapidly. However, when the mass fraction of transparent nylon exceeds 30%, the rate of increase in tensile strength gradually slows down. The reason for this phenomenon is that composite materials inherently exhibit rigidity. As the load-bearing capacity of the composite material increases, the polymer chains in the material do not have enough time to move, and the rate of structural entanglement decreases. Therefore, a noticeable "reaction lag" phenomenon occurs. Therefore, by adding transparent nylon, it is possible to significantly improve the tensile strength of the PA66/GF system.

*4.2.2 Notched impact strength analysis.* Testing the notched impact strength of samples was conducted using the XJ-300A cantilever impact testing machine. In the test, the machine's vibration mode was set to a single cantilever beam with an initial loading frequency of 0.5 Hz. The vibration frequency was then uniformly increased at intervals of 10 seconds until it reached 20 Hz. The temperature varied within the range of 40°C to 100°C. The formula for calculating the notched impact strength of the test specimen is provided below.

$$\sigma = \frac{C}{b \times d} \tag{4}$$

In the above formula, $C$ represents the energy absorbed by the specimen; $b$ represents the cross-sectional area of the specimen; $d$ represents the depth of the notch in the specimen.

The samples were placed on a simple-supported impact testing machine. The original thickness of the samples was 12 mm, and the remaining thickness after testing was 7.4 mm, with a width of 10.6 mm. Six samples were taken for each group, and the formula mentioned earlier was used to calculate the notched impact strength of the samples, with the results averaged. The test results are presented in Table 4.

**Table 4.** Impact strength of PA66/GF system with different transparent nylon contents.

| Numbering | Nylon Content /% | Notched Impact Strength $\left(kJ \bullet m^{-2}\right)$ |
|---|---|---|
| A1 | 0 | 5.2 |
| A2 | 10 | 5.6 |
| A3 | 20 | 7.3 |
| A4 | 30 | 8.4 |
| A5 | 40 | 6.8 |
| A6 | 50 | 1.9 |

Observing the table above, it can be seen that as the mass fraction of nylon is added to the composite material in the range of 0% to 50%, the notched impact strength of the material gradually increases before rapidly decreasing again. When the nylon content reaches 30%, the notched impact strength of the material reaches 7.3. The main reason for this phenomenon is that transparent nylon in the composite material can act to disperse the copolymer crystalline phase, gradually increasing the degree of deformation in the composite material matrix. Additionally, during the process of increasing the dispersed phase content from 0% to 20%, the number of fibers released by the dispersed phase also increases. Therefore, when the material is subjected to strong impact, stress gradually diffuses from the material's matrix to the fibers, forming a fiber-interlocking network. This reduces the impact of interface defects. However, when the content of transparent nylon continues to increase, the interface voids between the dispersed phase and the material matrix also increase. The size of these voids also becomes larger. When the material is subjected to external impact, it effectively suppresses the energy absorbed by the material at the interface. As a result, the notched impact strength of the samples decreases.

In summary, the research involved preparing PA66/GF composite systems using the method of incrementing blending, testing the experimental samples, and analyzing the relationship between the content of transparent nylon and material properties. The test results indicate that transparent nylon significantly enhances the flame-retardant performance of the composite material. The higher the content of transparent nylon, the higher the melting point and crystallization temperature of the PA66/GF system. Moreover, a higher content of transparent nylon also leads to increased tensile strength in the PA66/GF system. When the nylon content in the material reaches 30%, it can achieve the highest notched impact strength.

## 5. Conclusion

This study utilized PA66 and PA6T as base materials and incorporated a compatibilizer and toughening agent to create a modified PA66/GF composite material. With transparent nylon content as the variable, the research investigated the impact of different transparent nylon contents on the thermal and mechanical properties of the PA66/GF system. The test results demonstrated that this composite material significantly improved thermal and mechanical properties compared to pure nylon. The addition of transparent nylon further enhanced the overall performance of the composite material.

## References

[1] Ye Shibing, Xia Jianmeng, Feng Jiachun. "Performance Study of Halogen-Free Flame Retardant Reinforced PA66/PA6 Alloy" [J]. Synthetic Materials Aging and Applications, 2023, 52(01): 6-8+140.

[2] Yu Yunan, Wu Hongyuan, Yan Wenli. "Improvement of Nylon Reinforced Materials and Size Stability and Water Absorption of PA6/ABS" [J]. Plastics Industry, 2023, 51(02): 36-41+78.

[3] Li Fuhai, He Qichao, Wang Jian. "Mechanical Properties of Insert Injection Molded Continuous Carbon Fiber Reinforced Polyamide 6 Composite Materials" [J]. Plastics Industry, 2023, 51(01): 97-102.

[4] Liang Qingru, Ji Zhenyan, Dong Chunlei, et al. "Preparation of Epoxy-Functionalized Polyethylene-Octene by Radiation and Its Application in Toughening Modification of Nylon 6" [J]. Journal of Radiation Research and Radiation Processing, 2023, 41(01): 26-34.

[5] Jiang Aiyun, Liu Xuewen, Yan Panpan, et al. "Preparation and Properties of Glass Fiber Reinforced Nylon MXD6 Composite Materials" [J]. Engineering Plastics Application, 2023, 51(01): 17-21.

[6] Wu Yifeng, Zhang Yan, Li Aiqun, et al. "Experimental Study on Compression and Shear Performance of Nylon Fabric-Layered Rubber Bearings" [J]. Vibration and Shock, 2022, 41(18): 150-156.

[7] Wang Jinxi. "Research on the Properties, Production Process, and Modification of Nylon 6" [J]. Chemical Industry Times, 2022, 36(09): 21-24.

[8] Wu Yifeng, Xu Yubo, Si Mingfei, et al. "Experimental Study on the Mechanical Properties of Nylon Fabric-Reinforced Rubber Bearings" [J]. Building Structures, 2023, 53(03): 10-14+20.

[9] Zhu Minjie, Wei Longsha, Yan Yiwu. "Influence of Linear Epoxy-Based Surface Modifier on the Properties of Nylon/Carbon Fiber Composite Materials" [J]. Composite Materials Science and Engineering, 2022(11): 77-81.

[10] Wu Shuyan, Qiao Yunhe, Liu Yingjun, et al. "Influence of Rotor Speed on the Properties of Ethylene-Vinyl Acetate Copolymer/Ternary Copolymer Nylon Thermoplastic Vulcanizate" [J]. Synthetic Rubber Industry, 2022, 45(06): 511-515.

[11] Xiao Hao. "Research on the Properties of Polytetrafluoroethylene Blended Modified High-Temperature Nylon" [J]. Science and Technology Innovation and Application, 2022, 12(32): 66-69+73.

[12] Wu Yifeng, Si Mingfei, Lu Song, et al. "Experimental Study on the Mechanical Properties of Thick-Layer Rubber Bearings Reinforced with Nylon Fabric" [J]. Journal of Civil Engineering, 2022, 55(S1): 126-131.
[13] Li Hui. "Mechanical Properties Study of Cast Nylon Composite Material for Conveyor Roller" [J]. Shandong Coal Science and Technology, 2022, 40(10): 130-132+135.
[14] Jiang Tao. Study on properties of organically modified graphene oxide/nylon fiber [J]. Synthetic fibers, 2012,51(09):30-34.
[15] Wang Chenchen, Song Lifu, Bai Nan, et al. Effect of POE-G-MHA content on Compatibility, Structure and Mechanical Properties of POE/EPDM composites [J]. Plastics Industry, 21,49(12):56-61.

## AUTHOR INDEX